About Island Press

Since 1984, the nonprofit Island Press has been stimulating, shaping, and communicating the ideas that are essential for solving environmental problems worldwide. With more than 800 titles in print and some 40 new releases each year, we are the nation's leading publisher on environmental issues. We identify innovative thinkers and emerging trends in the environmental field. We work with world-renowned experts and authors to develop cross-disciplinary solutions to environmental challenges.

Island Press designs and implements coordinated book publication campaigns in order to communicate our critical messages in print, in person, and online using the latest technologies, programs, and the media. Our goal: to reach targeted audiences—scientists, policymakers, environmental advocates, the media, and concerned citizens—who can and will take action to protect the plants and animals that enrich our world, the ecosystems we need to survive, the water we drink, and the air we breathe.

Island Press gratefully acknowledges the support of its work by the Agua Fund, Inc., Annenberg Foundation, The Christensen Fund, The Nathan Cummings Foundation, The Geraldine R. Dodge Foundation, Doris Duke Charitable Foundation, The Educational Foundation of America, Betsy and Jesse Fink Foundation, The William and Flora Hewlett Foundation, The Kendeda Fund, The Andrew W. Mellon Foundation, The Curtis and Edith Munson Foundation, Oak Foundation, The Overbrook Foundation, the David and Lucile Packard Foundation, The Summit Fund of Washington, Trust for Architectural Easements, Wallace Global Fund, The Winslow Foundation, and other generous donors.

The opinions expressed in this book are those of the author(s) and do not necessarily reflect the views of our donors.

FIRE ECOLOGY IN ROCKY MOUNTAIN LANDSCAPES

FIRE ECOLOGY
IN
ROCKY MOUNTAIN LANDSCAPES

William L. Baker

Washington | Covelo | London

Library of Congress Cataloging-in-Publication Data

Baker, William L. (William Lawrence)
  Fire Ecology in Rocky Mountain landscapes / William L. Baker.
      p. cm.
  Includes bibliographical references and index.
  ISBN-13: 978-1-59726-182-1 (cloth : alk. paper)
  ISBN-10: 1-59726-182-3 (cloth : alk. paper)
  ISBN-13: 978-1-59726-183-8 (pbk. : alk. paper)
  ISBN-10: 1-59726-183-1 (pbk. : alk. paper) 1. Fire ecology—Rocky Mountains Region.
I. Title.
  QH104.5.R6B35 2009
  577.2′40978—dc22

                                                                              2008039908

Printed on recycled, acid-free paper

Manufactured in the United States of America

10 9 8 7 6 5 4 3 2 1

This book is dedicated to the forest reserve scientists H. B. Ayres, Henry Graves, John Jack, John Leiberg, George Sudworth, and F. E. Town and the geographer Henry Gannett. Under difficult conditions around 1900, these scientists not only documented the need for creating national forests, but also made remarkable maps, given the limited technology of the day, and recorded perceptive and systematic observations of the ecology of fire and vegetation.

Contents

Illustrations

Tables

Boxes

Preface

Years ago, while traveling by car in northern Australia, I waited alongside a road with other cars as an intense fire burned across the road in the overhanging crowns of the eucalyptus trees. After the fire crossed, we all drove on. The fire was as fascinating to watch as a snake or group of kangaroos crossing the road. There were no sirens, as it was too far away to be worth the trouble in this large, comparatively wild part of Australia. I think of that scene now from my current vantage point in the Rocky Mountains, an area of the United States that is in some ways comparable, in being lightly populated by humans and vulnerable to wildfires. The Rockies are one of the wilder parts of the United States, containing only about 3 percent of the U.S. population; substantial public lands, including wilderness and parks; and large areas of natural vegetation.

Here as elsewhere we continue to struggle with wildland fire. We only conditionally accept the place of fire in nature, even in locations far from homes. We still commonly attack wildfires with crews and aircraft even if fires are distant from valued resources. Despite substantial evidence to the contrary, we still misunderstand fire's aftermath as an emergency for plants and animals. In part this is because we EuroAmericans have not lived long enough in the region to have experienced and understood fire deeply, developed a viable relationship with fire, and instilled it in our collective cultural memory. In the past few decades, we have accumulated enough science and experience to accept, in concept, that fire is an essential process in Rocky Mountain ecosystems. However, we have yet to appreciate or fully understand fire in its severe manifestations, just as we are unlikely to appreciate the ecological importance of the grizzly when a big brown bear is charging down the trail in our direction.

A healthier relationship between people and fire must have a sound scientific basis to have lasting value, yet older stories and understandings about fire in Rocky Mountain ecosystems have been supplanted by new ones. A prevailing story, which no longer fits the evidence, is that fire was historically benign and manageable in fire-adapted, stable, resilient vegetation prior to EuroAmerican settlement but was disrupted by a period of misguided fire control, logging, livestock grazing, and other land uses. An associated redemption story suggests that fire can be reinstated and managed using science-based prescribed burning and mechanical treatments so it could again play its role in ecosystems without threatening homes, infrastructure, and valued natural resources. The story provides a compelling, peaceful vision of harmony and stability, as people would then be able to live near wildlands without fear. Wildlife and plants also would be sustained once vegetation was restored and functioning properly and fire well managed by the government.

However, science has shown that this story does not fit the evidence, except in limited areas. This book uses detailed scientific data to show that fire regimes in the Rockies were more generally dominated historically by infrequent episodes of large, often severe, difficult-to-control fires that burned under severe weather conditions, often during droughts. The evidence suggests that a different relationship would be more sensible, a relationship based on humility in the face of the power of nature, a desire to minimize our impacts on the natural world and keep as much wildness as possible, and the good sense to get our homes and infrastructure protected or out of fire-prone settings, as fire will eventually come our way. Fire is a force that shapes nature through overwhelming power, a force we cannot tame and with which we ultimately have to learn to live if we are finally to settle into the country.

This book could not have been completed without the help of many people. Foremost was my wife, Deb Paulson, who encouraged me and tolerated long periods of distraction and work. The University of Wyoming provided a sabbatical and the Department of Geography favorable arrangements that enabled me to complete the book on time. My graduate students were patient with me, even though I was distracted and scurrying. Interlibrary Loan at the University of Wyoming was fantastic at finding and quickly obtaining obscure materials. I appreciated a research grant from the Association of American Geographers that allowed me to visit the National Archives to find early materials. Many scientists reviewed parts or all of chapters, including Martin Alexander, Steven Buskirk, Amy Hessl, Dan Kashian, Natasha Kotliar, Dominik Kulakowski, William Massman, Michael Murray, Roger Ottmar, Rosemary Sherriff, Jason Sibold, and Thomas Veblen. I also appreciated the comments of five anonymous reviewers who made many helpful suggestions that improved the book. Barbara Dean of Island Press provided important inspiration, encouragement, and guidance throughout the process of writing the book. Erin Johnson, Sharis Simonian, and Barbara Youngblood of Island and Jill Mason of Masonedit, were very supportive and helpful with the many details. The time I had as an undergraduate at Oregon State University to read about fire with Fred Swanson was seminal, as was Bill Chilcote's big-picture ecology. Growing up in wild country with tall trees, in southeastern Alaska, forced us all to get up on a mountain to see at the landscape scale. Tom Vale deepened this landscape proclivity and educated me about people and nature while I was his doctoral student at the University of Wisconsin.

Unfortunately, I was not able to include everything others wanted in the book or to treat some matters in depth, or even cover them at all. The book generally omits treatment of the effects of fire on the physical environment, as well as soils, microbial ecology, nutrient cycling, and energy flows, and largely ignores fire's impacts on aquatic ecosystems, focusing instead on the terrestrial. These omitted subjects are of considerable importance but would have required a larger volume, more time than I had, and another author. The scientific literature is enormous and continuing to grow rapidly, and I attempted to cite and use as much as I could, but in the end literature had to be omitted to keep the size of the book reasonable. It is likely that some key studies also were simply not found or were omitted because readers would not be able to access them.

CHAPTER 1

Introduction

As I write this in the early summer of 2008, bark beetles are killing trees over large areas, concern about global warming is rising, and drought and fire are on our minds. It is a difficult time to study fire and vegetation across landscapes, as landscapes appear to be unraveling and time seems short. Yet if our ecosystems and our communities are to weather the fire increase that is likely to come with global warming and an expanding population, we need to begin redesigning our communities at the landscape scale immediately.

This book has several themes, introduced below and elaborated in the chapters, that build a case for an increasing landscape-scale approach to fire. Landscapes, which are land areas that range in size from hundreds of meters to kilometers, are a spatial scale on which both fire and people naturally function, which is part of our difficulty in getting along together. Perhaps by better understanding the natural history of fire and how fire affects plants and animals, we can redesign our land uses to better allow people and nature to live together in landscapes.

THEMES

In the past few decades, the world has become, in some senses, smaller and more comprehensible because of new information and technology, but it has also become more complex because this new information has more fully revealed the substantial spatial heterogeneity and temporal variability in the earth's systems. The rise of computers, geographical information systems, satellites, and other spatial technology has facilitated understanding at an expanded spatial scale. The academic discipline of landscape ecology expanded in the 1980s, and land management began including landscape-level approaches (Liu and Taylor 2002). Fire ecology also expanded from its roots in small-plot, detailed studies to the landscape scale (Romme 1982; M. G. Turner 1987).

Emerging from this expanded spatial scale over the past few decades has been awareness of the remarkable variability in fire behavior and fire effects across large landscapes, a theme of this book. Wildland fire in the Rocky Mountains is now understood as inherently a landscape phenomenon, as it is capable of burning across hundreds of thousands of hectares, spotting over rivers and even mountain ranges. Hundreds of fires can ignite in a single, fast-moving thunderstorm and, under drought and strong winds, coalesce into a enormous firestorm, as happened in 1910 (Koch 1942; S. Cohen and Miller 1978). Such firestorms can burn hundreds of hectares per hour with flames in forests as much as 100 meters high. However, during century-scale interludes between firestorms, fire may be nearly absent or might burn through the understory of a tall pine forest at such low intensity a person could jump over it.

Research on this variability, formally called the *historical range of variability* (HRV), seeks to understand how fire and vegetation varied spatially and temporally over the last few hundred to few thousand years (box 1.1). HRV research has changed our understanding of variability in fire and vegetation across landscapes and over time. This research required systematic use of tree rings, charcoal and pollen, and early historical records (chap. 5). Somewhat surprisingly, this understanding could not be obtained more simply from experience, even over a lifetime.

Human life spans are simply too short relative to the periods over which fire functions in Rocky Mountain landscapes, and this mismatch is another theme of the book. The time required to burn across an area equal to a landscape of interest, the *fire rotation*, has been found to be centuries in most Rocky Mountain landscapes (chaps. 6–9; table 5.7). Most difficult to appreciate is that the area burned during a fire rotation mostly accrues from one or more *episodes* of large, often severe fires after long *interludes* of mostly small, lower-severity fires that burn little total land area. Such interludes can last a lifetime, however, so the episodes can be completely missed. The interludes can be easily misconstrued as periods of fire suppression, and the episodes can appear to represent an abnormal fire increase; but episodes and interludes are normal in all fire regimes.

New research also suggests a significant synchronizing role of climate in fire episodes and interludes, through distant links, or *teleconnections* to sea surface and

BOX 1.1
Historical Range of Variability

Research on the historical range of variability (HRV) of fire and vegetation structure is fundamental to understanding ecosystems. HRV refers to "the ecological conditions, and the spatial and temporal variation in these conditions, that are relatively unaffected by people, within a period of time and geographical area" (Landres, Morgan, and Swanson 1999, 1180). HRV thus represents how vegetation is structured (e.g., tree density), how it varies spatially and temporally, and how fire functions (e.g., fire size, intervals) with little effect of people, except where people have been a significant structuring force. In the Rocky Mountains, the several hundred years prior to EuroAmerican settlement are most relevant to the HRV, but earlier periods are relevant as well. HRV represents the full spectrum of conditions, not a single fixed state.

Understanding the HRV is important, because if ecosystems are functioning under the HRV, it is likely that biological diversity and ecosystem services (e.g., nutrient cycling, forage production) will be sustained. Ecosystems can be intentionally displaced from the HRV to obtain products or services, but this usually requires external inputs and subsidies, such as water and nutrients or pest control. Ecosystems can also be unintentionally modified from the HRV by unsustainable land uses, such as overgrazing by livestock, control of fires, or land uses that favor invasive species. If the HRV is not understood, land uses may inadvertently alter an ecosystem in ways that reduce biological diversity and ecosystem services. Thus, those who value nature and want to minimize the impacts of land uses on species and ecosystems often use the HRV as a frame of reference for assessing land-use effects and for developing lower-impact land-use methods.

Restoration ecology can also use the HRV as a framework for setting ecological restoration goals (Shinneman, Baker, and Lyon 2008). Of course, environments change naturally, evolution continues, and it would be inadvisable to restore the exact conditions of one particular time under the HRV. However, restoring many of the conditions of the HRV is wise, even if the environment is changing, as the species, community structure, and ecological functions of the HRV provide the most promising reservoir of options for responding to environment change.

atmospheric conditions in the oceans (chap. 2). Detecting human impacts on fire regimes, given natural fluctuations, is thus exceedingly difficult, and this book will present evidence to challenge past interpretations that emphasize fire suppression as a central cause of changes in vegetation in the Rocky Mountains after EuroAmerican settlement (chap. 10).

The large, often severe wildland fires of the episodes are commonly perceived as ecological catastrophes. However, infrequent, large, severe fire was characteristic of most Rocky Mountain ecosystems under the HRV. Even the most severe fire is shown not generally to be a disaster for native animals and plants in Rocky Mountain ecosystems, as most animals and plants have adaptations that allow survival and recovery after fire (chaps. 3–4, 6–9). Animals commonly avoid mortality by moving away from fire or by surviving in burrows or other fire refugia, and many animals benefit from resources in the postfire environment (chap. 4). The usual vegetation response to fire is initially bare ground, then rapid recovery from surviving aboveground stems or underground roots, stems, and seeds, aided to a lesser extent by seed dispersal from nearby unburned areas (chap. 3). Many Rocky Mountain trees, and some nonsprouting shrubs, are slower to recover, and temporary openings from lags in regeneration were common under the HRV (chap. 6–9).

People like to tinker and fix things, and the appearance of the bare, burned, postfire environment may foster a desire to apply expensive seed to burned areas, plant trees, or control erosion. Rocky Mountain ecosystems are capable of full recovery after fire, and seeding, logging, and other past postfire actions will be shown to actually slow or permanently damage natural postfire recovery (chap. 10). New approaches can enhance natural recovery.

In fire history research, the transition from small plots to landscapes is not yet complete, and popular, but unsupported, ideas remain from the past. Large, infrequent, severe fires at the landscape scale under the HRV contrast with the idea of "frequent fire" that arose from past small-plot studies of fire history. Many past fire histories were detailed studies in small plots in purposely chosen locations. At that time, researchers commonly sought old-growth forests and old trees with multiple fire scars, as these were thought to lead to the most complete historical record of fire. Fire was thought to be relatively uniform across landscapes. Thus, it was not considered necessary or of interest to obtain an unbiased spatial sample of fire history across a landscape; the large spatial heterogeneity at the landscape scale was not generally known. Fortunately, new methods for sampling and analysis are available for both small plots and landscapes that overcome the limitations of past studies (chap, 5). The popular idea of frequent fire is shown generally not to hold when fire is studied systematically across large landscapes (chaps. 5, 7).

Threats to sustaining fire and vegetation in the Rocky Mountains in a manner consistent with the HRV are shown to be primarily land uses over the last century and emerging global warming (chap. 10). The concept of *fire exclusion*, a general term for decreased fire from a variety of land uses, is shown not to fit the history of land uses (chap. 10). Moreover, fire has continued and appears to be increasing in the Rocky Mountains over the last few decades. Instead of fire exclusion, the most significant human impact on Rocky Mountain fire regimes is shown to have been increased fire from logging, livestock grazing, roads, invasive species, and human-set fires (chaps. 10, 11). These land uses, combined with expanding wildland-urban interfaces, continue to add fire across landscapes during a time of global warming.

Restoring fire has become policy for some agencies, but in the past this focused on creating vegetation structure that could resist severe fire and grow efficiently. Low-severity fire that does not kill larger trees in forests has been the ideal, and mechanical thinning and low-severity fire are extolled for increasing tree growth, forage for livestock and wildlife, species diversity, and other attributes. A prevailing idea has been that ecosystems under the HRV were efficient machines structured by low-severity, frequent fire to produce optimum products and services, including pleasing forests with little undergrowth. Thus, restoration, in this view, requires gardening—fixing ecosystems to again have healthy individual plants and animals optimally producing products and services. Part of our difficulty with fire is in these contrasting views: *nature as a garden* tended by benign and frequent fire, often ignited by people or under their control, and *nature as wild landscape* subject to remarkable and uncontrollable variability in fire. Fire under the HRV in the Rocky Mountains is definitely shown (chaps. 3, 5–9) to have been highly variable and at times uncontrollable, producing and destroying vegetation structures with little regard for human ideas about efficiency.

New management policy is needed that recognizes the power of fire and our inability to fully control it (chap. 11). A promising approach is *wildland fire use*, the current term for allowing fire to play a more natural but guided role in ecosystems. Wildland fire use is expanding in areas where fire does not threaten valued resources. However, we also need to redesign our communities and individual houses to resist fire effectively. Developing sound policy and redesigning communities have been limited by the rapid in-migration and population growth that are feeding housing development in the interface with wildlands. A culture of individualism and private-property rights discourages restraint and careful planning in development. Our government is assuming the risk by providing costly fuel treatments and fire protection, an enabling behavior that leads to further development in a positive feedback loop, at a time when drought and fire are increasing. Yet all this can be changed by communities seeking to live sustainably with fire.

There has never been a better time for people to reflect on the places they live, understand the science of fire and vegetation in local ecosystems, and use science to help create landscapes that work for both people and nature. The idea of creating better landscapes to allow us to live with fire is a positive step, and practical ideas are available that can be implemented in time to prepare for increased fire (chaps. 11, 12).

THE ENVIRONMENTAL SETTING OF THE ROCKY MOUNTAINS

The geology, climate, and vegetation of the mountains that are the subject of this book substantially shape fire occurrence and behavior in the region.

Physiography, Geology, Climate

The Rocky Mountains are a series of ranges and intermontane valleys loosely connected along a 2,000-kilometer northwest-to-southeast axis across seven states (fig. 1.1). My boundary for the Rockies is derived from Fenneman (1931), Hunt (1974), and Bailey (2002). I began with Bailey's ecoregions but modified them in western Colorado to exclude the Uncompahgre Plateau and Grand Mesa, better treated as mountainous parts of the Colorado Plateau (Hunt 1974). I also connected the southern and central Rockies, as in Hunt (1974), and included the Black Hills, even though they are disconnected from the Rockies. I followed Fenneman and Hunt in excluding Oregon's Blue-Wallowa Mountains, in spite of their affinities with the Rockies. Although the Rockies extend into Canada, this book covers only the United States portion, an area of about 55 million hectares. I drew upon scientific literature within the Rockies and from some nearby or even distant areas.

The Rockies generally arose in the Laramide uplift of the late Mesozoic to early Cenozoic and typically have an uplifted igneous or metamorphic core with sedimentary rocks along the flanks or in intermontane basins (Thornbury 1965). The southern Rockies also have linear ranges oriented north-south that lead to contrasts in wind and precipitation on the two sides of the ranges. The exception is the east-west-oriented San Juan Mountains, composed of Miocene to Quaternary volcanics. The southern Rockies have large intermontane basins (e.g., Middle Park, San Luis Valley) with sagebrush, grasslands, and some salt desert vegetation. The central Rockies, also called the middle Rockies, are separated from the southern Rockies by a large basin. They are less co-

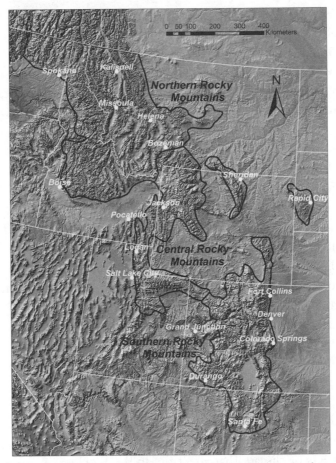

Figure 1.1. The Rocky Mountains in the United States. The boundary has been modified, as explained in the text, beginning with Bailey's ecoregions (Bailey 2002) from the *North American Environmental Atlas* Web site (http://www.nationalatlas.gov). The backdrop includes cities and 1-km-resolution digital elevation data, also from the *North American Environmental Atlas*.

hesive, having an east–west-oriented range, the Uintas, and a series of generally north–south-oriented ranges, along with the disjunct Bighorn Mountains. The northern section of the central Rockies is distinguished by the volcanic Yellowstone Plateau, which is less mountainous than the rest of the central Rockies, allowing for more continuous forests and potentially larger fire spread. The northern Rockies contain the central Idaho mountains, formed in a large body of intrusive igneous rocks characterized by broad mountains lacking distinctive orientation. North of this area is a series of steep, north–south-oriented ranges separated by trenchlike valleys. In northwestern Montana's Glacier National

Park, extensive mountain glaciation in layered sedimentary rocks led to striking topography within undistinctive north–south-oriented ranges.

The Rockies have a continental climate, except for a near maritime climate in the northwestern corner of the region. Precipitation increases and temperature declines with elevation, but topography also shapes local climate. The central and northern Rockies are most strongly influenced by westerly flow from the Pacific, which in winter leads to periodic cyclones and in summer to warm, dry conditions (V. L. Mitchell 1976). The southern Rockies in winter have predominantly southerly flow, interrupted by infrequent cyclonic storms. In summer, the North American monsoon, an influx of tropical moist air, increases lightning and thunderstorms from about July 1 to mid-September, particularly in the southern part of the southern Rockies (D. K. Adams and Comrie 1997). South of the latitude of Denver, some vegetation differs from that to the north, with bristlecone pine forests, a southwestern variant of mixed-conifer forests containing white fir, and Arizona fescue grasslands, perhaps related to the monsoon.

Climate diagrams show that the central and northern Rockies generally have peak monthly precipitation in April, May, and June and a dry period in July and August (fig. 1.2). In contrast, the eastern slope of the southern Rockies has peak precipitation in April and May with a longer dry period from June to early October. Farther south, the dry period is from April through June, with the North American Monsoon causing increased precipitation in July and August, followed by another dry period from September to November. Variation in the aspects of climate that affect fire in the Rockies, including drought and teleconnections with the Pacific and Atlantic, are discussed in detail in W. L. Baker (2003) and in chapter 2.

Vegetation

The Rocky Mountains have nearly the highest vegetation diversity in the United States (Wickham et al. 1995). The ecology of natural vegetation is reviewed for the Rocky Mountains as a whole (Peet 2000), the northern Rockies (Daubenmire 1943; Arno 1979; Habeck 1988), the central Rockies (Knight 1994), and the New Mexico and southern Colorado parts of the southern Rockies (Dick-Peddie 1993; D. E. Brown 1994). The distribution of trees and shrubs that dominate the vegetation is strongly shaped by elevation and topographic-moisture gradients, illustrated by gradient diagrams (fig. 1.3). Common and Latin names for major trees, shrubs, and graminoids are in appendixes A, B, and C, and fire adaptations of plants are discussed in chapter 2. Natural vegetation still dominates about 86 percent of the Rockies land area, including about

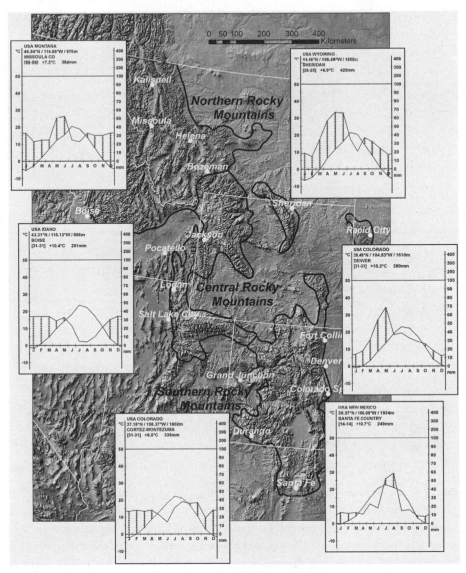

Figure 1.2. Climate diagrams for selected locations in the Rocky Mountains. Each diagram shows (1) the latitude, longitude, and elevation of the station and its name, years of record for temperature and precipitation (in brackets), mean annual temperature (°C), and total annual precipitation (mm) at the top between the axes; (2) a left axis and line graph for mean monthly temperature (°C), which is generally unimodal and centered on summer; (3) a right axis and line graph for mean monthly precipitation (mm), which is more irregular; (4) vertical shading between the two line graphs indicating moisture surplus; and (5) an unshaded area between the two line graphs indicating moisture deficit.

Source: The climate diagrams and data are from a public domain data set produced by the Oak Ridge National Laboratory, plotted using Climate Plot 32, supplied by Rivas-Martínez et al. (2002).

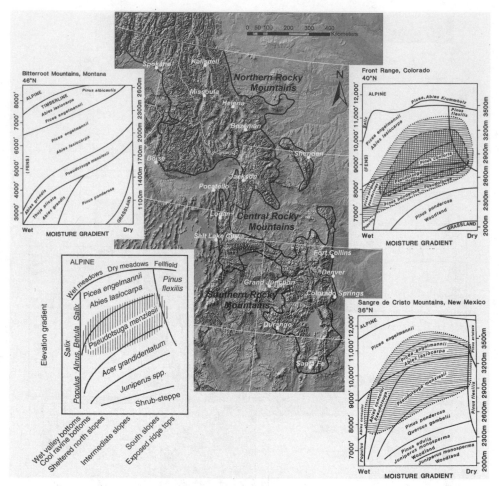

Figure 1.3. Gradient diagrams for four places in the Rocky Mountains. The lower left diagram is for northeastern Utah. Each diagram shows vegetation dominants versus elevation and a topographic-moisture gradient. Zones tilt down to the left, as vegetation types occur at lower elevations on sheltered slopes, and also are shifted down in elevation toward the north. Each place has some vegetation zones that are found across the Rockies and also some unique vegetation. Latin and common names for trees and shrubs in the diagrams are given in appendixes A and B.

Source: Gradient diagrams for the Bitterroot Mountains, the Front Range, and the Sangre de Cristo Mountains are reproduced from Peet (2000, p. 89, fig. 3.7) with permission of Cambridge University Press; the diagram for northeastern Utah is from J. N. Long (2003, p. 1093, fig. 5), used with permission of Heron Publishing.

equal areas of grasslands, shrublands, and subalpine forests but much more montane forest (table 1.1).

The montane zone (chaps. 6, 7) is the first forested zone above low-elevation semiarid grasslands and shrublands. Piñon-juniper woodlands (chap. 6) of Utah juniper and twoneedle piñon are extensive on the western slope of

Table 1.1. Major Vegetation Types in the Rocky Mountains

Vegetation Type	Area (ha)	Percentage of Area of Rocky Mountains
Montane woodlands & forests		
Piñon–juniper woodlands	2,140,600	3.9
Ponderosa pine–Douglas-fir forests	3,910,200	7.1
Douglas-fir forests	2,563,000	4.7
Moist northwestern forests	2,469,400	4.5
Mixed-conifer forests	5,497,300	10.0
Quaking aspen forests (partly subalpine)	1,497,900	2.7
Total montane woodlands & forests	18,078,400	32.9
Subalpine forests		
Lodgepole pine forests	4,468,300	8.1
Spruce-fir forests	5,384,900	9.8
Five-needle pine forests (whitebark, limber, bristlecone)	21,800	0.1
Total subalpine forests	9,875,000	18.0
Shrublands		
Salt desert shrubs (e.g., greasewood)	869,200	1.6
Sagebrush	6,628,700	12.1
Misc. shrublands (e.g., mahogany)	197,400	0.4
Mixed mountain shrubs	2,884,400	5.3
Total shrublands	10,579,700	19.4
Grasslands		
Great Plains grasslands (e.g., shortgrass prairie)	4,411,200	8.0
Plateau grasslands (e.g., Indian ricegrass)	352,000	0.6
Montane grasslands (e.g., bluebunch wheatgrass)	1,321,700	2.4
Subalpine grasslands (e.g., Thurber's fescue)	1,967,100	3.6
Total grasslands	8,052,000	14.6
Riparian	870,800	1.6
Agriculture, urban, or unvegetated	7,430,000	13.6

Source: Areas and percentages are from the Westgap map (GAP Analysis Program at http://gapanalysis.nbii.gov).

the southern Rockies, extending north to the Wyoming border, with oneseed juniper and twoneedle piñon most common on the eastern slope of the southern Rockies north to central Colorado. Utah juniper woodlands extend north in Wyoming in the foothills of the central Rockies. Rocky Mountain juniper can be codominant or dominant in the upper elevations of piñon–juniper woodlands, but also forms an extensive low-elevation woodland in the central and northern Rockies (Peet 2000). Ponderosa pine–Douglas-fir forests (chap. 7) can be pure ponderosa at the lowest elevations, a mixture in midelevations, and grade upward into pure Douglas-fir forests. Northwestern Montana, northern Idaho, and northeastern Washington have moist northwestern forests that can be dominated by western white pine, western hemlock, western red cedar, grand fir, or mountain hemlock. Above these forests in elevation are heterogeneous mixed-conifer forests, which I have divided latitudinally. Northern Rocky Mountain mixed-conifer forests are mixtures of Douglas-fir, western larch, and lodgepole pine, with lesser amounts of other trees. An upper-montane mixed-conifer forest occurs in the northern Rockies, south of the range of western larch, the central Rockies, and the northern part of the southern Rockies. This forest has Douglas-fir, lodgepole pine, limber pine, quaking aspen, and minor amounts of other trees. A southwestern mixed-conifer forest, from northeastern Utah to southern Colorado and northern New Mexico, has white fir, Douglas-fir, ponderosa pine, quaking aspen, blue spruce, and southwestern white pine. Finally, Utah, southern Colorado, and northern New Mexico have extensive stable quaking aspen forests in the montane and the subalpine zone.

The subalpine zone (chap. 8) has lodgepole pine forests from the northern Rockies into southern Colorado, and spruce-fir forests throughout the Rockies. Five-needle pine forests include whitebark pine in the northern Rockies and northern half of the central Rockies, limber pine throughout the Rockies but most commonly from Wyoming to central Colorado, and bristlecone pine from central Colorado to northern New Mexico.

Shrublands (chap. 9) are found throughout the Rockies. Salt desert shrublands occur on saline sites, particularly at low elevations, where greasewood, fourwing saltbush, Gardner's saltbush, shadscale saltbush, valley saltbush, spiny hopsage, winterfat, and other shrubs may each dominate or occur in mixtures. Sagebrush shrublands are common throughout the Rockies, with Wyoming big sagebrush at low elevations, mountain big sagebrush at higher elevations, basin big sagebrush in swales or on deeper soils, Bigelow sagebrush in rocky canyons and on outcrops, black sagebrush often on rocky uplands (particularly if calcareous), and little (low) sagebrush on upland sites with a buried impermeable horizon. Silver sagebrush occurs extensively on the eastern plains and foothills and also in cold basins in the mountains. Threetip sagebrush is particularly common in foothills of the northern and central Rockies. Miscellaneous shrublands include curlleaf mountain mahogany, a tall shrub or small tree that is most common on steep, rocky sites in the northern and central Rockies, and alderleaf mountain mahogany, which is common on rocky sites on the eastern slope in the ecotone with plains grasslands in Wyoming and Colorado. Mixed mountain shrublands, common in northern New Mexico and Colorado, particularly on the western slope, and in northeastern Utah include alderleaf mountain mahogany, big sagebrush, chokecherry, Gambel oak, mountain snowberry, skunkbush sumac, Utah serviceberry, bigtooth maple (in Utah), and a variety of other shrubs.

Grasslands (chap. 9) also occur throughout the Rockies. Great plains grasslands include (1) shortgrass prairies, dominated by blue grama and buffalograss, in northern New Mexico and eastern Colorado, and (2) mixed-grass prairies, dominated by little bluestem, needle and thread, and western wheatgrass, along the foothills of the southern Rockies and throughout the plains adjacent to the central and northern Rockies. Plateau grasslands, which have a mixture of James' galleta, needle and thread, and Indian ricegrass, typically occur on sandstone mesas on the ecotone between the Rocky Mountains and the Colorado Plateau in western Colorado. Montane grasslands include (1) greenleaf fescue and rough fescue grasslands in the northern Rockies, (2) bluebunch wheatgrass–Idaho fescue grasslands from the northern Rockies to the northern part of the southern Rockies, (3) needle and thread grasslands throughout the Rockies, (4) Parry's oatgrass grasslands, most common in the southern Rockies, and (5) Arizona fescue–mountain muhly grasslands in the southern Rockies. Subalpine grasslands include (1) tufted hairgrass meadows, particularly in the central and southern Rockies, (2) timber oatgrass grasslands, and (3) Thurber's fescue grasslands in the southern Rockies.

Riparian vegetation (chap. 6) includes low-elevation cottonwoods (plains cottonwood, Rio Grande cottonwood) and montane cottonwoods (black cottonwood, narrowleaf cottonwood), which may be mixed with, or occur in a mosaic with blue spruce woodlands in the southern Rockies. Quaking aspen and a variety of conifers can occur in riparian areas, particularly in the subalpine. Willow-dominated wetlands predominate on shallow slopes and alluvium in both the montane and subalpine, and wetlands dominated by sedges, rushes, and grasses also occur.

SUMMARY AND ORGANIZATION OF THE BOOK

The Rocky Mountains have among the richest diversity of vegetation and landscapes in the United States and enough latitudinal range to include substantially contrasting climates and fire settings. Wildland fire, the book will argue, operates at large spatial extents and over long periods but does so in episodes and interludes that lead to fluctuating mosaic landscapes. Native plants and animals have adaptations that allow their persistence in the face of fluctuating landscapes subject to fire, but people continue to struggle in living with fire in landscapes.

To help move thinking about fire to the landscape scale, I first review the factors that affect fire behavior (chap. 2) and some of the basic functional traits and responses of plants (chap. 3) and animals (chap. 4) to fire at the landscape scale, before explaining how scientists reconstruct the history of fire in land-scapes under the HRV (chap. 5). The particulars of fire and vegetation responses under the HRV in the major ecosystems of the Rockies are elaborated in chapters 6–9, before I attempt to disentangle the role of EuroAmericans in creating the landscapes and fire situations of today (chap. 10). Out of all of this, I hope that better sense can be made of the options available to us to begin creating landscapes that function better for both people and nature in the face of fire (chaps. 11, 12).

CHAPTER 2

Lightning, Fuels, Topography, Climate, and Fire Behavior

This chapter reviews the major factors that influence fire from ignition to spread across landscapes. Fire is a complex phenomenon because spatial and temporal variation in lightning, fuels, topography, and weather and climate is substantial and produces regional, landscape-scale, and local variation in fire.

LIGHTNING AND IGNITION

Lightning varies geographically in the Rockies, but annual density is generally less than three flashes per square kilometer, which is low relative to the U.S. maximum of more than nine flashes per square kilometer in Florida (Orville and Huffines 2001). Annual density increases from less than one-half flash per square kilometer in Idaho and northwestern Montana to three to four flashes per square kilometer in northern New Mexico (Huffines and Orville 1999; fig. 2.1), reflecting more importance of the North American monsoon toward the south (Reap 1986). Flash density also increases with elevation (Reap 1986). Lightning most likely to ignite fires is not the most intense but instead has a long, continuing current (LCC; Fuquay et al. 1967, 1972; Latham and Schlieter 1989), which occurs in only a small percentage of total flashes (A. R. Taylor 1974), most of which are negatively charged. Negatively charged lightning with LCC has a slightly higher probability of ignition than positive lightning has (Latham and Schlieter 1989). In Idaho, negative strikes were correlated with number of fires ($r = 0.49$) more than were positive strikes ($r = 0.24$); number of strokes per strike and strike intensity were only weakly correlated with fires

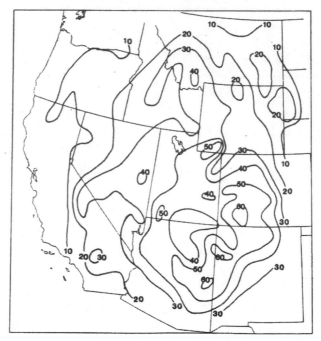

Figure 2.1. Geographical variation in lightning in the Rocky Mountains. Shown is the number of days with two or more lightning strikes, based on 1983–84 summers.
Source: Reproduced from Reap (1986, p. 791, fig. 7) with permission of the American Meteorological Society.

(r = <0.28; Benoit and Strauss 1994). Lightning characteristics are weakly correlated with ignition because ignition is limited by moisture and fuels (Rorig and Ferguson 1999).

Lightning storms that lead to fires typically have shorter rains before and after the fire than storms that do not ignite fires (Gisborne 1931; Fuquay, Baughman, and Latham 1979; Hall 2007). So-called dry lightning is favored in (1) high-base storms, (2) outside rain areas in wet storms, and (3) fast-moving storms with little precipitation in one place (Rorig and Ferguson 2002). Tall thunder clouds have the most lightning, and high base height favors precipitation evaporating before reaching the ground (Fuquay 1962), an effect enhanced by a dry atmosphere below the storm (Bothwell 2000). In the northern Rockies, lightning days with little rain and many ignitions are thus most common with low-level dry air, instability leading to convection, and high base height; days with few ignitions have rain reaching the surface (Rorig and Ferguson 1999, 2002).

Dry lightning may occur with a disturbance in the upper levels of the atmosphere moving around an upper-level ridge of high pressure (Bothwell

2000). Fast-moving frontal storms (e.g., cold air advancing on warm air) may also track across the United States, leaving a wave of ignitions (Komarek 1966). In the southern Rockies, monsoon flow with moist tropical air can promote ignitions if precipitation is limited (Komarek 1966). Lightning and ignitions may occur in episodes and spatial clusters, mirroring storms (Komarek 1968). In 2000, most of the more than a thousand ignitions in the central and northern Rockies traced to eight major lightning events on days with little rain and low humidity (Bothwell 2000). Lightning started 335 fires in one day in the northern Rockies (Barrows 1951a) and in 1940 about 2,000 fires in three days in two wilderness areas, where a hundred strikes led to 60 fires (G. A. Thompson 1964). Dry and wet lightning days differ, and indices of moisture and instability allow forecasting (Rorig and Ferguson 1999) by the National Weather Service's Storm Prediction Center (Bothwell 2000).

The shock wave of a lightning strike dislodges and fragments fuels (e.g., bark, needles, wood) and volatilizes flammable extractives (e.g., terpenes), allowing rapid ignition and commonly producing a fireball visible as a 1- to 4-meter-diameter "flare-up" at the base of a tree that lasts less than a second. A flare-up can collapse, igniting fine fuels (e.g., litter) at a tree base (A. R. Taylor 1974); however, of 11,835 fires in the northern Rockies in the 1930s, 34 percent of ignitions were in snags, 30 percent in *duff* (partly or fully decomposed organic matter below litter), 12 percent in down wood, 10 percent in green tree tops, and 8 percent in grass (Barrows 1951a). Rotten wood is among the most easily ignitable fuels (Stockstad 1979). Ignitions are more likely in deeper duff and in finer fuels with lower bulk density (Latham and Schlieter 1989).

These fuel effects lead to unequal ignition probability, with grasslands and sagebrush less likely than forests to ignite. The *ignition ratio*, the number of lightning strikes per fire start, is between 330 and 84 in grasslands, 144 in sagebrush-grass, 34 in lodgepole pine, between 29 and 42 in mixed conifer, between 24 and 34 in ponderosa pine, 20 in Douglas-fir, and 10 in logging slash, based on modern fire records (Meisner et al. 1994; Latham and Williams 2001). Western hemlock, grand fir, subalpine fir, and Engelmann spruce forests appear most ignitable, based on more fires per unit area than expected, whereas western white pine and lodgepole pine are less ignitable (P. M. Fowler and Asleson 1984). In a study by Fowler and Asleson (1984), older forests were somewhat more ignitable than expected, but there was no difference among forest density classes. Data from the 1930s and 1940s in the northern Rockies show that grand fir and western hemlock forests, along with grasslands, had the highest fire densities (Barrows 1951a). However, ignition ratio and *fire density* (number of fires per unit area) may be poor predictors of more meaningful measures of fire (e.g., fire rotation, mean fire interval). Probability of fire spread, controlled by fuels and weather, may substantially reshape initial patterns of ignition and fire density.

FUELS

This section reviews classification and inventory of fuels, how fuels vary spatially, and the processes that influence spatial and temporal variation in fuels. Rocky Mountain fuels are characterized by spatial heterogeneity and fluctuation, not uniform buildup.

Classifying and Inventorying Fuels and Their Potential for Fire

The Fuel Characteristic Classification System (FCCS; Ottmar et al. 2007) considers a *fuelbed*, a homogeneous vegetation unit on the landscape, to have potentially six horizontal strata and 18 fuelbed categories sharing common combustion properties (fig. 2.2a). Each stratum or category is characterized by variables affecting its contribution to fire behavior (Prichard et al. 2007; Riccardi, Ottmar, et al. 2007). For example, the canopy stratum includes percentage cover, height, crown base height, foliar moisture content, stem density, ladder fuels (small trees and shrubs that can allow fire to climb into tree canopies), and snags (Riccardi, Ottmar, et al. 2007).

The physical properties of fuels affect the intensity and rate of fire spread (table 2.1). *Heat content* controls potential energy release during combustion but does not vary much, except that pitch and other extractives have high values (Whelan 1995). *Loadings*, the mass of a particular fuel per unit area, represent the potential energy available per unit area from combustion, but other properties affect fuel consumption. More fuel will likely be consumed if (1) surface area–to–volume ratio is high (so initial fuel moisture is lower); (2) fuels are in shorter time-lag moisture classes, so moisture evaporates more easily; and (3) *packing ratio* is near optimum. At a low packing ratio, fire burns slowly because heat transfer between particles is insufficient; at a high packing ratio, fire burns slowly because of deficient oxygen (Burgan and Rothermel 1984). *Time-lag fuel moisture classes* reflect the time for dead fuels with a particular moisture content to adjust about 63 percent of the way to equilibrium with atmospheric conditions (Deeming, Burgan, and Cohen 1977). The actual time lag, however, rarely matches diameter limits exactly, because other fuel properties affect response (H. E. Anderson 1990).

Accuracy of data on fuels varies along a gradient from detailed field studies, which are most accurate but cover small areas, to remote sensing, which is the

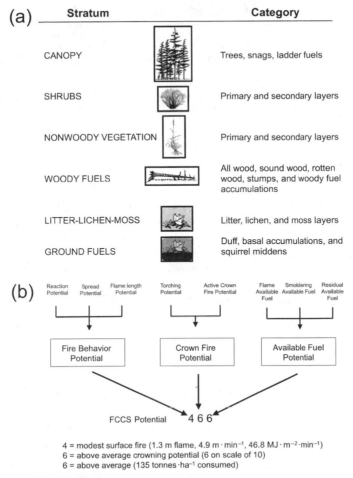

Figure 2.2. The Fuel Characteristic Classification System, including (a) fuelbed strata and categories and (b) fire potentials.

Source: Reproduced from Ottmar et al. (2007, p. 2387, fig. 3, and p. 2388, fig. 4) with permission of the National Research Council, Canada.

least accurate but covers large areas. Detailed field sampling (J. K. Brown, Oberheu, and Johnston 1982; J. H. Scott and Reinhardt 2002; Lutes et al. 2006) is time consuming, but yields highly accurate data. Faster, but less accurate, is to field match a particular fuelbed with photos of actual or synthetic fuelbeds whose properties are known (table 2.2; Keane and Dickinson 2007). Less accurate, but acceptable for some uses, are standard *fuel models* for average conditions. Thirteen standard models for use across the United States (Albini 1976; H. E. Anderson 1982) have been expanded to 40 (J. H. Scott and Burgan 2005), but FCCS had more than 250 fuelbeds in 2007 (Prichard et al. 2007), and the user can create custom fuelbeds or modify existing ones. Finally, remote sensing, reviewed below, can estimate fuels over large areas.

Table 2.1. Some Physical Properties of Fuels

Property	Symbol	General Meaning	Effects	Units
Heat content	h	Energy per unit of fuel	Potential energy and fire intensity per unit mass of fuel	MJ kg^{-1}
Loading	w	Total mass of fuel	Potential energy and fire intensity from total fuel	mT ha^{-1}
Fuelbed depth	δ		One factor in bulk density	cm
Surface area to volume	σ	Fuel surface area per unit volume of fuel	Rate of change in fuel moisture and temperature	m^2 m^{-3}
Particle density or specific gravity	ρ_p	Fuel mass per unit volume of fuel	Heat transfer into fuel particles and time to ignition	kg m^{-3}
Bulk density	ρ_b	Fuel mass per unit volume of fuel and air; also fuel loading or fuelbed depth	Affects rate of spread	kg m^{-3}
Packing ratio	β	Proportion of the fuelbed that is fuel (as opposed to air): bulk density/particle density (ρ_b/ρ_p)	As packing ratio increases up to an optimum point, reaction intensity increases, but above this, insufficient oxygen reaches the fuel and spread declines	ratio from 0.0 to 1.0
Time-lag fuel moisture			Time to move 63% toward equilibrium moisture	
1-hour	1-h	≤0.64 cm/≤0.25″ diameter		
10-hour	10-h	0.64–2.54 cm/ 0.25–1″ diameter		
100-hour	100-h	2.54–7.62 cm/1–3″ diameter		
1,000-hour	1,000-h	7.62–20.32 cm/3–8″ diameter		
>1,000-hour	>1,000-h	>20.32 cm/>8″ diameter		

Sources: Burgan and Rothermel (1984); Prichard et al. (2007); Riccardi, Prichard, et al. (2007).

Table 2.2. Photo Series for Estimating Fuel Loadings in Rocky Mountain and Nearby Ecosystems

Vegetation Type	Location	Sources
Montane grasslands	Northwest	Ottmar, Vihnanek, & Wright (1998)
Montane grasslands	C CO	Battaglia et al. (2005)
Sagebrush shrublands	Southwest	Ottmar, Vihnanek, & Regelbrugge (2000)
Sagebrush shrublands	Northwest	Ottmar, Vihnanek, & Wright (1998)
Sagebrush shrublands	MT	Ottmar, Vihnanek, & Wright (2007)
Mixed mountain shrublands	Rocky Mtns.	Ottmar, Vihnanek, & Wright (2000)
Piñon-juniper woodlands	Southwest	Ottmar, Vihnanek, & Regelbrugge (2000)
Ponderosa pine forests	C CO	Battaglia et al. (2005)
Ponderosa pine forests	MT	Fischer (1981b)
Ponderosa pine–juniper forests	MT	Ottmar, Vihnanek, & Wright (2007)
Ponderosa pine–western larch forests	MT	Fischer (1981b)
Douglas-fir forests	C CO	Battaglia et al. (2005)
Douglas-fir forests	MT	Fischer (1981b)
Moist northwestern forests	MT	Fischer (1981a)
Moist northwestern forests, harvested	ID	Koski & Fischer (1979)
Quaking aspen forests	C CO	Battaglia et al. (2005)
Quaking aspen forests	NW WY	J. K. Brown & Simmerman (1986)
Quaking aspen forests	Rocky Mtns.	Ottmar, Vihnanek, & Wright (2000)
Upper-montane mixed-conifer forests	C CO	Battaglia et al. (2005)
Lodgepole pine forests	C CO	Battaglia et al. (2005)
Lodgepole pine forests	N CO	M. E. Alexander (1978)
Lodgepole pine forests	Rocky Mtns.	Ottmar, Vihnanek, & Wright (2000)
Lodgepole pine forests	MT	Fischer (1981c)
Lodgepole pine forests, harvested	SE WY	Popp & Lundquist (2006)
Spruce-fir forests	C CO	Battaglia et al. (2005)
Spruce-fir forests	MT	Fischer (1981c)
Spruce-fir forests, harvested	SE WY	Popp & Lundquist (2006)

Note: CO = Colorado, ID = Idaho, MT = Montana, WY = Wyoming, C = central, N = north, SE = southeast. (Similar abbreviations are used in subsequent tables.)

The potential of standard fuel models to support various types of fires can be evaluated for a standard set of weather conditions to estimate relative fire hazard, a notion partly embedded in the National Fire Danger Rating System (Deeming, Burgan, and Cohen 1977), but expansion is warranted to include more factors (H. E. Anderson 1974). H. E. Anderson has suggested seven: (1) rate of spread, (2) rate of area growth, (3) fire intensity, (4) crowning potential, (5) firebrand potential, (6) spot fire potential, and (7) fire persistence. FCCS includes eight, grouped in relative indices (on a 0–10 scale) of surface fire behavior, crown fire, and available fuel (fig. 2.2b; Sandberg, Riccardi, and Schaaf 2007).

Spatial Variation in Fuels among Ecosystems and across Landscapes

Loadings by stratum for the FCCS fuelbeds available for the Rockies (table 2.3) illustrate common patterns among ecosystems. In forests, loadings are strongly shaped by trees, because as trees die and become snags, then sound wood, rotten wood, and ultimately duff (fig. 2.3), loadings move among these categories. The largest loadings in forests are thus in the trees, snags, large sound wood (1,000-hour), rotten wood, and duff (see table 2.3). The category with the largest value may mirror recent mortality agents; for example, lodgepole pine fuelbed 22, which lacks insect activity, has lower loadings of snags, and large sound and rotten wood, than does fuelbed 23, which has recent insects and disease. Duff has large loadings in almost all vegetation types. Other fine fuels nearly always have low loadings (i.e., less than 3 metric tons per hectare in each category) but highest spatial continuity, along with high surface area–to–volume ratios, enabling fire spread in forest understories. Loadings are lowest in grasslands, sagebrush, and piñon-juniper, intermediate in ponderosa pine and mixed-conifer forests, and highest in subalpine forests, except lodgepole pine.

Remote sensing has modest promise for mapping spatial variation in fuels (Keane et al. 2001). However, some key fuel parameters vary substantially over

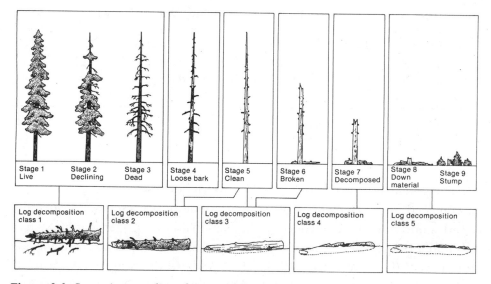

Figure 2.3. Stages in mortality of trees and decomposition of snags and down wood. *Source:* Reproduced from Maser et al. (1979, p. 80, fig. 44).

Table 2.3. Loadings for Strata and Categories in Fuelbeds for the Rocky Mountains

Fuelbed No.	Fuelbed Name	Trees	Snags	Shrubs	Non-Woody	Sound Wood				Rotten Wood	Stumps	Lichens, Moss	Litter, Duff
						1-hr	10-hr	100-hr	1,000-hr				
	RIPARIAN WOODLANDS												
1	Black cottonwood–Douglas-fir–quaking aspen	**78.4**	3.8	2.7	0.5	0.5	2.2	9.5	2.7	**32.7**	0.0	3.5	**101**
	PIÑON-JUNIPER WOODLANDS												
210	Piñon-juniper forest	**3.5**	**7.1**	1.4	0.3	0.5	1.1	2.7	0.8	1.1	0.0	**3.8**	1.4
	PONDEROSA PINE–DOUGLAS-FIR FORESTS												
27	Ponderosa pine–twoneedle piñon–Utah juniper	**9.5**	3.8	4.4	1.4	0.5	0.8	2.7	**5.4**	**24.5**	0.3	3.3	2.7
28	Ponderosa pine savanna	5.2	**20.1**	4.4	2.2	0.3	0.3	0.5	2.2	**9.5**	0.5	2.5	**9.8**
34	Interior Douglas-fir–interior ponderosa pine–gambel oak	12.0	**25.3**	1.9	0.5	0.3	4.1	10.9	27.4	9.5	0.3	3.8	**27.5**
67	Interior ponderosa pine–Douglas-fir	15.8	**47.4**	3.8	2.2	1.4	3.8	3.3	**16.3**	7.1	1.1	3.8	**57.4**
24	Pacific ponderosa pine–Douglas-fir	**15.8**	5.2	0.0	1.4	0.3	0.5	2.2	**7.6**	6.8	4.6	4.6	**13.6**
53	Pacific ponderosa pine	**26.1**	6.5	0.0	0.3	0.3	0.3	4.1	**29.9**	12.3	0.3	5.2	**50.9**
212	Pacific ponderosa pine (selection-logged)	14.2	**21.0**	0.0	0.3	2.7	8.7	8.2	**117**	7.4	1.9	2.5	**58.8**
4	Douglas-fir–ceanothus (clear-cut 15–20 yrs ago)	10.3	0.0	7.6	1.6	0.5	0.8	10.3	0.0	**64.0**	**35.9**	5.4	**29.9**
18	Douglas-fir–oceanspray (prescr. fire 5 yrs ago)	**20.7**	6.0	0.3	0.5	1.1	2.2	0.5	**27.5**	4.6	1.6	2.2	**32.7**

Table 2.3. (Continued)

Fuelbed No.	Fuelbed Name	Trees	Snags	Shrubs	Non-Woody	Sound Wood 1-hr	10-hr	100-hr	1,000-hr	Rotten Wood	Stumps	Lichens, Moss	Litter, Duff
	MIXED-CONIFER FORESTS												
29	Interior ponderosa pine–Engelmann spruce–Douglas-fir (selection-logged with slash left)	28.3	1.9	4.1	0.5	2.7	4.1	5.4	49.0	43.6	1.4	1.6	40.3
273	Engelmann spruce–Douglas-fir–white fir–interior ponderosa pine	28.0	96.1	0.3	0.3	0.8	4.1	8.2	25.9	6.8	0.5	7.4	46.0
208	Grand fir–Douglas-fir	30.8	234	1.9	0.3	1.4	4.4	9.0	45.2	8.2	0.3	3.3	63.7
286	Interior ponderosa pine–limber pine	10.3	18.2	1.4	0.3	0.5	2.7	2.7	24.5	46.3	0.3	7.4	34.6
	SUBALPINE FORESTS												
21	Lodgepole pine (wildfire)	1.6	28.9	0.0	0.0	0.5	2.7	6.8	23.1	2.5	0.0	4.1	17.7
22	Lodgepole pine (none)	11.4	0.5	0.0	0.5	1.1	6.0	7.6	25.9	4.9	0.0	4.1	35.4
23	Lodgepole pine (insects and disease)	10.3	77.9	0.8	0.5	0.5	3.0	4.4	52.8	18.0	0.0	3.8	17.7
59	Subalpine fir–Engelmann spruce–Douglas-fir–lodgepole pine	16.1	42.7	2.7	0.8	1.4	2.7	5.4	28.6	28.6	0.0	4.1	119
70	Subalpine fir–lodgepole pine–whitebark pine–Engelmann spruce	26.1	44.7	2.5	0.5	0.8	3.3	6.8	51.7	38.7	0.5	5.7	38.1
61	Whitebark pine–subalpine fir	31.6	84.9	2.2	0.5	0.3	1.1	4.6	49.0	67.8	1.1	5.4	42.7
42	Trembling aspen–Engelmann spruce	19.9	1.9	0.0	1.1	0.8	1.9	4.9	26.1	13.6	0.3	2.7	30.5

GRASSLANDS												
66 Bluebunch wheatgrass–bluegrass	0.0	0.0	**2.7**	0.0	0.0	0.0	0.0	0.0	0.0	**3.5**	**65.9**	
235 Idaho fescue–bluebunch wheatgrass	0.0	0.0	0.0	**5.4**	0.0	0.0	0.0	0.0	0.0	0.0	**0.3**	—
63 Showy sedge–alpine black sedge	0.0	0.0	0.0	**3.8**	0.0	0.0	0.0	0.0	0.0	0.0	**0.8**	—
131 Bluestem–Indian grass–switchgrass	0.0	0.0	0.1	**9.0**	0.0	0.0	0.0	0.0	0.0	0.0	**5.2**	—
57 Wheatgrass–cheatgrass	0.0	0.0	0.0	**0.3**	0.0	0.0	0.0	0.0	0.0	0.0	**0.3**	—
213 Wheatgrass–cheatgrass grassland	0.0	0.0	0.0	**0.5**	0.0	0.0	0.0	0.0	0.0	0.0	0.0	—
MIXED MOUNTAIN SHRUBLANDS												
216 Gambel oak–bigtooth maple	**50.6**	0.8	0.5	2.7	3.0	5.2	0.0	0.0	0.0	0.0	**6.5**	**37.0**
231 Gambel oak–juniper–ponderosa pine	**16.3**	5.2	2.7	0.3	1.4	1.4	5.4	**19.1**	5.4	0.3	5.4	**12.8**
SAGEBRUSH SHRUBLANDS												
56 Sagebrush shrubland	0.0	0.0	**3.3**	**1.9**	0.0	0.0	0.0	0.0	0.0	0.0	**0.3**	0.0
60 Sagebrush shrubland (burned 2 yrs ago)	0.0	0.0	**1.1**	**0.5**	0.0	0.0	0.0	0.0	0.0	0.0	**2.2**	0.0

Sources: Fuel Characteristic Classification System (Ottmar et al. 2007; Prichard et al 2007; Riccardi, Ottmar, et al. 2007).

Note: Table entries are in megagrams or metric tons per hectare. The highest three loadings within each fuel model are in bold. Loadings are omitted for several minor categories.

short distances (J. K. Brown 1981), making mapping difficult; and some, such as bulk density, require field measurements. A national map of fuel models was developed at a resolution of one square kilometer from satellite imagery with extensive ground-truthing (Burgan, Klaver, and Klaver 1998). Total loading was predicted with moderate accuracy (R^2 = about 0.7) from 1:15,840-scale aerial photographs of forests in northern New Mexico (K. Scott et al. 2002). Canopy properties, such as crown closure, cover, and bulk density were mapped with modest success (R^2 = 0.46–0.76) from Aster imagery (a satellite that is part of NASA's Earth Observing System) and derived vegetation maps (Falkowski et al. 2005) and with similar success (61–78 percent match with field data) from AVIRIS data (an airborne remote-sensing instrument) collected by NASA aircraft (Jia et al. 2006). Light detection and ranging (LIDAR), a laser-based remote-sensing technique, can accurately estimate canopy bulk density, canopy base height, crown volume, tree height, and fuels by height class, including identification of ladder fuels (Riaño et al. 2003; Andersen, McGaughey, and Reutebuch 2005; Skowronski et al. 2007) and mapping of fuel models (Mutlu et al. 2008).

Important surface fuels, such as large deadwood, cannot be estimated using only aerial imagery because of obstruction by tree crowns, but inclusion of biophysical factors in models allows indirect estimates (Keane et al. 2001). Modest success (55–71 percent of variance explained) was obtained in modeling key fuels (i.e., duff and litter depth and loadings by time-lag moisture classes) across the Black Hills using Landsat TM imagery, topographic variables, and vegetation types (Reich, Lundquist, and Bravo 2004). These methods were also effective in mapping fuel models (Reich, Lundquist, and Bravo 2004; Falkowski et al. 2005). The LANDFIRE project is developing 30-m national maps of the H. E. Anderson (1982) and J. H. Scott and Burgan (2005) fuel models and FCCS fuelbeds, as well as canopy cover, height, bulk density, and base height for forests, using Landsat and biophysical variables (Reeves, Kost, and Ryan 2006). Errors in these products require further research, as error accumulates across each mapping step, and some equations are unreliable above certain limits (Reeves, Kost, and Ryan 2006). Modeling using LANDFIRE fuel maps led to lower spread rates and less crown fire than observed in two fires (Krasnow 2007). Another method is to use field plots to develop equations for spatial prediction, avoiding remote sensing. This was used to map fuel models and canopy height, cover, bulk density, and base height in Boulder County, Colorado, but only 56–62 percent of variation in fuels was explained (Krasnow 2007).

Given the potential errors, indirect methods may not produce locally reliable data, but indirect methods do allow some understanding of the scale and pattern of variability in fuels. The ecological implications of these maps have yet to be elucidated. Fuels in the Black Hills, for example, show broad trends related to topography and fine-scale heterogeneity, yet to be explained (Reich, Lundquist, and Bravo 2004). A limitation of models of fuel loads, in addition to error, is that the processes that shape fuels are stochastic at a fine scale and thus only partly predictable from modeling (e.g., J. K. Brown and See 1981). Some fuels also can change rapidly, so that fuel maps can quickly become outdated.

Processes That Produce Spatial and Temporal Heterogeneity in Fuels

The complexity of processes and interactions that shape the "fuel complex" is illustrated by factors influencing flammability (fig. 2.4). Processes that change fuels (J. K. Brown 1975; Knight 1987) include (1) stand development and succession, (2) natural and human disturbance, (3) disease and parasitism, and (4) decomposition. These interact and vary with the setting. Loadings of duff, litter, grass and forbs, sound wood, and shrubs build up during primary succession from glacier to rock, talus, meadow, prairie, shrubland, krummholz, and forest, a

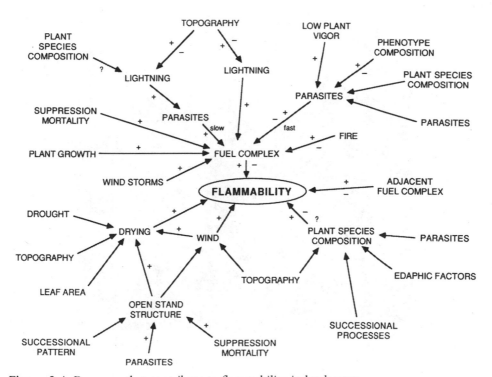

Figure 2.4. Processes that contribute to flammability in landscapes.
Source: Reproduced from Knight (1987, p. 66, fig. 4.4) with the kind permission of Springer Science and Business Media.

trend found using a chronosequence of 431 plots in Glacier National Park (Jeske and Bevins 1979).

Natural disturbances can produce dead fuels, consume fuels through fire, break up fuels, and change the properties and location of fuels. Although disturbances produce dead fuels, they may also change live fuels, with a complex net effect. Bark beetles, for example, may have complex effects on fuels and ultimately flammability (see box 4.2). Low-intensity fires can kill small trees and reduce litter and duff if moisture content is low, but they also can have little effect on fuels, killing few mature trees and reducing fine fuels—which can be rapidly replaced—by less than 50 percent (e.g., Lawson 1972; J. K. Brown 1975). High-severity fires, in contrast, may consume large fractions of fine, dead fuels (J. K. Brown and See 1981), but only about 24 percent of large deadwood was consumed in a 1996 high-severity fire in Yellowstone, leaving a large input of snags and deadwood (Tinker and Knight 2000; see also fig. 2.5). Insect outbreaks, blowdowns, fires, weather events, and diseases can periodically affect large areas, producing extensive mortality and dead fuels; in some cases, complex interactions favor extensive mortality (e.g., W. L. Baker and Veblen 1990).

Figure 2.5. Down wood 10 to 15 years after a crown fire in a lodgepole pine forest in the Bighorn Mountains, Wyoming.
Source: Reproduced from Town (1899, plate XXXIX). The original caption reads: "View in Bighorn Mountains, Wyoming, Hyattsville Road, East Fork of Big Goose Creek showing fallen timber ten to fifteen years after fire ran through it, with new forest growth slowly starting up."

Cycles of fungus, beetles, and fire that tend to synchronize fuels were found in lodgepole pine in Oregon but are unlikely across Rocky Mountain landscapes because disturbance, disease, and parasites have fine-scale, complex, interacting effects (Knight 1987; Lundquist and Negron 2000) that interrupt synchrony. Small patches (i.e., tens of meters in diameter) of altered fuels may arise from damage by lightning (A. R. Taylor 1964), small fires, blowdowns, insects, disease (Lundquist 2007), mistletoe, and other agents. Thirteen disturbance agents were found in a 482-hectare ponderosa pine forest in South Dakota (Lundquist and Negron 2000). Mistletoes, root rots, and other diseases usually expand slowly and laterally from infection centers (box 2.1), contributing to patchiness in fuels across landscapes (e.g., Lundquist 1995). Resulting fuel jackpots (concentrations of fuel), or areas with little fuel, influence fire behavior. Fine-scale variation in fire severity in the 2000 Jasper fire in ponderosa pine in the Black Hills was linked to prefire fine-scale variation in fuels, particularly tree size and density (Lentile, Smith, and Shepperd 2005, 2006).

All fuels do not build up after disturbances in a generally predictable manner. In the Rocky Mountains, data on fuel changes during succession after

---

BOX 2.1

Mistletoe and Fire

Mistletoes of the genera *Arceuthobium* and *Phoradendron* are native parasites of Rocky Mountain conifers and can reduce tree growth, can lead to branch or even tree mortality, and often promote abnormal branching. Mistletoe may raise the fire hazard by increasing resin near infection points, as well as needle and branch mortality, while brooms retained on the trees act as a ladder fuel. Ignited brooms can drop from trees and lead to torching of live trees or be carried aloft, producing a spot fire (Alexander and Hawksworth 1975). Low-severity fires may especially kill mistletoe-infected trees (Harrington and Hawksworth 1990), but Wicker and Leaphart (1976) suggest that low-severity fires were patchy under the HRV, allowing survival of some regenerating trees, which could infect other trees. High-severity fire kills mistletoe, but surviving trees or groups may also allow infection of postfire regeneration (Alexander and Hawksworth 1975; Kipfmueller and Baker 1998b). Ponderosa pine, Douglas-fir, and western larch can particularly survive fires and lead to infection of regeneration (Wicker and Leaphart 1976).

(continued on next page)

BOX 2.1
Continued

Mistletoe disperses through explosive dehiscence and birds, but infection moves slowly, about 10 meters per decade (Geils, Tovar, and Moody 2002). Given fire rotations of two to five centuries, expansion of mistletoe about 0.5 kilometer from surviving trees is feasible between fires. A patchy pattern of infection centers about 0.5 to 2 kilometers in diameter was found across a lodgepole pine landscape in Wyoming and may characterize healthy landscapes subject to high-severity fire (fig. B2.1.1). The population of 24 out of 28 birds that use ponderosa pine forests in Colorado, and bird diversity in these forests, increased with increasing mistletoe infection (Bennetts et al. 1996), also suggesting that mistletoe contributes to a healthy ecosystem.

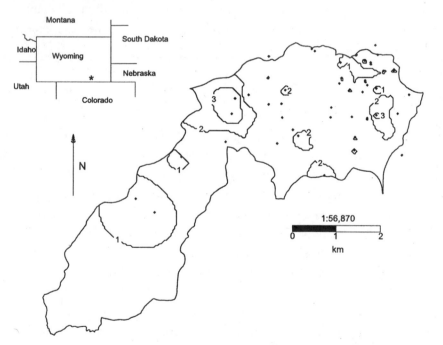

Figure B2.1.1. Mistletoe and fire. A 3,241-ha subalpine watershed in the Medicine Bow National Forest, Wyoming, showing the location of study plots (black diamonds) and contours representing dwarf mistletoe ratings: 1 = light, 2 = moderate, 3 = high.

Reproduced from Kipfmueller and Baker (1998b, p. 79, fig. 1), copyright Elsevier 1998, with permission from Elsevier Limited.

high-severity disturbance come entirely from forests. Fine dead fuels may generally build up after fire, until tree canopies are closed (e.g., the first 125 years in fig. 2.6a, f, g), primarily from needles, leaves, and twigs shed annually (J. K. Brown and See 1981). These fuels may then decline slightly as stands age, probably because decomposition exceeds accumulation (Romme 1982), but the trend is weak. Larger fuels do not build up consistently because of variable legacies from the prefire stand and stochastic processes that produce fuel regardless of stand age (J. K. Brown 1975; J. K. Brown and See 1981). Down wood can be a prominent legacy in some young postfire forests (fig. 2.5)—particularly if an old stand burns—but it may be a small legacy if a young stand burns or the fire is intense and consumes much wood. This variable legacy of down wood produces large scatter in 1,000-hour sound fuels as *stand development* (the process by which forest structure redevelops after a disturbance) begins (fig. 2.6d) and in 1,000-hour rotten fuels 80–120 years later (fig. 2.6e), when these fuels may commonly have rotted (P. M. Brown et al. 1998). Large deadwood and branchwood also accumulate from patchy small disturbances and from self-thinning, which can be a significant source of deadwood during stand development (Kashian et al. 2005). Thus, 10- to 1,000-hour fuels have moderate but quite diverse values in stands 100 to 400 years of age (fig. 2.6b–e).

Decomposition of fine dead fuels can occur within years, but decomposition of snags and large deadwood may proceed through stages that play out over decades (fig. 2.3), the rate of fragmentation and decay dependent on moisture, temperature, and other conditions (Harmon et al. 1986; Keane 2008). Snags can persist in Rocky Mountain forests for long periods on some sites (e.g., see figs. 8.10, 8.11). Logs on the forest floor may not decay fully for more than a century in subalpine forests (Brown et al. 1998), and some hard down wood in ponderosa pine forests can persist for centuries, surviving multiple fires (Ehle and Baker 2003; W. L. Baker, Veblen, and Sherriff 2007). Snags and down wood are not just fuels; they also provide important wildlife habitat and regeneration sites for plants, replenish soil nutrients, and maintain soil structure (Maser et al. 1979; J. W. Thomas 1979; Harmon et al. 1986). Snags and large wood are inherently patchy and fluctuate temporally (Tinker and Knight 2001; Harmon 2002; J. K. Brown, Reinhardt, and Kramer 2003).

Multiple interacting patchy processes leave few spatial and temporal consistencies in fuels. A buildup of total fuel loading may be found only in small samples—for example, in 11 plots in spruce-fir forests in one area (fig. 2.6i). A larger sample of 71 lodgepole-pine plots showed no trend (fig. 2.6h). The authors of a very large study of spruce-fir fuel plots ($n = 4,309$) found that "sometimes a particular cover type in a particular forest showed periodicity between

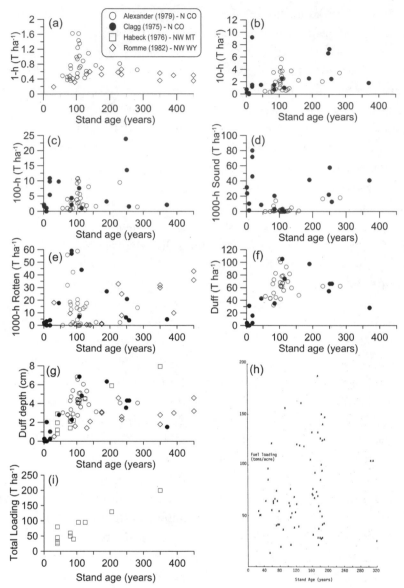

Figure 2.6. Trends in fuels in lodgepole pine (a–g) and spruce-fir forests (h–i). The trends are those during stand development after high-severity fires in Colorado, Montana, and Wyoming: (a) 1-hour time-lag fuel loading; (b) 10-hour time-lag fuel loading; (c) 100-hour time-lag fuel loading; (d) 1,000-hour time-lag sound fuel loading; (e) 1,000-hour time-lag rotten fuel loading; (f) duff loading; (g) duff depth; (h) total fuel loading in successional lodgepole pine forests in the Tolan Creek drainage, western Montana; and (i) total fuel loading in spruce-fir forests in the Selway–Bitterroot Wilderness, Idaho and Montana.
Sources: Graphs (a)–(g) are based on tables in Alexander (1979) and interpolated from graphs in Clagg (1975), Habeck (1976), and Romme (1982); (h) was reproduced from Mathews (1980, p. 25, fig. 1) with the permission of Ed Mathews, Missoula, Montana; (i) data were interpolated from a graph in Habeck (1976).

loading and stand age, but a consistent pattern was not evident" (J. K. Brown and See 1981, 7). They also found that very old forests may have high fuel loads, but "fuel quantities cannot be predicted from age alone in young, immature or mature stands" (p. 12). Change, fluctuation, and spatial heterogeneity in fuels, rather than fuel buildup, characterize Rocky Mountain ecosystems.

TOPOGRAPHY, FIRE BREAKS, AND FIRE CHANNELS

Major topographic effects on fire behavior arise from aspect, slope, slope position, and elevation. Studies of these effects use simple counts of fires, fire density (number of fires per unit area), fire frequency (number of fires per unit period), or area-based measures, such as fire rotation. Southerly facing slopes are generally drier and might have more and larger fires. An aspect effect on fire density was evident in some decades but weak in others (Barrows 1951a,b; Barrows, Sandberg, and Hart 1976; P. M. Fowler and Asleson 1984). Fire rotation and area burned under the HRV did not vary significantly among aspects in subalpine forests in the southern Rockies (W. L. Baker and Kipfmueller 2001; Buechling and Baker 2004). Forest fires were more frequent on south- than on north-facing slopes below 2,800 meters in elevation in northern Colorado (E. Howe and Baker 2003). A stronger aspect-related effect on fire at lower elevations is consistent with evidence that climate and fuel moisture differences among aspects are greater at lower elevations (Hayes 1941). However, this pattern may not hold if ecosystems differ among aspects.

Steeper slopes result in a generally drier environment than shallower slopes because of higher runoff and because steeper south-facing slopes receive higher insolation (Ryan 1976). Ignition probability is likely little affected by slope steepness, although very steep slopes have reduced ignitions (Barrows 1951b; Barrows, Sandberg, and Hart 1976). Steeper slopes may have more rapid fire spread and larger fires because of increased direct flame contact and forward heat transfer by convection and radiation (Barrows 1951b; Ryan 1976). Fires in the 1900s in Colorado were largest, on average, on intermediate (20–29 percent) slopes, but the greatest percentage of fires that became large was on very steep slopes (Ryan 1976). However, fire atlas data in the Selway-Bitterroot Wilderness of Idaho and Montana showed no significant relationship between fire frequency and slope steepness (Rollins, Morgan, and Swetnam 2002). Also, in a smaller Wyoming area, fires under the HRV did not burn more area on steeper slopes (W. L. Baker and Kipfmueller 2001).

Fires may commonly spread uphill because of enhanced heat transfer in an uphill direction and upslope winds during the day (Barrows 1951b). Ignitions and fire density in recent decades were highest at the top of slopes, but percentage of ignitions that became large was highest at the base of slopes (Barrows 1951a,b; Barrows, Sandberg, and Hart 1976; P. M. Fowler and Asleson 1984). Fires also can and do spread downhill. In a Wyoming lodgepole pine forest, fire scars on the downhill side of trees suggested that spread was often downslope, likely because of downslope winds during episodes of rapid spread (W. L. Baker and Kipfmueller 2001). Nighttime cold-air drainage also can produce strong winds from ridges down slopes and valleys (Barrows 1951b; Sturman 1987). A surface fire can back down a hill under these conditions, preheating and drying canopy fuels and allowing a rapid upslope crown fire if winds later shift (Barrows 1951b). This contributed to loss of 14 firefighter lives in the 1994 South Canyon fire in Colorado (B. W. Butler et al. 1998). If a temperature inversion forms over a valley at night, a low-level jet may produce very strong winds and rapid nocturnal fire spread on ridges exposed above the inversion (Baughman 1981).

As elevation increases, air temperature declines, precipitation and humidity increase, and fuel loadings increase (Ryan 1976). Length of the fire season also declines with elevation (Larsen 1925a). Mean fire size and number of large fires in the 1930s and 1940s were lower at higher elevations (Barrows 1951a). Highest fire density in the northern Rockies over the last century was typically in a middle zone about 1,000–2,000 meters in elevation (Barrows 1951a,b; P. M. Fowler and Asleson 1984; Rollins, Swetnam, and Morgan 2001). This zone may be a *thermal belt* with the driest and warmest conditions because of nighttime temperature inversions and cold-air drainage (Hayes 1941). However, the zone also may have the highest ignition probability because of opposing trends of increasing lightning toward higher elevations and increasing drought toward lower elevations (P. M. Fowler and Asleson 1984). The thermal belt may have high fire density, but it is not known whether more meaningful area-based measures (i.e., fire rotation) are short. Fire rotation in subalpine forests in Colorado was about 275 years at about 2,600 meters elevation and 350 years at about 3,400 meters elevation (Buechling and Baker 2004), but this may also reflect the predominance of lodgepole pine in the lower subalpine and spruce-fir in the upper subalpine (Sibold, Veblen, and González 2006).

Larger topographic features may also affect fire regimes. Western slopes of mountain ranges have more upslope wind and higher precipitation, while eastern slopes have more downsloping wind and less precipitation (Alington 1998; Sibold, Veblen, and González 2006). Late-day convection may counter downsloping winds on eastern slopes, limiting fire spread (Alington 1998). Fires under the HRV were more frequent but smaller, and fire rotations were longer on eastern than on western slopes in the southern Rockies (Alington 1998; Sibold, Veblen, and González 2006). However, these climatic effects may be only one factor. More flammable vegetation on western slopes in the northern Rockies may explain higher fire density on western than on eastern slopes (Barrows 1951b). More topographic complexity and fire breaks on the western slopes and more continuous forest with less topographic diversity on the eastern slopes of a range on the Idaho–Montana border led to larger fires and shorter fire rotations under the HRV on the eastern slope (M. P. Murray, Bunting, and Morgan 1998). Recent fires were similar in number but, on average, about twice as large on eastern as on western slopes in Colorado (Ryan 1976). Fires can burn across zones in some cases (Ehle and Baker 2003; Buechling and Baker 2004) but not others (Alington 1998; M. P. Murray, Bunting, and Morgan 1998).

Slope, aspect, and elevation effects on fire regimes may be most evident using measures based on counts (number of fires, fire density, fire frequency), but effects appear muted or lacking when more meaningful area-based measures are used (fire size, area burned, fire rotation). Counts are a poorer measure, since small fires can be numerous yet have little ecological impact. Counts suggest that more fires ignite in the middle-elevation thermal belt, on south-facing (as opposed to north-facing) slopes, near the top of slopes, and perhaps on the eastern slopes of mountain ranges. But area-based measures reveal little net topographic effect (fig. 2.7), because most ignitions go out without burning much area, and most of the burned area in fire regimes comes from the small percentage of fires that are large (chap. 5). These large fires typically burn across multiple topographic settings, may burn downhill as well as up, and are likely to become largest where fuels are continuous, winds are strong, and fire breaks are few.

A *fire break* is a physical or biological feature that can slow or stop fire spread, usually because it has no fuel. Low-fuel areas that slow or stop fire are sometimes called *fuel breaks*. Common natural fire breaks (fig. 2.8) include lakes, streams, ridgelines, rock outcrops, talus slopes, mass movements, and snow avalanche tracks. In modest weather, even moist vegetation (e.g., aspen forests, wetlands) can serve as a fire break. During extreme weather, shady forests can act as a fire break at times, because temperature and wind speed are lower and relative humidity and fuel moisture higher (Gisborne 1935). A backing spot fire can be an ephemeral but effective fire break during a fire, as during the 1967 Sundance fire in northern Idaho (H. E. Anderson 1968). Fire breaks can decrease fire sizes and lengthen fire rotation by limiting fire spread. Fire breaks thus limit large fires that can lead to landscape age-class homogeneity, favoring spatial heterogeneity in age classes across landscapes (W. L. Baker 2003).

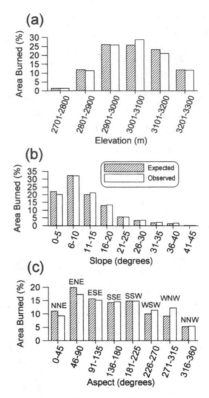

Figure 2.7. Percentage of total burned area in relation to topography. The area is a 3,241–ha subalpine forest in southeastern Wyoming, and the data ("Observed") are from 1680 to 1868 by classes of (a) elevation, (b) slope, and (c) aspect. "Expected" represents the percentage of land area found in each class.

Source: Reproduced from W. L. Baker and Kipfmueller (2001, p. 255, fig. 5) with permission of Blackwell Publishing.

Topographic effects on wind can be important (Schroeder and Buck 1970). Terrain features can channel and accelerate winds and fire spread or create turbulent eddies that increase spotting. Exposed slopes and ridges allow wind to fully affect spread, as in the 1967 Sundance fire in Idaho (H. E. Anderson 1968). Narrow passages can force convergence, increasing wind speed and fire spread; dissected terrain can increase vertical mixing and bring strong winds to the surface (Goens 1990). Canyons commonly increase winds and fire spread, as in the 1949 Mann Gulch fire in Montana, in which 13 firefighters died (Rothermel 1993). In the central and northern Rockies, canyons oriented southwest to northeast may align with winds from cold fronts, which are common in this area, and can channel up-canyon winds. Canyon mouths, where wind direction changes, often have turbulent eddies that can throw up firebrands. Up-canyon winds may collide with regional winds, creating turbulence that enhances fire

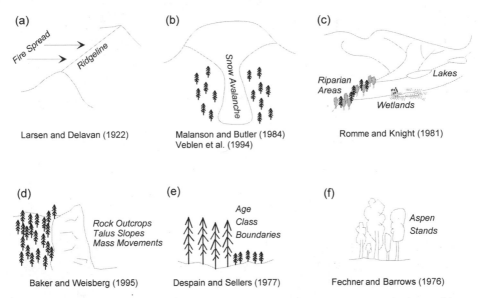

Figure 2.8. Potential fire breaks in Rocky Mountain landscapes: (a) ridgelines, (b) avalanche chutes, (c) wetlands and riparian areas, (d) rock outcrops, talus, and mass movements, (e) forest age–class boundaries, and (f) quaking aspen stands.
Source: Reproduced from W. L. Baker (2003, p. 139, fig. 5.6) with the kind permission of Springer Science and Business Media.

intensity, a contributor to explosive fire behavior and the 14 firefighter deaths at the 1994 South Canyon fire in Colorado (B. W. Butler et al. 1998).

WEATHER AND CLIMATE

Fires in the Rockies are promoted by drought, including high temperatures, strong winds, and unstable air, along with low precipitation, relative humidity, and fuel moisture, documented in many studies (W. L. Baker 2003). Finer fuels adjust more rapidly to drying, but as drought continues, moisture in large fuels can reach low levels (i.e., less than 10 percent). As more fuel becomes combustible, fuel continuity increases, allowing higher intensity and larger fires (Renkin and Despain 1992; M. G. Turner, Hargrove et al. 1994). Earlier, I reviewed effects of temperature, precipitation, humidity, other weather factors, and fuel moisture on fire in the Rockies (W. L. Baker 2003).

Research attention has shifted to regional and global climatic patterns that promote droughts and fires across the Rockies (table 2.4). Some patterns are derived from modern climate and fire data, which have limited length and may

Table 2.4. Links of Rocky Mountain Fires to Drought and Atmospheric and Oceanic Influences

Sources	Location	Period	Variable	Forest or Other Vegetation	Fire Severity	PDSI or precipitation conditions during fire year and preceding years[a]	PNA	ENSO	PDO	AMO
NORTHERN ROCKIES										
Collins, Omi, & Chapman (2006)	ID, MT	Modern	Area burned	All	All	Dry, −1 wet	—	El Niño	Pos.	Neg.
Morgan, Heyerdahl, & Gibson (2008)	ID, MT	Modern	Large fires	All	All	Dry	—	No	Pos.	—
Westerling et al. (2003)	ID, MT	Modern	Area burned	All	All	Dry	—	—	—	—
Knapp (1998)	S ID	Modern	Large fires	Sagebrush, grassland	High	Not nec. dry, −1 wet	—	—	—	—
Gedalof, Peterson, & Mantua (2005)	ID, W MT	Modern	Area burned	Montane	All	Dry	Yes	No	Pos.	—
Barrett, Arno, & Menakis (1997)	ID, MT	Modern, HRV	All fires	All	All	Dry	—	—	—	—
Westerling & Swetnam (2003)	ID, MT	Modern, HRV	Area burned	All	All	Dry	—	El Niño	Pos.	—
Kipfmueller & Swetnam (2000); Kipfmueller (2003)	ID, MT	Modern, HRV	Large fires	Subalpine	Gen. high	Dry, −1 dry	—	No	—	—
Heyerdahl, Morgan, & Riser (2008)	ID, MT	HRV	Small, large	Montane	All	Dry	—	No	No	—

CENTRAL ROCKIES

Study										
Collins, Omi, & Chapman (2006)	WY	Modern	Area burned	All	All	Dry, −1 wet	—	El Niño	Pos.	Neg.
Westerling et al. (2003)	WY	Modern	Area burned	All	All	Dry	—	—	—	—
Knapp (1998)	ID	Modern	Large fires	Sagebrush, grassland	High	Not nec. dry, −1 wet	—	—	—	—
Balling, Meyer, & Wells (1992)	NW WY	Modern	Area burned	Subalpine	All	Dry	—	—	—	—
P. M. Brown et al. (2008)	NE UT	HRV	Regional fire years	Montane	All subalpine	Dry, −2 wet	—	La Niña	Neg.	No
Schoennagel et al. (2005)	NW WY	HRV	Large fires	Subalpine	High	Dry	—	No	No	—

BLACK HILLS

McCutchan & Main (1989)	WY, SD	Modern	Large fires	Montane	All	Dry	—	—	—	—
P. M. Brown (2006)	WY, SD	HRV	Large fires	Montane	All	Dry	—	La Niña	Neg.	Pos.
Kitzberger et al. (2007)	WY, SD	HRV	% of sites	All	All	Dry	—	La Niña	Neg.	Pos.

SOUTHERN ROCKIES

Westerling et al. (2003)	CO, NM	Modern	Area burned	All	All	Dry, −1 wet	—	—	—	—
Collins, Omi, & Chapman (2006)	CO	Modern	Area burned	All	All	Dry, −1 wet[b]	—	La Niña	Neg.	Pos.
Westerling & Swetnam (2003)	WY	Modern, HRV	Area burned	All	All	Dry, −1 wet	—	La Niña	Neg.	—
Schoennagel et al. (2007)	NW CO	Modern, HRV	Large fires	Subalpine	High	Dry, −1 dry	—	La Niña	Neg.	Pos.
Kitzberger et al. (2007)	CO, NM	HRV	% of sites	All	All	Dry, −1 wet	—	La Niña	Neg.	Pos.
P. M. Brown & Shepperd (2001)	CO, WY	HRV	All fires	Montane	All	Dry, −1 wet	—	—	—	—
P. M. Brown & Wu (2005)	S CO	HRV	All fires[c]	Montane	All	Dry, −1–3 wet	—	La Niña	Neg.	—
Sherriff & Veblen (2008)	N CO	HRV	Large fires	Montane	All	Dry, −2 wet[d]	—	La Niña	Neg.	Pos.

Table 2.4. Continued

Sources	Location	Period	Variable	Forest or Other Vegetation	Fire Severity	PDSI or precipitation conditions during fire year and preceding years[a]	PNA	ENSO	PDO	AMO
Veblen, Kitzberger, & Donnegan (2000)	N CO	HRV	Large fires	Montane	All	Dry, −2–4 wet	—	La Niña	—	—
Buechling & Baker (2004)	N CO	HRV	Large fires	Subalpine	High	Dry	—	—	—	—
Schoennagel et al. (2005)	N CO	HRV	Large fires	Subalpine	High	Dry	—	La Niña	No	—
Sibold & Veblen (2006)	N CO	HRV	Large fires	Subalpine	High	Dry	—	La Niña	Neg.	Pos.

Note: PDSI = Palmer drought-severity index, PNA = Pacific–North American pattern, ENSO = El Niño–Southern Oscillation, PDO = Pacific Decadal Oscillation, AMO = Atlantic Multidecadal Oscillation. The modern period is the period after EuroAmerican settlement, and HRV is the period of the historical range of variability, generally before EuroAmerican settlement. The fire variable is important, as studies that use area burned or focus on large fires are of much higher value than studies that use all fires in understanding these links. Table entries indicate conditions during the year of the fire (PNA, ENSO, PDO, and AMO). A dash (—) indicates the effect was not analyzed.

[a]Column entries indicate moisture conditions, based on PDSI or precipitation, during the year of the fire (first entry in column) and preceding years (lag; second entry). If a second entry is given, the number indicates which preceding years, and the condition during those preceding years is indicated by general terms (e.g., wet).

[b]In western Colorado no one-year lag was found.

[c]Fires scarring less than two trees were excluded.

[d]An influence from a wet period two years prior to a fire year was found only in the lower montane (<2,100 meters in elevation).

be confounded with land-use effects. Patterns under the HRV are largely from pre-EuroAmerican fire history and climatic reconstructions.

Magnitude of drought in the fire season, measured by the Palmer drought-severity index (PDSI), an index of dryness or precipitation deficits, is correlated with area burned. Negative PDSI values, or dry years, correspond with large fire years and area burned throughout the Rockies (table 2.4). In low-elevation forests, high precipitation the year before or up to four years before a drought in both modern records and under the HRV (table 2.4) is thought to have increased live fine fuels, promoting larger, more intense fires in subsequent droughts, a logical inference not yet validated with fine-fuel data (Hessl, McKenzie, and Schellhaas 2004). Similar lags are not evident in subalpine forests, where fine fuels are generally high (table 2.4). Surprisingly, only one study (Kipfmueller 2003) found that a preceding dry year increased probability of fire, and extended drought is not linked to increased fire probability (table 2.4). In grassland and sagebrush in southern Idaho and northern Utah, most large fires occurred the year after high precipitation, but the fire year was either dry or wet (Knapp 1998). Widespread fire in mixed-grass prairies of the Great Plains may also have peaked during wet periods with abundant fuels (K. J. Brown et al. 2005).

Spatial and temporal patterns of drought in the United States link to several sources. Drought in the 1900s was correlated with atmospheric and sea surface temperature indices (explained below), particularly the Pacific Decadal Oscillation (PDO) and the Atlantic Multidecadal Oscillation (AMO; McCabe, Palecki, and Betancourt 2004; McCabe and Palecki 2006). The summer PDSI for 1645–1990, from tree ring records across North America, shows several periodicities (Fye et al. 2006): (1) a biennial period linked with winter Quasi-Biennial Oscillation (QBO), an alteration in tropical stratospheric winds; (2) a three- to six-year period linked with the El Niño–Southern Oscillation (ENSO); (3) a seven- to eight-year period linked with the North Atlantic Oscillation (NAO); (4) a decadal period possibly linked with solar variation; and (5) a bidecadal period linked with solar-lunar variation or the PDO. The response of area burned or large fires to atmospheric, oceanic, solar, and lunar forcing is often opposite in the northern and southern Rockies (table 2.4; chap. 10), mirroring a known climatic dipole (Dettinger et al. 1998; Westerling and Swetnam 2003).

The first ocean-atmosphere teleconnection found to be of importance to fire was with ENSO, which represents interannual variation between an El Niño with a warm eastern equatorial Pacific and a La Niña that is cool in the same area, cycling over a period of a few years (Enfield 1989). In the southern Rockies, winter precipitation and snowpack increased during El Niño and decreased during La Niña (Ropelewski and Halpert 1986). ENSO had less effect farther north, leading to average snowfall decreases (El Niño) or increases (La Niña) of about 20 percent (S. R. Smith and O'Brien 2001). ENSO had no effect on modern area burned in northern Idaho and western Montana (Gedalof, Peterson, and Mantua 2005), and it was not related to modern regional fire years in Idaho and western Montana (P. Morgan, Heyerdahl, and Gibson 2008). More area burned in El Niño years in parts of nearby northeastern Oregon under the HRV (Heyerdahl, Brubaker, and Agee 2002). However, in northeastern Washington, ENSO had weak and inconsistent effect on fire under the HRV (Hessl, McKenzie, and Schellhaas 2004), and the spatial synchrony of fires was unrelated to ENSO by itself (Kitzberger et al. 2007; Heyerdahl et al. 2008). In contrast, in low-elevation (less than 2,100 m) ponderosa pine forests in northern Colorado, years of widespread fire under the HRV typically occurred in a dry La Niña year two to four years after El Niño–enhanced spring precipitation (Veblen, Kitzberger, and Donnegan 2000; Sherriff and Veblen 2008). Similarly, large fire years in subalpine forests in this area were promoted under the HRV by La Niña in the year of the fire (Sibold and Veblen 2006; Schoennagel et al. 2007). Regional fire years in montane and subalpine forests in Utah also occurred in a dry La Niña year and, in mixed-conifer forests only, two to three years after wet years (P. M. Brown et al. 2008). ENSO thus may influence fire in the southern Rockies and the southern part of the central Rockies.

The two major Northern Hemisphere winter atmospheric patterns, the Pacific–North America Pattern (PNA) and the North Atlantic Oscillation (NAO), provide some ability to predict summer fire conditions from the vantage of the preceding winter. Droughts and hot, dry weather that lead to low fuel moisture and large modern fires are promoted by an upper-level ridge in the atmosphere over the western United States or by relatively uniform zonal flow (flow generally parallel to latitudinal lines) across the northern United States (Brotak 1983; Gedalof, Peterson, and Mantua 2005). The PNA is characterized by a deep winter Aleutian low linked to a blocking ridge over the northwestern United States and parts of the northern Rockies. This ridge diverts moisture to the north or south, leading to low snowpack in the central and northern Rockies and high snowpack in the southern Rockies (Changnon, McKee, and Doesken 1993). Low winter snowpack contributes to low fuel moisture and larger area burned the following summer in Yellowstone (Balling, Meyer, and Wells 1992). However, winter drought is weakly correlated with modern area burned the following summer in northern Idaho, unless drought persists through spring and summer (Gedalof, Peterson, and Mantua

2005). The winter PNA pattern is associated with large fire years the following summer in modern fire records along the west coast of the United States (Trouet et al. 2006). Although there is little advance predictive power, a ridge over the northwestern United States in summer is associated with modern large fire years in northern Idaho (Gedalof, Peterson, and Mantua 2005) and is linked to warm summer sea surface temperatures in the North Pacific (Liu 2006). The exceptional 1967 Sundance fire in northern Idaho, for example, burned during a period with a warm north Pacific (Finklin 1973).

The NAO is indexed by the strength and position of the subtropical anti-cyclone over the North Atlantic (Fye et al. 2006). A negative NAO is associated with a weak Bermuda high in the eastern Atlantic, which favors lower winter-spring moisture penetration and thus drier conditions in the central United States, including decreases in average PDSI (i.e., deeper drought) in the Rocky Mountains the following summer (Fye et al. 2006). An explicit effect on fire has not yet been studied.

Another key ocean-atmosphere link is the Pacific Decadal Oscillation, characterized by shifts over 20- to 30-year periods in Pacific sea surface temperatures poleward of 20 degrees N latitude (Mantua et al. 1997). The PDO is linked with variability in ENSO (Newman, Compo, and Alexander 2003). During a positive (warm) PDO, a PNA pattern is favored with a blocking ridge (high pressure) and warm, dry winters in the northern Rockies and Northwest and cool, wet conditions in the southern Rockies (Cayan 1996). The positive (warm) phase of the PDO had an enhancing effect, with a five-year lag on large regional fire years under the HRV in northeastern Washington (Hessl, McKenzie, and Schellhaas 2004) and a two-year lag on modern fires in the northern Rockies (Collins, Omi, and Chapman 2006). Lags suggest that the PDO influences longer-term fine-fuel conditions or surface energy relations linked to climate (Hessl, McKenzie, and Schellhaas 2004). However, the PDO was unrelated to fires in ponderosa pine forests in Idaho and western Montana under the HRV (Heyerdahl, Morgan, and Riser 2008) but was positive in seven of nine modern regional fire years (P. Morgan, Heyerdahl, and Gibson 2008). In northern Idaho, modern area burned is correlated with annual to interannual, rather than decadal, variability in the PDO (Gedalof, Peterson, and Mantua 2005). In the 20- to 30-year period of cool, wet winters during a negative (cool) PDO in the northern Rockies, fuel moisture may be high, damping the effects of drought (Collins, Omi, and Chapman 2006). Farther south, however, highest drought frequency in the 1900s was during negative PDOs (McCabe, Palecki, and Betancourt 2004). Large fire years under the HRV in subalpine forests in northern Colorado (Sibold and Veblen 2006; Schoennagel et al. 2007), in ponderosa pine forests in the Black Hills (P. M. Brown 2006) and northern Colorado (Sherriff and Veblen 2008), and in mixed-conifer forests in Utah (P. M. Brown et al. 2008) were during a negative (cool) PDO.

Another oceanic-atmospheric link of importance to fire is with the Atlantic Multidecadal Oscillation, a 65- to 80-year cycle in Atlantic sea surface temperatures north of the equator, where temperatures are shaped by heat transport in the Atlantic thermohaline circulation (Enfield, Mestas-Nuñez, and Trimble 2001; McCabe, Palecki, and Betancourt 2004). A positive (warm) AMO from about 1930 to 1960 was associated with warm, dry conditions and high drought frequency in the West (Enfield, Mestas-Nuñez, and Trimble 2001; McCabe, Palecki, and Betancourt 2004; Sutton and Hodson 2005). In much of the Rockies, a positive AMO had the highest drought frequency when the PDO was negative, but in the northern Rockies and northern plains, highest drought frequency occurred when the PDO was positive (McCabe, Palecki, and Betancourt 2004). In both montane and subalpine forests in northern Colorado, large fire years occurred preferentially under the HRV during a positive AMO with a negative PDO and a La Niña (fig. 2.9; Sibold and Veblen 2006; Sherriff and Veblen 2008). In contrast, in the northern Rockies, low PDSI had

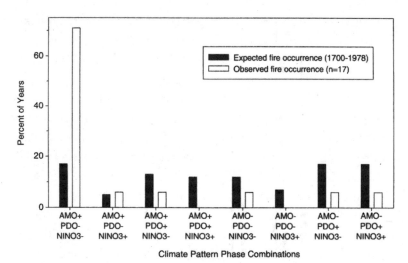

Figure 2.9. Teleconnections between subalpine forest fires and the Atlantic and Pacific. Shown is the "expected" percentage of documented years of large fires (more than 100 ha) in spruce-fir and lodgepole pine forests in Rocky Mountain National Park, Colorado, if years of large fires had been random relative to the climate pattern phase combinations between 1700 and 1978. "Observed" is the actual occurrence of years of large fires for each combination of positive (+) and negative (−) phases of the Atlantic Multidecadal Oscillation (AMO), Pacific Decadal Oscillation (PDO), and the El Niño–Southern Oscillation (NINO3).
Source: Reproduced from Sibold and Veblen (2006, p. 839, fig. 3) with permission of Blackwell Publishing.

strongest effect on modern area burned during a negative (cool) AMO. A negative AMO had lower modern drought frequency in the northern Rockies than did a positive AMO (McCabe, Palecki, and Betancourt 2004) but had more fire because a negative AMO may increase fine fuels in the summer-dry northern Rockies, promoting large fires when drought occurs (Collins, Omi, and Chapman 2006).

The onset of the North American Monsoon (NAM), its regular afternoon thunderstorms having most effect in the southern Rockies, leads to a decline in fire from about July 1 to mid-September. Ignitions continue, because of more lightning, but moist fuels lead to less burned area (Mohrle 2003). A strong winter PNA leads to a subsequent wet monsoon (R. W. Higgins, Mo, and Yao 1998). Both a positive (warm) PDO (Barlow, Nigam, and Berbery 2001) and negative (cool) AMO (Enfield, Mestas-Nuñez, and Trimble 2001) enhance the monsoon, decreasing fire in the southern Rockies (Collins, Omi, and Chapman 2006).

Seasonal pattern and length of the fire season strongly influence fire regimes. In the southern Rockies, the fire season is bimodal, peaking in June before the monsoon (Floyd, Romme, and Hanna 2000; Hall 2007) and in September–October after the monsoon (Ryan 1976). In contrast, in the northern Rockies, where the monsoon has less effect, the fire season is unimodal, spanning the period when mean air temperature is more than 10° centigrade and monthly precipitation is less than five centimeters (Larsen 1925a), and fires typically peak in July (Barrows 1951a). The average fire season is about 150 days in low-elevation forests and about half as long in subalpine forests (Larsen 1925a). In northern Idaho, midsummer conditions generally favor fire, and length of fire season explains modern interannual variation in area burned better than particular weather events do (Gedalof, Peterson, and Mantua 2005). A longer fire season, from lower winter precipitation and earlier springs, is linked to more large (>400 ha) fires throughout the West, but especially in mid-elevation forests in the northern Rockies, comparing 15-year periods before and after the mid-1980s (Westerling et al. 2006).

The well-known 22-year cycle of drought correlates with the sunspot cycle (J. M. Mitchell, Stockton, and Meko 1979; E. R. Cook, Meko, and Stockton 1997) and an 18.6-year lunar nodal-tide effect (Currie 1984). Statistical links of climate with solar variation are found in some studies (e.g., Millspaugh, Whitlock, and Bartlein 2000; Suh and Lim 2006), but not others (J. Moore, Grinsted, and Jevrejeva 2006). A physical link with solar variation, if a link exists, rests on amplification of small variations in solar output through effects on ultraviolet radiation, ozone, clouds, or winds (e.g., Suh and Lim 2006). In Idaho and Montana, number of lightning fires was correlated with sunspot numbers from 1915 to 1939 (Bumstead 1943). In bristlecone pine forests in southern Colorado, stand origins peaked near the Maunder sunspot minimum (W. L. Baker 1992b). Lunar effects on drought-related fire are unstudied. An alternative forcing that overlaps the periodicity of solar-lunar phenomena is the PDO, discussed previously.

Storms with strong winds that significantly affect burned area are relatively ephemeral (days, weeks), and their probability or importance may or may not correlate with seasonal, annual, or decadal indices of drought, atmospheric, and oceanic influences (Heyerdahl, Brubaker, and Agee 2002; Kipfmueller 2003). Almost all published studies of the behavior of large fires in the Rockies document or mention strong winds as a major reason the fire became large. Sustained winds of more than 40 kph gusting to more than 65 kph are associated with large fires in a variety of ecosystems in the Rockies (table 2.5).

Few studies have analyzed the synoptic or mesoscale conditions that explain the strong wind, but five general situations occurred. First, high- or low-level jet streams over fires contributed to rapid spread and extensive burned area in the Rockies, as in fires in the 1950s and 1988 (e.g., Schaefer 1957; Goens 1990). Second, late-season dry cold fronts, particularly in the central and northern Rockies, commonly produced gusty winds and extensive fire spread, as in Yellowstone and the northern Rockies in 1988 (Goens 1990; D. A. Thomas 1991). A cold front contributed to rapid fire spread and firefighter deaths at South Canyon, Colorado, in 1994 (B. W. Butler et al. 1998). A related pattern is the breakdown of an upper-level ridge, moving to the east, followed by a surface trough or surface low and dry air (Nimchuk 1983; Prevedel 2007), a pattern that contributed to firefighter deaths at Mann Gulch, Montana, in 1949 (Werth 2007). Third, dry downsloping winds, particularly on the eastern slope of mountains, likely contributed to the 1978 Ouzel fire and the 1996 Buffalo Creek fire in Colorado and may occur in extreme fire years in the Northwest (Gedalof, Peterson, and Mantua 2005). Finally, high pressure east of the Rockies can lead to strong, dry winds, as during part of the 2002 Hayman fire in Colorado (Finney et al. 2003) and some pre-EuroAmerican fires in southeastern Wyoming (W. L. Baker and Kipfmueller 2001).

FIRE BEHAVIOR

This section discusses the essential physical aspects of fire behavior, including fire spread, fireline intensity, crown fire initiation and spread, spotting, and glowing combustion. The ecological effects of fire are strongly shaped by these

Table 2.5. Wind and Some Large Rocky Mountain Fires

Year	Fire Name & Location	Sources	Winds (kph)	Synoptic Pattern	Approx. Fire Area (ha)	Vegetation
1910	N. Rockies, ID & MT	Koch (1942)	High	—	1,200,000	Many
1926	Quartz Creek, ID	Gisborne (1927)	High	—	7,290	Moist northwestern
1931	Freeman, ID	Jemison (1932)	20–30	—	>8,100	Moist northwestern
1936	Sundance, WY	Cochran (1937)	40	—	3,360	Ponderosa pine
1940	McVey, SD	A. A. Brown (1940)	Variable	—	8,900	Ponderosa pine
1949	Mann Gulch, MT	Rothermel (1993); Werth (2007)	22–35	Breakdown of upper-level ridge	?	Ponderosa pine–Douglas-fir
1955	Robie Creek, ID	Schaefer (1957)	—	Jet stream	3,365	Ponderosa pine
1967	Sundance, ID	Finklin (1973)	55 gusting to 85	Cold front	>20,240	Moist northwestern
1977	La Mesa, NM	Foxx (1984)	50	—	6,180	Ponderosa pine
1978	Ouzel, CO	Butts (1985)	65–80	Downsloping?	405	Subalpine
1984	North Hill, MT	Schwecke (1989)	65–80	—	10,930	Ponderosa pine–Douglas-fir
1985	Butte, ID	Rothermel & Mutch (1986)	16–24 gusting to 32	—	>10,700	Subalpine
1988	Canyon Creek, MT	Goens (1990)	25–50 gusting to 95	Cold front and low-level jet	96,000	Many
1988	Fayette, WY	Beighley & Bishop (1990)	Up to 48	—	15,590	Subalpine
1988	North Fork, WY	D.A. Thomas (1991)	Gusts to 95	Cold front	204,050	Subalpine
1994	Butte City, ID	B.W. Butler & Reynolds (1997)	40–65 gusting to 95	—	2,750	Sagebrush
1994	South Canyon, CO	B. W. Butler et al. (1998)	50 gusting to 72	Cold front	856	Piñon–juniper
1996	Buffalo Creek, CO	W. L. Baker, observation	Strong winds	Downsloping	4,818	Ponderosa pine–Douglas-fir
2000	Cerro Grande, NM	Paxon (2000); Crimmins (2006)	>80	Upper-level ridge	19,290	Ponderosa pine–Douglas-fir
2001	Moose, MT	S. W. Barrett (2002)	Strong winds	—	29,000	N. Rockies mixed-conifer
2002	Hayman, CO	Finney et al. (2003)	32 gusting to 82	Prefrontal, front, then high pressure	55,870	Ponderosa pine–Douglas-fir

Note: Table includes fires for which descriptions of their behavior have been published, along with the wind and synoptic conditions that contributed to fire spread over large areas; other factors, such as drought and unstable air, may also have contributed.

physical aspects. Case studies of fire behavior are presented in chapters 6–9 on Rocky Mountain ecosystems.

Fire Spread and Fireline Intensity

In theory, fires ignited at a particular point increase rapidly in spread rate at first, then increase more slowly, approaching a steady-state rate of spread within minutes in the case of a grass fire or hours in the case of a forest fire (Cheney 1981). Acceleration is often stepped as the fire engages increasing fuel sizes and as wind shifts, but acceleration becomes steadier as fireline width increases and a convection column develops that draws air into the fire (Cheney 1981). In the Rockies, acceleration is fastest for grasslands and sagebrush; a little slower for other shrublands and piñon-juniper woodlands; intermediate for Douglas-fir, ponderosa pine, and lodgepole pine forests; and slowest for mesic subalpine spruce-fir forests and moist northwestern forests (see fig. 10.7).

As a fireline moves across a fuelbed, a certain heat is released per unit area of ground and unit of time. Potential heat is limited by available fuel and by moisture and other conditions. The net rate of heat release, or *reaction intensity*, I_R, is:

$$I_R = \frac{dw}{dt} H \qquad (2.1)$$

where I_R is in kilowatts per square meter, dw/dt is the rate of consumption of fuel per unit area and unit of time in kilograms per square meter per second, w is the weight of fuel consumed per unit area in kilograms per square meter, t is the time in seconds, and H is the heat of combustion of the fuel in kilojoules per kilogram (Rothermel 1972). Rothermel estimated dw/dt using a maximum combustion rate, given properties of the fuel, that is adjusted for mineral and moisture damping of the reaction. Reaction intensity can be calculated for *flaming combustion* or for both flaming and *glowing combustion* (combustion without flames) by quantifying the fuel consumed in each process (R. A. Wilson 1990). The part of reaction intensity propagated forward in flames helps drive the moving fire front (Rothermel 1972; R. A. Wilson 1990); the vertical component leads to crown scorch and a convection column; and the downward component, particularly from glowing combustion, heats the soil.

If the front of the flaming combustion part of the fire, or fireline, moves more quickly through the fuelbed, flames will be higher and the fireline more intense. It may be useful to envision a moving fireline as analogous to a conveyor belt feeding fuel into a stove, in which fire intensity increases as fuel enters the stove faster. Thus, faster spread across a fuelbed leads to higher *fireline in-*

tensity. Formally, fireline intensity, *I*, is the rate of energy output, in kilowatts per meter, from a 1-meter-wide strip parallel to the fireline, extending over the *flame depth*, *D*, from the front to the back of the active (flaming) combustion zone (fig. 2.10a), calculated by:

$$I = Hwr \qquad\qquad (2.2)$$

where *H* is the heat of combustion of the fuel in kilojoules per kilogram, *w* is weight of fuel consumed per unit area in the zone of flaming combustion in kilograms per square meter, and *r* is the rate of spread in meters per second (Alexander 1982). *H* is nearly constant at about 18,700 kilojoules per kilogram, except that extractives (e.g., resins) have higher values (Whelan 1995). Rothermel (1972) defined *unit energy* or *heat per unit area* as *Hw* in equation 2.2. Heat per unit area is also the reaction intensity times the length of time the fire is resident within a unit area (Andrews 1986). Fireline intensity is thus the product of rate of spread and the energy in available fuel. A particular area can burn with a limited range of reaction intensities, which are constrained by unit energy, but with a large range of fireline intensities. Rate of spread is thus often the key variable controlling fireline intensity. Fuel consumption varies over perhaps a 10-fold range and rate of spread over a 100-fold range, so that fireline intensity may vary more than 1,000-fold, from 10 to at least 100,000 kilowatts per meter (Alexander 1982).

Fireline intensity can be estimated from (1) rate of spread and fuel consumption, (2) flame length, and (3) scorch height. These terms are explained below. In the first method, which uses equation 2.2, the fire's rate of spread is monitored and fuel consumption is estimated after the fire (e.g., J. K. Smith, Laven, and Omi 1993), a method validated in a laboratory test (R. M. Nelson and Adkins 1986). Using the second method, flame length (fig. 2.10a) is estimated visually during a fire, using known reference markers and photography (e.g., Britton, Karr, and Sneva 1977), although consistent estimation is difficult (V. J. Johnson 1982) because of fluctuating flames and intensity (Rothermel and Deeming 1980). If flame length can be measured, fireline intensity can be estimated. Fireline intensity (*I*), in kilowatts per meter, was found empirically (Byram 1959) to be:

$$I = 259.833 \ L_F^{2.174} \qquad\qquad (2.3)$$

where L_F is flame length in meters (Alexander 1982). This equation was validated in a laboratory test (Ryan 1981). P. H. Thomas (1963) derived a different equation:

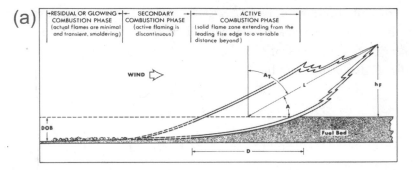

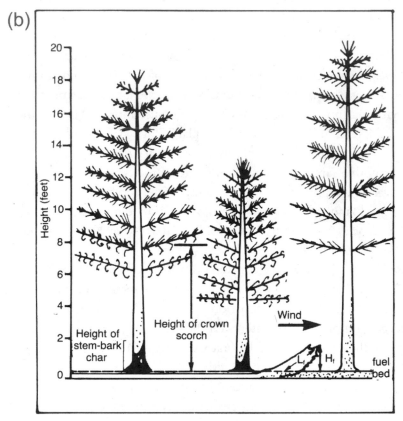

Figure 2.10. Cross section of a surface fire and its effects on trees. (a) The cross section of a wind-driven surface head fire with flame measurements, combustion zones, and consumption of the fuelbed. The fireline extends over the flame depth (D), and the flames themselves have a flame length (L), a flame height (h_F), a flame angle (A), and a flame tilt angle (A_T). Flaming and glowing combustion together result in a certain depth of burn (DOB). (b) Cross section of a wind-driven surface head fire showing flame length (L_f) and flame height (H_f), as well as the height of bark char and height of crown scorch.

Sources: (a) is reproduced from Alexander (1982, p. 351, fig. 1) with permission of the National Research Council, Canada; (b) is reproduced from Cain (1984, p. 17, fig. 1).

$$I = 230.5 \, L_F^{1.5} \tag{2.4}$$

Rothermel (1991b) suggested that equation 2.4 may better apply to crown fires and 2.3 to surface fires. The third method is to estimate fireline intensity from scorch height, h_s (fig. 2.10b), in meters, after a fire, using an equation from Van Wagner (1973), based on 13 experimental oak and pine fires in eastern Canada: $h_s = 0.1483 \, I^{0.67}$, where I is fireline intensity in kilowatts per meter (Alexander 1982). This can be rearranged to estimate fireline intensity from scorch height:

$$I = 17.5 \, h_s^{1.5} \tag{2.5}$$

Unfortunately, field tests show that these methods do not produce congruent estimates of fireline intensity. Equation 2.5 underestimated fireline intensity from equation 2.3 by a factor of about three in a field trial in Arkansas. Scorch height was generally more than twice flame length and more than four times the height of bark char (Cain 1984). In an experimental fire in quaking aspen in Colorado, fireline intensity, from equation 2.2, was underestimated by about six times using equation 2.3 (J. K. Smith, Laven, and Omi 1993). J. K. Smith et al. showed that equation 2.3 also underestimated fireline intensity, from equation 2.2, by about 15–63 times for experimental burns in ponderosa pine forests in northern Idaho, using data in Armour, Bunting, and Neuenschwander (1984). R. M. Nelson (1980) found that equation 2.3 underestimated the intensity of *head fires* (fires burning with the wind) in southern forests, and intensity was better estimated by $I = 462.481 \, L_F^2$, while intensity of *backfires* (burning against the wind) better fit $I = 145.190 \, L_F^{1.5}$. H. E. Anderson et al. (1966) derived equations to be $I = 54.445 \, L_F^{1.536}$ in lodgepole pine and $I = 103.191 \, L_F^{1.493}$ in Douglas-fir slash in Idaho. Thus, fireline intensity may vary with the square or three-halves power of flame length or scorch height, but none of equations 2.3–2.5 appears valid for general use. Equation 2.2 may be the most reliable, but it is difficult to use except in controlled fires. Further research is needed on flame length, scorch height, height of bark char, and fireline intensity in the Rocky Mountains.

Rothermel (1991b) brought fireline intensity, rate of spread, and flame length together in a fire characteristics chart that is useful for comparing fires (e.g., fig. 2.11). This chart has axes for rate of spread and unit energy, along with contours representing their product, fireline intensity, each contour line having corresponding estimates of flame length. An improved chart uses a horizontal axis of actual fuel consumption (Alexander, Stocks, and Lawson 1991), which is

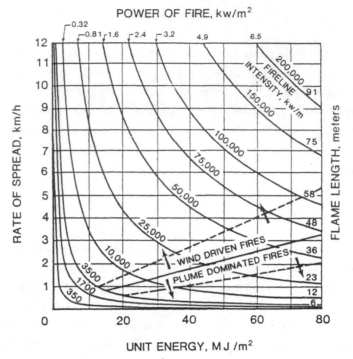

Figure 2.11. Fire characteristics chart for crown fires. The chart has contours for fireline intensity and corresponding estimates of flame length, from equation 2.4, given the fuel's unit energy and rate of spread. Illustrated is a line that approximately separates expected wind-driven and plume-dominated fires.
Source: Reproduced from Rothermel (1995, p. 263, fig. 1).

most useful where that information is known. Estimates made during actual fires can be plotted on the chart to provide comparisons.

Crown Fire Initiation and Spread

In vegetation with elevated shrub or tree crowns, fires can be categorized by which layers are combusted. A *surface fire* combusts only surface fuels below the shrub or tree canopy, a *passive crown fire* combusts surface fuels and torches individual stems or groups of stems, an *active crown fire* combusts both surface and canopy fuels, and the rare *independent crown fire* burns only in canopy fuels (Van Wagner 1977; J. H. Scott and Reinhardt 2001). A *ground fire* combusts organic soils (e.g., wetlands) by glowing combustion (Hungerford, Frandsen, and Ryan 1995).

Initiation of a crown fire from a surface fire can depend on availability of surface fuels for flaming combustion, vertical continuity of fuels up to the crowns, foliar moisture content of the crowns, and wind speed. Van Wagner

(1977) derived an equation for the fireline intensity that could lead to initiation of crown fire from a spreading surface fire, $I'_{initiation}$, in kilowatts per meter, an equation reformulated by Scott and Reinhardt (2001) as:

$$I'_{initiation} = \left(\frac{CBH(460 + 25.9FMC)}{100} \right)^{3/2} \tag{2.6}$$

where *CBH* is the lowest canopy base height above ground, in meters, with enough canopy fuels to allow fire to carry vertically through the rest of the canopy, and *FMC* is the foliar moisture content of the tree crowns, in percent of dry weight. However, M. G. Cruz, Alexander, and Wakimoto (2004) empirically modeled probability of crown fire initiation, using 71 experimental surface and crown fires that were set in Canada and Australia to measure fire behavior. Their logistic model, which correctly classified about 85 percent of fires as surface or crown fires, links crown fire initiation to wind speed, the fuel strata gap (the vertical gap from the top of the surface fuels to the bottom of the canopy or ladder fuels), and surface fuel consumption (fig. 2.12a).

The critical wind speed (U'_{10} = wind speed at 10 meters above open ground) needed to initiate crown fire is considered by M. G. Cruz, Alexander, and Wakimoto (2004) to be up to an order of magnitude or more too high in common model systems. BEHAVE (Andrews 1986); BehavePlus (Andrews, Bevins, and Seli 2005); FARSITE (Finney 1998); FlamMap (Finney 2006); and NEXUS (J. H. Scott 1999), based on equation 2.6 (J. H. Scott 2006), all underestimate the probability of crown fire initiation at a particular windspeed (M. G. Cruz, Alexander, and Wakimoto 2004). Scott (2006) showed that FlamMap has larger errors than NEXUS in matching the empirical Cruz et al. model for Douglas-fir, ponderosa pine, and lodgepole pine forests in the Rockies (fig. 2.12). The M. G. Cruz, Alexander, and Wakimoto (2004) model (fig. 2.12a) provides the only validated empirical model of crown fire initiation and is a significant advance, but all empirical models may not work as well outside conditions for which they were developed (M. G. Cruz, Butler, and Alexander 2006). The Fuel Characteristic Classification System includes algorithms for estimating crown fire initiation potential that are consistent with the M. G. Cruz, Alexander, and Wakimoto (2004) model (see Schaaf et al. 2007). Crown fire initiation and spread (CFIS) software (Alexander, Cruz, and Lopes 2006), which includes the M. G. Cruz, Alexander, and Wakimoto (2004) model, is the best available tool for estimating crown fire initiation.

Physically modeling the spread of crown fires is difficult because of fluctuating intensity and spread rates and inherent instability (B. W. Butler et al. 2004),

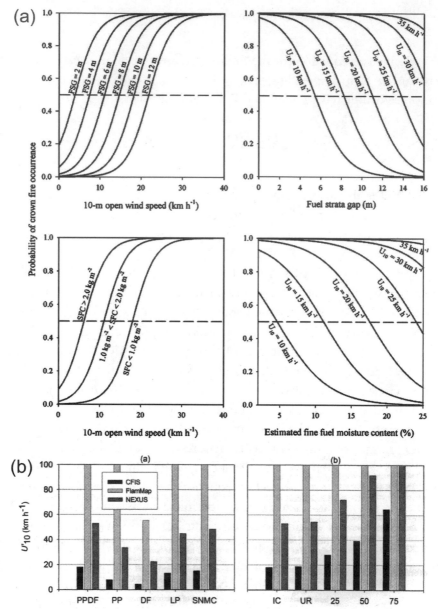

Figure 2.12. Crown fire initiation and spread. Shown are results based on the M. G. Cruz, Alexander, and Wakimoto (2004) logistic regression model: (a) general patterns from the model, showing the probability of crown fire occurrence given combinations of wind speed (U_{10} = wind speed 10 meters above open ground), fuel strata gap (FSG), and surface fuel consumption (SFC); and (b) the estimated critical wind speed (U'_{10}) at which active crown fire spread is possible, based on CFIS (M. G. Cruz et al. 2005), FlamMap (Finney 2006), and NEXUS (J. H. Scott 1999), for ponderosa pine–Douglas fir (PPDF), ponderosa pine (PP), Douglas-fir (DF), lodgepole pine (LP), and Sierra Nevada mixed–conifer (SNMP), and, in the right panel, for a ponderosa pine stand, given initial conditions (IC), understory removal (UR; i.e., removal of all trees less than 5 centimeters in diameter at breast height), and 25, 50, or 75 percent of initial basal area removed (25, 50, 75) by thinning from below.

Source: (a) is reproduced from M. G. Cruz, Alexander, and Wakimoto (2004, p. 649, fig. 3) with permission of the Society of American Foresters; (b) is reproduced from J. H. Scott (2006, p. 16, fig. 5).

including rapid changes in behavior that can lead to "blowup" fires with unexpectedly high spread rates. Infrared imagery of a prescribed crown fire in Alaska showed bursting convective plumes of heat shooting tens of meters horizontally, at speeds ten times greater than ambient winds, before they became buoyant and rose vertically (Coen, Mahalingam, and Daily 2004). Wind-driven crown fires typically have strongly bent convection columns and relatively high, steady rates of spread compared to plume-dominated fires, in which a convection column can lead to turbulent indrafts, increased radiation, and unexpectedly high spread rates (figs. 2.11, 2.13), including horizontal heat and flame bursts. Plume-dominated fires are more likely on slopes (Rothermel 1991a). High convection columns can also have rapid downbursts, triggered by evaporative cooling and precipitation, that can produce sudden exceptional winds and fire spread (Haines 1988; Rothermel 1991b).

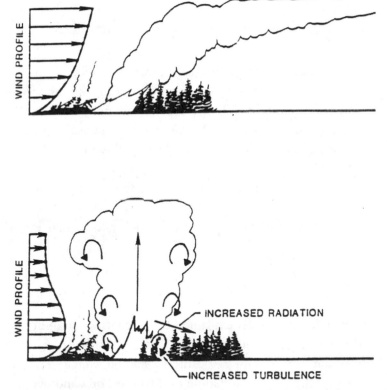

Figure 2.13. Wind-driven and plume-dominated crown fires. Crown fires may be either wind driven (top), with increasing winds aloft, or plume dominated (bottom), with stronger winds near the surface due to indrafts driven by a strong convection column.
Source: Reproduced from Rothermel (1991a, p. 254, figs. 1, 2) with permission of Richard Rothermel, Missoula, Montana.

Maximum spread rates of wind–driven active crown fires are typically 25–100 meters per minute (A. Cruz et al. 2005). The median rate of spread of 57 Canadian and American wildfires, including 5 in the Rockies, was about 35 meters per minute, or 2.1 kilometers per hour (Alexander and Cruz 2006). A set of 8 wind-driven crown fires in the Rockies spread on average 3.34 times as fast as predicted surface fire spread in the same fuels (Rothermel 1991b). Rothermel (1991b) modeled crown fire spread rate as a function of wind speed and surface fuel loading and moisture, bypassing crown fuel characteristics. This model consistently underpredicted the rate of spread of the 57 crown fires, with an average error of about 60 percent (Alexander and Cruz 2006). This model is included in BehavePlus (Andrews, Bevins, and Seli 2005), FARSITE (Finney 1998), FlamMap (Finney 2006), and NEXUS (J. H. Scott 1999).

A. Cruz et al. (2005) empirically modeled rate of spread of crown fires on flat terrain, based on 24 experimental crown fires in Canadian boreal forests, and derived the following equation:

$$CROS_A = 11.02U_{10}{}^{0.9} \ CBD^{0.19} \ e^{-0.17EFFM} \qquad (2.7)$$

where $CROS_A$ is the active crown fire rate of spread, in meters per minute; U_{10} is the wind speed at 10 meters above open ground, in kilometers per hour; CBD is canopy bulk density, in kilograms per cubic meter; and $EFFM$ is estimated fine-fuel moisture, in percent. Active crown fire spread is predicted if the criterion for active crowning (CAC) exceeds the following:

$$CAC = CROS_A/(3/CBD) \qquad (2.8)$$

If not, a passive crown fire rate-of-spread model, rather than equation 2.7, applies:

$$CROS_P = CROS_A \ e^{-CAC} \qquad (2.9)$$

The A. Cruz et al. (2005) model explained 61 percent of variability in the development data set. In the independent test of 57 North American wildfires that exhibited extensive crowning, the model had an average error of 51–60 percent, underpredicting spread for some of the fastest fires and over predicting for some low fuel-moisture conditions (Alexander and Cruz 2006). The model projects that wind speed most strongly affects spread rate, which also increases with CBD and declines as fine-fuel moisture rises (fig. 2.14). A minimum CBD of about 0.1 kilogram per cubic meter for active crown fire spread was found

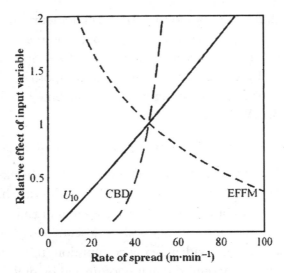

Figure 2.14. Factors affecting the rate of spread of crown fire. Shown are the relative effects of wind speed (U_{10}), canopy bulk density (CBD), and estimated fine-fuel moisture (EFFM) on the rate of spread of crown fires. The graphs are based on the M. G. Cruz, Alexander, and Wakimoto (2005) model under the following conditions: U_{10} = 28 kilometers per hour, CBD = 0.16 kilograms per cubic meter, and EFFM = 7 percent.
Source: Reproduced from Alexander and Cruz (2006, p. 3017, fig. 2) with permission of the National Research Council of Canada.

experimentally (M. G. Cruz et al. 2005, fig. 1e) and was also found in northwestern forests (Agee 1996). Complex, physically based models of crown fire initiation and spread that may allow even better understanding are under development and show promise, but challenges remain (B. W. Butler et al. 2004; M. G. Cruz, Butler, and Alexander 2006; M. G. Cruz et al. 2006). The empirical M. G. Cruz, Alexander, and Wakimoto (2005) model in CFIS currently provides the best available software (Alexander et al. 2006).

Spotting

Fires also spread by *spotting*, in which dislodged, burning fuel particles rise in the fire's convection column, potentially landing and igniting a spot fire ahead of the fireline. Large particles fall close to the fireline, and small particles burn out, so intermediate-size particles have the greatest potential for long-distance spotting. A model estimates maximum spotting distance, which is increased by higher wind speeds, flame heights, and flame duration and is modified by the species of tree and terrain situation (e.g., Albini 1983). Observed spotting distances have commonly been 0.5–1.5 kilometers in high-severity fires in a variety of settings in the Rockies (Cochran 1937; Small 1957; Beighley and Bishop

1990; Finney et al. 2003), with spotting to 6–10 kilometers in some cases (Gisborne 1935; Koch 1942) and even 16–19 kilometers in the 1967 Sundance fire in northern Idaho (H. E. Anderson 1968; Finklin 1973). Spot fires can increase rate of spread by bypassing low-fuel areas or if spotting distance is large (Alexander and Cruz 2006). Spotting has also contributed to firefighter deaths (Rothermel 1993; B. W. Butler et al. 1998).

Glowing Combustion and Heat Release to the Soil

Glowing or smoldering combustion is *pyrolysis* (thermal breakdown) and oxidation without flames (Hungerford et al. 1991). Smoldering begins in ignited wood after flaming stops or in a crack or depression in organic soils, and spreads into wood or downward into organic soils in a sequence of drying, pyrolysis, and onset of a reaction zone with glowing combustion (fig. 2.15). Fuel loadings in Rocky Mountain ecosystems often are dominated by duff and large woody fuels (table 2.3) consumed in glowing combustion after a fireline passes. Rate of

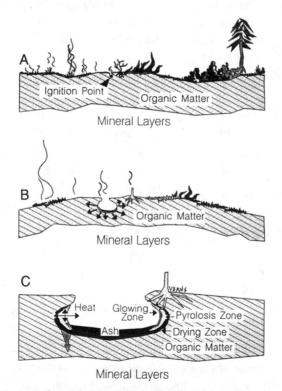

Figure 2.15. Glowing combustion in organic soils: (a) ignition after passage of a fireline; (b) concentric burn hole as combustion expands; (c) sequence of heating and drying, pyrolosis, and glowing combustion, along with heat release into the soil or into the atmosphere. *Source:* Reproduced from Hungerford, Frandsen, and Ryan (1995, p. 83, fig. 4).

spread of glowing combustion is typically two orders of magnitude lower than in flaming combustion, and maximum combustion temperatures are half or less, but with longer duration, leading to a potentially much higher total cumulative heating (heat loading) of the soil (Hungerford et al. 1991). However, much of the heat is either radiated upward away from the soil or escapes in convective currents that heat the atmosphere rather than the soil. For example, only about 30 percent of total heat released during a peat fire actually heats the soil (Hungerford, Frandsen, and Ryan 1995).

Aboveground fire effects of smoldering are dependent on *depth of burn* (fig. 2.10a), reflected in postfire depth of unburned organic material, exposure of mineral soil (Alexander 1982), and consumption of woody fuels. Plants and animals or their perennating structures (e.g., roots) can possibly survive flaming combustion but cannot survive near or within fuels undergoing glowing combustion, where temperatures may exceed 300° centigrade for long periods (Hungerford, Frandsen, and Ryan 1995). The total heat flux into the soil or an organism or its perennating structure determines the belowground biotic effects of fire. If organic soil layers do not ignite because of high moisture content, they may insulate the soil against heat from a passing fireline; but when duff or woody fuel in contact with the soil ignites, the potential heat flux into the soil is high (Hungerford et al. 1991, 1995).

Consumption of duff and large woody fuels by glowing combustion is variable and difficult to predict (Hungerford et al. 1991). Both moisture and mineral content limit glowing combustion in duff (Frandsen 1987), but moisture content is the best predictor of duff consumption. Based on 449 prescribed fires in the northern Rockies, the Southwest, and the Pacific Northwest, consumption was less in deeper duff and was negatively correlated with duff moisture (Reinhardt et al.1991). Duff consumption was less than 15 percent for average duff moisture above about 175 percent and was more than 50 percent when duff moisture was about 50 percent. Preburn loadings of large woody fuel were not important predictors of duff consumption. Consumption of large woody fuels is most shaped by moisture content, extent of rot, and horizontal and vertical arrangement (J. K. Brown et al. 1985). Empirical equations are used to model duff consumption, and the Burnup model (Albini et al. 1995; Albini and Reinhardt 1995, 1997) is used to model consumption of large woody fuels in the First Order Fire Effects Model (FOFEM; Reinhardt 2003). Burnup predicted woody fuel consumption well in a test of 68 prescribed fires in the northern Rockies and the Pacific Northwest (Albini and Reinhardt 1997). Another fuel consumption model is Consume (Prichard, Ottmar, and Anderson 2007), which links to the Fuel Characteristic Classification System (Ottmar et al. 2007).

A fire creates a *heat pulse*, which transfers heat to the soil by conduction and radiation, as well as causing movement of moisture and changes in phase, leading to complex effects on soil temperature (Steward, Peter, and Richon 1990; Campbell et al. 1995; Albini et al. 1996). A passing fireline can raise soil temperature, but glowing combustion of duff and large woody fuels may especially heat soil (Massman et al. 2003). Higher fuel loadings may often increase soil heating but not always (Massman, Frank, and Reisch 2008). As soil temperatures reach about 100° centigrade, most organisms and perennating structures are killed, and by about 300° centigrade, soil structure may be altered (DeBano, Neary, and Ffolliott 1998). Moisture generally increases heat transfer into the soil, but also prevents soil temperature from rising above about 90° centigrade in a moist soil layer, because the heat will evaporate moisture rather than raise soil temperature (Frandsen and Ryan 1986; Campbell et al. 1995).

Soil temperature data from fires in the Rockies are rare. Maximum temperature, at a depth of 2 centimeters, in prescribed surface fires in Colorado was lower in a medium-loading ponderosa pine forest (36° centigrade) than in a higher-loading forest (69° centigrade), a medium-loading grassland (222° centigrade), a high-loading grassland (260° centigrade), or beneath a slash pile (407° centigrade [Massman et al. 2003] or about 300° centigrade [Massman, Frank, and Reisch 2008]). Temperatures did not exceed 100° centigrade, sufficient to kill most organisms, at 5 centimeters or greater soil depth, except beneath the slash pile, where 100° centigrade was reached at more than 5 centimeters depth in one case (Massman, Frank, and Reisch 2008) and more than 10 centimeters depth in another case (Massman et al. 2003). Mathematical models have been derived for estimating soil temperature beneath surface fires and the depth at which lethal temperatures are reached (e.g., Steward, Peter, and Richon 1990). The Campbell et al. (1995) model is used in the First Order Fire Effects Model (Reinhardt 2003). However, instantaneous temperatures do not measure the time-integrated heat flux into organisms that kills tissues (see chap. 3).

SUMMARY

Lightning is most abundant in the southern Rockies and at higher elevations, but ignitions are favored wherever there is little rain before or after lightning, especially in forests and particularly those with ignitable snags, dry duff, or rotten wood. Multiple physical properties of fuels shape post-ignition fire behavior. Wood and duff are influential in having largest loadings, while fine dead fuels are most flammable and spatially continuous. Because of patchy and fluctuating processes, fuel buildup between fires is not a consistent trend. Large fires, which do most of the work in Rocky Mountain landscapes, appear insensitive to topography, except for fire breaks, physical features that modify wind, and differences in setting across mountain ranges. Drought is correlated with area burned and linked to atmospheric-oceanic conditions in the Pacific and Atlantic, but with contrasting trends in the summer-dry northern versus the summer-monsoon southern Rockies. Large fires often have strong winds (e.g., jet streams, cold fronts). Fireline intensity can vary more than 1,000-fold and is most strongly affected by variation in rate of spread, leading to large variations in fire severity, which affect plant and animal survival. New model systems can better predict crown fire initiation and spread, as well as heat transfer into soils.

CHAPTER 3

Fire Effects on Plants: From Individuals to Landscapes

The effects of fire on plants are partly a function of properties of the fire (e.g., fireline intensity), but they also depend on the ability of plants to survive or recolonize. Depending on how soon after a fire they develop, fire effects—immediate effects, postfire reorganization and recovery, or long-term adjustment—can be categorized as first, second, or third order, respectively (fig. 3.1). These time scales interact, however; reorganization and recovery affect the behavior of subsequent fires, as does long-term adjustment. Effects occur across levels of organization, from individuals to populations, communities or ecosystems, and landscapes (Pausas and Lavorel 2003). Lotan et al. (1981) and J. K. Brown and Smith (2000) reviewed general fire effects on plants, primarily at individual and community levels; other studies cover parts of the Rockies (table 3.1). Online information is available about effects on individual species (W. C. Fischer et al. 1996; USDA Forest Service 2006). Computer models can help analyze effects from individual mortality to burn patterns across landscapes (Reinhardt, Keane, and Brown 2001).

PLANT FUNCTIONAL TRAITS AND FIRE RESPONSES

Plant functional traits (Pausas and Lavorel 2003) that allow persistence differ among levels of organization—an individual tree may resprout, allowing it to survive, but effective seed dispersal may be needed to allow a species to persist across landscapes with large fires. Traits were not selected, in an evolutionary sense, solely by fire, since some (e.g., decay resistance) offer advantages against

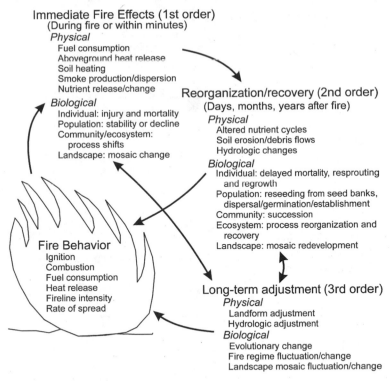

Figure 3.1. Fire effects on the physical and biological environment.
Source: Adapted from Schmoldt et al. (1999).

other selective forces (e.g., drought, disease). Traits also may vary among populations or environments. Thus, the following are generalities.

Trees

Rocky Mountain trees have several fire-relevant functional traits (table 3.2, appendix A, fig. 3.2). To persist, individual trees may either resist fire damage or suffer topkill and resprout. Only a few deciduous trees in three genera (*Alnus, Betula, Populus*) in the Rockies commonly resprout. Riparian trees in these genera may resprout when burned (fig. 3.2a), a trait that is also useful after floods. Several conifers are tall (greater than 40 m at maturity), but few are fire resisters with low-flammability foliage, low epiphyte loads, thick bark, and decay resistance. Decay resistance means trees can discourage disease by sealing scars from surface fires with resin or by regrowth (Loehle 1988). The ability to self-prune lower branches, as in ponderosa pine, means fire cannot easily climb into crowns; ponderosa pine also can recover from extensive crown scorch (Dieterich 1979). Thus, most resistant are the three tall, thick-barked, and decay-resistant conifers: Douglas-fir, ponderosa pine, and western larch.

Table 3.1. Regional Summaries of Fire Ecology in the
Rocky Mountains

Location	Sources
IDAHO	
Northern Idaho	J. K. Smith & Fischer (1997)
Central Idaho	Crane & Fischer (1986)
Eastern Idaho	A. F. Bradley, Fischer, & Noste (1992)
MONTANA	
Western Montana	W. C. Fischer & Bradley (1987)
Eastern Montana	W. C. Fischer & Clayton (1983)
UTAH	A. F. Bradley, Noste, & Fischer (1992)
WYOMING	
Northeastern Wyoming	Meyer, Knight, & Dillon (2005)
Western Wyoming	A. F. Bradley, Fischer, and Noste (1992)
Southeastern Wyoming	Dillon, Knight, & Meyer (2005)
COLORADO	
Western Colorado	Kulakowski & Veblen (2006)
Northern Colorado	Veblen & Donnegan (2005)

Population persistence, in contrast, is rare in Rocky Mountain trees. Only one, lodgepole pine, has a persistent *seed bank* (storing seeds in either canopy or soil), in the canopy, that can achieve population persistence after mortality of individuals, by immediately releasing seeds. Whitebark pine has delayed seed germination and a soil seed bank that lasts a few years (Tomback et al. 2001), but it is unclear to what extent the seed bank survives fire, allowing population persistence. Most resprouter, resister, and seed-banker trees in the Rockies have wind-dispersed seeds and can thus also function as community or landscape persisters (table 3.2), suggesting one strategy is not always effective.

Most Rocky Mountain trees are solely postfire dispersers, not resprouters, resisters, or seed bankers. These trees are commonly killed by fire because of low branches, flammable foliage, high epiphyte loads, thin bark that may even be flammable (e.g., Rocky Mountain juniper), and little decay resistance. If fire is infrequent, a viable strategy is to avoid spending energy resisting fire, accept mortality, and then disperse seed into burned areas (Loehle 1988). Dispersers include shade-tolerant community persisters and light-loving landscape persisters (table 3.2). Community persisters have traits that allow them to survive competition or take advantage of *facilitation* (one species modifying the environment in a way that favors another); landscape persisters rely on effective dispersal (Pausas and Lavorel 2003).

Community persisters are typical on relatively moist sites, and often their seeds germinate best on organic seedbeds (appendix A). High shade tolerance

Figure 3.2. Seven trees that demonstrate the functional strategies of trees (see table 3.2): (a) *resprouter*, narrowleaf cottonwood, after a fire in Great Sand Dunes National Park, Colorado; (b) *resister*, ponderosa pine, illustrating the thick bark that allows it to survive some fires; (c) *seed banker*, lodgepole pine, regrowing from serotinous seed released during the 1988 Yellowstone fires; (d) *long-lived shade tolerator*, western red cedar, growing in a moist northwestern forest in northern Idaho; (e) *short-lived shade tolerator*, subalpine fir; (f) *wind-dispersed light lover*, young quaking aspen, which resprouted after the 1988 Yellowstone fires; (g) *animal-dispersed light lover*, Utah juniper, growing on a rocky site in Colorado National Monument. *Source:* Photos by the author, except (e), which is reproduced with the permission of Dave Powell, USDA Forest Service (Bugwood.org).

Table 3.2. Rocky Mountain Trees by Level of Persistence and Functional Strategy

A. INDIVIDUAL PERSISTERS (I)	D. LANDSCAPE PERSISTERS (D)
1. Resprouters	6. Wind-dispersed light lovers
Riparian	Light seeded
Black cottonwood	Engelmann spruce
Narrowleaf cottonwood	Lodgepole pine
Plains cottonwood	Quaking aspen
Thinleaf alder	Western larch
Upland	Moderate seeded
Paper birch	Bristlecone pine
Quaking aspen	Douglas-fir
2. Resisters	Ponderosa pine
Douglas-fir	Western white pine
Ponderosa pine	7. Animal-dispersed light lovers
Western larch	Lower-elevation semiarid sites
B. POPULATION PERSISTERS (P)	Oneseed juniper
3. Seed bankers	Rocky Mountain juniper
Lodgepole pine	Twoneedle piñon
C. COMMUNITY PERSISTERS (C)	Utah juniper
4. Long-lived shade tolerators	Higher-elevation windy sites
Mountain hemlock	Limber pine
Western hemlock	Southwestern white pine
Western red cedar	Whitebark pine
5. Short-lived shade tolerators	
Grand fir	
Pacific yew	
Subalpine fir	
White fir	

Note: Shown are level of persistence (A–D) and functional strategy (1–7) relative to fire. Individual and population persisters may also be landscape persisters.

allows seedlings and saplings to grow beneath other trees or shrubs after high-severity fire. A subgroup is long lived and has lightweight wind-dispersed seeds. Longevity allows them to eventually outgrow shorter-lived trees in long intervals without fire. A second subgroup includes shorter-lived trees with medium-weight wind-dispersed seeds (grand fir, Pacific yew, subalpine fir, white fir), in which age of first seed is relatively long, 20 to 50 years, suggesting that energy is spent on early growth, not regeneration.

The other postfire dispersers are shade-intolerant, or light-loving, *landscape persisters.* One subgroup has wind-dispersed seeds that arrive early in postfire succession, germinate well on charred seedbeds, and has generally precocious seed production (i.e., taking less than 20 years), allowing further reseeding in the postfire environment. Seeds range from very light (Engelmann spruce, lodgepole pine, quaking aspen, western larch) to medium weight (Douglas-fir, ponderosa pine, Rocky Mountain juniper, western white pine). Seeds that weigh less than 100 milligrams are effectively dispersed by wind, although some are also dispersed by animals (Benkman 1995). Trees with medium-weight seeds are often found on drier sites, as larger seeds may allow seedlings to develop a taproot, enabling survival where moisture deficits are common (D. P. Turner 1985). A second subgroup may arrive later in succession, often depends on animals for dispersal, and generally lacks early seed production. These include low-elevation piñons and junipers and moderate- to high-elevation five-needle pines often found on windy, rocky sites.

Dominance of Rocky Mountain forests by community and landscape persisters that rely on dispersal is likely related to generally long fire rotations (see table 5.7). Where short-rotation, high-severity fire occurs (Bond and Midgley 2001), as in Mediterranean areas (e.g., California chaparral), 75–90 percent of dominants are resprouters and seed bankers (Pausas et al. 2004). Resprouting is common in angiosperms, which suggests that there is a phylogenetic limit on resprouting (Bond and Midgley 2001). Belowground resprouting structures require energy, thereby limiting height growth (Bond and Midgley 2001). Upland tree resprouters in the Rockies (paper birch, quaking aspen) are short lived and not tall, eventually losing in competition with longer-lived, taller conifers. Population persistence via seed banks also is difficult with the long fire rotations of the Rockies, because it is difficult to store viable seed a long time (Pausus et al. 2004). Lodgepole pine overcomes this by storing seed in sealed cones in the canopy.

Shrubs

Shrubs may reinvade burned areas via surviving or off-site seed or by resprouting from buds in various locations (fig. 3.3). The extent of shrub reseeding after fire is poorly known; information is available for only about half the taxa (fig. 3.4, appendix B). Of the 35 taxa for which we have information, reseeding from surviving seed is uncommon; the largest seed source of shrubs that do reseed is off-site (fig. 3.4a). Fewer shrubs (11 of 67 taxa total) are known to have soil seed banks that can survive fires, but in many of these, the seed appears stimulated (scarified) to germinate after fire. Best known of these are two ceanothus (redstem, snowbush), shrubs that also resprout (fig. 3.5), leading to a flush of growth and density or cover increase after fire. Several shrubs, known to reseed only rarely after fire, actively resprout. Shrubs likely do reseed after fire, but the extent is unclear in many cases.

In contrast to trees, about three-quarters of common shrubs in the Rockies resprout after fire, usually from the root crown or a rhizome, but occasionally from unburned buds on the stem or root or even an unburned branch (fig. 3.4b).

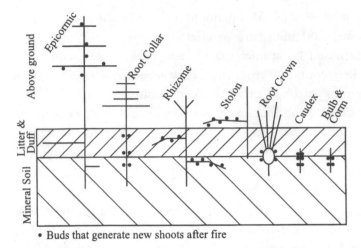

Figure 3.3. Plant parts that can regenerate new shoots after fire. Shown is the location of plant parts in and above the soil.
Source: Reproduced from M. Miller (2000, p. 16, fig. 2-3).

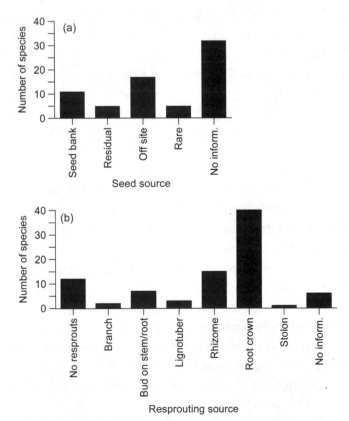

Figure 3.4. Seeding and resprouting in shrub regeneration after fire. Shown are (a) the seed source and (b) the resprouting source for common Rocky Mountain shrubs.
Source: Data are from the Fire Effects Information System, http://www.fs.fed.us/database/feis.

Figure 3.5. Snowbush ceanothus (*Ceanothus velutinus*).
Source: Photo by the author.

Lignotubers (large underground stem or root swellings that store energy and have buds) are common in California chaparral but rare in the Rockies; only Gambel oak (see fig. 9.8), green manzanita (which also grows in chaparral), and occasionally antelope bitterbrush are known to have lignotubers. Nearly all montane and subalpine shrubs of forests and woodlands in the Rockies for which we have information do resprout. Most of the 12 shrub species known not to resprout dominate low- to mid-elevation semiarid shrublands.

Resprouting success varies with shrub size, fire intensity, and climatic or other environmental factors (Vesk, Warton, and Westoby 2004). For example, 32 of 37 taxa (86 percent) for which we have information show reduced resprouting when fire intensity is higher (appendix B). The five shrubs without a fire intensity reduction are among those that flush or become temporarily more abundant after fire. Flushing may occur in about ten shrubs, or 15 percent of taxa, those that often dominate after high-severity fire in montane or lower-subalpine zones, particularly in the northern Rockies.

The time required for shrubs to recover after fire is quite variable, as shrubs occupy many environments and vary in recovery strategy (appendix B). Resprouting flushers may fully recover within two to five years, often aided by heat-stimulated seed banks in the soil. Some resprouting shrubs that do not flush, and lack a seed bank, recover nearly as fast, but most shrubs require 5 to

25 years, depending on fire intensity. However, the nonresprouting sagebrushes require from a half century to several centuries (chap. 9), as they are poor dispersers and may require infrequent wet periods to successfully germinate (e.g., Maier et al. 2001). At low elevations in semiarid shrublands, only about half the common shrubs resprout after fire; others rely on dispersal from off-site seed, as soil seed banks are apparently lacking. Germination and growth are limited for these species by infrequent episodes of favorable climate.

Graminoids

Graminoids include grasses, sedges, and rushes that dominate in forest or shrubland understories or form openings. About two-thirds of 34 common Rocky Mountain graminoids for which there is information have a *bunchgrass* life form (leaves and stems originating in a cluster or bunch from fibrous roots), and about half have a *rhizome* (horizontal belowground stem) or, rarely, a *stolon* (horizontal aboveground stem) (see figs. 3.6, 3.7). A few are bunchgrasses with short rhizomes. Rhizomes help avoid heating that may kill bunchgrasses or stoloniferous plants. However, bunchgrasses may resprout from surviving basal buds or tillers if fire intensity is low. If killed, they must reestablish from a soil seed bank or off-site seed. In regenerating after fire, bunchgrasses that resprout are most common, followed by graminoids that resprout from rhizomes; seeding is primarily from off-site seed (fig. 3.6a).

Graminoids do not have to regrow a permanent aerial stem after fire, as shrubs and trees have to, but fast recovery may still be generally limited to rhizomatous species or where fire is low in intensity and bunchgrasses survive. Response of graminoids varies considerably with fire intensity, environmental setting, and other factors (appendix C). However, the Rockies have roughly equal numbers of graminoid taxa that are generally stimulated or unchanged after fire and graminoid taxa that are generally depressed or, at best, unchanged (fig. 3.6b, appendix C). The bunchgrass life form dominates those that are generally depressed by fire and recover slowly, and the rhizomatous life form dominates those that are generally stimulated by fire and recover quickly. If fire intensity is high enough to kill bunchgrasses, recovery by seed may require one to three decades. If bunchgrasses survive, or graminoids are rhizomatous, recovery may occur within two to five years, or even the year after fire, in some cases.

Forbs

Information on the response of individual *forbs* (nongraminoid herbaceous plants, frequently with showy flowers) is limited, perhaps because there are many of them. Forbs usually are vulnerable to topkill, if not complete mortality,

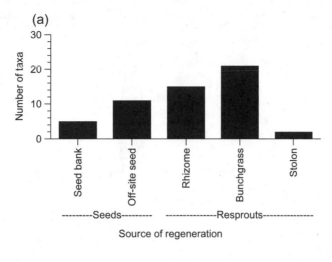

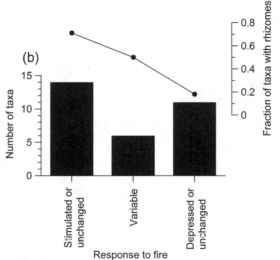

Figure 3.6. Graminoid regeneration and response to fire. Shown are (a) the seed source and resprouting source for common Rocky Mountain graminoids and (b) the response to fire by these graminoid taxa.

Source: Data are from the Fire Effects Information System, http://www.fs.fed.us/database/feis.

by fire, since aboveground foliage is often below flame height. Forbs survive in microsites (e.g., rocky areas with low fuel), resprout from surviving underground structures, or reseed from soil seed banks or off-site seed. Survival from underground structures varies with fire intensity, season of fire, and type and depth of structure. Season of fire affects ability to resprout, which depends on seasonally varying belowground carbohydrate storage and moisture and other stresses (Flinn and Pringle 1983).

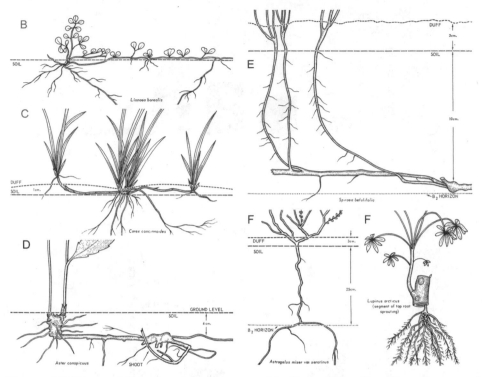

Figure 3.7. Types of forbs in relation to potential effects of fire. B = fibrous roots and stolons, C = rhizomes above mineral soil, D = rhizomes 1.5–5 cm below mineral soil, E = rhizomes 5–13 cm below mineral soil, and F = taproots. McLean omitted a diagram of type A.
Source: McLean (1969, p. 121, fig. 1). Copyright 1969, reprinted with permission of the Society for Range Management.

Most important for survival is depth of the resprouting structure. The relationship is imperfect, suggesting that species also differ in heat tolerance and survival potential (Flinn and Pringle 1983). However, neither type of rhizome sheath nor surface-to-volume ratio of rhizomes apparently affects tolerance to heat. In southern British Columbia, McLean (1969) classified 30 common plants, including 17 forbs of the Douglas-fir zone, by rooting depth and potential survival (fig. 3.7). Kuntz (1982) analyzed the response of some of these to prescribed fires in mountain big sagebrush in Idaho, and Antos, McCune, and Bara (1983) studied the response to wildfire in a montane grassland in Montana. McLean thought types A, B, and C, which featured only fibrous roots or fibrous roots and a stolon or rhizome in the top 1.5 centimeters of soil, would be susceptible to most fires. Kuntz and Antos et al. did find that mat-forming and fibrous-rooted forbs declined after fire. These kinds of species would probably be damaged in fires in lodgepole pine forests, as the 1988 Yellowstone fires charred soil to a mean depth of 0.58 centimeter in light surface fires and 1.36 centimeters in crown fires (M. G. Turner, Romme, and Gardner 1999). McLean considered type D, which had rhizomes 1.5–5.0 centimeters deep, intermediate in survival potential, and types E and F, with rhizomes or a taproot below 5 centimeters, likely to survive even high-severity fires. Kuntz found that tap-rooted forbs, in general, and forbs with bulbs, in cool fires, were not adversely affected.

Scattered studies suggest that fire commonly topkills forbs, but many then resprout; some germinate from a soil seed bank, some survive in low-fuel microsites and reseed from surviving plants, and some reseed from off-site seed. Survival and resprouting from rhizomes or roots appear most common. Fourteen of 18 forbs growing a year after the 1988 Yellowstone fires in lodgepole pine forests were resprouts from rhizomes, perhaps aided by dispersal; four others came from off-site dispersal, and only one species came from a seed bank (J. E. Anderson and Romme 1991). The seed banker American dragonhead (*Dracocephalum parviflorum*) also may be common in or may dominate the early postfire environment in Douglas-fir forests (Lyon 1971; Crane, Habeck, and Fischer 1983). Another common seed bank forb is streambank wild hollyhock (*Iliamna rivularis*), which may dominate early succession after fire in quaking aspen forests (Bartos et al. 1994). Some common postfire forbs, such as fireweed (*Chamerion angustifolium*) have an effective dual strategy: survivors resprout and mature rapidly, producing abundant wind-dispersed seed and saturating the postfire environment (Armour, Bunting, and Neuenschwander 1984; Bartos et al. 1994; M. G. Turner et al. 1997).

Nonvascular Plants: The Biological Soil Crust

Nonvascular plants, plants without specialized water-conducting tissues, may be scattered across soil surfaces or in the soil or organized into a microcommunity, commonly called *biological soil crust* (also called *cryptogamic, microphytic,* or *microbiotic crust*). The crust is a three-dimensional community of *cyanobacteria* (blue-green algae), green algae, mosses, and lichens formed in and on top of the soil (fig. 3.8). In the Rockies, soil crusts are most common on semiarid sites, where they play key roles in fixing nitrogen, reducing soil erosion and runoff, and shaping plant regeneration (Belnap and Lange 2001).

Fire damages crust, but some components recover quickly, others slowly. Green algae species composition was similar to unburned composition as soon as a month after fire in sagebrush in Washington state (Johansen, Ashley, and Rayburn 1993) and within three years after fire in mixed mountain shrubs in Utah (Johansen, Javakul, and Rushforth 1982). Photosynthetic capacity,

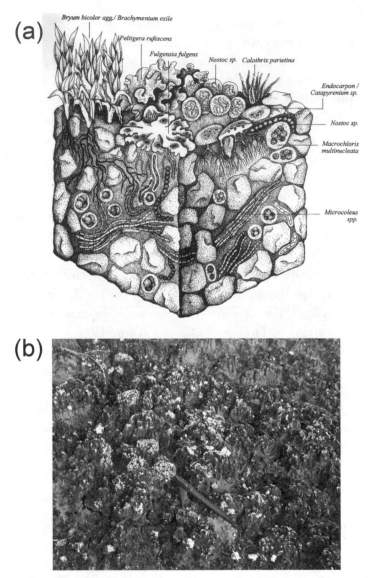

Figure 3.8. Biological soil crust. (a) shows a diagram of a biological soil crust with typical colonizers. *Bryum* is a moss; *Macrochloris* is an alga; *Peltigera*, *Fulgensia*, and *Endocarpon* are lichens; and *Nostoc*, *Calothrix*, and *Microcoleus* are cyanobacteria. (b) is a photograph of biological soil crust in a piñon-juniper woodland in Colorado.

Source: (a) is from Belnap and Lange (2001, p. 4, fig. 1.1), © Springer Science, reproduced with kind permission of Springer Science and Business Media; (b) is a photo by the author; note pencil for scale.

however, recovered only about halfway within three years after fire in Wyoming big sagebrush on the Snake River plains (Rychert 2002); it may generally recover by five years (Johansen 2001). Some fire-adapted mosses may recover on burned Wyoming big sagebrush sites in ten years, but taller mosses may take longer (Hilty et al. 2004). Lichens can reappear early but generally require longer to recover, perhaps 13 to 35 years (Johansen 2001). Better developed, older crust may require more than 50 years to recover in burned piñon-juniper in Utah (Evangelista et al. 2004). In contrast, wildfire in Hell's Canyon Palouse grasslands (bluebunch wheatgrass and Idaho fescue) led to only minor declines in mosses and lichens, most likely because burns were fast and of low intensity (Bowker et al. 2004). However, in a related montane grassland in Montana, mosses and lichens had not recovered within three years after wildfire (Antos, McCune, and Bara 1983).

FIRE EFFECTS: FROM INDIVIDUALS TO LANDSCAPES

Fire effects vary with level of plant organization. Individual plants may survive if they can avoid significant injury during combustion, but plant populations persist through survival of individuals or regeneration from seed banks. Communities and ecosystems undergo succession after fire, and the landscape structure of a mosaic of patches is reshaped by fire.

Effects on Individual Plants: Injury and Mortality

Plants suffer immediate injury or mortality as fuel is consumed. The amount of combustion of the plant itself and the heat output from nearby fuel combustion control the extent of injury to a plant (R. E. Martin 1963; Vines 1968; Dickinson and Johnson 2001). Both intensity and duration of heat affect the extent of injury. Lethal temperatures are lower for growing plants than for dormant plants or seeds and vary among species; 44–68° centigrade for short periods can be lethal to most growing plants (Kayll 1968; Levitt 1980; Seidel 1986), and 60° centigrade is often considered lethal for tissues (Vines 1968; Levitt 1980). However, lower temperatures for long duration may cause tissue death as readily as higher temperatures of short duration. Formally, the temperature, t, that can kill is inversely proportional to the logarithm of duration, T, of that temperature: $t = a - b \log T$, where a and b are constants (R. E. Martin 1963; Levitt 1980). Most meaningful is the total heat received while flames pass or glowing combustion occurs. This *time-integrated heat flux*, long known to be important (Fahnestock and Hare 1964), is strongly related to depth of tissue damage in trees (Bova and Dickinson 2005). If fireline intensity is relatively constant, fire *residence time* (the time that fire is burning at a particular point) is more easily observed in the field and can serve as a proxy for time–integrated heat flux.

Injury or mortality can occur from direct ignition and combustion or from indirect impacts of total heat transfer to plant crowns, boles, or roots. The foliage of forbs, graminoids, short shrubs, and small trees may be within the flame height of wildfires, and thus those plants may be directly combusted or suffer lethal internal temperatures. Taller shrubs and trees may also be combusted if a fire becomes a crown fire. If fire intensity is lower, in surface fires in forests or in fires in shrublands or grasslands, some plants may survive in low–fuel refugia, such as rocky areas or interspaces between shrubs (Bowker et al. 2004). Thus, plants survive fires only if they can escape lethal temperatures, if flames are below the plant's canopy, fire intensity in the vicinity is low, survival structures are protected in the soil, or the plant has other fire–resistance traits (e.g., thick bark).

Crowns, if not ignited, can still be damaged by the plume of hot gases transferring heat into the canopy by convection and radiation from flaming combustion in a surface fire (Dickinson and Johnson 2001). If canopy foliage is thin, offering little resistance to heat transfer from the plume, the height at which tissue necrosis in the crown (*crown scorch*) is expected is:

$$z = (jI\,(T_p - T_o))\,I^{2/3} \qquad\qquad (3.1)$$

where z is *scorch height* (in meters), j is a constant estimated from field data, T_p is temperature of the plume (in degrees centigrade), T_o is ambient temperature, and I is fireline intensity (in kilowatts per meter; Van Wagner 1973; Dickinson and Johnson 2001). Fireline intensities of 100 kilowatts per meter, for example, lead to scorch heights of about 10 meters. A more complex equation estimates heat transfer into crown components that are larger or have bark or other structures (e.g., hardened cones) that resist heat transfer (Dickinson and Johnson 2001). Percentage of crown scorch (crown foliage killed), rather than scorch height, is a better indicator of tree mortality (Ryan and Reinhardt 1988; Sieg et al. 2006).

Woody plant parts of sufficient size, whether boles, branches, or roots, may resist heating of the cambium to lethal temperatures because of their volume and the insulation provided by bark. The *cambium* is the thin, actively growing layer inside the bark; when it reaches lethal temperature, it no longer conducts fluids. Bark thickness is the resistance characteristic protecting the cambium, but bark properties (e.g., thermal conductivity, heat capacity) control heat transfer. In general, flame residence time needed to kill the cambium is a linear function of the square of bark thickness, so small increases in bark thickness greatly increase resistance to injury from fire (Gutsell and Johnson 1996). Bark may or may not actually catch fire (Vines 1968) and typically remains on the stem for several years after cambium is killed in that area, eventually falling off and leaving a fire scar. Complete girdling by cambium death kills the plant, but so can scarring around part of the stem (Ryan and Frandsen 1991). Stem scarring increases as ambient temperature rises, as fireline intensity and flame residence time increase and if resistance by the tree is insufficient (Vines 1968). Higher ambient temperature (e.g., summer) means less temperature rise is needed to reach lethal internal temperature.

Rates of spread of typical surface fires suggest flame residence times on bark might be too short to kill the cambium, particularly if bark is thick. But, as the fireline reaches the stem, flame vortices may be set up that increase residence time and temperature of flames on the leeward side of the trunk (fig. 3.9), if the bole is large enough and the fire is not fast moving (Gutsell and Johnson 1996). Leeward triangular fire scars thus are typical, since temperature is highest near the base and declines upward. Flame residence time is increased by twice the tree diameter (*m*) divided by rate of spread (*m/s*). Small trees, if not directly combusted, may be killed around their bole, as they are too small to produce vortices and differential heating on their leeward side.

Damage to stems may result not only from flaming combustion as the fireline passes, however, but also, perhaps more so, from glowing combustion after the fireline passes (Ryan and Frandsen 1991). Glowing combustion may continue in duff at tree bases for hours after flaming combustion, increasing temperatures in the cambium between stems and roots. Ryan and Frandsen (1991) found depth of burned duff correlated with probability of cambium mortality, with duff-depth consumption greater than 19 centimeters likely to cause cambium death. Contrary to the idea that larger trees with thicker bark are more likely to survive fires, larger trees produce deeper duff that burns longer, leading to higher cambium mortality (Ryan and Frandsen 1991). Finer roots that grow up into duff between surface fires may contribute to tree mortality when surface fires and glowing combustion occur (Swezy and Agee 1991).

Glowing combustion also heats the soil, potentially injuring or killing belowground plant structures (e.g., taproots, rhizomes) and microorganisms. Low-intensity surface fires may not elevate soil temperatures to lethal levels for plant tissues at 5 centimeters of depth (Neary et al. 1999). However, Ryan and Frandsen (1991) observed temperatures above lethal levels at 2 centimeters of depth for over ten hours after prescribed fire led to glowing combustion in duff in a

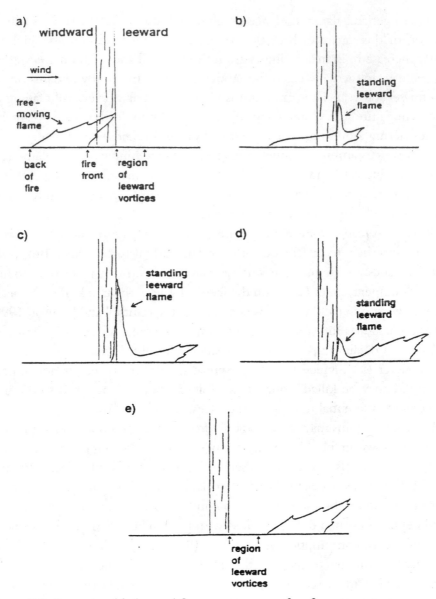

Figure 3.9. Formation of a leeward flame vortex as a surface fire passes a tree.
Source: Reproduced from Gutsell and Johnson (1996, p. 169, fig. 4) with permission of the National Research Council of Canada.

ponderosa pine forest. Under burning slash piles or slow-moving fires with deep surface fuels, soils can reach lethal temperatures greater than 100° centigrade at greater than 20 centimeters of depth (Neary et al. 1999). Plant roots are concentrated in the top layers of soil, so damage from glowing combustion is possible in many cases.

Survival of Rocky Mountain trees after fires has been measured to develop equations to predict mortality from fires and estimate which trees may die (appendix D). Typically, both stem resistance and crown damage are included, using either bark thickness or its proxy, tree diameter, along with percentage of the crown scorched. The First Order Fire Effects Model (FOFEM) uses Ryan and Reinhardt's (1988) equations and data and other information to predict tree mortality (Keane, Reinhardt, and Brown 1994). A simple two-variable model (fig. 3.10) does reasonably well at predicting mortality for seven western conifers (Ryan and Reinhardt 1988). A more complex model includes intercept terms for each species, which shifts the response surface down (increasing

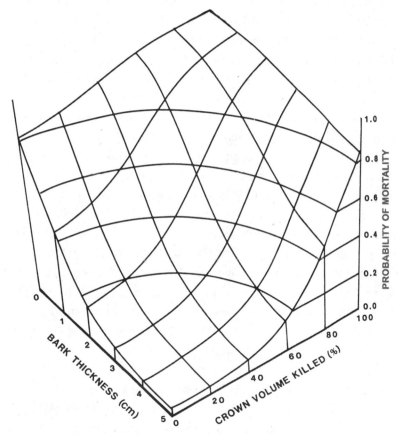

Figure 3.10. Probability of mortality of seven western conifers after fire. Probability of mortality is shown to be a function of crown volume killed and bark thickness. This is the pooled probability of mortality for all seven conifers; individual conifers have surfaces that are similar but elevated or lowered (see text for details).

Source: Reproduced from Ryan and Reinhardt (1988, p. 1294, fig. 2) with permission of the National Research Council of Canada.

probability of mortality) for Engelmann spruce, subalpine fire, and western hemlock and up (decreasing probability of mortality) for Douglas-fir, western larch, and western red cedar (Ryan and Reinhardt 1988). The model suggests that mortality increases sharply for bark thickness of less than 2 centimeters or if more than 60 percent of the crown volume is scorched (fig. 3.10).

Effects on Populations: The Role of Seed Banks

The immediate effects of fire on populations are a consequence of mortality or survival of individuals after fire. Survival of individuals, except trees, is high in many ecosystems, as many species resprout. Populations may also survive because of erratic fire behavior (e.g., wind shift) or in low-fuel refugia (e.g., rocky areas) within fire perimeters. Populations that suffer complete topkill may also survive via either canopy or soil seed banks. A soil seed bank consists of viable seeds left in the litter and on or below the soil surface (D. L. Clark 1991). Seed banks fluctuate, as some seed is transient, persisting only until germination the year after dispersal, while other seed is stored and remains viable for longer periods, sometimes a century or more (D. L. Clark 1991).

Among Rocky Mountain trees, the only known consistent seed bank species are lodgepole pine, which has a canopy seed bank, and whitebark pine, which has a soil seed bank. About one-third of Rocky Mountain shrubs for which there is information have soil seed banks (fig. 3.4), and some of these have seeds that are stimulated by heat, leading to a postfire flush of seedlings (appendix B). Fewer graminoids appear to have effective soil seed banks (fig. 3.6). There is no single comprehensive source on seed banks in Rocky Mountain forbs, but D. L. Clark's (1991) study is very thorough, covering many species. He found that many forbs have soil seed banks.

Soil seed banks are found in all Rocky Mountain ecosystems (D. L. Clark 1991). Seed bank density and richness tend to be low in dry montane grasslands and dark conifer forests; highest in wetlands and aspen forests; and intermediate in moist grasslands, shrublands, and open conifer forests (e.g., Douglas-fir, whitebark pine). Seed bank composition is highly similar (i.e., 81–94 percent) to that of aboveground vegetation in montane and subalpine nonforested ecosystems, suggesting stable communities; but in many forests (Whipple 1978; D. L. Clark 1991) and in shortgrass prairies near the mountains (Coffin and Lauenroth 1989) the similarity is often only 20–35 percent.

Viable seeds are concentrated in top layers of the soil, where they are vulnerable to fires. Seed density of some species is highest in the litter (Pratt, Black, and Zamora 1984), but seeds of others are concentrated in the top 2 centimeters (Moore and Wein 1977; Pratt, Black, and Zamora 1984) or top 5 centime-

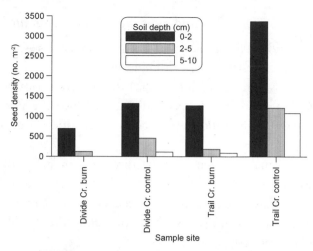

Figure 3.11. Seed density in soil at three depths after fire at two sites in Yellowstone National Park. Shown is seed density in a spruce-fir burn (Divide Cr. site three to four years after fire) and a lodgepole pine burn (Trail Cr. site four to five years after fire) and matching unburned controls in Yellowstone National Park. Note that the 5- to 10-cm depth in the Divide Cr. burn had no seed; thus the bar is missing.
Source: Data are from M. A. Wood (1981).

ters (Kramer and Johnson 1987) of mineral soil. Seed densities were highest in lodgepole pine and spruce-fir forests in the top 2 centimeters of soil (fig. 3.11).

Given these shallow depths, it is unsurprising that fires generally decreased seed bank density by about 20 percent and species richness about 15 percent in nonforested ecosystems, but a 77–78 percent seed density reduction and a 58–72 percent species-richness reduction occurred in conifer forests (M. A. Wood 1981; D. L. Clark 1991). Still, even severely burned forests had seeds sufficient for revegetation, if only marginally (D. L. Clark 1991). More emergence from the seed bank occurred because of stimulation by smoke than because of stimulation by heat (Abella, Springer, and Covington 2007). Seeds of many species survived temperatures of 100° centigrade, but high heat tolerance (to 150° centigrade) was common only in annual and biennial forbs (D. L. Clark 1991). The soil seed bank of the invasive annual, cheatgrass, was reduced 50 percent after fire in Wyoming big sagebrush in Utah but quickly regained even higher density (Hassan and West 1986; see fig. 9.6). As succession proceeds in conifer forests, seed bank density typically declines, but richness of species does not (D. L. Clark 1991). Forests are more vulnerable than nonforested ecosystems to changes in species composition after fire, at least where survival of individuals through underground structures or in fire refugia is low and seeds are a primary means of population persistence.

Effects on Communities and Ecosystems: Succession after Fire

The immediate response to fire is a shift from mortality to resprouting, germination, and dispersal. Resprouting can begin even a few weeks after spring or summer fires (Antos, McCune, and Bara 1983; Floyd, Romme, and Hanna 2000). Germination from seed banks and dispersal into the postfire environment also may begin quickly. Postfire succession gets under way as soon as the fire cools.

Postfire successional research includes direct observations of burns, using adjacent unburned forest as a control, and study of a set of burns differing in time since fire, a *chronosequence*. Chronosequences are a weaker source, as they assume sites are physically similar and have the same history, which is seldom true (E. A. Johnson and Miyanishi 2008).

Direct postfire observations of burns showed that resprouters dominated, as a percentage of the early (years 1 to 5) postfire flora in lodgepole pine (14 species, or 74 percent of flora; J. E. Anderson and Romme 1991) and Douglas-fir forests (10 species, or 71 percent of flora; Crane, Habeck, and Fischer 1983), as well as in sagebrush (10 species, or 77 percent of flora; L. D. Humphrey 1984). Seed bank species, in contrast, were low but variable, with one, three, and zero species, respectively; species arriving from off-site seed dispersal were equally variable, with four, one, and three species, respectively. Early domination by resprouters is expected, given the many shrubs, graminoids, and forbs capable of resprouting (appendixes B, C). In many Rockies ecosystems, native annual forbs and graminoids may germinate from seed banks and codominate with resprouting perennials the first few years after fire. A flush of annuals is known in lodgepole pine (M. G. Turner et al. 1997), aspen (Bartos et al. 1994), ponderosa pine forests (Merrill, Mayland, and Peek 1980), piñon-juniper woodlands (Shinneman 2006), and sagebrush (Kuntz 1982).

The ensuing successional process is incompletely understood. Early ideas remain as a backdrop to an expanded, but still incomplete theory of postfire succession. Frederick Clements (1916) envisioned a holistic, orderly, facilitative progression of stages leading to a climax vegetation in balance with regional climate. Groups of species, organized in discrete stages, or *seres*, were thought to invade after fire and modify the environment in ways (e.g., by adding organic matter to soil) that would make it unfavorable for them but facilitate arrival of the next stage, leading eventually to the climax. Many of Clements's ideas were challenged. Gleason (1917) argued that succession is a gradient in time along which species are arrayed independently, rather than in discrete seres. Climaxes in balance with factors other than climate were documented (Tansley 1935).

Drury and Nisbet (1973, 362), in a reaction that was reductionistic (explained by individual components, in contrast to Clements's holistic approach), proclaimed that "a complete theory of vegetational succession should be sought at the organismic, physiological or cellular level, and not in the emergent properties of populations or communities." However, the reductionistic perspective was challenged as too simplistic, because it omitted positive interactions among species, as well as evidence that some species do modify environments during succession (Finegan 1984).

A reductionist-based but broader framework has emerged more recently, emphasizing species' life history and physiology, multiple processes (e.g., dispersal, herbivory), and interactions (e.g., competition, allelopathy) that may shape succession (Pickett, Collins, and Armesto 1987; Pickett and McDonnell 1989). Succession is considered a combination of community-level site availability, species availability, and species performance (fig. 3.12). Sites become available at the community level following a fire (the effects of size and dispersion of fires are discussed later). Species differ in availability because of differing survival abilities (e.g., resprouting), propagule pools (i.e., seed banks), or dispersal to the burned site. Once present, species' performance is shaped by constraints of the postfire environment, including resource availability (e.g., nutrients) and environmental stresses (e.g., drought), as well as species' autecology, including functional traits (e.g., rapid early seeding) and ecophysiology (e.g., growth rate). Species also must contend with competition, chemical interactions (e.g., allelopathy), and losses to consumers. Positive interactions (e.g., nurse plants) that may be facilitative are neither emphasized nor excluded in the model.

Seed dispersal agents and dispersal distances are poorly known for Rocky Mountain plants, except trees (see table 3.2). In general, wind dispersal is favored over animal dispersal in dry environments and among taller plants, such as canopy trees, where mean wind speed is high (Howe and Smallwood 1982). In the Rockies, tall canopy trees are generally wind dispersed, and shorter trees, particularly in more stressful settings, are predominantly animal dispersed. The distance that seeds may be dispersed by wind from a forest edge into an opening increases as the height of trees increases and as seed mass declines (Greene and Johnson 1996). Density of dispersed seed typically declines sharply with distance, following a negative exponential curve (e.g., fig. 3.13). For most trees, little seed is dispersed beyond about 150 meters from a forest edge; however, rare updrafts can disperse light seeds up to several kilometers (Greene et al. 1999). Rocky Mountain trees vary substantially in density of seed dispersed by wind with distance from a source. For example, western hemlock seed density at 100 meters from a source may still be 40 percent of seed density at the edge of the source,

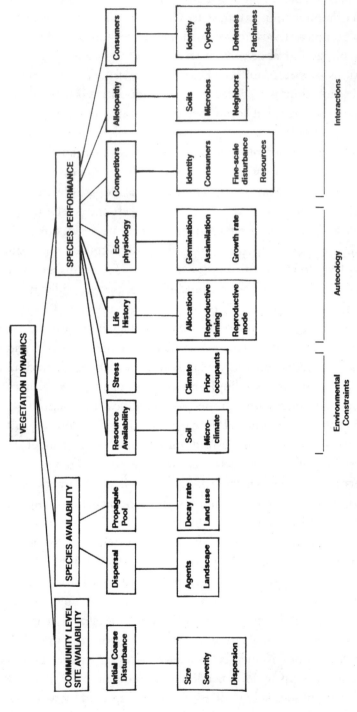

Figure 3.12. Factors affecting plant succession. Succession is shown as a function of community-level site availability, species availability, and species performance.

Source: Reproduced from Pickett and McDonnell (1989, p. 244, fig. 2), copyright Elsevier 1989, with permission of Elsevier Limited.

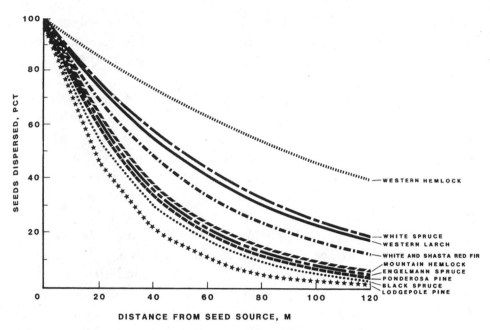

Figure 3.13. Percentage of tree seeds dispersed vs. distance from seed source. The graphs are logarithmic curves fitted to empirical data from unpublished sources.
Source: Reproduced from McCaughey, Schmidt, and Shearer (1986, p. 51, fig. 1).

whereas lodgepole pine seed density may be less than 5 percent at that distance. Western hemlock and western larch stand out as the best wind-dispersed trees in the Rockies; other trees are much poorer. Dispersal by animals may be important, however, particularly among trees that dominate lower-elevation semiarid and high-elevation windy sites (table 3.2). Clark's nutcracker, for example, can disperse seeds 10 kilometers or more (Tomback 2001).

Species' performance in the postfire environment varies with resources and stresses on the burned site. Succession may be slower on drier sites or if nutrients are limiting. Succession rates were correlated with nutrient levels and were twice as high on mesic lower slopes as on drier south-facing slopes in a Colorado subalpine forest (Donnegan and Rebertus 1999). Similarly, recovery of sagebrush after fire may be contingent on favorable moisture (Maier et al. 2001).

Several mechanisms, including facilitation, inhibition, and tolerance or their combinations, may shape succession (Pickett, Collins, and Armesto 1987). Clements's (1916) *facilitation*, discussed earlier, is also called *relay floristics*, as early arrivers make the environment unfavorable for themselves but more favorable for later arrivers. *Inhibition*, also called *initial floristics*, emphasizes that survivors or first-arriving plants may control succession by preventing other species from dominating (Egler 1954). The *tolerance* idea is that late-successional species tolerate the conditions (e.g., low resource levels) imposed by early-successional species but eventually outgrow them to dominate (Pickett, Collins, and Armesto 1987).

In the Rockies, all these mechanisms and both scales of explanation (holistic and reductionistic) may help us to understand succession; the mechanisms and scales of explanation that are relevant may vary in importance with time since fire, species traits, fire severity, and environment. Strong evidence supports the initial-floristics mechanism in the early postfire environment in montane and subalpine forests, where survivors (especially shrubs, graminoids, and forbs) may resprout or germinate from seed banks within the first one to two years after fire (Lyon and Stickney 1974; J. E. Anderson and Romme 1991; Doyle et al. 1998). The reductionistic perspective, that life history traits may explain succession, is supported in general after fire in Rocky Mountain forests (Kessell and Potter 1980) and in sagebrush, but in the latter the mechanism is tolerance, not inhibition (L. D. Humphrey 1984). Late-successional trees, particularly in moist forests in the northern Rocky Mountains, include shade-tolerant trees that, in their life history and ecophysiology, also may represent tolerance. Resprouting shrubs, which dominate after high-severity fires in many forests in the Rockies, can sometimes act as *nurse plants*, providing shade, nutrients, or other resources that enhance survival and growth of young trees; this is an example of a facilitation mechanism (M. E. Floyd 1982; M. H. Jones 1995; Liang 2005). High-severity fire could favor species with seed banks, since seeds are more heat tolerant than underground resprouting structures are, as well as species that are off-site dispersers (Stickney 1990; Doyle et al. 1998), a trend found in the boreal forest (Wang and Kemball 2005).

Because of multiple processes, succession after fire may include divergent pathways, which may become stabilized, at least temporarily, in alternative states or may ultimately converge (Pickett, Collins, and Armesto 1987). For example, lags in tree regeneration can occur on dry slopes in low-elevation ponderosa pine forests, leading to openings that persist a century or more before trees recover (Kaufmann et al. 2001). Lags in tree regeneration for decades or centuries may also occur at high elevations because of infrequent favorable climate and competition with dense postfire herbaceous vegetation (Stahelin 1943). Succession may be delayed or may favor especially fire-resistant trees (e.g., western larch) or dense shrublands, if forests burned in high-severity fires reburn before most trees are able to mature and develop sufficient fire resistance (Larsen 1925b). Human land uses, such as livestock grazing, may degrade the postfire environment, altering succession and changing the resulting forest (W. L. Baker 1991).

Species diversity varies over succession. Most studies are of *alpha diversity*, diversity at the plot or stand level, and *species richness*, number of species per unit area. In chronosequences of aspen, lodgepole pine, and spruce-fir forests, richness was highest early in succession, from years 4 to 25, and declined as stand closure occurred (fig. 3.14) 50 to 100 years after fire (Kleinman 1973; D. L. Taylor 1973; Peet 1978). Richness rose again in late succession as the canopy reopened (Peet 1978). In postfire observations, richness in a burned lodgepole pine forest was about the same one year after high-severity fire as in adjoining unburned forest, but richness was 40 percent above that in the unburned forest by three years after fire (Striffler and Mogren 1971). A chronosequence in mountain big sagebrush showed richness was highest two years after fire, with little trend in richness 3 to 36 years after fire (L. D. Humphrey 1984).

Fire severity also affects postfire species richness. In Yellowstone's lodgepole pine forests, richness in high-severity burned areas one year after the 1988 fires was about half as high as in moderate-severity burned areas (J. E. Anderson and Romme 1991). Similarly, richness was highest three and five years after fire in

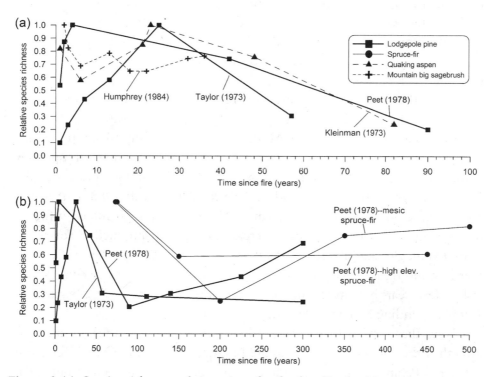

Figure 3.14. Species richness and time since fire for four Rocky Mountain ecosystems. Shown are data for (a) the first 100 years, and (b) 500 years. Species richness is normalized to the fraction of the highest richness reached at any point during postfire succession. *Sources:* Data are from Rocky Mountain National Park (Peet 1978), Yellowstone National Park (D. L. Taylor 1973), southeastern Idaho (L. D. Humphrey 1984), and northern Utah (Kleinman 1973).

areas burned by light surface fires, intermediate in severe surface fire areas, and lowest in crown fire areas (M. G. Turner et al. 1997). Before-and-after (year 1) comparisons of effects of low-severity prescribed fires in northern Colorado showed no significant change in richness in ponderosa pine, Douglas-fir, aspen, or aspen-conifer forests or in Wyoming big sagebrush (Petterson 1999). Thus, high-severity fire may injure or kill some species, reducing richness temporarily, but richness may rise to prefire levels or above them within a few years or up to about 25 years after fire. Richness may recover quickly, even by the year after fire, if the fire is low severity in forests or even high severity in sagebrush, suggesting that most species survive fires and quickly resprout or germinate from seed banks.

Productivity also changes during succession. Managers have intentionally burned vegetation to shift production from woody plants to palatable grasses and forbs, and evidence suggests that fires do this but only for limited periods. Annual net primary productivity (ANPP) is the net rate (gross productivity minus community respiration) at which biomass (in kilograms) is accumulated per unit area (square meters) per unit of time (year). Ecosystems can restore ANPP within a few years after fire (R. A. Reed et al. 1999). Primary effects of fire, then, are short-term declines in ANPP after fire as plants initially recover and regrow to maturity, accompanied by shifts in the components of the community (e.g., forbs, grasses) that contribute to ANPP.

In the Rockies, the ability of many grasses, forbs, and shrubs to resprout after fire, and the inability of most upland trees and some shrubs (e.g., big sagebrush) to do so, determines how ANPP gets shifted initially. Many studies have found temporary increases in herbaceous production after fires in aspen and conifer forests, sagebrush, mixed mountain shrublands, and grasslands (e.g., Kleinman 1973; Merrill, Mayland, and Peek 1980; Shariff 1988; Bartos et al. 1994; J. G. Cook, Hershey, and Irwin 1994). However, high-intensity fires, in dry periods or on dry sites, can decrease herbaceous production for several years (Mueggler and Blaisdell 1958; Schwecke and Hann 1989; J. G. Cook, Hershey, and Irwin 1994). Herbaceous production increase, if it occurs, may last only a few years but up to one or two decades (Kleinman 1973; Bartos et al. 1994), depending on rate of recovery of woody plants.

Trees that survive fires, typically low-intensity fires, may show increased growth if they are not seriously injured (Wyant, Laven, and Omi 1983; Reinhardt and Ryan 1988), but often trees suffer some crown scorch or bole damage, and growth declines (D. L. Peterson et al. 1991). Even if individual trees increase in growth, mortality of other trees may lead to net decline in growth at the stand level for several years after fire (Reinhardt and Ryan 1988; D. L. Peterson et al. 1991). Trees, of course, can recover in forests burned completely in high-severity fires, and maximum productivity in recovering forests is usually reached 40–100 years after fire or other disturbance in lodgepole pine forests (J. A. Pearson, Knight, and Fahey 1987; Kashian et al. 2005) and near or before 125 years in spruce-fir forests (Aplet, Smith, and Laven 1989). Productivity may increase rapidly in the first 100 years in whitebark pine forests, continuing slowly upward after that (Forcella and Weaver 1977).

Effects on Landscapes

The idea that natural disturbance and succession play out spatially across a landscape as a mosaic of seral stages developed long ago (Watt 1947), but it was formalized by Whittaker and Levin (1977). With the rise of landscape ecology in the 1980s, a *shifting mosaic* of disturbance patches, varying in size, severity, time since fire, and spatial pattern, became a common part of successional concepts (Pickett and McDonnell 1989; fig. 3.12).

The shifting mosaic is from multiple fires (box 3.1), but each fire may be heterogeneous, contributing to the mosaic (fig. 3.15). Smaller fires may be more homogeneous, perhaps even of a single severity, but larger fires may be more diverse, with a high-severity crown fire component, possibly a high-severity surface fire component, and a low-severity surface fire component (e.g., M. G. Turner, Hargrove, et al. 1994). Often some fraction of the area within a burn perimeter contains surviving single trees or islands of trees that escape burning (figs. 3.15, 6.8). A scorch zone may border the high-severity part of the burn area. Spot fires may occur, particularly if there is a high-severity component to the fire. Even when the fire is entirely low severity in a forest and tree survival is high, some fraction of unburned area is typical due to shifts in wind or low fuel loads (W. L. Baker and Ehle 2001). Some individual trees may torch, producing a high-severity patch and leaving a small opening. Thus, a single large fire can leave many patches across a landscape. The 1988 fires, for example, left a mosaic of greater than 90,000 distinct patches in Yellowstone National Park (M. G. Turner, Hargrove et al. 1994). As subsequent fires occur, they modify the mosaic produced by past fires, leading to a kaleidoscopic, shifting mosaic of patches (box 3.1).

The shifting mosaic of patches that make up a fire regime can be characterized by attributes of individual patches such as fire severity, patch size, patch shape, and extent of the scorch zone or other types of edge, as well as attributes of the overall mosaic, including landscape diversity and age-class structure (see also chap. 5). Patch size affects plants because large patches may exceed short-term plant dispersal distances, but they also provide an open, less shaded

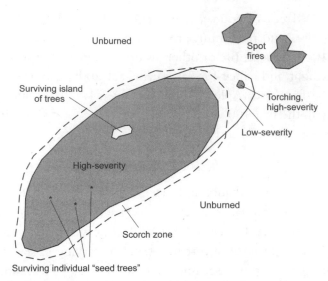

Figure 3.15. Patchiness potentially produced by a single fire. Patchiness sources include variation in fire severity, surviving islands of trees, and single-seed trees, as well as torching and spot fires and a scorch zone around areas of high severity.
Source: Diagram by the author.

environment that could favor some tree seedlings. Narrow patch shapes and complex boundaries lead to shorter distances from unburned vegetation, facilitating postfire seed dispersal (M. G. Turner et al. 1997). A surprising finding after the 1988 fires that burned about 230,000 hectares in Yellowstone was that all moderately burned area and 75 percent of crown fire area was within 200 meters of unburned vegetation (fig. 3.16), a distance that may allow eventual seed dispersal and tree regeneration. In Grand Teton National Park, distance to unburned forest influenced which trees dominated after fire (Doyle 2004). The scorch zone in burned ponderosa pine forests may be an area within which postfire tree regeneration is favored (Bonnet, Schoettle, and Shepperd 2005). Landscape heterogeneity in fire effects leads to *spatial contingency in postfire succession* (particularly for trees), in which the pattern and composition of nearby surviving trees lead to a diversity of postfire successional patterns in forests, an idea first formulated by Whitford (1905) and elaborated in later studies (chaps. 6–9).

FIRE SEVERITY

Fire severity refers to the magnitude of fire effects on biological and physical components of ecosystems, or simply "the degree of environmental change

BOX 3.1
Fluctuating Landscapes

Until the 1980s, landscapes subject to fires and other disturbances were thought to have a shifting mosaic of patches that was stable over some large area. W. S. Cooper (1913) thought that the state of a small patch of forest might fluctuate widely over time, but that on some larger area the landscape fraction in each age class might remain fixed as disturbances shift, producing a *shifting-mosaic steady state* (Bormann and Likens 1979). In the Boundary Waters Canoe Area (BWCA) in Minnesota, small areas did fluctuate widely (fig. B3.1.1c–f), but even on the whole 404,000-hectare BWCA, a steady state was lacking and fluctuation was large over two centuries (compare fig. B3.1.1b to a; W. L. Baker 1989b). Similarly, a steady state has been lacking in a 130,000-hectare lodgepole pine landscape in Yellowstone over the last three centuries (fig. B3.1.1g; Romme and Despain 1989a). In both areas, extensive fires in the 1700s produced large areas of young forest (e.g., LP0 in fig. B3.1.1g), which recovered and aged until mature forest dominated, in the late 1800s in the BWCA and the late 1900s in Yellowstone, when large fires occurred again.

In nearly all fire regimes studied to date, *episodes* of infrequent large fires are followed by *interludes* of recovery and small fires. Fire regimes worldwide have fire-size distributions that are inverse-J shaped, with exponentially more small fires than large ones (chap. 5). Landscapes fluctuate substantially, because the large fires occur in infrequent episodes. In the interludes between them much of the landscape ages (e.g., the peak in the sequence of landscape age structures in fig. B3.1.1b), even though the matrix is peppered with small fires (W. L. Baker 1989b). Infrequent large fires control the landscape dynamics throughout a fire rotation, because a very high percentage of burned area is from the infrequent large fires. The small interlude fires do perforate the recovering and aging matrix, increasing the diversity of age classes and decreasing connectivity of the matrix; however, the small interlude fires account for very little total burned area and cannot prevent the general recovery and aging of the matrix.

Landscapes subject to fires or other large disturbances (e.g., insect outbreaks) thus have inherent fluctuation, characterized by episodes of large disturbances, followed by recovery and aging. Fluctuation in landscapes is inherent in fire regimes, including fluctuations in landscape diversity

(continued on next page)

BOX 3.1

Continued

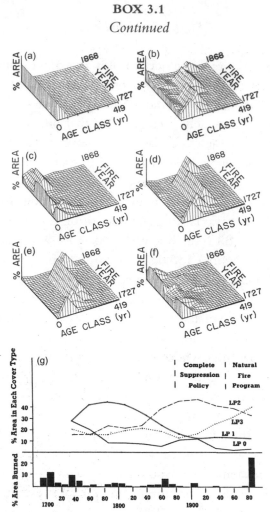

Figure B3.1.1. Fluctuating landscapes. Temporal fluctuation in age-class structure in the Boundary Waters Canoe Area (BWCA), Minnesota, and Yellowstone National Park, Wyoming: (a) theoretical steady-state mosaic for the BWCA with little temporal variation in the fraction of forests in each age class (six 70-year classes from 0 to 419 years); (b) actual temporal pattern on the whole BWCA with a cohort of young forests in 1727 that aged until extensive fires in the 1860s shifted old forests back to young age classes; (c–f) actual temporal patterns on four sample subareas, each representing $\frac{1}{16}$ of the whole BWCA; (g) percentage of a 130,000-ha area in Yellowstone National Park burned each decade (bottom) and percentage of the area covered by successional stages from early postfire (LP0) to mature (LP3) lodgepole pine forest. Periods of fire management policy (e.g., natural fire program) are indicated above the line graphs.

Sources: Parts (a–f) reproduced from W. L. Baker (1989b, p. 27, figs. 2 and 3), with permission of the Ecological Society of America, © 1989 Ecological Society of America; part (g) reproduced from Romme and Despain (1989a, p. 696, fig. 2), with permission of the American Institute of Biological Sciences, © 1989 American Institute of Biological Sciences.

BOX 3.1

Continued

(Romme 1982), connectivity, and age–class structure. Variation in intervals between fires (chap. 5) adds to these fluctuations. Several implications follow from this inherent landscape fluctuation:

- *Episodes of large, often severe fire and interludes of small, low-severity fire are normal.* Infrequent episodes of large, often severe fire, at intervals near the length of the fire rotation account for most of the rotation (chap. 5). During the interludes, which also last nearly as long as the rotation, the rotation appears very long, fires are small, and landscapes become patchier. Long interludes with small, less severe fires followed by episodes with large, more severe fire are normal in all fire regimes.
- *Observed change is typically recovery from the last episode of fire, not directional change.* Many Rocky Mountain vegetation types are subject to crown fires, but some recover quickly relative to the fire rotation, others slowly (see table 5.7). In fast-recovery types, most of the fire rotation is spent in mature, fully developed condition, although vegetation may still age. In slow-recovery types, however, much of the rotation is spent recovering from the last large fire. During recovery, vegetation undergoes stand development, including self-thinning and increasing size and cover of shrubs and trees. Rocky Mountain conifer landscapes that appear to be changing in this way most often are simply recovering from the last large, often severe fire (Kulakowski, Veblen, and Drinkwater 2004; Zier and Baker 2006). Given the large fluctuation inherent in landscapes, it is exceedingly difficult to detect actual directional change.
- *Normal long interludes of recovery and periods of fire suppression appear identical.* Both have relatively little fire and an appearance of a long fire rotation; both have increasing tree or shrub cover and density, along with similar fuel changes; and both have early-successional species replaced by later-successional species. In Yellowstone before the 1988 fires, the rate of burning was so low as to give the impression that fires were not occurring because of suppression, but that was simply an interlude (Romme and Despain 1989a). Natural recovery has commonly been misinterpreted as fire suppression.

(continued on next page)

BOX 3.1
Continued

- *Large fires and recovery may be synchronized across large areas and vegetation types.* Large, infrequent fires commonly occur in regional drought years with drought promoted by teleconnections to Atlantic and Pacific atmospheric and oceanic conditions (chap. 2). Fires in drought years can become large enough (see table 5.2) to burn across multiple vegetation types and synchronize recovery for long periods across large land areas.

- *Services and processes from ecosystems, valued by people, can change rapidly.* Wildlife habitat, erosion and water quality, extractive resources such as timber and livestock forage, even the safety of communities in the wildland–urban interface, and a variety of other processes and services naturally fluctuate. For example, fluctuation in landscape diversity, measured by richness or evenness of the number of types of patches, has significant effects on plant, bird, and mammal populations (Romme and Knight 1982). The fluctuation is potentially disruptive, however, changing rapidly over short intervals when large fires occur after long periods of stability.

- *Fires that burn large land areas play an important evolutionary role in forests.* Native disease, insect, and parasite (e.g., dwarf mistletoe) populations are reduced across large land areas, alleviating selection pressure by these agents that could otherwise strongly shape host tree genetics. Reduction of extensive forest cover to random, small, remnant unburned patches also allows a random genetic subset of the former extensive tree population to dominate postfire environments, likely the major source of random genetic drift in Rocky Mountain trees (G. E. Howe 1976), which contributes to adaptation to changing environments.

caused by fire" (Key and Benson 2006, LA-6). Severity has focused on qualitative measures of tree mortality in forests: *low-severity fires* are generally surface fires that kill few or no canopy trees; *moderate-severity fires* kill some fraction of canopy trees (e.g., 25–75 percent); *high-severity fires* are either crown fires or intense surface fires that kill most canopy trees; and *mixed-severity fires* kill trees

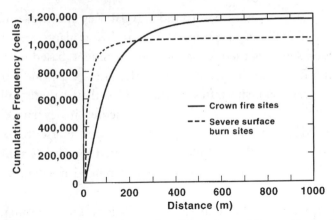

Figure 3.16. Distance to nearest unburned or lightly burned forest. Shown is the cumulative frequency distribution of the distance of cells, in a raster GIS map of severe surface fires and crown fires, to the nearest unburned or lightly burned forest edge after fires in Yellowstone in 1988. The graph shows that most of the points in these burns are within 100 meters (surface fires) to 300 meters (crown fires) of unburned or lightly burned forest where seeds are available for regeneration.

Source: Reproduced from M. G. Turner, Hargrove, et al. (1994, p. 734, fig. 2) with permission of Opulus Press.

extensively in some areas and lightly or not at all in others. Usually, the fire effects that occur immediately, rather than those that occur during reorganization and recovery, are evaluated (fig. 3.1). National programs, such as the Burned Area Emergency Rehabilitation (BAER) program (USDI Bureau of Land Management et al. 2006), loosely assess and map fire severity immediately after fire without clear and consistent criteria but perhaps with an emphasis on the effects on soil (Safford et al. 2008).

A tree-centric or soil-focused perspective ignores other ecosystem components, which may be affected differently. Fires that kill trees may cause little animal mortality, as many animals have effective escape mechanisms (chap. 4). Understory plants often have foliage and crowns in the flaming combustion zone, even in so-called low-severity forest fires, and thus commonly are combusted and topkilled. This equals high severity to the aboveground parts of those plants, yet nearly all resprout or have soil seed banks or dispersal mechanisms, allowing recolonization after fire. Also, severity to canopy trees may be unrelated to severity to physical properties and biology of soils (Odion and Hanson 2008; Safford et al. 2008). Fire thus affects thousands of components of ecosystems (Key and Benson 2006), each potentially differently.

To better assess severity requires measuring effects on more components of ecosystems. At least the three major layers in an ecosystem are proposed: canopy, soil and soil surface, and understory vegetation (Roberts 2004). A more elaborate system, the composite burn index (CBI; Key and Benson 2006), evaluates scorch, char, fuel consumption, damage to living plants, and the nature of ash and exposed soil. It is used for landscape assessment, based on a set of plots in which magnitude of change in specific features (e.g., duff consumption) within five vertical strata (substrates, vegetation less than 1 meter tall, vegetation 1–5 meters tall, intermediate trees, big trees) is scored in severity classes. These are then averaged and summed to determine the index, but scores of each component can be used directly. CBI plots are also used to validate remote-sensing evaluations of fire severity, such as delta or differenced normalized burn ratio (dNBR).

BAER mapping may be focused on soil, whereas CBI evaluates multiple components; but neither evaluates change relative to the HRV. Missing is recognition of the force-resistance relationship between the physical process of fire and the functional traits of plants and animals that allow survival or response to fire (Key and Benson 2006), reviewed here and in chapters 2 and 4. A strongly applied force effectively resisted does not have severe effects. Rocky Mountain plants nearly all have means to persist at the individual, population, community, or landscape scale (e.g., the trees in table 3.2). Is the burning of an aspen forest, which commonly only topkills stems, "severe" when resprouts often appear within a few weeks after fire, allowing individual aspen stems and full clones to persist? Even though many trees lack traits to achieve persistence of individual stems, or even a local population, they persist at the landscape scale by accepting mortality and reinvading by dispersal. This strategy has apparently been successful for generations, so why is temporary loss of trees from a burned area considered "severe" or an emergency?

I suggest that nearly all current measures of fire severity are really just measures of degree of change within a specific period, not measurement of fire severity, which should evaluate current change relative to degree of change from fires under the HRV or relative to a measure of potential recovery. To create an index of fire severity requires estimates of degree of change under the HRV, estimated from reconstructions (chap. 5) or fires in reference areas (e.g., wilderness). Then a ratio of the degree of change for a recent fire and the median change values from reference areas or reconstructions would estimate fire severity, as would current recovery rates and patterns relative to rates and patterns of recovery under the HRV. CBI and other indices measure degree of change after fire, not ecosystem damage. Similarly, BAER characterizes effects on soil and potential erosion but does not evaluate these relative to historical fires.

This is an important distinction, since severity has the connotation of damage. Are our ecosystems truly damaged by wildfire, as seems implicit in BAER and other approaches? Evidence presented in this book suggests that ecosystems are temporarily changed by fire, to varying degrees, but seldom to a degree that prevents the natural recovery that likely occurred under the HRV. In this ecological sense, wildland fires would not usually be considered severe, even if high intensity, and certainly not an emergency, as implied by BAER. However, developing an index of fire severity and changing the terms would require further scientific research and some consensus; thus, for this book I will continue to use traditional qualitative severity terms rather loosely.

MODELS OF SUCCESSION AND LANDSCAPE CHANGE IN RESPONSE TO FIRE

Fire effects that play out across large landscapes over decades to hundreds of years are not easily analyzed without models (Reinhardt, Keane, and Brown 2001). Models allow estimation of long-term consequences of management before large areas are committed to a particular action. For example, models can assess current landscape condition relative to behavior under the HRV so that we can understand the impacts of current or projected fire management policy (Keane, Parsons, and Hessburg 2002).

One set of early fire models used in the Rockies was not spatially explicit (i.e., the actual locations and configurations of trees, communities, and vegetation patches were not used) and focused instead on succession as a process of transition among a set of states. States typically represented community dominants in successional stages. Transitions between states were based on the functional traits or "vital attributes" of dominant species (see, for example, table 3.2, appendix A). Transitions after a particular interval or fire severity depended on which dominant species were able to respond (fig. 3.17). The most complex models of this type were those of Kessell and colleagues, which were calibrated with empirical data for use across landscapes of Glacier National Park, Montana (Cattelino et al. 1979; Kessell 1979). Vital attributes models have been developed conceptually for much of the northern Rockies, have been presented in regional summaries (table 3.1), and now commonly are part of larger models (e.g., LANDSUM: Keane, Parsons, and Hessburg 2002).

Another thread of development focused on detailed population and physiological processes, based on modifications of the JABOWA physiological model (Botkin, Janak, and Wallis 1972). For example, Keane, Arno, and Brown (1989, 1990) and Keane, Morgan, and Menakis (1994) developed a FIRESUM model to study fire in ponderosa pine–Douglas-fir forests and the impact of fire

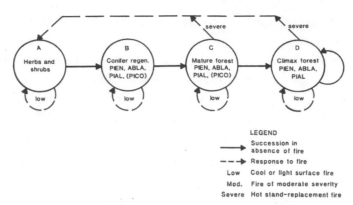

Figure 3.17. Example of a simple graphical successional model. The model is of four stages (A–D) in a subalpine forest with Engelmann spruce (PIEN), subalpine fir (ABLA), white-bark pine (PIAL), and lodgepole pine (PICO), based on vital attributes of these trees in response to low-severity fire and severe (high-severity) fire. PICO in parentheses means it may or may not occur.
Source: Reproduced from Bradley, Fischer, and Noste (1992, p. 73, fig. 32).

exclusion on the decline of whitebark pine. Keane, Ryan, and Running (1996) later merged FIRESUM with an ecosystem model to form a spatial biogeochemical fire model, FIRE-BGC, which they applied in Glacier National Park to understand how fire affects forest productivity, nitrogen cycling, and other ecosystem processes. Models such as FIRE-BGC have mostly been used in research applications due to their complexity.

Nonspatial state-and-transition models of forest succession have been used to assess the effects of fire exclusion in U.S. landscapes. Fire regime condition classes (FRCC) initially characterized departure of a particular area of vegetation from the successional state expected under the HRV (Schmidt et al. 2002). Later, the LANDFIRE program used a state-and-transition computer model, VDDT, to estimate the fraction of the landscape in a particular successional state under the HRV, which could then be compared with current conditions (Long, Losensky, and Bedunah 2006).

Advances in computer technology have made it possible for models to become spatially explicit. The location and configuration of patches of vegetation can be modeled and temporal variation in the shifting mosaic predicted. Spatially important species-level processes, such as seed dispersal, known to play important roles in succession, can be included. Effects of a spatially varying mosaic on wildlife can also be modeled (e.g., Potter and Kessell 1980).

Complex spatially explicit computer models are valuable but have significant limitations that are not generally understood. The algorithms or equations that make up the model often are approximations or are theoretical, lacking sufficient field data and measurements. For example, RMLANDS used a fire-size distribution from lodgepole pine forests in Yellowstone to model piñon-juniper woodland dynamics in western Colorado, because data from the woodlands were not available (William Romme, pers. comm., 2005). Some fire processes are poorly understood or even contested—for example, the relative importance of fuel and fire weather in shaping fire regimes is a topic of scientific debate (W. L. Baker 2003), but models typically adopt one perspective on this debate. Some processes have even been modeled incorrectly; for example, data on mean composite fire intervals have commonly been used as though they represent fire rotation, an error that can lead to large miscalculations (chaps. 5, 10). The frequency of large fires, which account for much of the burned area, is largely uncertain because evidence is available for only a few events. Potential errors and uncertainties implicit in model equations and limited empirical evidence seldom are carried through into explicit uncertainties about model predictions (Lo 2005). Moreover, most fire models have been only partially validated for their intended purpose (Rykiel 1996).

Models remain valuable tools, but too commonly they are applied to make predictions and as a basis for assessments, even though they have not yet been shown to be technically valid for such purposes (W. L. Baker 1999). Skepticism about modeling results is warranted, particularly when models are complex, including hundreds of equations and parameters. An appropriate use of complex computer models is to generate ideas and hypotheses about the impacts of fire on spatial and temporal scales that exceed our ability for direct study. Simpler, more easily validated models with fewer components are in order when valid predictions are needed.

SUMMARY

Most Rocky Mountain trees are killed by fire and rely on postfire seed dispersal to persist, a viable strategy where fire rotations in forests are generally long, as in the Rockies. In contrast, about three-quarters of common shrubs resprout after fire. About equal numbers of graminoids resprout well as resprout poorly. Forb survival and resprouting are increased by greater rooting depth. Seed banks are rare in Rocky Mountain trees and uncommon in graminoids, but about one-third of shrubs and many forbs have them. Seeds are concentrated in the top layers of soil, where they are vulnerable to reduction by soil heating, particularly by high-intensity fires in forests. Postfire succession in most ecosystems in the

Rockies is controlled by rapid resprouting of shrubs, graminoids, and forbs. Tree regeneration that relies on seed dispersal is contingent on trees in nearby unburned forests, leading to varied outcomes, and may be slow, including multidecadal lags. At the landscape scale, fire creates a mosaic of disturbance patches that fluctuates. Infrequent, large, severe fires are followed by interludes with small, often less severe fires. Since plants in the Rockies have functional traits that allow survival and recovery, regardless of fire intensity or size, fire is rarely an emergency for them.

CHAPTER 4

Fire Effects on Animals: From Individuals to Landscapes

As with plants, animals experience immediate fire effects, postfire reorganization and recovery, and long-term adjustment (fig. 3.1). Fire effects on animals also occur across levels of organization, from individuals to populations across landscapes (Pausas and Lavorel 2003). Broad reviews of fire effects on animals cover the international (Komarek 1969; Lyon et al. 1978) and United States literature (J. K. Smith 2000). Reviews of specific groups include birds and mammals (Bendell 1974), small mammals (Ream 1981), birds in North America (Saab and Powell 2005a,b), amphibians in North America (Pilliod et al. 2003), amphibians and reptiles (Russell, Van Lear, and Guynn 1999), and soil fauna and microbes (Ahlgren 1974; Neary et al. 1999). Kotliar et al. (2002) review the effects of fire on birds in western coniferous forests, and Saab et al. (2005) review the effects on birds in the Rockies. A review of the response of mammals to forest fires in boreal forests (Fisher and Wilkinson 2005) is relevant to subalpine forests in the Rockies.

Understanding animal responses to fire is difficult because of animals' mobility, hierarchical habitat selection and patterns of resource use, as well as the background temporal and spatial variability in animal populations. Mobility allows individual animals, unlike plants, to modify their response to fire. Habitat selection can be hierarchical (Kotliar, Reynolds, and Deutschman, 2008), including spatial scales from the regional to meters (fig. 4.1). For example, elk and American bison selected particular parts of a large winter range based on snow and forage, then at the scale of hundreds of hectares chose topographic settings that favored south-facing slopes and burned areas, finally favoring particular vegetation types at a fine (1-hectare) scale (S. Pearson et al. 1995).

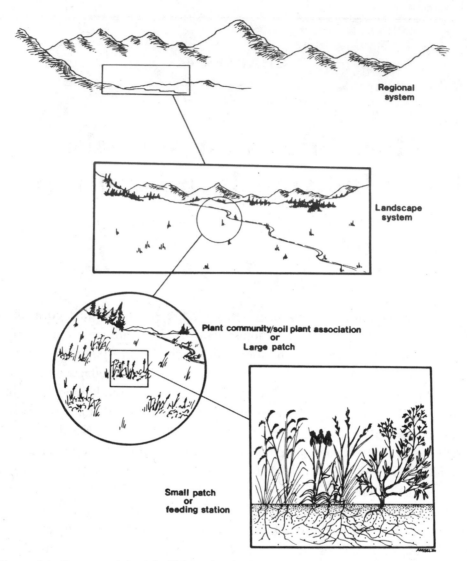

Figure 4.1. Conceptual model of hierarchical selection of habitat by herbivores.
Source: Reproduced from Senft et al. (1987, p. 790, fig. 1), copyright 1987; reprinted with permission of the American Institute of Biological Sciences.

Animal populations also may fluctuate from year to year or place to place, and large sample sizes may be needed to detect directional change above this background variability (Kotliar, Kennedy, and Ferree 2007). Some animals may track ephemeral resources (e.g., rain-induced pulses of annual seed-producing plants, bark beetles), so their populations are not spatially fixed (Rotenberry and Knick 1999). In such cases, absence after fire may be unrelated to the fire. Migratory bird and mammal populations may be regulated by mortality during migration or in the winter, so their absence or lower density after fire on a summer range could be from off-site mortality rather than fire effects. In contrast, high site tenacity may result in some migratory birds returning to a breeding site for several years, even if it is unsuitable after fire (Kotliar et al. 2002). To address these problems in assessing animal responses to fire, Petersen and Best (1999) suggested that studies need unburned controls or preburn data, multiple burns, and postfire monitoring for a period proportional to the life spans of the organisms and rate of recovery of vegetation. Detecting the response to fire of short-lived passerine birds in slowly recovering sagebrush required four to five years of postfire data. Well-replicated studies with sufficient sample sizes, control areas, and adequate duration of postfire monitoring are rare, and the summaries in this chapter should thus be interpreted cautiously.

Rather than just presence or absence, it is also important to know whether animals found after fire are reproducing there at higher rates or instead suffering increased mortality (Saab and Powell 2005b). Such population responses are poorly known. Nest success appeared higher for early-successional cavity-nesting birds in burned versus unburned ponderosa pine forests (Saab and Powell 2005b). In contrast, sage and Brewer's sparrows were only as productive in a mosaic burn as in unburned sagebrush (Petersen and Best 1987).

FIRE EFFECTS: FROM INDIVIDUALS TO LANDSCAPES

Individual animals may survive if they avoid injury during combustion. However, populations and communities are affected by postfire movements and vegetation succession, and animals may be sensitive to landscape changes from fire.

Immediate Fire Effects on Individual Animals

Immediate effects on animals are shaped by physical attributes of the fire (e.g., rate of spread), as well as the prefire animal community and the ability of animals to survive or recolonize after fire. Animal mobility allows escape, but other adaptations (e.g., burrowing) allow survival (Handley 1969). Flames, heat, and asphyxiation may injure or kill animals. In spite of the danger, some animals seem calm near fires, easily maneuvering to avoid flames (Howard, Fenner, and Childs 1959; Singer and Schullery 1989); those with limited mobility (e.g., snails) are more likely to suffer mortality (fig. 4.2).

Animals can avoid injury or mortality by fleeing or taking refuge in wet or low-fuel areas or in burrows (fig. 4.3), where they appear to tolerate short-term

Figure 4.2. Dead snails in an aspen grove burned in the 1988 Yellowstone fires. The snails (*Oreohelix*) are an example of mortality suffered by animals with low mobility.
Source: Reproduced from Beetle (1997, p. 7).

Figure 4.3. Some examples of animal fire refugia: (a) prairie dog mound in a mixed-grass prairie, which can be used as a burrow but also as an aboveground low-fuel fire refugium; (b) rock outcrop, in piñon-juniper woodland, with cracks that allow animals to escape flames; (c) a willow-dominated wetland that may remain unburned except during drought.
Source: Photos by the author.

temperatures of 59–63° centigrade without mortality (Howard, Fenner, and Childs 1959). However, the lethal temperature for mammals declines as relative humidity rises, limiting their ability to evaporatively cool. Animals in burrows often survive if they are more than 5 centimeters belowground, particularly in burrows with multiple entrances (Lawrence 1966). But they may suffer mortality, even 15 centimeters belowground, if they are beneath dead wood concentrations with long-duration heating (Howard, Fenner, and Childs 1959). Burrows can serve as refugia for many species, including insects and arachnids, snakes, lizards, and mammals (Russell, Van Lear, and Guynn 1999). Survival analogs of soil or canopy seed banks of plants include insect eggs, aquatic life stages of amphibians, and other dormant animal forms in soil or water (e.g., Branson 2005).

Direct animal mortality has been observed in fires in many ecosystems, particularly where fires spread at high rates or produce low-lying smoke. Birds may be killed by smoke inhalation or be struck by vehicles if disoriented by smoke and flames, and young in nests are especially vulnerable (McEneaney 1989). A chaparral fire in Southern California killed wood rats, rabbits, mice, and voles via asphyxiation or heat (Chew, Butterworth, and Grechman 1959). Meadow voles suffered some direct mortality in fire in tallgrass prairie in Nebraska, but many survived by fleeing or by taking refuge in pocket gopher mounds and in burrows 5–12 centimeters belowground (Geluso, Schroder, and Bragg 1986). Pygmy rabbits largely survived fire in Wyoming big sagebrush in Idaho where the fire burned in a mosaic (fig. 4.4) but suffered high mortality where fire burned more completely and rapidly (Gates 1983). Tree-dwelling small mammals (e.g., squirrels) likely suffered high mortality in mixed-severity fire in piñon-juniper and ponderosa pine forests in northern New Mexico (Guthrie 1984). The 1988 Yellowstone fires killed about 250 elk, more than half in a single set of several harem groups that panicked in the face of a fast-moving fire, but that mortality was less than 1 percent of total elk in the park (Singer et al. 1989). Fire in tallgrass prairie in Nebraska killed several species of snakes (Geluso, Schroder, and Bragg 1986), but many reptiles and amphibians are able to escape fire by moving to burrows, holes, or cracks in rocky areas (Russell, Van Lear, and Guynn 1999; Pilliod et al. 2003). Grassland fires may kill some insects, although if fires are late season, insect eggs in the soil may survive (Branson 2005). Seasonal timing of fire also can allow amphibians to survive if they are in burrows, as was likely for New Mexico's Jemez Mountains salamander, a state threatened endemic species, in high-severity fires (Cummer and Painter 2007). If amphibians are aboveground, survival may be higher for aquatic life stages in moist fire refugia (Pilliod et al. 2003); however, wildfires that burned across

Figure 4.4. A mosaic of burned and unburned Wyoming big sagebrush in Idaho.
Source: Photo by the author.

small Montana streams killed Rocky Mountain tailed frogs and westslope cut-
throat trout (*Oncorhynchus clarki lewisi*; Pilliod et al. 2003).

Predators may capture fleeing animals, adding to mortality. Ferruginous
hawks and other predators followed smoke plumes into Yellowstone as the 1988
fires were burning and were observed in concentrations hunting small mammals
(McEneaney 1989; Singer and Schullery 1989); this was also observed during
fires elsewhere (Komarek 1969). Ravens, cranes, and herons have been seen
seeking fleeing prey ahead of firelines burning across meadows (McEneaney
1989).

Reorganization and Recovery after Fire

Survival, rather than mortality and recolonization, is the primary means of population persistence in most animals, but survival of individuals does not ensure
population persistence in burned areas. Survivors may move to more favorable
habitat in unburned or less severely burned areas. Birds moving out of burned
areas may elevate populations in nearby unburned areas (Wauer and Johnson
1984; Smucker, Hutto, and Steele 2005). Deer mice increased in a tallgrass
prairie burn in Kansas from both immigration by individuals moving through
or resident in the adjoining unburned landscape and reproduction of survivors

in the burn (Kaufman, Gurtz, and Kaufman 1988). Woodpeckers may also immigrate, attracted to beetles that attack fire-injured trees. Thus, postfire movement increases the complexity of animal responses to fire relative to the simpler response of resprouting, seed germination, and comparatively slow dispersal of plants.

THE FIRST FIVE YEARS AFTER FIRE

Animals may take advantage of short-term food resources after fire. In the first three months after the 1988 Yellowstone fires, large numbers of common ravens, bald eagles, bears, and coyotes scavenged carcasses of elk killed in the fires (Singer et al. 1989). Opening of serotinous lodgepole cones by heat led to a short-term flush of seed-eating birds, including red crossbills, pine grosbeaks, pine siskins, and Clark's nutcrackers (McEneaney 1989). A flush of mobile insects (e.g., bark beetles) and insect-eating birds often occurs in the first few years after forest fires in the Rockies (box 4.1). Do postfire beetle outbreaks lead to further fires that then precondition landscapes for another beetle outbreak, in a positive feedback loop (see box 4.2)?

Predation may be higher in postfire environments, reducing some prey populations. Mortality of pygmy rabbits increased 66 percent in burned sagebrush in Idaho, mostly from predation, leading to population elimination 20 months after fire (Gates 1983). Postfire predation may have further decreased deer mice in sagebrush after fire in northwestern Colorado (Olson et al. 2003). Increased snow depths in burned forests in Glacier National Park may also increase predation of elk and white-tailed deer relative to predation in unburned forests (Singer 1975).

Arachnids (spiders, mites, ticks), insects, earthworms, and mollusks that live on low vegetation, in the litter, or near the soil surface often suffer short-term reductions after fires, but occasionally their populations are unchanged or increased. Fire in mixed-grass prairies on the edge of the Black Hills, South Dakota, substantially reduced surface-dwelling insects for at least two years (Forde 1983). Fire in sagebrush in southeastern Washington reduced ground beetles (Coleoptera; Rickard 1970). Fire in Wyoming big sagebrush in Idaho led to reduced ants (Hymenoptera) and grasshoppers (Orthoptera) two to three years postfire (R. A. Fischer, Reese, and Connelly 1996). In contrast, Nelle et al. (2000) observed no effect of fire on dry weight of ants, beetles, grasshoppers, and miscellaneous invertebrates 1–14 years after fire in mountain big sagebrush, except that beetles increased in year 1. Ants commonly survived high-severity fire in underground nests in piñon-juniper woodlands in Colorado (Kendall 2003). Arachnids (mostly spiders and mites) on the soil surface remained significantly

BOX 4.1
Fire, Insects, and Birds

Trees struck by lightning (Schmitz and Taylor 1969) or damaged by small fires, disease, or physical disturbances support beetles (table 4.1) at endemic levels (Logan et al. 1998). Bark beetles may attack some trees that have been injured and weakened by low-severity fires in ponderosa pine forests (W. C. Fischer 1980; Safay 1981). After high-severity fire, dead trees are typically attacked by wood-boring beetles, while surviving trees in patches inside the fire perimeter and in the crown-scorch margin often are attacked by bark beetles, more so as basal damage and crown scorch increase (Amman and Ryan 1991; Rasmussen et al. 1996; H. D. W. Powell 2000; Cunningham, Jenkins, and Roberts 2005). Some wood-boring beetles can locate fires tens of kilometers away by orienting to smoke or by infrared sensing of heat (McCullough, Werner, and Neumann 1998). Insect attacks on fire-injured trees typically cause less mortality than the fire injury itself (Rasmussen et al. 1996), but one-third to two-thirds of trees that initially survive fire in the scorch zone may eventually die from the fire or from insects (Connaughton 1936; Rassmussen et al. 1996). Beetles that reproduce in fire-injured trees can spread into unburned forests, killing some live trees (Rasmussen et al. 1996; K. Gibson, Lieser, and Ping 1999), but the extensive mortality of large outbreaks has not apparently begun with beetles attracted to fire-injured trees. Postfire beetle-caused tree mortality has been viewed from a pest and timber perspective, reflected in terms such as "infested."

However, fire is a natural process that maintains the diversity of native species, including beetles (McCullough, Werner, and Neumann 1998; Kendall 2003) and a suite of birds (fig. 4.6). Other insects in several guilds, including ants and mobile insects such as flies (Diptera), may also increase after fire. In piñon-juniper woodlands in Mesa Verde National Park, Colorado, the number of insects and their diversity were higher in years 2–3 after high-severity fire than in unburned woodlands (Kendall 2003). Woodpeckers rare in unburned forests may be abundant for a few years after high-severity fires (Blackford 1955; Koplin 1969; Saab, Dudley, and Thompson 2004; Vierling, Lentile, and Nielsen-Pincus 2008). These include black-backed woodpeckers, which focus on wood-boring beetles (H. D. W. Powell 2000), and American three-toed woodpeckers, which

BOX 4.1
Continued

consume bark beetles or wood-boring beetles (Pfister 1980; Hutto 1995; Kotliar, Reynolds, and Deutschman, 2008). Hairy and downy woodpeckers and northern flickers may also be common in early postfire forests (Harris 1982; Caton 1996; Smucker, Hutto, and Steele 2005; Vierling, Lentile, and Nielsen-Pincus 2008). Woodpeckers, particularly the three-toed, can control small beetle outbreaks and significantly reduce larger outbreaks (Koplin 1969; Fayt, Machmer, and Steeger 2005). Postfire beetles and beetle-dependent birds often decline by years 2–3 after fire (Furniss 1965; H. D. W. Powell 2000; Cunningham, Jenkins, and Roberts 2005), persisting for four to five years at times (Rasmussen et al. 1996; Gibson et al. 1999; Saab et al. 2007). Aerial insectivores, including Lewis's woodpecker (Saab and Vierling 2001), bluebirds, and flycatchers, also typically increase in burned forests (Saab et al. 2007; Vierling, Lentile, and Nielsen-Pincus 2008) until tree regeneration leads to canopy closure two to five decades later (D. L. Taylor and Barmore 1980). From the standpoint of a suite of native insects and birds, high-severity fire is not detrimental.

reduced in spruce-fir forests three to four years after high-severity fire, relative to unburned forests, although populations of beetles and flies (Diptera) were elevated (M. A. Wood 1981). In lodgepole pine forests in Yellowstone, arachnids (mites and springtails) on soil surfaces were reduced for one to two years (Christiansen and Lavigne 1996), particularly in high-severity fires, and remained reduced for four to five years (M. A. Wood 1981). Arachnids and soil insects are likely to recover as litter builds back up (Ahlgren 1974; Christiansen and Lavigne 1996). Land snails were generally killed in aspen forests in the 1988 Yellowstone fires (fig. 4.2) and had not recolonized six years later (Beetle 1997). Earthworms are usually reduced by fires and the subsequent drier environment (Ahlgren 1974).

Among small mammals, deer mice commonly increase and dominate, or are slightly reduced but still dominant, after fire in many low- to moderate-elevation ecosystems in and near the Rockies. These include salt desert shrublands (Groves and Steenhof 1988), sagebrush shrublands (McGee 1982; W. E. Moritz 1988), mixed-grass prairies (Forde 1983), piñon-juniper woodlands (Guthrie 1984), ponderosa pine forests (C. E. Bock and Bock 1983), and moist

BOX 4.2
Beetles and Fire: A Positive Feedback Loop?

Beginning with attack or outbreak, bark beetle effects on fire were hypothesized to be a potential positive feedback loop thought to promote subsequent fires because of increased dead fuels (Hopkins 1909; J. K. Brown 1975; Amman and Schmitz 1988; Amman and Ryan 1991; Parker and Stipe 1993). After the resulting fire, a cohort of even-aged trees grows up and simultaneously becomes susceptible to insect attack as the trees reach maturity (Amman and Schmitz 1988), and the loop repeats. Fire-scarred trees may also be attractive to bark beetles, which kill initially surviving trees and may expand to kill live trees in adjoining forests, also leading to fire. Parts of this loop tested in the Rockies suggest that a new model is needed. Romme et al. (2006) reviewed beetles and fire in the southern Rockies.

Fuels after a beetle outbreak and a fire may be similar, except for finer fuels. Large deadwood is also common 10–15 years after high-severity fires (fig. B4.1.1c) and after blowdowns. Finer fuels (litter, 1-hour fuels, perhaps 10-hour fuels) may be higher after a beetle outbreak than after a fire, which consumes many of these fuels, but this difference disappears in a few years. Fires are not very likely in the first few decades after high-severity fire in forests (chaps. 5–9), except for the higher reburn potential in the northern Rockies (chaps. 6, 8). High-severity fires in the Rockies appear relatively insensitive to fuels in subalpine forests (chaps. 2, 8) but may be more sensitive to fuels at lower elevations. Escaped slash fires in logged areas (chap. 10) suggest that increased fire could follow beetle outbreaks if fires are ignited by people.

However, the positive feedback loop may not commonly occur, because fuel changes from beetle outbreaks can be quite variable. Variability arises in part from differences in beetle-caused mortality in a stand, particularly if a stand is a mixture of nonhost species, which may increase after host trees are killed (Dordel, Feller, and Simard 2008). Also, diverse changes occur in fuels, in part because of variable mortality, but also because of contrasting trends among fuels. In theory, crown fire potential might increase for one to three years while leaves remain on dead trees (fig. B4.1.1a), but then decrease because canopy bulk density needed to carry a crown fire remains low for one to two decades after leaf fall

BOX 4.2

Continued

(Knight 1987). As dead trees fall, opening the canopy, understory fuel moisture may decline (J. K. Brown 1975), but growth of moist understory vegetation may counter this trend (Knight 1987; Reid 1989; Page and Jenkins 2007a). Regenerating understory trees and shrubs (fig. B4.1.1b) also represent ladder fuels (H. J. Lynch et al. 2006). Five to 30 years after a mountain pine beetle outbreak in lodgepole pine, large dead fuels may reach high levels, increasing potential fire intensity (J. K. Brown 1975; Armour 1982; Page and Jenkins 2007a). Modeling suggested a higher rate of spread and fireline intensity and a lower probability of active crown fire 20 years after a mountain pine beetle outbreak in lodgepole pine than in nonoutbreak stands; this was due not to fuel loading differences but to greater midflame wind speeds in beetle-killed stands (Page and Jenkins 2007b; Jenkins et al. 2008). Fuel loadings may later decline, not change, or increase further, depending on forest type (Armour 1982; Dordel, Feller, and Simard 2008).

In contrast to logical expectations and modeling studies, beetle outbreaks in subalpine forests often did not actually affect fire occurrence or extent, probably because fuels are not generally limiting to fire in these forests. Extent of fires did not increase a few years after a 1940s spruce beetle outbreak (69,500 hectares; Bebi et al. 2003) or a late 1990s spruce beetle outbreak (84,497 hectares; Kulakowski and Veblen 2007) in Colorado, when dead needles were thought to enhance flammability. Fire density during four decades after a 1940s spruce beetle outbreak actually was lower in the outbreak area below 3,000 meters elevation than in unaffected forests and did not differ above this elevation (Bebi et al. 2003). A finer-scale study (4,600 hectares) in the same area found that no severe fire occurred in beetle-affected areas for 54 years after the outbreak (Kulakowski, Veblen, and Bebi 2003). A low-severity fire affected more area outside the beetle outbreak, suggesting that postoutbreak growth of moist understory vegetation reduces flammability (Kulakowski, Veblen, and Bebi 2003). In Yellowstone, areas lightly or moderately affected by beetles 5–17 years prior to fire were more likely to escape burning in one study (M. G. Turner, Romme, and Gardner 1999), but another study found repeated

(continued on next page)

BOX 4.2
Continued

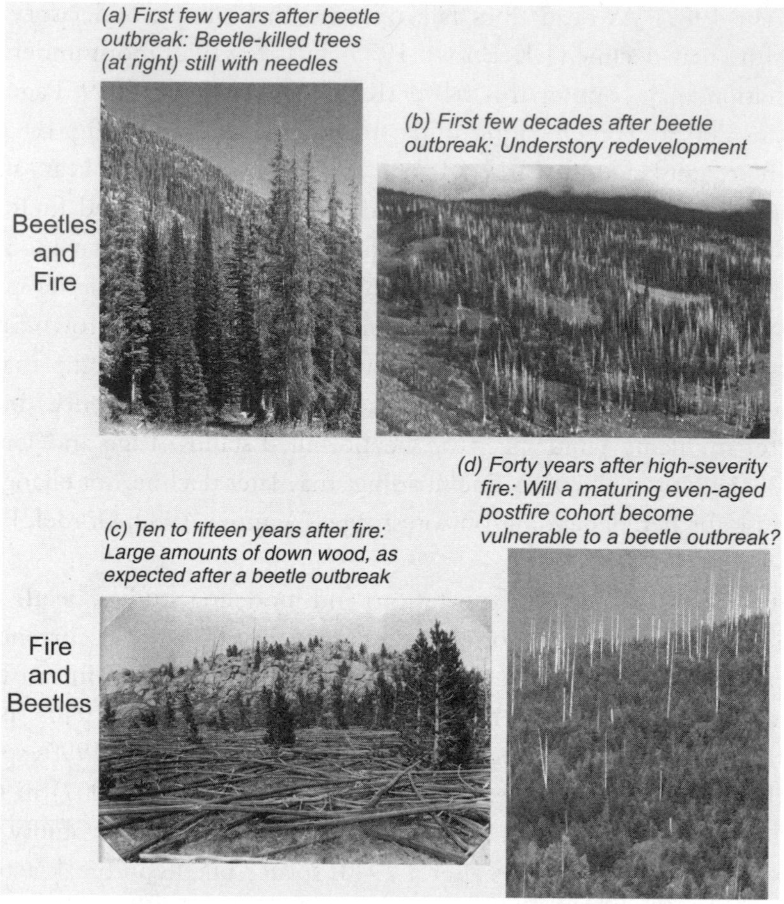

Figure B4.1.1. Beetles and fire: a positive feedback loop? (a) Needles may persist on a tree killed by bark beetles for a period of one to three years and are thought to temporarily increase flammability; (b) as an understory redevelops, potential fire intensity is increased by down wood but decreased by growth of relatively moist understory vegetation; (c) large amounts of down wood also occur after fire; (d) 40 years after fire an even-aged cohort of trees has replaced most of the standing snags, and a century after fire a mature even-aged forest may be becoming more vulnerable to a beetle outbreak.

Sources: Photos by the author except (c) is a photo by the author of an original photo at the National Archives, FRB, p. 33. no. 440: "Remains of fire and new growth. E. Fk. of Big Goose Creek on Hyattsville road," Bighorn Mountains, Wyoming, likely taken by F. E. Town, ca. 1898. A companion photo of the same area says it was taken 10 to 15 years after the fire.

BOX 4.2
Continued

beetle activity 13–16 years before fire increased odds of burning—though only by about 11 percent (H. J. Lynch et al. 2006).

Similarly, empirical studies showed that if fire does occur in subalpine forests, severity may or may not be increased by prior beetle outbreaks. Fire severity can be higher if beetle-attacked stands are older, have higher horizontal and vertical fuel continuity, and have more dead and down large woody fuels (M. G. Turner, Romme, and Gardner 1999; Bigler, Kulakowski, and Veblen 2005). In Yellowstone, areas lightly or moderately affected by beetles 5–17 years prior to fire were more likely to burn lightly, but areas with severe beetle damage were more likely to burn severely (M. G. Turner, Romme, and Gardner 1999). However, in a drought year, severity of a subalpine Colorado fire was only slightly increased in stands affected by the 1940s spruce beetle outbreak, and other variables (e.g., stand structure, previous fire) had more effect (Bigler, Kulakowski, and Veblen 2005). Nearby, fire during drought was also not more severe in areas affected by bark beetles a few years earlier (Kulakowski and Veblen 2007).

Another part of the positive feedback theory suggested that even-aged tree regeneration after large disturbances (e.g., fire, beetle outbreak, drought) might encourage a large beetle outbreak decades to more than a century later (fig. B4.1.1d), which is consistent with beetle responses (Raffa et al. 2008), but empirical evidence is mixed. Extensive mountain pine beetle outbreaks in the northern Rockies in the late 1900s followed extensive late-1800s fires (Schmid and Parker 1990). Widespread western spruce budworm and Douglas-fir beetle outbreaks in the Colorado Front Range in the late 1900s may have been favored by late-1800s fires (Hadley and Veblen 1993). The effect of late-1800s fires is unclear in other areas because early logging, lower-severity fires, and a wet period were at least as important as fires in creating stand conditions favorable to late-1900s spruce budworm outbreaks (Anderson et al. 1987; Swetnam and Lynch 1989). Moreover, extensive old postfire lodgepole pine and spruce-fir forests (Kipfmueller and Baker 2000; E. Howe and Baker 2003; Buechling and Baker 2004) suggest that beetle outbreaks often did not occur when forests reached susceptible ages after extensive fires of past centuries.

(continued on next page)

BOX 4.2
Continued

A related inference that patchy landscape mosaics may resist beetle outbreaks is not well supported. Holland (1986, 20) suggested that "fires . . . had historically created mosaics of forest vegetation, with a wide distribution of age, species diversity and size. These mosaics usually prevented the relatively large insect and disease infestations found today." This suggestion was repeated by others (e.g., Lundquist and Negron 2000; Raffa et al. 2008). However, simulation in Canada suggested that small fires, which would create more landscape heterogeneity, actually increased traversability by mountain pine beetles in lodgepole pine forests; while extensive fires decreased beetle movement (Barclay et al. 2005). This pattern could have an age threshold, however. Young postfire spruce-fir forests had little susceptibility to spruce beetle outbreaks for 50–100 years after fire but were susceptible after crossing this age threshold (Veblen et al. 1994; Bebi et al. 2003; Kulakowski, Veblen, and Bebi 2003; Bigler, Kulakowski, and Veblen 2005). Extensive beetle-related mortality occurred in older stands in landscapes with generally resistant postfire forests (Veblen et al. 1994; Bebi et al. 2003). Beetles can move across kilometers, and it is not clear that beetle spread to susceptible forests is discouraged by small intervening areas of young forest or by heterogeneous forests in general (e.g., Hughes 2002; Barclay et al. 2005).

Under the HRV, Rocky Mountain forests likely had periods of both heterogeneity and homogeneity (see box 3.1), and large outbreaks of bark beetles (table 4.1) are within their HRV (Schmid and Mata 1996). Spruce budworm–Douglas-fir beetle outbreaks in the San Juan Mountains, Colorado, were regionally synchronous for several hundred years and have not increased in severity (Ryerson, Swetnam, and Lynch 2003). Recent spruce beetle outbreaks in the southern Rockies have not approached the extent of an outbreak in Colorado in the late 1800s (W. L. Baker and Veblen 1990). However, it is likely that global warming is enhancing beetle outbreaks, including a recent mountain pine beetle outbreak, in lodgepole pine in the Rockies and Canada that appears larger than outbreaks known since the early 1900s (Raffa et al. 2008). The limited available evidence suggests that fires will not be substantially changed in intensity or extent.

northwestern montane forests (Stout, Farris, and Wright 1971). Deer-mice are omnivorous, and insects and surviving seeds or seeds of postfire annual plants favor them (M. A. Wood 1981; Halford 1981; McGee 1982). Deer mice may decline one to two years after fire in some cases, as vegetation regrows and other animals recolonize (C. E. Bock and Bock 1983; Forde 1983). Occasionally, other species (e.g., yellow-pine chipmunk) increase in the early postfire environment (M. A. Wood 1981).

Many other mammals may decline in early postfire environments until vegetation regrows. Townsend's ground squirrel and American badger were more than twice as abundant on unburned salt desert shrubland as in burns a year after fire (Groves and Steenhof 1988). The least chipmunk preferred unburned sagebrush two years after fire (W. E. Moritz 1988). Prairie voles and thirteen-lined ground squirrels also preferred unburned mixed-grass prairie (Forde 1983). Southern red-backed voles were reduced for at least four to five years in burned subalpine forests (M. A. Wood 1981). White-tailed deer avoided burned areas in Montana (Keay and Peek 1980). Moose, often considered fire followers (Bendell 1974), avoided burned forests after the 1988 Yellowstone fires (Tyers and Irby 1995). Thousands of elk and bison, faced with little forage on the burned winter range the first winter, migrated out of Yellowstone; those that stayed suffered high mortality (Singer and Mack 1999).

The short-term response of amphibians and reptiles to fire is almost unknown in the Rocky Mountains. After extensive fires in 1999–2005 in Glacier National Park, Montana, Columbia spotted frog and long-toed salamander populations in wetlands were little affected, but Rocky Mountain tailed frog larvae declined in streams, and boreal toads apparently responded favorably to changes in upland terrestrial environments (Hossack and Corn 2006).

SUCCESSIONAL TRENDS

Later stages of postfire succession may also favor certain animals. Cavity-nesting birds may decline as snags fall and the forest regrows (Pfister 1980) but be common in old-growth forests where snags are again abundant. Birds favored by midsuccession may include cordilleran and dusky flycatchers in coniferous forests and warbling vireos and dusky flycatchers in aspen (Kotliar et al. 2002). In subalpine forests in Wyoming, southern red-backed voles were common in old-growth spruce-fir forests with substantial understory plant cover but were reduced in mature spruce-fir forests with little understory cover (Nordyke and Buskirk 1991). Species associated with mature or old forest (e.g., northern goshawk [Graham et al. 1997]; hermit thrush [D. L. Taylor and Barmore 1980; S. J. Henderson 1997]; mountain chickadee, ruby-crowned kinglet, Swainson's thrush [S. J. Henderson 1997; Kotliar et al. 2002]) may not use postfire forests burned at high severity until canopy closure or a mature or old-growth condition is reached.

Reduced species diversity of birds and small mammals in the first few years after fire can be lacking, short lived, or longer lasting and can be followed by a rise above prefire levels. Reduction in bird species richness lasted only one year in ponderosa pine forests burned in a mix of severities in northern New Mexico (Wauer and Johnson 1984) and in burned subalpine forests in Yellowstone (D. L. Taylor and Barmore 1980). Species richness was similar before, and 1–2 years after all severities of the 2000 Cerro Grande fire in New Mexico (Kotliar, Kennedy, and Ferree 2007). Stout et al. (1971) suggested that reductions in richness from about 16 to 6 small mammals might be sustained for 5 to 10 years after fire in moist northwestern forests. In Yellowstone forests, after a first-year decline, species richness of birds and mammals rose above prefire levels to peak about 25 years after fire, then declined, reaching a low when canopy closure occurred, after about 50 years (D. L. Taylor 1971, 1973; D. L. Taylor and Barmore 1980; fig. 4.5). Eight years after fire, Roppe and Hein (1978) found seven species in a burned lodgepole pine forest and five in unburned forest, consistent with Taylor's findings. Bird and mammal richness fluctuated between two to six species during about 75 years of postfire succession in sagebrush (W. E. Moritz 1988). In burned Wyoming big sagebrush, diversity of small mammals declined

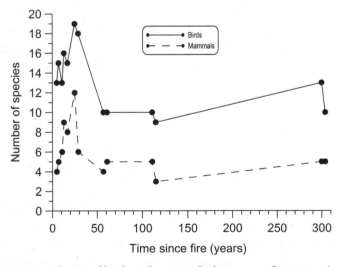

Figure 4.5. Species richness of birds and mammals during postfire succession. Shown is a chronosequence representing postfire succession in lodgepole pine forests of Yellowstone National Park.
Source: Based on data in D. L. Taylor (1971).

the first 5 years after fire but recovered or exceeded diversity in unburned areas by year 7 (Olson et al. 2003). Mean richness was 2.23 in burned and 7.54 in unburned big sagebrush, from 1–14 years or more after fire in several Rockies states (Welch 2002).

Fire Severity and Effects on Animals

Animal responses to fire often vary with fire severity. Birds are best studied (fig. 4.6). Severity effects vary because different postfire forest structures result (Smucker, Hutto, and Steele 2005). Low- to moderate-severity fires favored several birds, but the favored species differed in New Mexico (Kotliar, Kennedy, and Ferree 2007), Montana (Smucker et al. 2005), and the Black Hills (C. E. Bock and Bock 1983), perhaps because of differing study designs or other factors (Kotliar, Kennedy, and Ferree 2007). Low-severity fire may have little effect (Saab et al. 2007). Breeding bird density increased for seven species the year after low-severity fire in the Black Hills but remained elevated for only one species in year 2 (C. E. Bock and Bock 1983). Low-intensity fire in ponderosa pine–shrubland ecotones in Rocky Mountain National Park had little effect on many birds but had adverse effect on shrub nesters (e.g., green-tailed towhee), particularly those that use nonsprouting shrubs (Jehle, Savidge, and Kotliar 2006). Moderate-intensity fires, in contrast, can affect bird composition and density for several years (D. L. Taylor and Barmore 1980; Wauer and Johnson 1984).

High-severity fires can have some longer-lasting effects. Red squirrels disappeared from severely burned ponderosa pine forests in northern New Mexico, and montane voles and long-tailed voles, favored by grass vegetation, became common (Guthrie 1984). Red squirrels and the northern flying squirrel disappeared from severely burned subalpine forests (M. A. Wood 1981). Large patches of high-severity fires may reduce or eliminate birds that nest in live foliage (e.g., western tanager), but these birds may increase after low-severity fires (Smucker, Hutto, and Steele 2005). In moist forests in northern Idaho, shrews were absent 3 years after high-severity fire, probably because of limited dispersal ability and lack of litter in the burn (Stout, Farris, and Wright 1971). Some microtines (e.g., meadow vole, southern red-backed vole) may be reduced or absent for 5 to 10 years in burned, moist, northwestern forests (Stout, Farris, and Wright 1971). Voles, the chief food of American martens, were also absent from subalpine forests 10 to 15 years after high-severity fire (Koehler and Hornocker 1977). Vagrant shrews, western heather voles, deer mice, and long-tailed voles were more abundant and southern red–backed voles less abundant in lodgepole forests 5 to 10 years after high-severity fires than in unburned forests (P. R. Davis 1976; Roppe and Hein 1978).

High severity

Woodpeckers
American three-toed CO, NM
Black-backed BH, MT
Downy BH, MT
Hairy BH, MT, NM
Lewis's BH, ID
Northern flicker BH, MT, NM
Red-headed BH

Flush of insects
Wood-boring beetles
Bark beetles
Diptera (flies)
Predatory insects

Mammals
Deer mouse
Mule deer

Other birds
Hammond's flycatcher NM
House wren NM
Mountain bluebird MT
Olive-sided flycatcher MT
Townsend's solitaire MT
Western bluebird NM
Western wood-pewee NM

Low to moderate severity

American robin BH, MT
American three-toed
  woodpecker CO
Black-headed
  grosbeak NM
Chipping sparrow MT
Dark-eyed junco BH, MT
Hermit thrush MT
Mountain bluebird BH
Plumbeous vireo BH
Spotted towhee NM
Steller's jay NM
Virginia's warbler NM
Western tanager BH, MT
Yellow-rumped warbler BH

Figure 4.6. Birds and some mammals favored by low-, moderate-, and high-severity fire. Shown are ponderosa pine–Douglas-fir forests burned in the upper Bitterroot Valley, Montana, in 2003, near the study area of Smucker, Hutto, and Steele (2005). Fires, particularly high-severity ones, are typically followed by a short-term flush of insects that feeds a suite of woodpeckers and other birds.

Sources: Based on studies in the Rocky Mountains, including Wauer and Johnson (1984) and Kotliar, Kennedy, and Ferree (2007) in northern New Mexico (NM); Kotliar, Reynolds, and Deutschman (2008), in central Colorado (CO); Saab and Vierling (2001) in Idaho (ID); Hutto (1995) and Smucker, Hutto, and Steele (2005) in Montana (MT); and C. E. Bock and Bock (1983) and Vierling, Lentile, and Nielsen-Pincus (2008) in the Black Hills, South Dakota (BH). Photos by author.

High-severity fires in shrublands also affect animal use. High-severity fire in sagebrush led to herbaceous-dominated vegetation avoided by sagebrush-obligate birds, such as Brewer's sparrows and sage thrashers, except where unburned shrubs remained nearby; however, grassland-associated birds, such as western meadowlark, vesper sparrow, and horned lark, were not reduced and even increased in some cases (Castrale 1982; C. E. Bock and Bock 1987; Kerley and Anderson 1995; Welch 2002; Knick, Holmes, and Miller 2005). Vagrant and cinereus shrews did not recolonize by 2 years after high-severity sagebrush fire (McGee 1982). Sagebrush-obligate birds remained reduced 7 to 9 years after fire (Kerley and Anderson 1995).

Although high-severity fire leads to more sustained changes in animal communities than does low-severity fire (Smucker, Hutto, and Steele 2005; Kotliar, Kennedy, and Ferree 2007), high-severity fire favors many species (fig. 4.6). In forests, a suite of birds is favored by high-severity fire (Smucker, Hutto, and Steele 2005; Kotliar, Kennedy, and Ferree 2007). Native insects (see table 4.1) also are attracted to dead or injured trees and provide food for these birds (see box 4.1), which commonly appear, sometimes at low density (Smucker,

Table 4.1. Bark- and Wood-Boring Beetles Common in Rocky Mountain Forests

Host	Primary Bark Beetles	Secondary Bark Beetles	Wood-Boring Beetles
Douglas-fir	*Dendroctonus pseudotsugae* (Douglas-fir beetle)	*Pseudohylesinus* sp.	Buprestidae Cerambycidae
Engelmann spruce	*Dendroctonus rufipennis* (spruce beetle)	*Ips pilifrons* *Scierus* sp.	Buprestidae Cerambycidae
Lodgepole pine	*Dendoctonus ponderosae* (mountain pine beetle)	*Ips pini* (pine engraver) *Dendroctonus valens* (red turpentine beetle) *Pityophthorus confertus* *Pityophthorus knechteli*	Buprestidae Cerambycidae
Ponderosa pine	*Dendoctonus ponderosae* (mountain pine beetle) *Dendroctonus brevicomis* (western pine beetle)	*Dendroctonus valens* (red turpentine beetle)	Buprestidae Cerambycidae
Twoneedle piñon		*Ips confusus* (piñon engraver beetle)	Buprestidae Cerambycidae

Sources: Primarily Amman and Ryan (1991) and Schmid and Mata (1996).
Note: Primary bark beetles may attack and kill live trees, secondary bark beetles typically attack weakened or recently killed trees, and wood-boring beetles primarily bore into dead trees.

Hutto, and Steele 2005), after high-severity fire in all forests in the Rockies, in-cluding ponderosa pine and Douglas-fir forests (Blackford 1955; Wauer and Johnson 1984; Hutto 1995; S. J. Henderson 1997; Kotliar et al. 2002; Saab, Dud-ley, and Thompson 2004; Gentry and Vierling 2007; Kotliar, Kennedy, and Fer-ree 2007; Vierling, Lentile, and Nielsen-Pincus 2008). Less specialized species, including mule deer and deer mice, are attracted to high-severity fires, and a va-riety of species use high-severity fires in the period before canopy closure (e.g., D. L. Taylor and Barmore 1980). Bird diversity, bird density, and number of bird guilds were higher 5 to 10 years after high-severity fire in lodgepole pine than in unburned forests in Wyoming (P. R. Davis 1976). Some species, such as the blue grouse, may use dense patches of regenerating Douglas-fir and ponderosa pine after high-severity fire (R. R. Martinka 1972).

Fire Effects on Animals at the Landscape Scale

Landscape mosaics of postfire successional stages appear to be an important component of habitat for many animals. Larger mammals and birds that can move kilometers (e.g., northern goshawk [Graham et al. 1997]) interact with landscape mosaics on a daily basis. Mosaics of successional stages, including early postfire and old growth, benefit the American marten (Koehler and Hornocker 1977) and one of its major prey, the southern red-backed vole (Nordyke and Buskirk 1991), as well as the northern goshawk and the moose (Gordon 1974). Salt desert shrublands (Groves and Steenhof 1988), Wyoming big sagebrush (W. E. Moritz 1988), and mountain big sagebrush (McGee 1982) in which fire left a mosaic of burned and unburned areas (fig. 4.4) had higher postfire small-mammal richness and density than solid burned areas did. If the burned area from small to moderate fires is contiguous, rather than patchy, across a winter range landscape, elk use and survival may be higher during se-vere winters (M. G. Turner; Wu et al. 1994). Unburned areas in a mosaic serve as fire refugia, increasing postfire recolonization and recovery of small mammals (McGee 1982; Longland and Bateman 2002). However, unburned patches in forests may also increase survival and recolonization of burns by nest predators, such as red squirrels, weasels, and bullsnakes, adversely affecting cavity-nesting birds postfire (Saab, Dudley, and Thompson 2004); thus, mosaics are not univer-sally beneficial to animals.

Although mosaics are known to be important for animals, less is known about which mosaic attributes are important (Longland and Bateman 2002). Mule and white-tailed deer may concentrate use in edges of burns (McCulloch 1969) and prefer adjacencies, such as burned openings with low snow cover and nearby unburned forest (Keay and Peek 1980). Canada lynx may select den sites in mature forest near travel corridors or early-successional forests where snowshoe hares are common (Lyon et al. 1994). Juxtaposition of burned and unburned forest may also favor olive-sided flycatchers and Townsend's solitaires, and a number of other birds are common near edges (Kotliar et al. 2002). Pine siskins and gray jays may concentrate in burn edges in lodgepole pine forests, and species richness may be elevated in the edge (Roppe and Hein 1978). However, positive effects on bird density were not found in narrow burn edges in lodgepole pine and spruce-fir forests in Yellowstone National Park (Pfister 1980).

Burn size and the pattern of linkages of burned and unburned forest also affect species. Some small mammals with limited dispersal ability (e.g., shrews) may only slowly recolonize large severe burns (Stout, Farris, and Wright 1971), but birds that move freely, such as the American three-toed woodpecker, may rapidly move out of large landscape areas and concentrate in burns (Blackford 1955). Tree swallow and white-crowned sparrow densities were significantly associated with larger burns in the northern Rockies, but plumbeous vireo and Townsend's solitaire had a significant negative association with larger, high-severity patches (Hutto 1995). Landscape linkages that facilitate animal movement can include recent burns or corridors of continuous old forest. Both appear important to movements of bighorn sheep in Colorado (Wakelyn 1987) and white-tailed deer in Montana (Freedman and Habeck 1985). Fishers and American martens prefer continuous forest for travel, particularly in winter (Lyon et al. 1994).

A temporally dynamic mosaic of postfire successional stages with a diversity of tree densities and patterns, associated with variation in fire severity, may influence a variety of species. The snowshoe hare, for example, is most abundant one to three decades after high-severity fires in boreal forests (Fisher and Wilkinson 2005), and Canada lynx populations track hare population fluctuations (Fox 1978; Lyon et al. 1994). Birds associated with postfire forests, such as mountain bluebirds, American three-toed and black-backed woodpeckers, and birds that prefer mature forest, such as ruby-crowned kinglets (D. L. Taylor and Barmore 1980), may fluctuate substantially in abundance in response to the irregular temporal pattern of episodic large, high-severity fires (Romme and Knight 1982). Species that can move across large land areas can modulate such fluctuations by moving to appropriate habitat elsewhere. For example, American three-toed, black-backed, and Lewis's woodpeckers may move among burn patches across large landscapes (Saab, Dudley, and Thompson 2004).

To manage animals at the landscape scale, their responses to specific attributes of landscape mosaics must be known, and mosaics can be overemphasized.

Fire regimes under the HRV did not provide identical burn patches at uniform intervals. In the Rockies under the HRV (see box 3.1), extensive burn patches from infrequent large fires, rather than fine-grained mosaic burns, influenced animals and their habitats for decades and even centuries (Romme and Knight 1982; Smucker, Hutto, and Steele 2005). During interludes between large fires, smaller fires covering little total land area were peppered across landscapes, allowing species that were dependent on fires, and could find and move to them, to persist until the next major fire.

FIRE EFFECTS ON SPECIAL SPECIES

In this section, I highlight the responses to fire by some well-known species: grizzly bears, elk, deer, American bison, bighorn sheep, pronghorn, and sage-grouse.

Grizzly Bear

Grizzly bears generally survived the Yellowstone fires of 1988. Only 2 of 21 radio-monitored bears with home ranges inside the burned areas disappeared, and one may have been killed in the fires (Blanchard and Knight 1990; French and French 1996). Three of 21 bears remained within the perimeter as the fires burned; three moved outside; and 13 moved into burned areas after the fire passed, often while fires remained active (Blanchard and Knight 1990). Bears fed on the carcasses of ungulates killed in the fires, along with plants, insects, and other foods. From 1989 to 1992, a larger sample of bears used the 1988 burns in proportion to, or in a lower proportion than, their availability, suggesting no preference for burns; however, rates of movement of many bears were lower, suggesting more plentiful foods, such as root crops, than before the fires (Blanchard and Knight 1996).

There may be long-term effects of severe fires on bears. The 1988 fires burned about 30 percent of the most important whitebark pine area in Yellowstone (Blanchard and Knight 1996). Seeds of this tree are a key source of high-energy food needed before hibernation (Arno 1986), and loss of trees is especially deleterious to bears in years of low cone crops (Blanchard and Knight 1996). Yellowstone is typical of the southern part of current grizzly range, but at the moister, northern edge of its Rocky Mountain range more shrubs with edible fruits occur (Zager 1980; Blanchard and Knight 1996). There, burns 35 to 70 years old were much more productive of fruit-bearing shrubs used by grizzlies than were mature and old-growth forests (Zager 1980).

Elk

Their mobility allows elk generally to avoid mortality from advancing fires, but they may face forage deficiency on burned ranges the first winter after fire. Direct mortality of elk in the 1988 Yellowstone fires was less than 1 percent of the population, but 24–37 percent of the population died the next winter. Winter mortality was mostly from malnutrition stemming from the 1988 drought and deep winter snow that made some forage inaccessible, combined with elk harvest outside the park (Singer, Coughenour, and Norland 2004). The fires burned 22 percent of the park's winter range, contributing to the winter mortality (Vales and Peek 1996; Singer and Harter 1996) and the unusual migration of 54 percent of the population to lower-elevation, unburned winter range outside the park (Singer, Coughenour, and Norland 2004). During milder winters, large contiguous burned areas, particularly outside winter ranges, may affect elk populations little, but patchy fires concentrated on winter ranges could elevate mortality in average or extreme winters (M. G. Turner, Wu et al. 1994).

As vegetation recovers, burned areas might be attractive to elk, possibly favoring population increases. Elk pellets suggested increased use, relative to unburned areas, one year after fire in Wyoming big sagebrush in Idaho (W. E. Moritz 1988) and in sagebrush in Montana, where increased use continued up to nine years after fire (Van Dyke, Debocr, and Van Beek 1996). Elk appear to have sought moderate- to high-severity burns in ponderosa pine forests on winter ranges in northern New Mexico (Rowland et al. 1983). However, a ground-based study of winter ranges in Yellowstone, during the second and third winters after the 1988 fires, found elk use was no different between burned and unburned pairs of sagebrush and forest sites (Norland, Singer, and Mack 1996). Aerial surveys the same years found 15–22 percent higher use of burned sagebrush and grassland in general and 52 percent higher use in the driest burned sagebrush (Singer, Coughenour, and Norland 2004), although this is a small part of the winter range (S. Pearson et al. 1995). Burned forests on the winter range were avoided because of deep snow and less forage (Singer and Harter 1996). In another ground-based study of elk and American bison, use was 6–33 percent higher in burned than in unburned areas in winter 3, but not much different in winter 4, except 5–6 percent higher in March (S. Pearson et al. 1995). At higher elevations on summer ranges in Yellowstone, pellet densities suggested that use was no different between burned and unburned pairs of meadows, forest edges, and forests the second summer after fire (Norland, Singer, and Mack 1996).

If these modest preferences for burns in winter led to increased elk survival or reproduction, it was difficult to detect in population estimates after the fires, and many elk left the burned winter range. Northern range estimates, which include elk in the park and migrating out of the park, were 19,316 the winter just prior to the fires; 12,328 in winter 1 after the fires; 15,805 in winter 2; 10,287 in winter 3; and 15,587 in winter 4 (Lemke, Mack, and Houston 1998). The winter 2 estimate fits with a fire contribution to population increase, but the winter 3 count does not, nor does the winter 4 count, when only slightly higher use of burns occurred (S. Pearson et al. 1995) while the population increased almost 50 percent. Also, after the fires, 2,000–4,500 elk began wintering outside the park at the northern end of the winter range (Lemke, Mack, and Houston 1998). This movement could be due to an attraction to less snowy sites with accessible forage, but, if so, burned vegetation was not sufficiently attractive to overcome weather effects. The evidence supports Pearson et al. (1995, 753): "Fire effects on northern Yellowstone ungulates are likely to be relatively short-lived, and in the long term, may have minimal impact on population dynamics compared to winter conditions." This is also consistent with a more recent synthesis (L. L. Wallace et al. 2004).

In contrast to Yellowstone, large changes in elk populations may be linked to episodes of high-severity fires in moister forests in northern Idaho and Montana, where taller resprouting shrubs and regenerating trees, rather than grasslands, dominate burned winter ranges. Elk increased dramatically in this area from 1925 to 1935 (Gaffney 1941; McCulloch 1955; C. J. Martinka 1974), peaking 20 to 25 years after large high-severity fires (e.g., 1910; McCulloch 1955; Leege 1968; C. J. Martinka 1974), although other management techniques (e.g., predator control) are confounded with this possible fire effect (McCulloch 1955). Elk are favored here by a diversity of forbs, grasses, and shrubs, with nearby cover found in shrub fields near young conifers (C. J. Martinka 1974; Irwin and Peek 1983). In a drier area in Glacier National Park, elk were attracted in winter to spruce forests, grasslands, and low-density lodgepole pine forests thought to be a result of thinning fires (Singer 1979). In the southern Rockies, elk on summer ranges, attracted to the combination of increased diversity of forage plants with nearby cover, preferred burned lodgepole pine to unburned forests five to ten years after fire (P. R. Davis 1977; Roppe and Hein 1978).

Deer

White-tailed and mule deer are found throughout the Rockies, but white-tailed deer have a spotty distribution in Colorado. Where they co-occur, white-tailed deer appear to favor sites with more cover than do mule deer (Keay and Peek 1980), and they may especially use river corridors (Freedman and Habeck 1985), as in Colorado.

White-tailed deer avoided large, open, early-successional areas following replacement fires, especially in winter when snow was deep (Singer 1975), preferring sites with shallow snow and short distance to live cover (Singer 1975; Keay and Peek 1980). They also preferred unburned forests over burned forests and grasslands (Keay and Peek 1980). They did use small burned areas in a Douglas-fir forest (Singer 1975). Freedman and Habeck (1985) thought that white-tailed deer rely on landscape-scale connections between riparian areas, with abundant browse, and nearby mature or old-growth upland forests with low snow depths but also low browse due to periodic low-intensity fires. However, surface fires likely were less frequent than these authors suggested (chap. 5), and most preferred browse species (Keay and Peek 1980: Saskatoon serviceberry, Scouler willow, dwarf rose, mallow ninebark) also recover fully within just a few years after low-intensity fire (see appendix B). Surface fire could not have kept browse height low.

Mule deer, in contrast, preferred ponderosa pine and Douglas-fir forests burned in a surface fire with scattered, individual-tree torching, over unburned forests or north-slope forests with substantial high-severity fire (Keay and Peek 1980). Preferred forests had intact canopies but lower average distance to cover than sites preferred by white-tailed deer. Mule deer used lower-density parts of about 40-year-old ponderosa pine–Douglas-fir forests recovering from high-severity fire in western Montana, making heavy use of snowbush ceanothus (fig. 3.5), which is abundant after high-severity fire in this area (Klebenow 1965). Mule deer also used five- to ten-year-old burns with dead trees more than unburned lodgepole pine forests in northern Colorado and southern Wyoming (P. R. Davis 1977; Roppe and Hein 1978). Mule deer use in northern Arizona piñon-juniper that burned in a high-severity fire was higher in the burned area than in unburned woodland in winter, and highest within 200 meters of unburned edge (McCulloch 1969). Mule deer used burned Palouse grasslands more than unburned grasslands one year after fire in the Snake River Canyon and also increased use in unburned areas on burn edges (C. A. Johnson 1989).

American Bison

The American bison response to fire in Yellowstone has been studied along with that of elk (M. G. Turner, Wu et al. 1994; S. Pearson et al. 1995; L. L. Wallace et al. 1995). Bison numbered about 2,500 before the 1988 fires (L. L. Wallace et al. 2004), and only 9 died in the fires, in episodes of rapid, high-intensity spread (Singer et al. 1989), although they suffered some mortality in the severe winter after the fires. Although an additional 569 bison were killed when they moved north out of the park (M. S. Boyce and Merrill 1991), this movement pattern preceded the fires (Meagher 1989). Altogether, bison populations increased by about a third after the fires, in part the result of increased forage on burned areas (Singer and Mack 1999).

On the Great Plains adjoining the Rockies, some bison were killed in fast-moving grassland fires in the 1870s and 1880s (Ross 1947; A. W. Thompson 1947). Plains bison may select burned patches or watersheds at the landscape scale, and particular vegetation patches within burns at a finer scale (e.g., Shaw and Carter 1990; Vinton et al. 1993), but this fine-scale selection may be absent in Yellowstone (L. L. Wallace et al. 1995). Bison in Wind Cave National Park in mixed-grass prairie sought prairie dog colonies across landscapes, then fed preferentially on the margins of colonies where forage quality is higher; however, after an off-colony prescribed fire, cow-calf herds selected for the burn (selection was 2.46 where 1.0 is random), and selection for the colony declined (to 1.95; Coppock and Detling 1986).

Bighorn Sheep

Bighorn sheep are typically associated with native montane and subalpine grasslands, usually on steep, rocky sites with considerable topographic relief and nearby escape cover (Wakelyn 1987; DeCesare and Pletscher 2006). Bighorns do also graze in grassy postfire shrublands and open postfire forests (Peek, Riggs, and Lauer 1979; Riggs and Peek 1980; DeCesare and Pletscher 2006). In Colorado, actual use of or attraction to burns was not studied; however, prescribed fire in montane grassland and mixed mountain shrubland improved the quality of winter, but not spring, diets of bighorn sheep (Hobbs and Spowart 1984). Winter diet improvements came largely from more consumption of green forage, which was obscured by litter on unburned sites. Unfortunately, two of the three primary green winter forages were nonnative grasses, Kentucky bluegrass and cheatgrass. Similarly, in Glacier National Park, bighorn sheep preferred open, recovering, postfire subalpine forests, burned 41 years prior to study, as well as a lawn, since both provided green winter forage (Riggs and Peek 1980). After a prescribed fire in Wyoming big sagebrush in Idaho, bighorn sheep grazed bluebunch wheatgrass at a higher rate during winters, especially years 1–2 after fire (Peek, Riggs, and Lauer 1979). It is not known whether postfire plants were preferred because they were greener in winter, but lack of litter accumulation likely was important (Peek, Riggs, and Lauer 1979; Hobbs and Spowart 1984).

Additional benefits of fire for bighorn sheep include increased visibility, which can lower predation and increase foraging efficiency (Wakelyn 1987), and opening of migration corridors from winter to summer ranges (Peek et al.

1985; Wakelyn 1987). However, Peek et al. caution that bighorn sheep response to fire is not always positive, because forage is lost in the first postfire winter, important forage plants may not resprout, forage gains may be short lived, other animals (including livestock) may usurp forage gains, and bighorn populations may have limited ability to respond to fire because of disease or other factors.

Pronghorn

Pronghorn are fast and usually can avoid mortality in fires, as reported in an observation of an 1874 fire on the northeastern Colorado plains (Ross 1947), but they may suffer from loss of forage if fires are large. Wyoming big sagebrush was more than 90 percent of the pronghorn diet from midfall through midspring in sagebrush in southeastern Idaho, and forbs were most of the summer diet; thus, fire may negatively affect winter habitat but could increase summer habitat value (Gates 1983). After further study of the same area, Moritz (1988) found pronghorn to generally use burns 2–77 years old more than unburned controls in March, June, and September; larger (i.e., greater than 1,000 hectares) and older burns (i.e., more than 35 years since fire) and burns with cheatgrass dominance were less preferred. Gates thought that small fires in large areas of sagebrush would not be detrimental to pronghorn, but both authors indicated that large fires could limit winter habitat value. Burned forests after the 1988 Yellowstone fires facilitated a few pronghorn movements to new parts of the park (M. D. Scott and Geisser 1996). Fire did not burn the sagebrush areas that are year-round range for about 80 percent of the park's pronghorn population, but they did burn substantial parts of a summer range dominated by mountain big sagebrush and grassland, used by about 20 percent of the population. Overall, there was no selection by pronghorns for burned over unburned summer range, although there was preference for burns in some parts of the summer range. Pronghorn populations declined by about 21 percent after the first winter, but the fires may not have been the cause, as they continued to decline through 1996 from multiple causes (Singer and Mack 1999).

Sage-Grouse

Fire is generally not beneficial for sage-grouse (Knick, Holmes, and Miller 2005; Beck, Connelly, and Reese, in press), but effects differ for leks (typically open areas where male grouse strut and breeding occurs), nesting areas, brood-rearing areas, summer range, and winter range. Response of greater sage-grouse to fire has been studied, but information is lacking for Gunnison sage-grouse. Fire generally kills most sagebrush taxa but can burn in a mosaic, leaving unburned area (chap. 9).

Fire might increase available lekking areas or use of existing areas. Male grouse did strut in some small burns in several cases (Schlatterer 1960; Connelly, Arthur, and Markham 1981; Gates 1983; W. E. Moritz 1988), but not another (Connelly et al. 2000). Fire led to short-term declines (year 1 or 2 postfire) in male attendance at leks in Colorado relative to the immediate prefire year, but attendance recovered by year 3 (L. A. Benson, Braun, and Leininger 1991). In Idaho, fire in and near lek areas decreased the number of leks and male attendance at remaining leks for five years postfire relative to unburned controls (Connelly et al. 2000). At higher elevations in Idaho, wildfires did not decrease male attendance at leks (R. C. Martin 1990).

Sage-grouse hens require adequate cover of sagebrush for nesting (Klebenow 1973). Where fire burned nesting areas in Colorado, sage-grouse nested only in unburned sagebrush patches (L. A. Benson, Braun, and Leininger 1991). Grouse usually avoided nesting in burns; however, five unsuccessful nests did occur in burns less than 20 years old in Oregon (Byrne 2002). Sage-grouse also did not nest in 8- to 9-year-old burned areas in Wyoming big sagebrush in Wyoming (Kerley 1994). Height and cover of mountain big sagebrush and screening cover from grasses needed for nesting were available in burns older than 25 years postfire in Oregon (McDowell 2000) and older than 20 years in Idaho (Nelle, Reese, and Connelly 2000).

Brood-rearing areas may occur in low-density sagebrush (R. C. Martin 1990) or a mosaic of dense sagebrush and openings that offer insects and forbs (Klebenow 1973). Thus, Klebenow and later Gates (1983) thought fire in a mosaic might enhance brood-rearing habitat by increasing forbs and insects while maintaining escape cover. However, Klebenow later was uncertain that this expected benefit would actually accrue (Klebenow and Beall 1978). In xeric Wyoming big sagebrush areas in Idaho, prescribed fire did create a mosaic but did not increase forbs, led to declines in some insects important to grouse in years 2 and 3 postfire, and had no effect on use by males, females, or broods (R. A. Fischer, Reese, and Connelly 1996). Similarly, fire did create a mosaic but did not increase forbs or grouse use in another Wyoming big sagebrush area in Idaho (Clifton 1981). Brood-rearing and broodless females in Oregon avoided 89 percent and 77 percent, respectively, of available burned area, suggesting that fire in little sagebrush and Wyoming big sagebrush provides no value for nesting, brood-rearing, or broodless females (Byrne 2002).

At higher elevation, some positive effects may sometimes occur. After wildfires in mountain big sagebrush in Idaho, forb cover and brood use increased, but only in the most heavily burned areas (R. C. Martin 1990). In contrast, Nelle et al. (2000) found no increase in forb cover in burns 1–14 years postfire in mountain big sagebrush in Idaho, possibly due to postfire drought and live-stock grazing, and only observed an increase in insects during the first postfire year. Specific forbs used by hens and chicks increased, and forage quality improved; but some insects declined in years 1 and 2 postfire in mountain big sagebrush in Oregon (McDowell 2000). Another study in Oregon in mountain big sagebrush–bitterbrush found increased forb cover and diversity and more of one forb used by grouse, but no effect on other primary foods, including insects in years 1–2 postfire (Pyle and Crawford 1996). Neither Oregon study monitored grouse use. In a simulation study, Pedersen et al. (2003) showed that fire was generally detrimental to sage-grouse in mountain big sagebrush areas, only increasing populations if fires burned very little area (i.e., 1 percent of the area once per 60 years).

After brood rearing, grouse may move to summer ranges. Fire did not change the timing, distance, or direction of movement of female grouse from breeding and nesting areas to summer ranges (R. A. Fischer et al. 1997). Little is known about sage-grouse use of burns on summer ranges. Kerley (1994) found sage-grouse used a Wyoming big sagebrush burn, eight to nine years postfire in Wyoming, less than expected during summer, based on availability. Fire is generally detrimental on winter ranges (W. E. Moritz 1988), where sage-grouse depend heavily on live sagebrush (Robertson 1991) and most sagebrush taxa are killed by fire (chap. 9).

SUMMARY

Fires may kill some animals, especially smaller and less mobile animals that live in litter or vegetation near the soil surface (e.g., arachnids, insects, earthworms, mollusks), particularly if fires are high intensity and have high rates of spread. Predators may attack fleeing animals as the fire burns, and scavengers will take advantage of animals killed by fire. Most animals survive by moving away from the fire or taking refuge in burrows, in cracks in rock outcrops, or in moist or low-fuel areas. Fire severity affects bird use of the postfire environment. After high-severity fires in forests, a flush of mobile insects, including wood-boring and bark beetles, and an open canopy provide food for woodpeckers and aerial insectivores. Among mammals, the deer mouse is common in the first one to two years after fire in many ecosystems. Some small mammals and birds are reduced until vegetation regrows a few years after high-severity fire. Postfire reduction can last more than a decade for some species (e.g., some sagebrush-obligate birds). Early-successional grass or shrub vegetation with some regenerating trees offers favorable habitat for many animals, including blue grouse, bluebirds, grizzly bears, bighorn sheep, elk, and deer. Although they show some attraction to recent burns in particular places, some large herbivores (e.g., bison, elk, pronghorn) appear regulated more by winter severity than by fires. Fires, regardless of severity, are certainly not catastrophes for most animal populations, but they are also not universally beneficial to animals. Species that use the postfire environment appear affected by specific attributes of the temporally fluctuating spatial pattern and extent of the landscape mosaic of burned patches (e.g., unburned areas, scorch zone, burn size). Not just any fire or any mosaic is beneficial to animals. However, animals, unlike plants, can more easily move across landscapes to seek or avoid specific burns or unburned areas.

CHAPTER 5

Fire Regimes and Fire History
in Landscapes

Fires may be small and of low severity in some areas and infrequent, large, and of high severity in others, reflecting differences in *fire regime*, the set of properties of fires over a relatively uniform time and space. Vegetation primarily shapes fire regimes, but they also vary with environment, as from low to high elevations (Keane et al. 2003). This chapter discusses how fire regimes are characterized and how fire history scientists reconstruct fire regimes under the HRV; it also presents a classification of fire regimes for the major vegetation types of the Rockies.

CHARACTERIZING FIRE REGIMES

Fire regimes are characterized by attributes of patches of burned area (W. L. Baker 1992b; table 5.1) within a particular landscape (usually thousands of hectares) over a particular time period (usually several hundred years or more). Attribute distributions, such as a fire-size distribution, are derived by either tallying the number of fires (fig. 5.1) or summing the area of fires in classes. Counts of fires in small plots were common in the past, but area-based distributions (e.g., fig. 5.2) are needed for landscapes, as fires vary too greatly in size to be simply counted. Summary parameters (e.g., mean fire size) can also be calculated to compare fire regimes. Empirical distributions from data are directly useful, but data can also be fitted to theoretical distributions to help understand processes influencing fires, to compare among sites or fire regimes, and for computer simulation. In this section, the key attribute distributions for fire size, fire interval, and fire severity are discussed, along with summary parameters.

Table 5.1. Major Attributes of Individual Fire Patches

Attribute	Meaning
Size of patch	The land area burned.
Size of patch interior	The land area within the fire perimeter but sufficiently removed from the perimeter to be relatively free of physical and biological effects from the unburned area.
Size of patch edge	The land area within the fire perimeter but sufficiently close to the perimeter to be strongly influenced by the physical and biological attributes of the unburned area.
Size of patch scorch zone	The land area of this zone, if present; the zone can result from radiant heating rather than flaming combustion, or from low-intensity fire, but plants may lose leaves and even be killed.
Shape of patch	The shape of a patch is often measured by a shape index.
Intensity	The energy output of the fire that burned the patch.
Severity	The ecological damage caused by the fire.
Time since fire	The elapsed time since the last fire within the burned area.
Spatial location	The setting of the fire.
Perimeter	The length of the border between burned and unburned vegetation.
Orientation	The compass direction of the main axis of the fire patch.

Source: W. L. Baker (1992a).

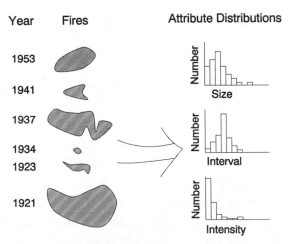

Figure 5.1. A sequence of fires and their corresponding attribute distributions. The sequence is hypothetical to illustrate the concept of attribute distributions.
Source: Reproduced from W. L. Baker (1992a, p. 183, fig. 1) with the kind permission of Springer Science and Business Media.

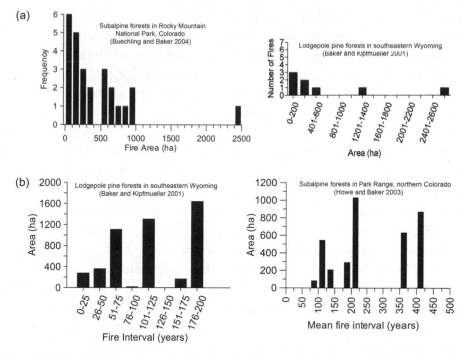

Figure 5.2. Empirical attribute distributions for fires in Rocky Mountain subalpine forests: (a) fire–size distributions, and (b) area–based fire–interval distributions. Note that size distributions are truncated at both ends, as small fires are difficult to detect and fires may exceed study area sizes.

Source: Adapted from graphs originally presented in a slightly different form in the cited publications.

Fire-Size, Interval, and Severity Distributions

Fire-size distributions, which show number of fires by fire size, are usually inverse-J shaped, with many more small than large fires (figs. 5.2a, 5.3). Recent data for the Rockies show that many fires ignite but few spread far. The reasons for this are poorly understood (M. A. Moritz et al. 2005; Cui and Perera 2008) but include the following: (1) fuels may have insufficient loading or are discontinuous; (2) weather often leaves fuels too moist or without sufficient wind for fire spread; (3) fire spread is stopped by a fire break (fig. 2.8); and (4) fire is put out by rain, snow, or people.

A power-law distribution (in which the probability of a fire greater than a particular size is the inverse of the size raised to a power) commonly fits fire-size data (M. A. Moritz et al. 2005; Cui and Perera 2008). However, the large, rare fires in the tail of the distribution often fit poorly (Alvarado, Sandberg, and Pickford 1998). For example, a power-law model did not fit, or fit over only a limited range of the historical data for six sites in North America, including two national forests in Idaho (W. J. Reed and McKelvey 2002). One solution is for large fires to be fit separately by an extreme-value distribution (e.g., a truncated Pareto) designed for infrequent events (Strauss, Bednar, and Mees 1989; Alvarado et al. 1998; Cumming 2001). Truncation with a minimum size (e.g., 200 hectares) makes ecological sense, as larger fires account for most of the burned area (see the next section) and behave differently than small fires do (M. G. Turner, Hargrove et al. 1994). Recently, truncated fire-size data from around the world have been found to fit well to a specific power-law model from physics and engineering, known as the probability-loss-resource (PLR) model (M. A. Moritz et al. 2005). Power-law distribution of fire sizes means that landscape mosaics, including in the Rockies, commonly fluctuate because large fires are infrequent and, between them, the many small fires burn little land area (box 3.1).

Fire-interval distributions, which show the amount of land area burned after particular intervals since the last fire, might be expected to have a single peak if fuel buildup led to a peak in ignition success, but available area-based empirical distributions for the Rockies instead have slight, if any, trends, lacking clear peaks (fig. 5.2b; see also W. L. Baker 1989a). This suggests fuel buildup only weakly shapes fire in the Rockies (chap. 2). Point-based interval distributions have often been fitted, commonly by the Weibull model (W. L. Baker 1989a; Grissino-Mayer 1999), although other models may better fit biologically expected trends (McCarthy, Gill, and Bradstock 2001). Fire-severity distributions show that large fires vary in severity, likely because of variation in fuels, weather, and topography (see fig. 7.1), but no attempts have been made to fit theoretical models.

Fire Rotation, Mean Fire Interval, Annual Probability of Fire, and Fire Frequency

Parameters and derivative measures from these data and distributions are important. The *fire rotation* or *fire cycle* is the expected time to burn an area equal to the area of a landscape of interest (Romme 1980), typically estimated by summing areas of fires over a time interval. Since fires vary in size, an area basis, rather than a frequency or count basis, is needed to measure the rate of burning; and fire rotation is the essential measure. Fire rotation can be estimated at any spatial extent, but to adequately estimate it for a fire regime requires an extent exceeding the largest fire and a period exceeding the fire rotation, as the rare, large fires that account for most of the fire rotation can be missed in short periods (chap. 10). In the Rockies, this generally means 100- to 500-year periods and tens of thousands of hectares (table 5.2). The formula follows:

$$FR = \frac{t}{\sum_{i=1}^{n} a_i \Big/ A} \tag{5.1}$$

where *FR* is fire rotation in years, *t* is period of observation in years, a_i is area of fire *i* of *n* total fires observed over period *t*, and *A* is total area of the landscape. For example, in a 10,000–hectare landscape, if 42 fires burn 15,000 hectares in 300 years, the rotation is 200 years.

Fire rotation within subareas, in a landscape having a uniform fire rotation, may be less or more than 200 years during short periods, because fires are patchy in time and space. In the long run, rotation in all parts of a landscape with a uniform fire rotation would tend toward 200 years. However, if analysis shows that there are distinct regimes and rotations within subareas of the landscape, it is appropriate to subdivide the landscape accordingly. Patchiness of fires also means that estimates of fire rotation from small areas, particularly with long rotations, such as in subalpine watersheds, may be very imprecise, as they may only reflect a few fires.

If the fire rotation for a landscape is 200 years, a point in the landscape will experience a fire every 200 years on average, which is the *point mean fire interval*. The *population mean fire interval* or *landscape mean fire interval*, the average mean fire interval across all points in the landscape, is equal to the fire rotation for the landscape (W. L. Baker and Ehle 2001). Population mean fire interval can be estimated from a set of valid point samples (see "Statistically Valid Sampling of Fire History in Landscapes" below), but a common measure, *mean composite fire interval*, is not a valid estimator of either point or population mean fire interval, as explained later in this chapter. W. J. Reed (2006) thought population mean fire interval was not equal to the fire rotation, but his formula incorrectly assumed that fire areas are summed until the time a last fire makes total burned area reach or exceed the area of the study area. This creates a possible inequality but is not how fire rotation is estimated. Instead, fire areas over a period are summed, and the fraction of study area burned in the period is used (equation 5.1). If this correct method is used, population mean fire interval and fire rotation will be equal.

The inverse of fire rotation is *annual probability of fire*, a decimal fraction. A fire rotation of 143 years is an annual probability of fire of $1/143 = 0.007$ (table 5.3). Annual probability of fire is the average fraction of a landscape expected to burn each year, or the probability, on average, that a point will burn each year. The inverse of point mean fire interval is *point fire frequency*, or number of fires per period (e.g., seven fires per 1,000 years). Fire rotation and population mean

Table 5.2. Years with Known Large Fires in the Rocky Mountains (1850–2007)

Fire Year & Name	Location(s)	Approx. Area (ha)	Sources
Ca. 1846	N Rockies	Unknown	Barrett, Arno, & Menakis (1997)
1851	CO	Extensive	Romme, Veblen et al. (2003); Sibold, Veblen, & González (2006)
Ca. 1856	N Rockies	Unknown	Barrett, Arno, & Menakis (1997)
1868 or 1869	C CO, W MT	Unknown	Jack (1900); Barrett, Arno, & Menakis (1997)
1879	W CO	Extensive	Hough (1882); Ensign (1888); Peirce (1915); Sibold, Veblen, & González (2006)
1880	C & W CO	Extensive	Sargent (1884); Ensign (1888); Jack (1900); Sibold, Veblen, & González (2006)
1889	N Rockies	Extensive	H. B. Humphrey & Weaver (1915); J. A. Larsen (1925b); Gaffney (1941); Gabriel (1976); Barrett, Arno, & Menakis (1997); Barrett (2000)
1890	N Rockies	Unknown	Kipfmueller & Swetnam (2000)
1894	SD & WY	>50,000	Graves (1899)
1895	N Rockies	Unknown	Kipfmueller & Swetnam (2000)
1910	N Rockies	1,200,000	H. B. Humphrey & Weaver (1915); J. A. Larsen (1925b); Gaffney (1941); Koch (1935, 1942); Wellner (1970); Gabriel (1976); Cohen & Miller (1978); Barrett, Arno, & Menakis (1997); Barrett (2000, 2002); Pyne (2001)
1919	N Rockies	>550,000	Koch (1935); Klebenów (1965); Wellner (1970); Barrett, Arno, & Menakis (1997)
1926	N MT	>20,000	
1929 Halfmoon	N MT	44,500	Gisborne (1931); Koch (1935); Sneck (1977); Barrett, Arno, & Menakis (1997); Barrett (2002)
1931 Quartzburg	W ID	18,220	Connaughton (1936); Wellner (1970)
1934	N Rockies	Unknown	Koch (1935); Kipfmueller & Swetnam (2000)
1967 Sundance	N ID	22,300	H. E. Anderson (1968); Finklin (1973); Lyon (1969); Lyon & Stickney (1974); Stickney (1986); Stickney & Campbell (2000)
1979	N Rockies	Unknown	Kipfmueller & Swetnam (2000)
1988 Fayette	NW WY	15,500	Beighley & Bishop (1990)

Table 5.2. Continued

Fire Year & Name	Location(s)	Approx. Area (ha)	Sources
1988 Yellowstone	WY, ID, MT	690,000	Rothermel, Hartford, & Chase (1994); many others
1988 Canyon Creek	W MT	101,215	National Interagency Fire Center (2006)
1989 Lowman	SW ID	18,600	National Interagency Fire Center (2006)
1992 Foothills	ID	104,049	National Interagency Fire Center (2006)
1994 Corral Creek	ID	46,826	USDI Bureau of Land Management (2004)
1994 Custer	MT	52,632	USDI Bureau of Land Management (2004)
1994 Idaho City complex	ID	62,348	National Interagency Fire Center (2006)
1994 Porphyry South	ID	34,282	USDI Bureau of Land Management (2004)
1994 Rabbit Creek	ID	59,271	USDI Bureau of Land Management (2004)
1996 Cox Wells	ID	88,664	National Interagency Fire Center (2006)
1999 Mule Butte	ID	56,241	National Interagency Fire Center (2006)
2000 Bear	MT	55,311	USDI Bureau of Land Management (2004)
2000 Burgdorf Junction	ID	23,290	USDI Bureau of Land Management (2004)
2000 Cerro Grande	N NM	19,284	Paxon (2000); Crimmins (2006)
2000 Clear Creek	ID	87,838	National Interagency Fire Center (2006)
2000 Diamond	ID	60,636	USDI Bureau of Land Management (2004)
2000 E. Idaho complex	ID	77,915	National Interagency Fire Center (2006)
2000 Flossie complex	ID	33,385	USDI Bureau of Land Management (2004)
2000 Jasper	W SD	33,920	Lentile (2004); Lentile, Smith, & Shepperd (2005, 2006)
2000 Kate's Basin	C WY	55,749	National Interagency Fire Center (2006)
2000 Maynard	ID	18,219	National Interagency Fire Center (2006)
2000 Mussigbrod complex	MT	34,413	National Interagency Fire Center (2006)

Table 5.2. Continued

Fire Year & Name	Location(s)	Approx. Area (ha)	Sources
2000 Salmon–Challis	ID	73,927	National Interagency Fire Center (2006)
2000 Sheep Mountain	W WY	>20,000	National Interagency Fire Center (2006)
2000 Shell Rock	C ID	31,151	USDI Bureau of Land Management (2004)
2000 Valley complex	W MT	118,247	National Interagency Fire Center (2006)
2001 Moose	N MT	28,745	USDI Bureau of Land Management (2004)
2002 Hayman	C CO	57,773	USDI Bureau of Land Management (2004); Romme, Kaufmann et al. (2003); Romme, Veblen et al. (2003)
2002 Missionary Ridge	S CO	29,604	USDI Bureau of Land Management (2004)
2002 Ponil	N NM	39,430	USDI Bureau of Land Management (2004)
2003 Hobbie	MT	15,520	Rocky Mt. Geographic Science Center (2007)
2003 Robert	MT	21,353	Rocky Mt. Geographic Science Center (2007)
2003 Wedge Canyon	MT	21,705	Rocky Mt. Geographic Science Center (2007)
2005 Valley Road	ID	16,556	Rocky Mt. Geographic Science Center (2007)
2006 Derby	MT	90,514	Rocky Mt. Geographic Science Center (2007)
2006 Heaven's Gate	ID	17,242	Rocky Mt. Geographic Science Center (2007)
2006 Rattlesnake Complex	ID	18,229	Rocky Mt. Geographic Science Center (2007)
2007 Ahorn	MT	21,276	Rocky Mt. Geographic Science Center (2007)
2007 Bridge	ID	17,684	Rocky Mt. Geographic Science Center (2007)
2007 Cascade Complex	ID	122,419	Rocky Mt. Geographic Science Center (2007)
2007 Castle Rock	ID	19,224	Rocky Mt. Geographic Science Center (2007)
2007 Chippy	MT	40,117	Rocky Mt. Geographic Science Center (2007)

Table 5.2. Continued

Fire Year & Name	Location(s)	Approx. Area (ha)	Sources
2007 East Zone Complex	ID	86,921	Rocky Mt. Geographic Science Center (2007)
2007 Fool Creek	MT	24,307	Rocky Mt. Geographic Science Center (2007)
2007 Krassel Complex	ID	19,133	Rocky Mt. Geographic Science Center (2007)
2007 Meriwether	MT	17,181	Rocky Mt. Geographic Science Center (2007)
2007 Neola North	UT	17,745	Rocky Mt. Geographic Science Center (2007)
2007 Rattlesnake	ID	73,771	Rocky Mt. Geographic Science Center (2007)
2007 Red Bluff	ID	21,429	Rocky Mt. Geographic Science Center (2007)
2007 Shower Bath Complex	ID	24,216	Rocky Mt. Geographic Science Center (2007)
2007 Skyland	MT	18,828	Rocky Mt. Geographic Science Center (2007)

Note: Large fires are defined here as those greater than 15,000 hectares in area. See Barrett et al. (1997) for earlier fire episodes, not dated to a single year, in the northern Rockies.

Table 5.3. Partitioning Fire Parameters by Fire Type or Vegetation Type in a Mosaic

Partition	Fire Rotation (yrs)	Annual Probability of Fire	Population Mean Fire Interval (yrs)	Fire Frequency (fires per 1,000 yrs)
All fires	143	0.007	143	7
Partition by fire type				
Crown	500	0.002	500	2
Surface	200	0.005	200	5
Partition by vegetation type				
Spruce-fir forest	333	0.003	333	3
Subalpine grassland	250	0.004	250	4

Note: Data are hypothetical. Annual probability of fire is 1/fire rotation. Population mean fire interval equals the fire rotation. Fire frequency is equal to annual probability of fire but is expressed relative to a period of years (e.g., 1,000 years). Note that the sum of the annual probabilities of fire in each partition equals the annual probability of all fires, and the sum of the fire frequencies in each partition equals the fire frequency of all fires.

fire interval are equal but have different meanings; the inverse of each, annual probability of fire and point fire frequency, are also equal but have different meanings. Mean fire interval and fire frequency are often used interchangeably, but that is incorrect.

These four measures can be estimated for all fires or partitioned by fire severity, vegetation type, parts of the landscape, or other criteria (table 5.3), although the option of partitioning these measures was not recognized in a recent review (Finney 2005). Partitioning requires estimating the contribution to the overall measure from each part of the partition. All four measures can be partitioned, but annual probability or fire frequency is simplest, since probabilities or frequencies in partitions sum to all-fire values. Corresponding fire rotations are the inverse of annual probability of fire.

Variability in fire size, interval, and severity is shown by attribute distributions (e.g., fig. 5.2) or summary parameters, such as standard deviation. If empirical data are fit to theoretical distributions, parameters of those distributions can also measure variability. Variability is of direct interest, but it is also needed to estimate confidence intervals and test hypotheses.

An Example Using Recent Data from the Rocky Mountains

To illustrate these calculations, consider a data set of 56,350 fires from 1980–2003 (USDI Bureau of Land Management 2004) on federal and Indian lands totaling 41.25 million hectares, about 75 percent of the area of the Rockies. These fires burned about 3.6 million hectares in 24 years, or about 150,000 hectares per year, an estimated fire rotation of 275 years (equation 5.1) and population mean fire interval of 275 years. Annual probability of fire was $1/275$, or 0.0036, and fire frequency about 3.6 fires per 1,000 years. Since this 24-year period is less than 10 percent of the fire rotation, the modern fire regime is not well characterized (chap. 10), but some aspects, while tentative, are interesting.

The fire-size distribution in this sample is inverse-J shaped, likely matching a power-law model. About 96 percent of the 3.6 million hectares that burned in 24 years was from the 2 percent of total fires ($n = 1,122$, or 47 per year) greater than 200 hectares in area. In contrast, the 55,228 fires (98 percent of fires) less than 200 hectares in area (fig. 5.3) accounted for only about 4 percent of total area burned, a pattern noted in other studies (Strauss, Bednar, and Mees 1989; Holmes et al. 2004). About half the burned area came from only 50 fires (about 0.1 percent of 56,350 total fires), about 2 per year, that were greater than or equal to 15,000 hectares, concentrated in the central and northern Rockies (fig. 5.4). Since 2003, there have been 18 additional fires greater than 15,000 hectares in the Rockies, 14 of them in 2007 (table 5.2). Even if fires were re-

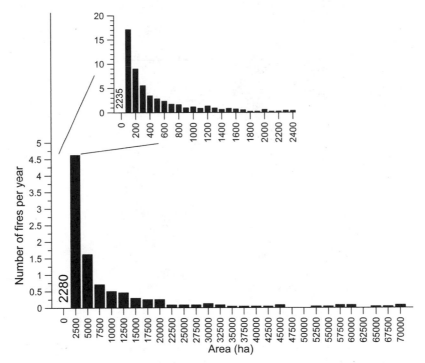

Figure 5.3. Frequency distribution of fire sizes in the Rocky Mountains. The graph is based on 56,350 fires from 1980 to 2003 on federal and Indian lands covering about 75 percent of the Rocky Mountains. It shows fires in the whole data set 0–70,000+ ha, and a close-up of fires 0–2,500 ha.
Source: Data from USDI Bureau of Land Management (2004).

duced over the last century (chap. 10), these data suggest that the essence of pre-EuroAmerican fire regimes can be understood by studying only large fires.

Variability, Fluctuating Landscapes, and Area-Based Measures of Fire under the HRV

Major sources of evidence used to understand fire under the HRV are brought to bear in reviewing the details of fire history by vegetation type in chapters 6–9, but together the evidence is compelling that fire under the HRV was also characterized by inverse-J fire-size distributions, as well as variability in interval and severity distributions, as in recent data, reviewed above. There is little doubt that variability was inherent in fire attribute distributions and landscape structure in the Rockies under the HRV, with significant ecological implications (box 3.1).

Variability in fires suggests that counting fires and using frequency-based measures are inappropriate, since fires differ. Since the rise of landscape ecology,

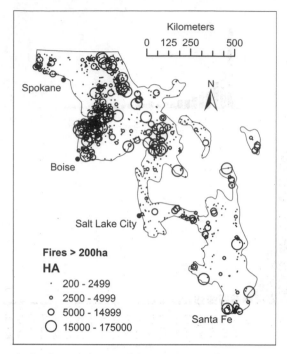

Figure 5.4. Recent large fires in the Rocky Mountains. Shown are fires greater than 200 ha in area from 1980 to 2003, using the same data set as in figure 5.3.

geographical information systems, remote sensing, and other spatial technologies, it has become feasible and essential to use area-based measures (e.g., area burned, fire rotation) to understand and characterize fire regimes.

Research on the relationship of fires and climate over large land areas cannot always use area-based measures. Using all fire-scar records, as some studies have done (table 2.4) is likely to identify the climatic conditions that promote ignitions and small fires, rather than the large fires that most shape landscapes. However, fire-climate studies can approximate area-based measures by quantitatively separating and analyzing the large fires that do most of the work in landscapes, as some studies did (table 2.4). These can be identified by bounding contiguous land areas that have evidence of a particular fire year (see "Estimating Fire-Regime Parameters from Landscape Data" below) or by selecting and analyzing the fires that burned many sites in the same year.

RECONSTRUCTING FIRE REGIMES

Historical records, direct observations of fires near EuroAmerican settlement, and extant evidence of past fires all allow reconstructions of fire regimes under

Table 5.4. Types of Evidence Used in Analyzing and Reconstructing Fire Regimes

Type of Evidence	Time Period in Rockies	Example of a Study
Systematic historical records		
Government agency records	Generally after 1900	K. F. Higgins (1984)
General Land Office survey	Late 1800s to early 1900s	M. D. Andersen & Baker (2006)
Forest reserve reports and maps	Ca. 1900	W. L. Baker, Veblen, & Sherriff (2007)
Early forest exams	Early 1900s	Ogle & DuMond (1997)
Early maps	Late 1800s to early 1900s	Kulakowski, Veblen, & Drinkwater (2004)
Historical observations		
Early explorers, settlers	Ca. 1600–1900	C. T. Moore (1972)
Early photographs	Late 1800s to early 1900s	Veblen & Lorenz (1991)
Tree rings		
Fire scars	Usually the last few centuries	Heyerdahl, Miller, & Parsons (2006)
Stand origins, tree age	Usually the last few centuries	Floyd, Romme, & Hanna (2000)
Deadwood, tree death	Usually the last few centuries	Ehle & Baker (2003)
Growth anomalies	Usually the last few centuries	Rubino & McCarthy (2004)
Charcoal and pollen	Usually post-Pleistocene	Millspaugh, Whitlock, & Bartlein (2000)

the HRV, a focus of this section. Primary sources are historical records, tree rings, and charcoal and pollen (table 5.4).

Systematic Historical Records

To help us understand fire over the last century, fire atlases and maps maintained in agency offices can be analyzed (Rollins, Swetnam, and Morgan 2001). *Fire atlases* are sets of maps of fires often maintained in agency offices; they record major fires but may miss some fires, particularly in the early twentieth century (C. E. Gibson 2006; Shapiro-Miller, Heyerdahl, and Morgan 2007). Gibson (2006) compiled fire atlas data from the late 1800s across 14 national forests and about 14 million hectares in the northern Rockies. Other available computer records include point locations and attributes of fires since the 1970s (e.g., USDI Bureau of Land Management 2004), polygon records since 2000 (http://geomac.usgs .gov; http://rmgsc.cr.usgs.gov/outgoing/geomac), and lightning locations (Orville and Huffines 2001). These have allowed geographic analyses of fire (Ryan 1976; Rollins, Swetnam, and Morgan 2001) and ignition (e.g., K. F. Higgins 1984) in the Rockies.

Early systematic records provide information about the HRV. The General Land Office (GLO) surveys provide locations and information on fires and vegetation in the late 1800s. For example, the GLO surveys show that fire was not a primary agent limiting tree invasion into forest openings in southeastern Wyoming (Andersen and Baker 2006). *Forest reserve reports* done about 1900 by government scientists provide maps and descriptions of forest fires and forest structure in the Rockies. W. L. Baker et al. (2007) used the reserve reports to document historical fires and forest structure in ponderosa pine–Douglas-fir forests in the Rockies. Similarly, early examinations of timber and other resources on the national forests provide systematic assessments and measurements, sometimes including maps, by trained personnel (e.g., Ogle and DuMond 1997). Early maps in forest exams or forest reserve reports can be scanned, georectified, and compared to modern remote sensing data or maps to understand changes, such as recovery after early fires (e.g., Kulakowski, Veblen, and Drinkwater 2004).

Historical Observations

Reports of early explorers and settlers and some scientific sources not focused on fires or vegetation may also provide information but were not collected systematically and have greater limitations and potential biases than systematic historical records. For example, W. L. Baker (2002) showed that early settlers and explorers were surprisingly unaware that lightning can ignite fires and commonly assumed that observed fires were set by Indians or EuroAmericans. Such early sources can be collected and analyzed systematically to enhance their value (e.g., C. T. Moore 1972), but they have significant inherent limitations. More objective information comes from comparing historical photographs and modern retakes. Veblen and Lorenz (1991) compared a large set of early photographs of forest structure and disturbances to modern rephotographs from the same locations in the Colorado Front Range.

Tree Ring Reconstructions

Parameters of fire regimes under the HRV can be reconstructed from extant evidence of past fires preserved in fire scars, stand or tree ages, dates and causes of tree death from standing dead or down wood, and growth anomalies in wood (table 5.4). This evidence is left across landscapes depending on the type and pattern of fire (fig. 5.5) and is differentially preserved over time. More research is needed on how modern fires leave evidence, so that methods can be developed for efficient, unbiased, and accurate reconstruction of fire regimes (W. L. Baker and Ehle 2001). Methods determine the extent, type, and pattern

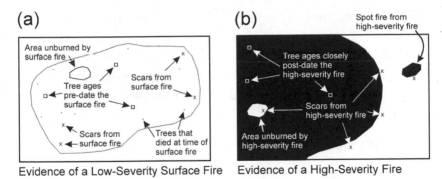

Figure 5.5. Pattern and types of evidence left by (a) low-severity surface and (b) high-severity fires in forests.
Source: Diagram by the author.

of evidence gathered and ultimately what is to be learned. Reviews of fire history methods include Arno and Sneck (1977), McBride (1983), E. A. Johnson and Gutsell (1994), Fall (1998), and W. L. Baker and Ehle (2001). As suggested above, emphasis is needed on spatial reconstruction and area-based measures.

LANDSCAPE RECONSTRUCTION OF FIRE REGIMES

The landscape scale is the appropriate scale to reconstruct fire regimes, as most of the burned area accrues from large fires (i.e., greater than 200 hectares). A contiguous area larger than the largest fire size is needed to fully reconstruct the fire-size distribution. Given fire sizes of 50,000–100,000 hectares over the last century (table 5.2), contiguous land area of tens of thousands of hectares is needed. Also, fires need to be reconstructed for a period of at least a full fire rotation, generally meaning one or more centuries in the Rockies. Recent large fires may erase evidence of previous fires, precluding reconstruction over a sufficient period (Barrett, Arno, and Key 1991). It is appropriate to avoid recently burned areas if the only goal is to understand the fire regime under the HRV. If a goal is also understanding the fire regime since EuroAmerican settlement, it would be biased to avoid recently burned areas or seek areas with older forests. In the Rockies, areas that are large enough often will include a mosaic of vegetation types (e.g., Buechling and Baker 2004), which can be left pooled or separated later in analysis.

The need for large areas with evidence undisturbed by logging or other land uses means large preserves or a set of small remnants is often studied, which can be biased or incomplete samples of a region. For some vegetation types (e.g., ponderosa pine forests), undisturbed landscapes may be lacking, and reconstruction from limited evidence is necessary. In both cases, it is essential to estimate the population to which the available sample applies, make unbiased choices from the full extent of usable evidence, and analyze potential bias in the available sample (W. L. Baker and Ehle 2001). For example, sampling only remnant, unlogged old forests (e.g., Grissino-Mayer et al. 2004) means recent high-severity fires likely will be missed and surface fires overemphasized, a bias that should be made explicit.

STATISTICALLY VALID SAMPLING OF FIRE HISTORY IN LANDSCAPES

Within a landscape to be sampled, a statistical sample of small plots or transects mapped using a geographical positioning system is needed to identify and map the extent of fires and variation in fire severity. To reconstruct parameters of fire regimes in landscapes, statistically valid spatial sampling is essential (Lorimer 1985; E. A. Johnson and Gutsell 1994). A valid sample is needed for choosing or placing (1) landscape samples within a region, (2) plot or transect samples within a landscape, and (3) samples of evidence within a plot or transect.

One effective method of placing plots or transects across a landscape is to obtain a stratified random sample or census of spatially delimited subareas. If possible, the landscape is divided into subareas based on biophysical settings and structural variability in vegetation visible in remote sensing imagery (e.g., Buechling and Baker 2004; fig. 5.6a). Subareas can also be defined by forest types, moisture settings, or environmental parameters or can be subdivided into equal-area units (e.g., Houston 1973; Arno and Sneck 1977). Then, either the entire set is censused, using plots or transects in each subarea (e.g., Buechling and Baker 2004), or a random sample of subareas in each stratum is sampled. Other valid spatial sampling methods, essential if variation in forest structure cannot be seen and mapped, are complete random sampling or systematic sampling using a grid of points (e.g., Haining 1990; fig. 5.6b).

The size of subareas or spacing of grid points needs adjustment for spatial autocorrelation and expected fire sizes (J. G. Fall 1998), emphasizing large fires that account for most of the burned area. Large fires create spatial autocorrelation at scales up to their typical maximum extent (table 5.2). Autocorrelated samples are not independent and should be conflated into a single estimate up to the spatial extent of significant autocorrelation (Polakow and Dunne 1999). Fire intervals were spatially autocorrelated among samples less than 2 kilometers apart in a subalpine landscape in Wyoming (W. L. Baker and Kipfmueller 2001). Spatial autocorrelation limits the number of sampling points worth placing into a landscape. In the Rockies, where fires greater than 100–200 hectares

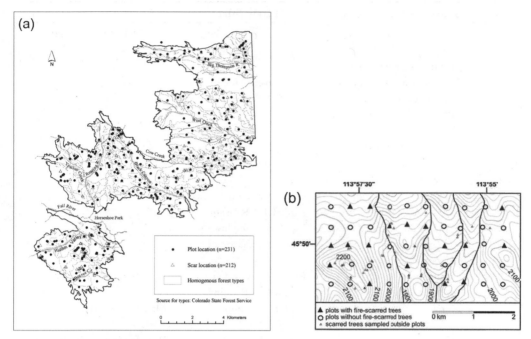

Figure 5.6. Examples of statistically valid landscape sampling methods. These include (a) a stratified random sample in Rocky Mountain National Park, Colorado, with strata that are forest structural categories and a single plot at the centroid of each polygon; and (b) a grid-based set of sampling plots in southwestern Montana.

Sources: Part (a) is reproduced from Buechling and Baker (2004, p. 1262, fig. 2) with permission of the National Research Council, Canada; part (b) is reproduced from Heyerdahl, Miller, and Parsons (2006, p. 108, fig. 1), © Elsevier 2004, with permission from Elsevier Limited.

account for most of the burned area (fig. 5.3), plot or transect spacing of 1–2 kilometers, or subareas averaging greater than 100 hectares, would facilitate identification of large fires and deemphasize small fires. Where remote sensing or field observation suggests more complexity, sampling on a finer grid or opportunistic sampling of fire scars or tree ages could be added between grid points (e.g., Heyerdahl, Miller, and Parsons 2006).

Many past studies used biased sampling, purposely choosing locations of landscapes, plots, and evidence to sample (*targeted sampling* [W. L. Baker and Ehle 2001]), or did not obtain a sufficient sample (table 5.5). Researchers using targeting often sampled primarily in landscapes containing old forests, seeking

Table 5.5. Common Sampling Biases and Limitations in Past Fire History Studies

No.	Source of Sampling Bias	Likely Effect	Sources
1	Biased sample selection (targeting)		W. L. Baker & Ehle (2001)
1a	Landscapes within regions		
	Landscapes with old trees, many scars	High-severity fire missed; MFI/FR too short	
1b	Plots/transects within landscapes		
	Plots with old trees, many scars	High-severity fire missed; MFI/FR too short	
1c	Trees within plots/transects		
	Old trees with multiple scars	High-severity fire missed; MFI/FR too short	
	"Recorder" trees	Incomplete fire history	
	Trees with open scars	Recent fires favored?	
2	Inadequate fire-scar sample		
2a	Inadequate fire-scar sample size	Inaccurate & wide range of MFI/FR estimates	W. L. Baker & Ehle (2001)
2b	Fire scars not cross-dated	MFI/FR too short	Madany, Swetnam, & West (1982)
3	Inadequate age sample (e.g., <5 trees/plot)		W. L. Baker & Ehle (2003)
3a	Insufficient age data to estimate severity	Poor fire severity estimate	
3b	No age data or age data not used	Fire severity unknown	
4	Biased fire-interval estimation		
4a	Composite fire intervals used	MFI/FR too short	W. L. Baker & Ehle (2001); W. L. Baker (2006a); Kou & Baker (2006b)
4b	Censored fire intervals omitted	MFI/FR too short	Polakow & Dunne (1999); W. L. Baker & Ehle (2001)

Note: Included is the likely impact on estimates of mean fire interval (MFI), fire rotation (FR), and fire severity.

long records of fire. However, as Lorimer (1985, 201) said, "One cannot sample only old-growth stands . . . and still obtain an accurate estimate of regional disturbance frequency, because the omission of young stands will lead to a biased estimate." Targeting led to fire-scar records that greatly underestimated fire since EuroAmerican settlement compared with fire atlas records from the same area (Heyerdahl et al. 2008). Other targets have been concentrations of fire-scarred trees or trees with multiple scars. Continuing use of targeted sampling (e.g., Grissino-Mayer et al. 2004) is unfortunate, as it is impossible to correct for biased sampling, and the resulting data are thus almost useless (E. A. Johnson and Gutsell 1994). Imagine estimating the fraction of child abusers in a human population by sampling only in prisons, a similar targeting. Statistically invalid sampling has persisted even though identified as a serious problem 20 years ago (Lorimer 1985) and again about a decade ago (E. A. Johnson and Gutsell 1994). To focus attention on valid studies, sampling biases of each study (table 5.5) are identified as studies are discussed in chapters 6–9.

Plot or Transect Samples and the Evidence

Evidence about fire regimes in a plot or transect containing woody plants potentially includes (1) fire scars, (2) dates of origin of live stems, (3) dates of death of standing dead and down stems, and (4) growth anomalies in stems. Studies that used fire scars or stand-origin dates were distinguished in the past, but it is now widely recognized that it is essential to acquire both and compare them locally. A fire scar by itself or a set of fire scars in a small area provides little or no evidence about the severity of a fire. At the location of a scar, the fire is on the surface, but scars are commonly found on the margin of high-severity fires, where surface fire is a tiny part of the fire (fig. 5.5b). The evidence needed to document a spreading surface fire is two or more separated, cross-dated scars from the same year, with intervening trees that predominantly predate the fire (fig. 5.5a). Similarly, to document that a fire scar dates nearby high-severity fire requires that trees near the scar be found to closely postdate the fire (fig. 5.5b). Thus, each scar needs to be compared with nearby tree ages and ages across the landscape, since surface fire in one area may be high severity elsewhere. Similarly, each tree age needs to be compared with scars across the area. Where field counts of tree ages or sizes suggest an even-aged stand, it is essential to search for unburned islands, surviving trees, or the margin of the stand, where scars that date the fire are likely. The relationship between a particular fire and forest structure cannot be determined if fire-scar and age-structure data are pooled across a landscape (e.g., Fulé et al. 2003; P. M. Brown and Wu 2005), as this prevents the detection of patches of high-severity fire at particular locations in the landscape, potentially misrepresenting the fire regime (W. L. Baker 2006a).

Plots or transects need to be large enough to include sufficient evidence about fire history at a point, but no larger, because of competing errors. The first error is omission of fires that affected the point, an error decreased by adding scarred trees (Falk 2004). The second is inclusion of fires that did not affect the point, an error increased by adding more scarred trees (Kou and Baker 2006b).

Stand-origin and other evidence are subject to the same errors. Plot area or transect length should thus be large enough to detect fires that affected a point, but no larger (J. G. Fall 1998).

For fire-scar evidence, the reason to sample an area around a point is that each fire only scars a fraction, the *scarring fraction*, of trees that actually had flames. Individual fires scarred 9–44 percent of ponderosa pine trees that received fire, which means 3–12 trees, whether scarred or not, around a point are expected to record all the fires affecting the point (table 5.6). A fire accumulation curve has been used to estimate needed sample size (W. L. Baker and Ehle 2001), which is often about 10 scarred trees to find most of the fires (fig. 5.7), but this curve may overestimate, as it does not show which fires actually affected a point. Based on scarring fraction, very small plots or transects are needed; if tree density is 120 trees per hectare, only about one-tenth hectare will have the 12 trees needed to detect all fires, given the lowest known scarring fraction (table 5.6).

An area around a point is also needed to obtain an adequate sample of stem ages and other evidence relevant to the point. Generally, 10–20 dates from the largest stems are required to estimate a stand-origin date, assuming the point was burned in high-severity fire (Kipfmueller and Baker 1998a), but more complete age structure, not just the largest stems, is needed if fires, varying in severity, affected the stand. Standing dead and down wood can be cross-dated to

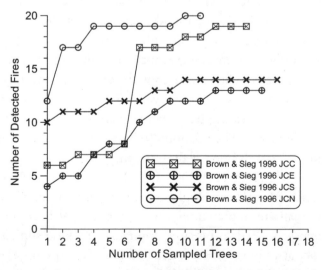

Figure 5.7. An accumulation curve for number of detected fires. The data are for four stands (JCC, JCE, JCS, and JCN) in ponderosa pine in the Black Hills.
Source: Data from P. M. Brown and Sieg (1996).

Table 5.6. Estimates of Scarring Fraction in Ponderosa Pine Forests

Sources	Fire Severity	Location	Scarring Fraction	Minimum Sample Size
J. M. Miller & Patterson (1927)	Low	S OR	0.090	12
Lachmund (1923)	Low	California	0.125	8
Morris & Mowat (1958)	Low	NC WA	0.200	5
Lentile (2004)	Low	Black Hills, SD	0.236	5
J. M. Miller & Patterson (1927)	Moderate	S OR	0.290	4
Lentile (2004)	Moderate	Black Hills, SD	0.437	3

Note: Studies are arranged in order of increasing values from field studies of single fires. Scarring fraction, which varies from 0.0 to 1.0, is the fraction of live, surviving trees on which a fire scar is formed after fire. Minimum sample size is the minimum number of trees that need to be sampled so an individual fire is recorded on at least one tree, given the scarring fraction. This is the inverse of the scarring fraction, rounded up. Omitted are studies that report only the cumulative fraction of trees scarred by a series of fires.

determine whether death dates are related to fires or other events; this also estimates fire severity (Ehle and Baker 2003). Growth anomalies in tree rings include traumatic resin canals and postfire increases or declines in growth (Rubino and McCarthy 2004; Druckenbrod 2005), which may not identify fire to a single year but can supplement mapping from fire scars (A. H. Taylor 2000).

Evidence gathered in the plot or transect should be censused or statistically sampled, but screening is needed. Fire scars must be distinguished from damage by other agents (Gara et al. 1986). Scars that are rotten, degraded, or physically difficult to remove cannot be used. Scars that have healed over may be missed, although the all-tree fire-interval method (see "The All-Tree Fire-Interval Method" below) requires census and dating of scarred and unscarred trees, allowing some detection of healed scars (Kou and Baker 2006a). Only larger down wood and standing dead trees are worth sampling to estimate timing of tree death, as cross-dating is ambiguous without sufficient rings (e.g., 50 years; Ehle and Baker 2003). Some types of wood might be differentially preserved, but the bias in the pool of dead wood is unknown.

In the past, in sampling fire scars, researchers commonly targeted multiple-scarred or previously scarred trees (*recorder trees*), as initial scarring was thought to enhance recording of subsequent fires. However, that idea was untested. W. L. Baker and Ehle (2003) found that a targeted sample of multiple-scarred trees identified more one-tree fires, which have little ecological significance, than a nontargeted sample but did no better at finding larger fires and missed 18 percent of the fires in the stand, including a significant high-severity fire and 30 percent of the ancient fires. Also, if recorders work, a particular fire should show up more often on recorders than as first scars, but 24 fires (40 percent of total fires) showed up on both recorder trees and as first scars; 62 percent of the scars documenting these fires were first scars, while only 38 percent were scars on recorders. Moreover, 32 percent of 60 fires found in a sample occurred only as first scars, while 28 percent of fires were only on recorders (W. L. Baker and Ehle 2003). Also, 62 percent of 154 scars in a complete census were first scars, and only 38 percent were scars on recorders. Finally, through simulation modeling, a separate study found that targeting multiple-scarred trees tends to omit the longer intervals that occur in a stand, because these are represented by either a single scar or no scar (Kou and Baker 2006a). Thus, multiple-scarred trees and recorder trees do not preferentially record fires, but instead contain a biased and incomplete sample of fires and fire intervals in a stand. Censuses and systematic or random samples are currently the only valid sampling approaches for the fire-scar record (W. L. Baker and Ehle 2003).

Evidence must also be gathered that allows discrimination of fire from other agents, which might also explain stem ages, dead wood, or other evidence. Primary agents in the Rockies are insects, drought and wet periods, and blowdown. Specific insects must be known, as they leave differing evidence. Mountain pine beetles may leave distinctive scars (Gara et al. 1986), and they also bring a fungus that leaves a blue stain in the wood (Ehle and Baker 2003). Drought and wet periods are identified in regional reconstructions (e.g., E. R. Cook et al. 2004) and local chronologies. Blowdown might be distinguished in the field by whether down wood is preferentially oriented (Ehle and Baker 2003). Little is known about how long evidence from agents is detectable, and more than one agent may coincide temporally, confounding causation (W. L. Baker, Veblen, and Sherriff 2007).

Evidence is extracted and brought to a laboratory for processing. Trees can be dated and growth anomalies identified on cores of wood extracted with an increment borer. Where allowable, wedges can be extracted from scar areas using a chain saw, causing minimal damage (fig. 5.8). Sectioned live trees may experience slightly higher mortality than neighbors that have not been sectioned (e.g., 8 vs. 2 percent [Heyerdahl and McKay 2001]). Sections of standing dead and down wood can also be removed. Manual saws can be used but make it difficult to avoid damaging the tree unless the scar is near the outer part of the stem. Increment borers can be used on trees with only a few scars (Barrett and Arno 1988; Sheppard, Means, and Lassoie 1988; Means 1989). More scars and more fires are found when scars are sampled at stem bases (Dieterich and Swetnam 1984). Methods for processing scars and cores are well established (Stokes and Smiley 1968). It is essential to cross-date scars, because it is otherwise im-

Figure 5.8. Collecting a fire-scar wedge from a tree.
Source: Reproduced from Arno and Sneck (1977, p. 13, fig. 5).

possible to document that scars from points across a landscape come from one fire (Madany, Swetnam, and West 1982). Analysis of the scar position in a tree ring allows estimation of the season of fires (Baisan and Swetnam 1990).

THE COMPOSITE FIRE-INTERVAL METHOD FOR PROCESSING FIRE-SCAR DATA

Small-plot or transect studies have commonly calculated *composite fire intervals* (CFIs), in which a composite or pooled list of fire years found in a plot or transect is made and intervals are calculated between listed fire years (Dieterich 1980). However, calculating intervals is not an essential step; they can be omitted and the list used in a landscape reconstruction (see "Estimating Fire-Regime Parameters from Landscape Data" below).

Sampling to obtain a list of fires that burned a point is sensible, but for estimating mean fire interval, it typically has included too many trees. The composite is subject to the two errors discussed earlier, omission of fires that burned the point and inclusion of fires that did not burn the point. Since most fires in a fire regime are small and very numerous, as mentioned previously, adding sampling area adds small fires—often representing new fire years—each of which, because small, has a high probability of not having burned the point. Sampling areas for fire history studies have commonly been larger than needed to correct for omission, and thus are subject to inclusion errors. Known scarring fractions may require only up to 12 trees to correct for omission (table 5.6), often obtainable in 0.1- to 0.5-hectare areas. Of 101 fire history plots in ponderosa pine forests in the West, all were greater than 1 hectare, and 95 percent were greater than 10 hectares (Kou and Baker 2006b). Thus, typical sampling areas likely have large inclusion errors, and each erroneous inclusion is often a new fire year, as small fires occur often.

Inclusion errors greatly affect CFI estimates, because mean CFI rapidly declines toward 1.0 as added sampling increases fire years in the list, narrowing mean intervals (Arno and Petersen 1983; W. L. Baker and Ehle 2001; Falk, Palmer, and Zedler 2007). This decline is an undesirable property. Imagine if estimated mean length of noses in a population of people declined rapidly as more noses were measured. CFI might accurately estimate point mean fire interval only if just enough trees were sampled to offset omission errors but no more, as extra trees add inclusion errors. No method exists to determine how many trees are needed at a point, as the slope of decline in CFI with added sampling varies with how scars are selected and with fire-size and interval distributions that are unknown a priori (Kou and Baker 2006b). Some researchers suggested that a decline in CFI with added spatial extent is not a flaw but an attribute of a fire regime (Falk, Palmer, and Zedler 2007). However, unless methods of selecting scarred trees are identical, trends may be little related to an attribute of fire regimes. Moreover, the likely attribute is that, as more area is sampled, more small fires are found. Small fires do little work in landscapes, as mentioned, and are likely of minor significance. To offset the effects of small fires, CFI can be restricted to larger fires (e.g., scarring more than 25 percent of trees), but the restriction for accurately estimating point mean fire interval is inconsistent, and often insufficient, so that correction is still needed (Kou and Baker 2006b).

An added problem is that CFI studies typically omit incomplete intervals. Fire-interval data may be censored because some intervals are incomplete, as study periods seldom begin and end with fires. Complete intervals would be as long as, or longer than, the incomplete intervals and are thus often the longest intervals in a stand. Omitting incomplete intervals biases CFI toward estimates that are too short relative to the point mean fire interval (Polakow and Dunne 1999; W. L. Baker and Ehle 2001; Kou and Baker 2006a).

The upshot is that *CFIs do not accurately estimate the point mean fire interval*, even though this is commonly assumed (see table 10.9). Many small fires and large inclusion errors are central reasons, as are targeting of particular points and omission of incomplete intervals. The error in estimating point mean fire interval from CFI is often large but variable because errors vary. An empirical study (Van Horne and Fulé 2006) and a simulation study (R. A. Parsons et al. 2007) of CFI accuracy only compared different methods of estimating CFI and did not include the necessary comparison of CFI with mean fire interval and fire rotation. However, a study that did include these comparisons showed that mean CFI estimates commonly are too short relative to point or population mean fire interval or fire rotation (Kou and Baker 2006a,b). Analysis (W. L. Baker and Ehle 2001) and empirical comparison with mean fire interval and fire rotation show mean CFI is too low by a factor of 3.6 to 16.0 (W. L. Baker 2006a). Thus, if mean CFI is 10 years, then mean fire interval and fire rotation may be between 36 and 160 years (see also fig. 5.9). A landscape characterized as having

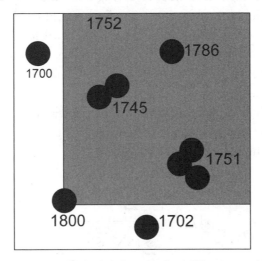

Area is 1 hectare. Each circle has a radius of 5 meters and thus an area of 0.008 hectare, representing a fire about the size of one tree. Shaded area is a fire that burned 80 percent of the 1-hectare area or 0.8 hectare.

Mean composite fire interval (Mean CFI)
    7 fires and 6 intervals, so Mean CFI = 100/6 = **16.7 years**

Fire rotation (FR) and population mean fire interval (PMFI)
    4 fires each burning 0.008 hectare = 0.032 hectare
    1 fire burning 0.016 hectare
    1 fire burning 0.024 hectare
    1 fire burning 0.8 hectare
    Total area burned is 0.87 hectare in a 1.0 hectare area
    Using equation 5.1, FR and PMFI = 100 years/0.87 = **114.9 years**

Figure 5.9. Why mean composite fire interval does not estimate mean fire interval or fire rotation. A hypothetical example with six small fires and one large fire (labeled by year) in a 1-ha sample area. Note that only the large fire burned a point at the center.

"frequent fire" (e.g., mean CFI = 10 years) may actually have a mean fire interval that is moderate to long, certainly not warranting the term "frequent fire." The CFI method is flawed, cannot be fixed, has contributed to significant misunderstanding of fire history, and warrants replacement.

The All-Tree Fire-Interval Method

A new method, the *all-tree fire-interval* (ATFI) method, has been shown by simulation to produce accurate and unbiased estimates of mean fire interval and fire rotation for small plots or transects (Kou and Baker 2006a). The ATFI estimate does not decline as plot area or number of sampled trees increases, but instead more accurately estimates the mean fire interval and fire rotation. ATFI includes all intervals, including incomplete intervals. It requires an estimate of scarring fraction, the fraction of trees receiving a fire that also scars, which can be estimated from field sampling in recent fires (e.g., Lentile 2004). Data are available that can be used in bracketing estimates until local data can be obtained (table 5.6). ATFI is estimated by sampling tree ages and fire scars in a random sample or census of all trees, both scarred and unscarred, in a plot or transect. Trees can be treated as a single population with consistent scarring fraction or can be divided into classes with differing scarring fractions, based on tree diameter, bark thickness, species, previous scarring, and other criteria (Kou and Baker 2006a). The formula is as follows:

$$ATFI = \frac{\sum_{i-1}^{m} TA_i}{\sum_{j=1}^{n} (SN_j/SF_j)} \tag{5.2}$$

where *ATFI* is mean fire interval estimated by the all-tree fire-interval method, TA_i is the age of sample tree i of m total sample trees, SN_j is the number of fire scars for *SF* (scarring fraction) class j of n total *SF* classes, SF_j is median *SF* for *SF* class j, and SNj/SF_j is an estimate of total number of fires experienced by trees belonging to *SF* class j (Kou and Baker 2006a).

Estimating Fire-Regime Parameters from Landscape Data

A list of fires from a small plot or transect need not be used to estimate fire-regime parameters (e.g., mean fire interval) but can instead be treated as simply point data on fire presence or absence, with fire-regime reconstruction done at

the landscape scale. Estimates in small plots or transects provide local information, but this is much less important than landscape estimates, since the fires
that account for most of the properties of fire regimes are much larger than
small plots or transects. Only landscape reconstructions allow estimation of a
fire-size distribution and the area-based fire-interval and severity distributions
that are needed. Methods for using small plot or transect data to reconstruct parameters of fire regimes across landscapes vary in quality, reflected in relative
ratings, used later in assessing fire history studies (chaps. 6–9).

A *high* relative rating goes to studies using *fire-year maps* (fig. 5.10b), reconstructed by circumscribing evidence from a particular fire year (e.g., stand origins, fire scars, tree deaths, growth anomalies). All methods have error, but by
far the best available landscape method is to reconstruct maps of past fires,
whose areas can then be summed and used with equation 5.1 to estimate fire
rotation and used directly to estimate attribute distributions. Boundary estimation is a potentially large source of error in fire-year maps. One method places
boundaries equidistant between points with and without evidence, unless fire
breaks are more likely boundaries, and uses estimated fire direction (e.g., A. H.
Taylor 2000; Beaty and Taylor 2001), possibly from scar directions (W. L. Baker
and Kipfmueller 2001). Minimum convex polygons can also be drawn around
evidence (e.g., Heyerdahl, Miller, and Parsons 2006) but appear to underestimate fire area (Shapiro-Miller, Heyerdahl, and Morgan 2007). A comparison
found that inverse distance weighting was more efficient and accurate at reconstructing boundaries than were indicator kriging, Thiessen polygons, or experts
(Hessl et al. 2007). Fuzzy-membership methods also may be effective (Jordan,
Fortin, and Lertzman 2005). In high- or variable-severity regimes, some
boundaries may still be visible and can be mapped (e.g., Buechling and Baker
2004). The estimated area of each fire in fire-year maps needs correction for (1)
missing evidence because of later fires, (2) areas the fire burned for which there
is no evidence, and (3) unburned area inside a perimeter (e.g., W. L. Baker
2006a). Recent fires can be reconstructed better, because less area is burned
over by later fires. Earlier fires can be bracketed using minimum and maximum
fire-extent maps, using differing assumptions.

Fire-year maps allow estimation of fire rotation, using equation 5.1, but
also fire-size distribution, severity distribution, and other distributions, including area-based fire intervals (fig. 5.2). Means and other measures of central
tendency and variability can be calculated. Fire-year maps can be overlain in a
GIS to map mean fire interval or rotation (E. Howe and Baker 2003; fig.
5.10c), which can be analyzed relative to topography and other landscape
features.

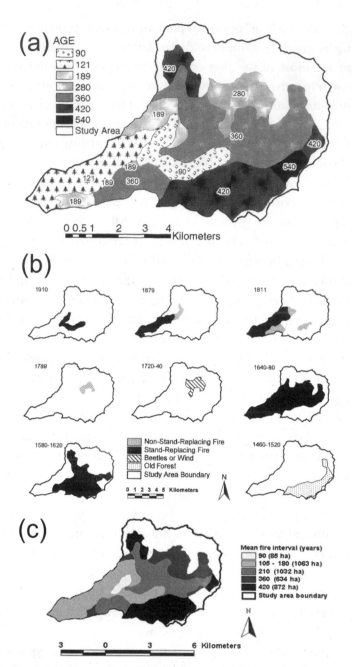

Figure 5.10. Fire history reconstruction across a subalpine forest landscape. The study area is in the Park Range, northern Colorado. The reconstruction, covering 1580–2000, includes (a) a stand-origin map showing stand age in the year 2000; (b) a fire-year map showing the estimated original extent of each fire, beetle, or wind event; and (c) spatial variation in mean fire interval.

Source: Reproduced from E. Howe and Baker (2003, p. 807, fig. 8a; p. 802, fig. 4; and p. 803, fig. 5c) with permission of Wiley-Blackwell Publishing Ltd.

A *medium* rating is given for methods that rely directly on *landscape snapshots* (observations or reconstructions of the state of a landscape at a particular moment, without reconstructions of past fire extent), including (1) mean or median of a set of stand-origin dates from an adequate, unbiased statistical sample across a landscape (e.g., Romme et al. 2001); (2) fire rotation and other parameters from landscape snapshots, using a theoretical relationship (Van Wagner 1978; E. A. Johnson and Gutsell 1994); (3) fire sizes from a stand-origin map and equation 5.1, rather than fire-year maps; and (4) total burned area reported or estimated over some period and equation 5.1. Landscapes fluctuate, even with constant fire rotation (box 3.1), so using landscape snapshots is more imprecise (W. L. Baker 1989a) than reconstructing fire-year maps that span centuries. Consider a snapshot of the Yellowstone landscape a year before the 1988 fires, when old forests dominated, and a year after, when young forests dominated, which would yield two quite different estimates of fire rotation. Stand-origin maps (fig. 5.10a) underestimate fire size, leading to a low estimate of fire rotation. Estimates of total burned area over a period (e.g., three-quarters burned in 250 years), without annual data, are imprecise.

A medium rating is also given to an aspatial estimate of actual fire areas, in which number of plots or transects recording a fire measures the fraction of the landscape burned by that fire (A. H. Taylor and Skinner 1998):

$$A_i = A \cdot \left(\frac{N_i}{N_t - N_b} \right) \qquad (5.3)$$

where A_i is the area of fire i, A is the total area studied, N_i is number of sample plots or transects with evidence of fire i, N_t is the number of sample plots or transects in the study area, and N_b is the number of sample plots or transects burned over by later fires, so not available to record fire i. These data are then used with equation 5.1. The method ignores the spatial configuration of points, differences in area represented by a point, likely direction of fire spread, occurrence of natural fire breaks, and unburned areas within perimeters and is inherently less precise than using the same data to estimate fire-year maps.

A *low* rating is assigned to fire rotation from (1) median, mean, or maximum of a small or nonrandom sample of stand-origin dates; or (2) time required for fuel buildup or development of mature forest. These are inherently very imprecise.

Rocky Mountain shrublands and grasslands lack fire scars, except on trees nearby or scattered within these areas. Evidence that a fire burned across an area is a cross-dated fire scar from the same year on both sides of the area (fig. 5.5). However, evidence is more typically available for only one margin, requiring corrections to estimate how much of the fire in the forest might have also burned in the shrubland or grassland (W. L. Baker 2006a; W. L. Baker, in press; chap. 9).

Charcoal and Pollen

Charcoal, aided by pollen in lake sediments, wetlands, small hollows in uplands, and soils can be counted and dated to reconstruct fire history over millennia (Patterson, Edwards, and Maguire 1987; MacDonald et al. 1991; Higuera, Sprugel, and Brubaker 2005). Charcoal amount varies with type of plant and intensity of fire, and charcoal suffers breakage, differential transport by wind and water, and potential mixing after deposition (Patterson, Edwards, and Maguire 1987; J. S. Clark 1988). The best records at the landscape or watershed scale are from larger charcoal particles in cores from annually laminated lakes with deep sediments (e.g., J. S. Clark 1988). Similar methods have been used with some success in small hollows in uplands in the Pacific Northwest (Higuera, Sprugel, and Brubaker 2005).

In the Rockies, annually laminated lakes are lacking, and their absence requires estimation of age by interpolating among dated points along a sediment core (e.g., Millspaugh, Whitlock, and Bartlein 2000). Charcoal is then counted in samples along the core, and charcoal concentration (particles per cubic centimeter) is converted to charcoal accumulation rate by adjusting for sedimentation rate (fig. 5.11). Charcoal accumulation rate is then divided into background and peaks (fig. 5.11). *Fire-episode frequency* is the number of peaks per some period (e.g., 1,000 years).

Simulation (Higuera et al. 2007) and analysis (Marlon, Bartlein, and Whitlock 2006) suggest that background charcoal may reflect area burned and regional trends, while peaks may represent local fires within source areas greater than 30,000 hectares (Higuera et al. 2007). Peaks may represent more than one fire (Millspaugh, Whitlock, and Bartlein 2000; Brunelle and Whitlock 2003) and likely primarily represent high-severity fires (Millspaugh and Whitlock 1995; Higuera, Sprugel, and Brubaker 2005). The validity of focusing on macrocharcoal to identify local fires was tested by comparing charcoal records with tree ring records of fire in the same watershed (Millspaugh and Whitlock 1995). However, in another calibration, the method identified too many peaks relative to actual local fires, and the full size-distribution of charcoal better distinguished actual local fires (Asselin and Payette 2005). Comparison of fire-episode frequency over the last few hundred years (see table 8.3) with estimated fire rotations over those periods (chaps. 6–9) hints that episode frequency may inconsistently estimate fire rotation, and further calibration research is warranted.

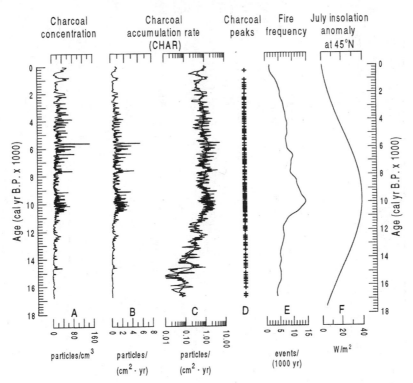

Figure 5.11. Charcoal data and fire history for a lake in Yellowstone National Park. Shown is fire episode frequency reconstructed for Cygnet Lake in Yellowstone National Park. *Source:* Reproduced from Millspaugh, Whitlock, and Bartlein (2000, p. 212, fig. 2) with permission of the Geological Society of America.

ROCKY MOUNTAIN FIRE REGIMES

Assigning fire regimes to vegetation types is not fully appropriate because regimes represent interactions among vegetation, topography, and climate (Keane et al. 2003). In the past, fire regimes in the Rockies were categorized by mean fire interval and fire severity. Fire severity is considered nonlethal or low; mixed; or stand-replacement or high (P. Morgan et al. 1996; J. K. Smith 1998). One scheme used overall intervals and severity (J. K. Smith 1998); another considered whether fires are lethal to canopy trees, with separate intervals for high- and low-severity fires (Barrett and Arno 1991). A six-class scheme had a nonlethal class, three mixed-severity classes, and two crown-fire classes, distinguished by MFI (Barrett 2004). A national scheme had five regimes (Schmidt et al. 2002; http://www.landfire.gov).

I classified fire regimes in the Rockies by fire rotation, rate of recovery, and typical pattern of topkill. Rotations do not form natural categories but may be

Table 5.7. Classification of Rocky Mountain Fire Regimes

Fire Regime	Rocky Mountain Ecosystems	High-Severity Fire Rotation (yrs)	Low-Severity Fire Rotation (yrs)
Short-rotation, fast-recovery, high-severity	Montane grasslands	50–115	None
	Subalpine grasslands	50–170	None
	Mixed mountain shrublands	100	None
	Dry mixed-grass prairies	140–160?	None
Short-rotation, slow-recovery, high-severity	Moist northwestern forests—grand fir	100–200	Rare
	Quaking aspen forests	>140	Rare
Short-rotation, slow-recovery, variable-severity	Upper-montane mixed-conifer forests	100–200	Unknown
	Southwestern mixed-conifer forests	130–200	Unknown
	Lodgepole pine forests	135–280	1,500–10,000 or more
	Northern Rockies mixed-conifer forests	150–250	Unknown
Medium-rotation, fast-recovery, high-severity	Shortgrass prairies	>200?	None
Medium-rotation, slow-recovery, variable-severity	Ponderosa pine–Douglas-fir forests	300?	60–300?
	Douglas-fir forests	400?	200?
Medium-rotation, slow-recovery, high-severity	Mountain big sagebrush shrublands	150–300	None
	Wyoming big sagebrush shrublands	200–350	None
	Little (low) sagebrush shrublands	>200	None
	Spruce-fir forests	175–350 (700)	1,500–10,000 or more
	Moist northwestern forests—upland western red cedar–western hemlock	200–300	Rare
	Whitebark pine forests	250	Rare
	Bristlecone pine forests	300	Rare
	Limber pine forests	350	Rare

Long-rotation, fast-recovery, high-severity	Salt desert shrublands	>500	None
Long-rotation, slow-recovery, high-severity	Piñon-juniper woodlands	290–600	Rare
	Moist northwestern forests—wet site western red cedar–western hemlock	>500	Rare

Note: Classification is based on fire rotation, recovery, and fire severity. Fire rotations are short (50–200 years), medium (200–400 years), or long (>400 years); recovery is fast (less than 25 percent of fire rotation) or slow (more than 40 percent of fire rotation); fire severity is low, variable, or high. The low-severity rotation is for both discrete surface fires and the surface-fire component of variable-severity fires. Similarly, the high-severity rotation is for both discrete high-severity fires and the high-severity component of variable-severity fires. Rotation estimates in parentheses represent extremes, and estimates with question marks are uncertain. Evidence for each estimate, as well as sources, is given in chapters 6–9. No estimates are available for curlleaf mountain mahogany shrublands, plateau grasslands, xeric low-elevation ponderosa pine–Douglas–fir forests, or riparian.

short (50–200 years), medium (200–400 years), or long (greater than 400 years). Rate of recovery is the percentage of the fire rotation needed to recover prefire composition and structure. Two categories are (1) fast (less than 25 percent of fire rotation), and (2) slow (40 percent or more).

Only eight combinations occur in the Rockies (table 5.7). The classification is based on few studies in some cases. Long-rotation, slow-recovery, high-severity regimes dominate wet northwestern forests and low-elevation semiarid woodlands, where fuels are sparse or patchy and ignition limited (W. L. Baker 2003). A medium-rotation, fast-recovery, high-severity regime occurs in short-grass prairies with low fuel loads. A medium-rotation, slow-recovery, variable-severity regime occurs in montane ponderosa pine and Douglas-fir forests. Medium-rotation, slow-recovery, high-severity regimes occur in mountain and Wyoming big sagebrush, little (low) sagebrush, upland moist northwestern forests, and subalpine spruce-fir and five-needle pine forests. Short-rotation, high-severity regimes have fast recovery in grasslands and mixed mountain shrublands and slow recovery in grand fir and aspen forests. A short-rotation, variable-severity, slow-recovery regime occurs in mixed-conifer and lodgepole pine forests.

Fire regimes can be predicted by models that link regimes to topographic variables and vegetation (Keane et al. 2003; Sherriff and Veblen 2007), which is useful or even essential if fire history evidence is unavailable or has been destroyed or altered. It may also be possible to directly estimate fire-regime parameters (e.g., mean fire interval) or the variables that influence fire regimes (e.g., fuel loads). Reasonable success (about 50–80 percent accuracy), for example, was obtained in predicting mean fire interval, as well as discrete categories of fuel and fire severity, when modeling was aided by ground-truth data (Rollins, Keane, and Parsons 2004).

SUMMARY

Understanding fire under the HRV requires reconstructing fire regimes across landscapes using historical records, tree rings, charcoal, and pollen. Past methods focused on obtaining long fire records from targeted patches of old-growth forest or long sediment cores in lakes. However, to accurately reconstruct fire regimes across landscapes requires unbiased sampling and area-based (e.g., fire rotation), rather than frequency-based (e.g., mean composite fire interval), measures. The best landscape method reconstructs fire-year maps from mapped evidence (fire scars, tree ages, death dates of trees, growth anomalies in tree rings), brackets uncertainties and potential errors in the maps, and uses the maps to estimate fire rotation and mean fire interval, as well as important attribute distributions, including fire-size, fire-interval, and fire severity distributions. I classified fire regimes in the Rockies into eight categories, based on fire rotation, rate of recovery, and pattern of topkill by fire, but all Rockies fire regimes are likely characterized by infrequent large fires that do most of the work of fire.

CHAPTER 6

Fire in Piñon-Juniper, Montane Aspen, Mixed-Conifer, Riparian, and Wetland Landscapes

This chapter covers a set of miscellaneous ecosystems that occur primarily in the montane zone, including low-elevation piñon-juniper woodlands found generally below the ponderosa pine zone, quaking aspen forests and several types of mixed-conifer forests found generally above the ponderosa pine zone, and riparian and wetland vegetation that occurs across all elevational zones. These ecosystems greatly enhance landscape diversity in the Rocky Mountains and influence the pattern and behavior of large wildland fires. Fire behavior and fire history, fire effects and succession, and land-use effects and restoration are discussed for each ecosystem.

PIÑON-JUNIPER WOODLANDS

Piñon-juniper woodlands, which are characterized by relatively short conifers, have received less fire research than other forests in the Rockies, in part because they are more difficult to study. Research has recently expanded and changed our understanding of fire and forest structure in these woodlands by showing that they often are hundreds of years old, tend to burn infrequently, and were rarely affected by low-severity fire.

Fire Behavior and Fire History

Fire spread in these piñon-juniper woodlands may be limited unless winds are strong (W. L. Baker 2003), as fuels often are in large size classes, and fine fuels sparse and discontinuous (Omi and Emrick 1980). Photo series are available for fuel estimation (Ottmar, Vihnanek, and Regelbrugge 2000). Drought, disease, and beetles caused piñon mortality after 2002 in and near southern Colorado (Breshears et al. 2005), but resulting fuels may decompose in a few decades (Kearns, Jacobi, and Johnson 2005). Ignitions are seldom *holdover fires*, fires that smolder in deadwood for long periods before spreading (Omi and Emrick 1980). Most go out, but if ignitions are followed by strong winds, rapid spread and large fires may occur (Hester 1952). At higher elevations, shrubs provide fuel continuity and can be flammable if dry, leading to potentially intense, rapidly spreading fires (Hester 1952), as at South Canyon, Colorado (fig. 6.1c), where, as mentioned previously, 14 firefighters died (B. W. Butler et al. 1998).

Fires in piñon-juniper woodlands across the western United States since EuroAmerican settlement have nearly all been high severity but have left unburned patches due to natural fire breaks, low fuel loads, or wind shifts (W. L. Baker and Shinneman 2004). Floyd et al. (2000) thought natural fire breaks, such as rock outcrops and cliffs, might explain persistence of old woodlands on Mesa Verde, but old woodlands were not limited to these sites on the nearby Uncompahgre Plateau (Eisenhart 2004; Shinneman 2006). Spot fires can be

Figure 6.1. Old-growth piñon-juniper woodlands. Shown are (a) comparatively dense woodlands with abundant down deadwood and a low-shrub understory; (b) trees with low branches in open woodlands at low elevations on typically rocky sites, neither setting conducive to low-severity surface fires; and (c) high-severity fire in piñon-juniper at South Canyon near Glenwood Springs, Colorado.

Source: Photos by the author.

ignited kilometers from a fire line, and fires can burn directly across clearings more than 10 meters wide (Hester 1952). In Colorado, large fires are most common in drought years, before monsoon onset in July (Omi and Emrick 1980; M. L. Floyd, Hanna, and Romme 2004).

Some have suggested that woodlands formerly were more savanna-like, maintained by low-severity surface fires (e.g., West 1999). Savannas in the southwestern United States have a low density of junipers in a matrix of perennial grass conducive to fire spread (McPherson 1997), although evidence of low-severity fire is meager (W. L. Baker and Shinneman 2004; Romme et al. 2008). However, in the Rockies, most woodlands near EuroAmerican settlement were denser than these savannas (see "Fire Effects and Succession" below) and had discontinuous grassy understories, shrub dominance (fig. 6.1a), or sparse trees on rocky sites (fig. 6.1b), none of which are conducive to widespread low-severity fire (W. L. Baker and Shinneman 2004; Eisenhart 2004; Shinneman 2006). Evidence of limited spreading low-severity fire was found in the upper ecotone with ponderosa pine forests in northern New Mexico (Allen 1989; Huffman et al. 2008), at higher elevation than savannas. In Idaho, some evidence was found of patchy low-severity fires (Burkhardt and Tisdale 1976). Evidence of widely spreading low-severity fires is lacking away from the ecotone in most piñon-juniper woodlands. Fire scars, which can indicate low-severity fires, were absent in woodlands on Mesa Verde (Floyd, Romme, and Hanna 2000; Floyd, Hanna, and Romme 2004), on the Uncompahgre Plateau (Eisenhart 2004; Shinneman and Baker, in press), and in southeastern Colorado (Tonnesen and Ebersole 1997). Old trees with low branches and abundant large deadwood also suggest low-severity fires were rare (Eisenhart 2004; Shinneman 2006); they were likely rare in piñon-juniper across the West (W. L. Baker and Shinneman 2004; Romme et al. 2008).

Pre-EuroAmerican fires were high severity on Mesa Verde and the Uncompahgre Plateau in southwestern Colorado (Erdman 1970; Floyd, Romme, and Hanna 2000; Floyd, Hanna, and Romme 2004; Shinneman 2006) and also were observed in southwestern Idaho (Burkhardt and Tisdale 1976). On the Uncompahgre Plateau, past high-severity fire was more common at high elevations (fig. 6.2; Eisenhart 2004; Shinneman 2006). At low elevations, drought may have played a significant role in maintaining low tree density (Eisenhart 2004; Shinneman 2006). Fire rotation for high-severity fires, which dominated under the HRV in piñon-juniper woodlands in the Rockies, is estimated at 290–600 years (table 6.1), and old-growth or ancient forests commonly dominated these woodland landscapes (Eisenhart 2004; M. L. Floyd, Hanna, and Romme 2004; Huffman et al. 2008; Shinneman and Baker, in press).

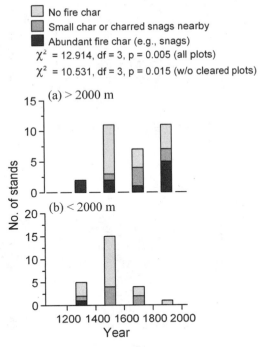

Figure 6.2. Stand origins and fire evidence by elevation in piñon–juniper woodlands. Data are from the Uncompahgre Plateau in western Colorado. Chi-squared values indicate that stand-origin distributions differ between low and high elevations. Evidence of char from past fires is also presented.

Source: Reproduced from Shinneman (2006, fig. 5), copyright 2006, with permission of Douglas J. Shinneman.

Fire Effects and Succession

Studies of fire effects are not numerous (table 6.2). Where shrubs were common, they may resprout beginning a few weeks after high-severity fire. Otherwise, the year after fire naturally has sparse cover (fig. 6.3a), a normal postfire response, as plants resprout or germinate from the seed bank. An initial flush of annuals and then resprouting perennial forbs and graminoids typify years 2+ after fire (Erdman 1970; Shinneman 2006). Species richness, for example, doubled from year 1 to 2 postfire (K. R. Adams 1993), and native forb and graminoid cover increased rapidly in years 2–4 postfire (Shinneman 2006). A shrub stage (fig. 6.3b) can last a century, with scattered trees slowly infilling and full recovery requiring greater than 200 years (Goodrich and Barber 1999) or as long as 300 years (Erdman 1969, 1970; Omi and Emrick 1980).

Tree age structures (table 6.3) and chronosequences (table 6.2) suggest that tree populations were also temporally variable, because of disturbance and

Table 6.1. Fire History Studies in Piñon-Juniper, Montane Aspen, and Mixed-Conifer Forests

Sources & Study Limitations, by Forest Type & Quality of Information	Location	No. Scar Plots	No. of Scarred Trees	Age-Structure Est. of Severity	Qual. Est. of Fire Size	No. Stand Origins	Stand-Origin Map	Fire-Year Maps	High-Severity Rotation (years) and Quality of Estimate (in parentheses)[5]
PIÑON-JUNIPER									
High									
Eisenhart (2004)	WC CO	136	0	High	No	136	No	No	No est.
M. L. Floyd, Romme, & Hanna (2000)	SW CO	1	0	High	No	2	No	Yes	400 (H)
M. L. Floyd, Hanna, & Romme (2004)	SW CO	0	0	High	Yes	8	Yes	No	400 (H)
Huffman et al. (2008; Canjilon)	N NM	1	61	High	Yes	106	Yes	No	290 (H)
Medium									
Shinneman & Baker (in press)[6]	WC CO	59	0	High	No	59	No	No	400–600 (M)
Low									
Burkhardt & Tisdale (1976)[0,1b,2a,3b]	SW ID	4	50?	No data	No	0	No	No	No est.
QUAKING ASPEN (STABLE)									
Medium									
Romme et al. (2001)[7]	SW CO	1	0	High	No	76	No	No	>140 (M)
MOIST NORTHWESTERN FORESTS									
Medium									
Barrett (2000)[2b,4a,4b]	C ID	<67	?	High, low	Yes	<67	No	No	230 (L)
Barrett, Arno, & Key (1991; maritime-influenced sites)[2b,4a,4b]	NW MT	6	<272	High	Yes	<230	Yes	No	261 (L)
Low									
Arno & Davis (1980)[2b,4a,4b,8]	N ID	2	38	High, moderate, low	No	?	No	No	No est.
Rapraeger (1936)[0,2b,3b]	N ID	1	?	High, low	No	0	No	No	No est.

Study	Location			Severity					Interval
Barrett & Arno (1991); J.K. Brown et al. (1994); Douglas-fir/grand fir)[9]	C ID	<77	?	High, low	No	?	No	No	119 (L)
Barrett & Arno (1991); J.K. Brown et al. (1994); western red cedar)[9]	C ID	<77	?	High	No	?	No	No	197 (L)
Antos & Habeck (1981)[9]	W MT							No	100–200 (L)
McCune (1983)[2a,7]	W MT	0	0	High, moderate, low	No	427	No	No	55–70 (L)
NORTHERN ROCKIES MIXED CONIFER									
Medium									
Barrett, Arno, & Key (1991; dry sites)[2b,4a,4b]	NW MT	8	<272	High, low	Yes	<230	Yes	No	140–240 (L)
Sneck (1977); K. M. Davis (1980)[2b,4a,4b]	NW MT	13	103	High, moderate	No	13	Yes	Yes	No est.
Barrett (2000)[2b,4a,4b]	C ID	<67	?	High, low	Yes	<67	No	No	230 (L)
Low									
Gabriel (1976)[2b,4a,4b,10]	NW MT	1	222	High, moderate, low	No	>22	No	Yes	150–200 (M)
Arno, Scott, & Hartwell (1995; plots F1,F2)[0,1b,2a,2b,4a,4b]	W MT	2	≤8	High, low	No	2	No	No	No est.
Arno, Smith, & Krebs (1997; plot L5)[0,1b,2a,2b,4a,4b]	W MT	1	>3	Low	No	1	No	No	No est.
Tesch (1981)[0]	W MT	?	?	High, low	Yes	2	No	No	No est.
UPPER–MONTANE MIXED CONIFER									
High									
N.T. Korb (2005; mesic)[4a,4b]; Sherriff (2004); Sherriff &	SW MT	1	18	High, low	No	40	Yes	Yes	117–186 (M)[6]
Veblen (2006; mixed conifer samples)[4a,4b]	N CO	5	81	High, moderate, low	No	9	No	No	No est.

Table 6.1. Continued

Sources & Study Limitations, by Forest Type & Quality of Information	Location	No. Scar Plots	No. of Scarred Trees	Age-Structure Est. of Severity	Qual. Est. of Fire Size	No. Stand Origins	Stand-Origin Map	Fire-Year Maps	High-Severity Rotation (years) and Quality of Estimate (in parentheses)[5]
Veblen, Kitzberger & Donnegan (2000; mixed conifer samples)[4a,4b]	N CO	~7	~110	No data	No	0	No	No	No est.
Low									
Houston (1973)[2a,2b,3b,4a,4b]	NW WY	13	47	No data	Yes	0	No	No	No est.
Clements (1910)[0,2a,2b]	N CO	?	?	High, moderate, low	No	?	No	No	No est.
Sherriff, Veblen, & Sibold (2001)[0,1b,1c,3b,4a,4b]	N CO	4	46	No data	No	0	No	No	No est.
Donnegan, Veblen, & Sibold (2001)[0,1b,1c,3b,4a,4b]	C CO	3	80	No data	No	0	No	No	No est.
SOUTHWESTERN MIXED CONIFER									
Medium									
Margolis, Swetnam, & Allen (2007)[0]	NM, CO	182	74	High	Yes	12		Yes	No est.
Wu (1999)[0,1b,1c,4a,4b]	S CO	11	88	High, moderate, low	No	9	No	No	130–210 (M)[6]
Low									
Grissino-Mayer et al. (2004)[0,1b,1c,3b,4a,4b]	S CO	3	42	No data	No	0	No	No	No est.
Alington (1998)[1c,2a,3a,4a]	S CO	16	49	No data	Yes	29	No	No	No est.
Allen et al. (2008)[0,3a,4a]	N NM	2	65	No data	No	1	No	No	No est.
Touchan, Allen, & Swetnam (1996)[0,3b,4a,4b,8]	N NM	3	96	No data	No	0	No	No	No est.

Notes: Studies were conducted in and near the Rocky Mountains and are ranked in column 1 specifically by the quality of information for understanding fire history across landscapes: *High* = Researchers sampled contiguous landscape areas and (a) included an adequate set of scarred trees with cross-dated fires, providing valid evidence about fire spread across the landscape; and (t) dated tree ages in a series of plots across a landscape, so fire severity and replacement fire rotation, or only fire rotation, could be determined. *Medium* = Researchers sampled contiguous landscape areas but (a) did not or could not cross-date fire scars; (b) did not collect or did not use tree age data, so fire severity is not known; or (c) did not sample contiguous landscape areas but did date tree ages in a series of plots across a multiple-landscape area, so fire severity and replacement fire rotation could be determined. *Low* = Study had multiple limitations, including (a) plots scattered across multiple landscapes or a small number of contiguous plots in a single landscape, so they provide only small-plot estimates of fire history; (b) biased sampling, low sample size, no age data, or other limitations (see table 5.5); (c) no analysis of fire spread among sampled trees or sample plots; or (d) no presentation of data. Studies are arranged from north to south within each forest type. Footnotes 1–4 use the same codes as in table 5.5, where citations are given explaining each limitation.

[0]Not landscape scale; small plots scattered many km apart or very few (e.g., 1–3) plots within a landscape.

[1a]Biased sample targeting of landscapes with old trees or many fire scars.

[1b]Biased sample targeting of plots with old trees or many fire scars.

[1c]Biased sample targeting of individual trees that are old, have many fire scars, or are thought to preferentially record fire scars.

[2a]Inadequate fire-scar sample size (e.g., <10 trees per plot).

[2b]Fire scars not cross-dated.

[3a]Inadequate age structure data to allow estimation of fire severity (e.g., a few trees per plot).

[3b]No age structure data to allow estimation of fire severity, or age data not used.

[4a]Biased fire-interval estimate based on composite fire intervals or individual tree fire intervals.

[4b]Biased fire-interval estimate due to omission of censored fire intervals.

[5]Fire rotation estimate based on high, medium, or low quantity/quality of evidence: (H) = fire areas from reconstructed fire-year maps and equation 5.1 (high); (M) = estimated mean or median from a histogram of stand-origin dates or from fire sizes estimated from a stand-origin map or from total burned area over some interval (medium); (L) = estimated mean from a small set of dated plots or based on the time required for fuel buildup and development of mature forests (low).

[6]My derivation of the estimate; see chapter text for explanation.

[7]No fire-scar data verifying that stands originated after fire.

[8]No analysis of fire spread.

[9]Data not presented.

[10]Data pooled across vegetation types.

Table 6.2. Studies of Postfire Responses of Plants in Piñon-Juniper, Montane Aspen, and Montane Mixed-Conifer Forests

Sources and Type of Observation by Vegetation Type	Location	Time since Fire (yrs)	Severity			Variable Studied				
			L	M	H	I	G	R	D	P
PIÑON-JUNIPER										
Observations/plots after fire										
K. R. Adams (1993)	SW CO	1–3				X	X		X	
Chronosequence studies										
Shinneman (2006)	W CO	1–10				X	X	X		
Erdman (1969, 1970)	SW CO	2–387				X	X		X	
Goodrich & Barber (1999)	NE UT	38–100				X			X	
QUAKING ASPEN										
Observations/plots after fire										
DeByle, Urness, & Blank (1989)	SE ID	1–2		X						
Bartos, Brown, & Booth (1994)	NW WY	1–3, 6, 12	X	X	X	X	X	X		X
J. K. Brown & DeByle (1987)	SE ID, NW WY	1–4	X	X	X			X		
J. K. Brown & DeByle (1989)	SE ID, NW WY	1–5				X	X	X		X
Bartos, Mueggler, & Campbell (1991)	NW WY	1–6						X		
Kay (2001)	NW WY	5–14		X				X		
Hessl & Graumlich (2002)	NW WY	5–15	X	X	X			X		
E. J. Larsen & Ripple (2003)	NW WY, SW MT	ca. 13, 66		X				X		
Chronosequence studies										
Gruell & Loope (1974)	NW WY	1–121		X				X		
MOIST NORTHWESTERN FORESTS										
Observations/plots after fire										
Stickney (1986)	N ID	1–10				X	X	X	X	
Humphrey & Weaver (1915)	N ID	2, 26				X	X	X		
Leiberg (1900b)	N ID	2–3						X		
J. A. Larsen (1925b)	N ID	4, 13				X	X	X		
Daubenmire & Daubenmire (1968)	N ID	33, 54				X	X	X		
Chronosequence studies										
McCune & Allen (1985)	W MT	48–240				X		X		
Habeck (1968)	NW MT	<50 to ca. 400				X	X	X	X	
Antos & Habeck (1981)	NW MT	Young to old				X	X	X		
J. A. Larsen (1929)	N ID	Young to old				X	X	X		

Table 6.2. Continued

Sources and Type of Observation by Vegetation Type	Location	Time since Fire (yrs)	Severity			Variable Studied				
			L	M	H	I	G	R	D	P
NORTHERN ROCKIES MIXED CONIFER										
Observations/plots after fire										
Crane, Habeck, & Fischer (1983)	W MT	1, 2, 5				X	X	X	X	
Habeck (1970)	NW MT	3, 34, 41	X	X	X	X		X		
UPPER-MONTANE MIXED CONIFER										
Observations/plots after fire										
J. K. Brown & DeByle (1987)	SE ID, NW WY	1–4	X	X	X			X		
J. K. Brown & DeByle (1989)	SE ID, NW WY	1–5				X	X	X		X
Doyle (1997, 2004)	NW WY	ca. 1–20	X	X				X		
Keyser, Smith, & Shepperd (2005)	W SD	4	X		X			X		
Peirce (1915)	NE WY	ca. 34				X		X		
Chronosequence studies										
S. J. Henderson (1997)	SW MT	≤22 to ≥200				X	X			
Kleinman (1973)	N UT	1–82				X	X	X	X	X	X
SOUTHWESTERN MIXED CONIFER										
Observations/plots after fire										
Walker (1993)	N UT	2, 3		X	X			X		
S. J. Henderson (1997)	SW MT	≤22 to ≥200		X			X			
Kleinman (1973)	N UT	1–82				X	X	X	X	X	X

Note: Studies were conducted in or near the Rocky Mountains. Authors were not specific about whether pure aspen forests were stable or successional. Table entries are ordered by time since fire and whether observations were of one or more fires or a chronosequence of fires (a set of fires ordered by time since fire). L = low severity, M = moderate severity, H = high severity, I = individual species, G = species groups (i.e., forbs, graminoids, shrubs), R = tree regeneration or density (or other tree population variables), D = diversity, P = production or growth (e.g., annual basal area increment).

episodic regeneration. Junipers may regenerate before piñons after fires (Erdman 1970; Barney and Frischknecht 1974; Omi and Emrick 1980) by a century or more on the Uncompahgre Plateau (Shinneman 2006). Junipers depend less on nurse plants and have greater drought tolerance and ability to compete with postfire plants (Erdman 1970; Chambers, Vander Wall, and Schupp 1999). Postfire piñon regeneration is favored by sagebrush or other shrubs that serve as nurse plants (Erdman 1970; M. E. Floyd 1982) and by wet periods (Eisenhart 2004; Shinneman and Baker, in press). Woodlands 100–200 years old often have a bell-shaped age structure, as a cohort of trees recovered

Figure 6.3. After high-severity fire in piñon-juniper woodlands. Shown after high-severity fire on the Uncompahgre Plateau in western Colorado are (a) sparse plant cover during year 1; and (b) recovered understory plant cover, including shrub dominance, about 10 years after fire.
Source: Photos courtesy of Douglas J. Shinneman.

after disturbance; older woodlands often are uneven aged, because drought, insects, and disease have periodically killed trees in the maturing woodland, and tree regeneration followed (Eisenhart 2004; Shinneman and Baker, in press).

Piñon is favored by relatively high moisture, juniper by drought, enabling large shifts in tree density and composition over time. Piñon expansion on the Uncompahgre Plateau began in the late 1700s after a long period of drought (Shinneman and Baker, in press). A wet period in the early 1900s also favored piñon regeneration (Eisenhart 2004; Shinneman and Baker, in press). However, density of piñons may decline rapidly, as it did during droughts in the early 2000s (Breshears et al. 2005; Mueller et al. 2005). Tree density also increases during recovery after high-severity fire (Huffman et al. 2006). Thus, fluctuations in tree density in these woodlands commonly reflect climatic fluctuations (Eisenhart 2004; Shinneman and Baker, in press) or recovery from disturbance (Erdman 1970; Huffman et al. 2006) rather than land-use changes (Romme et al. 2007).

These woodlands were highly spatially variable under the HRV. Pre-EuroAmerican tree density across the piñon–juniper zone on the Uncompahgre Plateau averaged 117 trees per hectare for Utah juniper and 128 trees per hectare for twoneedle piñon, totaling 245 trees per hectare, but it varied from near 0 to 1,176 trees per hectare; density was highest in upper elevations and lowest near the ecotone with sagebrush and grasslands (Shinneman 2006). Similar large variability was found in northern New Mexico in the ecotone with ponderosa pine, where pre-EuroAmerican tree density averaged 479 trees per hectare but varied from near 0 to 1,125 trees per hectare in a 409-hectare area (Huffman et al. 2006). Large spatial variability likely reflects patchy disturbances and regeneration and environmental variation.

Land-Use Effects and Restoration

Past reviews of piñon–juniper dynamics attributed tree density increases to fire exclusion and livestock grazing (e.g., West 1999; R. F. Miller and Tausch 2001; R. F. Miller et al. 2008). In a western Colorado study, densities of seedling and sapling piñons (less than 5 centimeters in diameter) were about three times as high in livestock-grazed woodlands as in reference areas, but most likely from loss of competing native forbs and grasses rather than fire exclusion (Shinneman and Baker, in press). The fire rotation under the HRV was long in piñon–juniper and has likely been little changed. Spatial lags (see box 10.1) mean that fire exclusion could have affected little land area had it occurred, and a general decline in fire is not even detectable for the Rockies (chap. 10).

Piñon–juniper invasion into adjoining sagebrush and grasslands has occurred but is uncommon on the Uncompahgre Plateau, where it was studied in detail (Eisenhart 2004; Manier et al. 2005; Shinneman 2006). Earlier work in Idaho (Burkhardt and Tisdale 1976) suggested that fire limited piñon–juniper invasion into sagebrush. Reanalysis of fire history in sagebrush (W. L. Baker 2006b) and piñon–juniper (W. L. Baker and Shinneman 2004) showed that fires could kill some invading trees, but fire rotations were too long for it to be a primary control, and fire has not declined since EuroAmerican settlement in sagebrush (chap. 9).

Dominance of burned woodlands by the invasive annual cheatgrass is a significant threat, particularly after smaller fires if cheatgrass is common and if the biological soil crust, native plant diversity, and cover of perennial forbs and graminoids have been reduced by overgrazing by livestock (Shinneman 2006). Postfire seeding reduced cheatgrass when considered across a suite of semiarid ecosystems (Beavers 2001) but had no effect or increased cheatgrass after fires in piñon-juniper woodlands on the Uncompahgre Plateau (Shinneman 2006; Getz and Baker 2008). Improved native seed mixes competitive with cheatgrass need development before postfire seeding is likely to be effective (Getz and Baker 2008; see also chap. 11).

Thinning woodlands to reduce tree density has been suggested as a focus of restoration by some (West 1999; Brockway, Gatewood, and Paris 2002; Landis and Bailey 2005). However, in the Rockies, tree density was spatially and temporally variable under the HRV (as mentioned previously), fire exclusion did not affect tree density, and livestock grazing primarily increased density of small trees (less than 5 centimeters in diameter). Thus, thinning and burning are not general restoration needs. Restoring native understory plants that livestock grazing reduced and controlling invasive species are the primary restoration needs (Shinneman, Baker, and Lyon 2008).

MONTANE QUAKING ASPEN FORESTS

Quaking aspen forests are not only beautiful but also are an important landscape feature, providing important wildlife habitat in the Rocky Mountains. Aspen paradoxically can serve as a fire break at times, but it is also remarkably well adapted to even high-intensity fires. There has been concern that aspen may be declining, in part because of fire exclusion, but recent research suggests otherwise.

Successional and Stable Aspen Forests

Quaking aspen, if present, may dominate the first century or more after high-severity fire in mixed-conifer forests. Aspen can resprout or sucker quickly from surviving lateral roots (Schier and Campbell 1978). Relatively pure, apparently stable aspen forests dominate large areas, particularly lower elevations (i.e., less than 2,800 m) on southerly-facing slopes in the central and southern Rocky Mountains (Romme et al. 2001). Stable aspen forests have few or no conifers and often an uneven aspen age structure with ongoing suckering. Stability could be disturbance maintained if a series of high-severity fires removed

(a)

(b)

Figure 6.4. A mosaic of stable aspen and successional aspen-conifer forests. The photographs are from central Utah in (a) about 1923, and (b) 2006. Stable aspen is visible as a comparatively bare area on the slopes of the background ridge. Successional aspen is visible as mixed conifers in the foreground, spruce-fir in the background.
Sources: Photo (a) is reproduced from F. S. Baker (1925, plate 1); (b) is by the author.

conifer seed sources (Romme et al. 2001), but stable and successional aspen can occur in a mosaic (fig. 6.4), so environment may also be important. This section focuses on stable aspen; successional aspen is discussed in the mixed-conifer sections later in this chapter and in chapter 8.

Fire Behavior and Fire History in Stable Aspen

High-intensity fire in conifers often becomes a surface fire and goes out a short distance into aspen, which thus can act as a fire break (Fechner and Barrows 1976; DeByle, Bevins, and Fischer 1987). However, aspen were burned by

Figure 6.5. Aspen stand burned in the 2002 Missionary Ridge fire in southwestern Colorado, north of Durango.
Source: Photo by the author.

intense surface fires and were not fire breaks in the 2002 Missionary Ridge fire in Colorado (fig. 6.5). Fires in Wyoming also burned through aspen (Gruell and Loope 1974). Nonetheless, these forests are less flammable than conifer forests; also, lightning ignitions are rare (Fechner and Barrows 1976), and aspen burned at much lower rates than expected, based on their area, in a severe fire season in Colorado (Bigler, Kulakowski, and Veblen 2005).

Fine-fuel loads in aspen vary substantially, depending on abundance of shrubs and forbs, but loadings of larger fuels (e.g., down wood) depend on local history of disturbance and disease (J. K. Brown and Simmerman 1986). Conifers and certain shrubs (e.g., common juniper) can torch in fires, increasing fire-line intensity (J. K. Brown and DeByle 1987), and, as aspen mixes with conifers, potential intensity may increase (DeByle, Bevins, and Fischer 1987). Fires burn the most area in aspen in July and August, during the most severe weather, and in October after leaf fall (Fechner and Barrows 1976; Ryan 1976), when vegetation is cured and more flammable.

Flames more than a half meter high commonly kill aspen (J. K. Brown and Simmerman 1986; J. K. Brown and DeByle 1987), so surface fire may be stand replacing, as in the Missionary Ridge fire (fig. 6.5). Fires can leave surviving individuals or groups of aspens or conifers in wet areas or natural fuel breaks inside a fire perimeter (E. Howe and Baker 2003), or on a burn margin where fire intensity is reduced and light surface fire occurs (J. K. Brown and DeByle 1987; E. Howe and Baker 2003). Keyser et al. (2005) found about 30 percent of aspen clones had burned at high severity and 30 percent at low severity, with only 25–75 percent mortality in the 2000 Jasper fire in the Black Hills. However, this was in a ponderosa pine landscape with more low-severity fire than may typify aspen forests. High-severity fire, perhaps often as intense surface fire, was likely the dominant type of fire under the HRV in aspen and aspen-conifer forests, based on even-aged aspen after fires in the 1800s in the Rockies (table 6.3) and early records (e.g., F. S. Baker 1918).

Only one study in the Rockies, in stable aspen forests in southwestern Colorado, provides landscape-scale fire history, from stand-origin dating in 76 plots, and the high-severity fire rotation was estimated at about 140 years (Romme et al. 2001). However, this study assumed that aspen cohorts regenerated after fire, but aspen cohorts can also originate from nonfire processes (W. L. Baker, Munroe, and Hessl 1997; Kurzel, Veblen, and Kulakowski 2007). Thus, the fire-driven part of the 140-year rotation may be longer, and I recorded the rotation as more than 140 years (table 6.1).

Light surface fires in aspen were probably uncommon under the HRV. Houston estimated the mean fire interval to have been 20–25 years from conifers on the northern Yellowstone winter range and thought nearby aspen had frequent fire (1973, 1115). However, lack of cross-dating (chap. 5) and the probability that many fires in conifers would have gone out in aspen (Fechner and Barrows 1976) mean that the fire rotation was likely much longer than 20–25 years in aspen. Meinecke (1929) found that about one-third of 240 sample trees in a Utah canyon had fire scars from 1771 to 1903. In the same area, F. S. Baker (1925, 19) found 43 trees with scars from 1863 to 1873 and suggested that "small, light fires occurred at intervals of 7 to 10 years" prior to EuroAmerican settlement. This suggests a fire rotation many times longer than 7–10 years, as fires were small (chap. 5). In contrast, thousands of hectares of aspen lacked fire scars in Montana (D. Lynch 1955), southwestern Colorado (Romme et al.

1996), and northwestern Colorado (Kurzel, Veblen, and Kulakowski 2007). Fire scars were found in only about 5 percent of 140 stands in Colorado and southern Wyoming and affected only about 25 percent of trees in stands with scars (Shepperd 1981). Under the HRV, light surface fires were likely occasionally frequent and small in particular places, but more commonly were absent or rare, probably having a very long fire rotation overall.

Kay (1997) thought ignitions by Indians explained pre-EuroAmerican fires that led to today's abundant aspen forests. He argued that lightning strikes and fires are rare today during times that aspen will readily burn, before May 15 and after September 15, when aspen is leafless and understory plants are cured. However, lightning fires occur and even reach a peak in October in aspen in the southern Rockies (Fechner and Barrows 1976; Ryan 1976). Aspen also commonly have burned in recent summer fires, as at Missionary Ridge (fig. 6.5) and in Yellowstone (Romme et al. 1997). Moreover, available evidence suggests that Indians burned primarily in low-elevation camping and travel areas, not in aspen forests (W. L. Baker 2002). Finally, late-nineteenth-century fires initiated large expanses of aspen forests in a climatic period favoring extensive fire (chap. 2). Similar fires in droughts in the last two decades (table 5.2) have again initiated aspen.

Fire Effects in Stable Aspen

Aspen stems are susceptible to fire because of thin bark (fig. 6.5). Topkill may be nearly complete from surface fires that char the bark less than 0.5 meter high on stems, but trees can also be killed by radiant heat without bark charring (J. K. Brown and DeByle 1987). Mortality of stems may continue for more than five years after low-intensity prescribed fires (J. K. Brown and DeByle 1989). After fire, aspen commonly regenerate by adventitious shoots (suckers) from relatively small (less than 2 centimeters in diameter) underground lateral roots 5–10 centimeters below the surface, but as deep as 20 centimeters if heating kills shallower roots (Schier and Campbell 1978). Low-intensity, spring, prescribed fire led to more suckers than did hotter fall fires in one case (Bartos, Mueggler, and Campbell 1991), but higher sprout density occurred after higher- than after lower-intensity wildfire in another case (Keyser et al. 2005), and no relationship was found in still others (J. K. Brown and DeByle 1989; Bartos, Brown, and Booth 1994).

Aspen were thought to have reduced suckering because of apical dominance by aging stems (Frey et al. 2003), which led to research on regenerating aspen with fire. After wildfires (Patton and Avant 1970; Kay and Wagner 1996; Keyser et al. 2005) or prescribed fires (J. K. Brown and DeByle 1989; Walker

1993; Bartos et al. 1994), sucker densities sometimes initially rose above levels in unburned controls but usually declined to prefire levels or below a few years later (J. K. Brown and DeByle 1989; Bartos et al. 1994; Kay 2001). Sucker density was no different 5–15 years after 12 prescribed fires in northwestern Wyoming (Hessl and Graumlich 2002) or 11 years after the 1988 Yellowstone fires in 23 stands (Ripple et al. 2001) than in controls.

Where sucker density initially rose after fire, native herbivores and livestock often reduced sucker densities and kept stems too short to escape browsing (Walker 1993; Bartos et al. 1994; Kay 2001; Hessl and Graumlich 2002; Keyser et al. 2005). Suckers were significantly taller in some debris piles from jackstrawed conifers killed in fires, because of protection from browsing (Ripple 2001), but a larger study found no net effect (Forester, Anderson, and Turner 2007). Two prescribed burns in northwestern Wyoming, with light herbivory, maintained high sucker density and height through year 6 after fire (Bartos, Mueggler, and Campbell 1991). Other stands with low herbivory have regenerated to tree height after fires since the 1930s (Kleinman 1973; Gruell and Loope 1974). Successful aspen regeneration after fire may occur if herbivory is low, but fires actually hasten aspen decline where herbivory is high (Bartos et al. 1994; Kay 2001).

Landscape and regional stand-age structures for more than 100 stands in several states (table 6.3) indicate stable aspen can be even aged, with regeneration in the first few years after fires or other events (Kurzel, Veblen, and Kulakowski 2007). Stable stands 60 years or older were often uneven aged (Gruell and Loope 1974; Betters and Woods 1981), with evidence of continuous or episodic regeneration. Regeneration may follow episodes of favorable climate, disease or insect damage (Alder 1970; Betters and Woods 1981; Hessl and Graumlich 2002; Kurzel, Veblen, and Kulakowski 2007), or low herbivory (W. L. Baker, Munroe, and Hessl 1997) but can be lacking for decades. For example, 20 percent of 10 stable stands in northwestern Colorado (Kurzel, Veblen, and Kulakowski 2007) and about 20 percent of 91 stable stands in northern Colorado (Kashian, Romme, and Regan 2007) lacked regeneration in the past 60–90 years. Stable aspen in northwestern Colorado commonly regenerated without fire and thus do not appear fire maintained; an alternative process is cohort-senescence dynamics, in which synchronous overstory dieback is followed by even-aged regeneration (Kurzel, Veblen, and Kulakowski 2007).

Understory production increased in Wyoming for up to 12 years after fire in aspen, with greater increases after moderate- and high- than low-severity burns (Bartos et al. 1994). Forbs were most of the production and were initially reduced but were 5–14 percent higher 12 years after, than before, fire (Bartos et al. 1994). In another study, in Idaho and Wyoming, understory production

Table 6.3. Studies of Tree Age Structure and Stand Origins in Piñon–Juniper, Montane Aspen, and Montane Mixed-Conifer Forests

Sources by Scale, Type of Dating, and Forest Type	Location	No. of Trees	Stand Age (yrs)	No. of Stands	No. of Cores per Stand
LANDSCAPE SCALE—STAND AGE					
Piñon-juniper woodlands					
Eisenhart (2004)	W CO	2,319		136	~17
Shinneman (2006); Shinneman & Baker (in press)	W CO	~1,300		28	~40
Tonnesen & Ebersole (1997)	SE CO	385		2	190–195
Stable quaking aspen forests					
Hessl & Graumlich (2002)[a]	NW WY	774		28	25–50
Betters & Woods (1981)[a]	NW CO	500		5	75–150
Quaking aspen forests—status unknown					
Gruell & Loope (1974)	NW WY	245		16	5–34
Northern Rockies mixed-conifer forests					
Tesch (1981)	W MT	125		2	53–72
Arno, Scott, & Hartwell (1995: plots F1, F2)	W MT	244		2	109–135
Arno, Smith, & Krebs (1997: plot L5)	W MT	172		1	172
Upper-montane mixed-conifer forests					
N. T. Korb (2005)	SW MT	?		40?	?
Hadley & Veblen 1993	N CO	1437		7	95–379
Mast (1993: NBH1, NBM1)	N CO	125		2	56–69
Sherriff (2004); Sherriff & Veblen (2006)	N CO	~1,000		9	67–164
Veblen & Lorenz (1986)	N CO	630		8	25–80
Southwestern mixed-conifer forests					
Wu (1999)	SC CO	>500?		9	>40?
Touchan, Allen, & Swetnam (1996)—successional aspen	N NM	183		4	~45
REGIONAL SCALE—STAND AGE					
Stable quaking aspen forests					
Kurzel, Veblen, & Kulakowski (2007)	NW CO	508		10	41–63
Alder (1970)[a]	UT, AZ	3,168		44	144

Upper-montane mixed-conifer forests					
Kurzel, Veblen, & Kulakowski (2007)—successional aspen	NW CO	107		2	47–60
Southwestern mixed-conifer forests					
Kleinman (1973)—successional aspen	N UT	1,280		6	160–240
LANDSCAPE SCALE–STAND ORIGIN					
Piñon-juniper woodlands					
Eisenhart (2004)	W CO		40 to >700	136	~17
Shinneman (2006)	W CO		50 to >700	56	20–40
Stable quaking aspen forests					
Hessl & Graumlich (2002)[a]	NW WY		95–165	29	25–50
Kulakowski, Veblen, & Kurzel (2006)	NW CO		100–200	12	15–21
Kashian, Romme, & Regan (2007)	N CO		<150	27	8
A. E. Smith & Smith (2005)	W CO		<60 to >140	9	~11
Romme et al. (2001)	SW CO		120 to > 240	76	20
Quaking aspen forests—unknown status					
Krebill (1972)	NW WY		<40 to >160	100	2
E. J. Larsen & Ripple (2003)	NW WY, MT		10 to >140	180	≤9
Ripple & Larsen (2000)—Warren plots	Yellowstone		10–160	20	1
Ripple & Larsen (2000)—their own data	Yellowstone		20–190	57	<2
Romme et al. (1995)	Yellowstone		120–170	15	5–10
W. L. Baker, Munroe, & Hessl (1997)	Rocky Mt. NP		17–115	23	5–10
Upper-montane mixed-conifer forests					
Kulakowski, Veblen, & Kurzel (2006)—successional aspen	NW CO		100–200	9	15–21
Kashian, Romme, & Regan (2007)—successional aspen	N CO		<150	24	8
A. E. Smith & Smith (2005)—successional aspen	W CO		<60 to >140	22	~11

Table 6.3. Continued

Sources by Scale, Type of Dating, and Forest Type	Location	No. of Trees	Stand Age (yrs)	No. of Stands	No. of Cores per Stand
REGIONAL SCALE–STAND ORIGIN					
Stable quaking aspen forests					
Shepperd (1990)[a]	CO, S WY		<40 to >150	140	?
Quaking aspen forests—unknown status					
Mueggler (1989)[b]	UT, ID, WY		<30 to >200	713	3

Notes: Studies were conducted in or near the Rocky Mountains. Stand ages are approximate, as ages are seldom provided. Type of dating varies. A stand-age distribution is a complete sample (usually truncated above a minimum tree size) of tree ages within a plot, with multiple plots across a landscape (areas up to thousands of ha) or a region (areas up to millions of ha). A stand-origin distribution, in contrast, is a sample of only the suspected oldest trees within a plot, with multiple plots across a landscape or region. Stand-age distributions allow estimation of within-stand processes affecting tree regeneration (e.g., whether the stand is uneven aged or even aged). Studies are arranged from north to south within each category. For aspen forests, "stable" refers only to stands generally lacking conifers and thus unlikely to be successional to conifer forests, not whether suckers are sufficient to lead to long-term persistence. "Status unknown" means the author(s) did not determine whether a stand was stable or successional.

[a] Author(s) did not often identify whether stands were stable or successional; this is my determination based on the data presented in the study and is thus tentative.

[b] Author(s) did not distinguish montane and subalpine stands, but only part of the total sample is montane.

increased the year after fire and roughly doubled the first 5 years after fire, mostly due to forbs (J. K. Brown and DeByle 1989). Fire-stimulated forbs, with seed banks or deep roots, included fireweed (*Chamerion angustifolium*), streambank wild hollyhock (*Iliamna rivularis*), and American dragonhead (*Dracocephalum parviflorum*). These plants greatly increased and were a substantial fraction of forb production in some stands 5–12 years after fire, while other forbs had not returned to prefire levels (J. K. Brown and DeByle 1989; Bartos et al. 1994). These authors found shrubs were generally reduced during the first 5–6 years or even up to 12 years after fire.

Land-Use Effects and Restoration in Stable Aspen

Livestock grazing reduced fine fuels in some aspen forests in the late 1800s and early 1900s (F. S. Baker 1925) and can still nearly eliminate ignition and spread (J. K. Brown and Simmerman 1986). However, aspen burns poorly even if fine fuels are abundant, and evidence is lacking that loss of fine fuels has significantly decreased fire in aspen over large areas.

Some have predicted the decline and eventual demise of aspen, arguing that the tree depends on disturbance to avoid conifer dominance and to regenerate successfully (Kay 1997; Bartos and Campbell 1998). However, this is relevant only for successional aspen, discussed later in this chapter and in chapter 8. Stable aspen can regenerate without high-severity disturbance, documented by the uneven age structures discussed previously, and is not generally being replaced by conifers, based on repeat mapping, repeat photography, and age-structure analysis (Romme et al. 2001; Manier and Laven 2002; Kulakowski, Veblen, and Drinkwater 2004; Kulakowski, Veblen, and Kurzel 2006; Zier and Baker 2006; Kashian et al. 2007; Kurzel, Veblen, and Kulakowski 2007; table 6.3). For example, 78 percent of 42,600 hectares that were unburned and aspen dominated in 1898 in northwestern Colorado remained so in 1998, reflecting stable aspen, and 85 percent of 11,600 hectares that were burned aspen in 1898 remained aspen dominated in 1998 (Kulakowski, Veblen, and Drinkwater 2004).

Moreover, the notion that disturbance has declined in aspen is premature. Fires burned little area in aspen forests over a 13-year period (DeByle, Bevins, and Fischer 1987), but that is not enough time to reliably assess change (chap. 10), as the aspen fire rotation exceeded 140 years (Romme et al. 2001). Much of a rotation occurs in episodes of large fires at intervals near the rotation (box 3.1). Between episodes, the rotation normally appears quite long. Reliably detecting change in rotation in aspen requires more than a century of data, not one to two decades. It is thus too early to suggest a need for harvesting or burning aspen forests to offset declining disturbance.

MOIST NORTHWESTERN FORESTS

Moist northwestern forests occupy limited areas in the northern Rockies where a maritime-influenced climate is found. These forests naturally can have relatively high fuel loads for the Rockies yet occur where summers can be quite dry, leading to the most explosive potential fire behavior in the Rockies. These forests have been significantly altered by land uses and introduced disease and are in need of protection and restoration.

Fire Behavior and Fire History

Moist northwestern forests are subject to an unusual fire hazard because of high fuel loads and summer drought (Daubenmire and Daubenmire 1968; Shiplett and Neuenschwander 1994). Explosive fire behavior in these forests is illustrated by the 1967 Sundance fire in northern Idaho, which burned 22,300 hectares in nine hours, about 2,475 hectares per hour (Anderson 1968), and the 1931 Freeman Lake fire, which burned more than 8,100 hectares in 12.5 hours, about 650 hectares per hour (Jemison 1932). The regional summaries in table 3.1 have more on fire history, fuels, fire effects, and succession.

These forests vary substantially in tree composition, moisture conditions, and fuel loads from drier grand fir forests with comparatively light fuels to moister western red cedar and western hemlock forests with heavy fuels. Trees were rather dense in grand fir forests around 1900, varying from 2,500–7,500 trees per hectare in young forests to 500–750 trees per hectare in forests about 100–150 years old, but more open forests also occurred on rocky sites (Leiberg 1900b). The understory was commonly shrubby and had substantial litter and fallen, decaying trees. Mature forests on wettest sites had 750–1,250 trees per hectare. Leiberg (1897, 59) described western white pine forests in 1895 in northern Idaho (text in brackets added by W. L. Baker):

> The stand of forests is very close; there is a vast amount of vegetable débris, decaying trees, fresh and old windfalls piled upon one another, broken-off tree tops, and young trees bent over by the snow and forming impenetrable thickets. Very little grass, more often none at all, grows on the ground, which is heavily covered with a humus reeking with moisture and topped off with a growth of mosses and liverworts. Multitudes of fungi are everywhere . . . the number of trees per acre is always considerable, but varies widely. A fair estimate per acre for the bottoms of the canyons would be 600 to 700 trees [1,500–1,750 trees per hectare], with diameters from 25 cm. to 60 cm. (10 to 24 inches), and 2,000 to 3,000 trees [5,000–7,500 trees per hectare], with di-

ameters from 15 cm. to 25 cm. (6 to 10 inches); of saplings there are often tens of thousands . . .

These forests were subject to high-severity fires during episodic droughts, as well as infrequent small, lower-severity fires (Shiplett and Neuenschwander 1994). Fire severity may have been highest on wind-exposed slopes and in mid-elevation thermal belts from temperature inversions, with severity lower and fires patchier on moister north-facing slopes and stream bottoms (Arno and Davis 1980; Antos and Habeck 1981). Moist stands may have been fire breaks at times, with fires burning into the edge and scarring a few trees before going out (W. C. Fischer and Bradley 1987).

High-severity fires occurred at moderate rotations of from one to two centuries in dry grand fir forests to two to three centuries in upland western red cedar and western hemlock forests (table 6.1). Wettest western red cedar–western hemlock forests along streams and near seeps likely burned in high-severity fires at long intervals (e.g., greater than 500 years; Arno and Davis 1980). McCune (1983) suggested a short rotation, based on an atypical series of high-severity fires (chap. 5). Low-severity fires occurred but were likely infrequent and patchy because of moist fuels (references in table 6.1). Western white pine may commonly have been killed by smoldering combustion in deep duff after a flaming front passed (Leiberg 1899b). The landscape at any instant may have been dominated by forests from a few large, high-severity fires (Barrett, Arno, and Key 1991; Barrett 2000) or from patchy smaller fires (Arno and Davis 1980).

Fire Effects and Succession

Plant composition after fire appears to be set by *initial floristics*—that is, it is shaped by surviving plants on site and seeds that arrive quickly from unburned areas (McCune and Allen 1985; Stickney 1986; Shiplett and Neuenschwander 1994). H. B. Humphrey and J. E. Weaver (1915) listed early postfire plants, including trees, two years after the 1910 fires. Stickney (1986) monitored 18 plots for a decade after the 1967 high-severity Sundance fire in northern Idaho. He found 103 plant species the year after fire; 84 were still present at year 10, when 139 were found. Of the 139, 46 percent had survived and resprouted, 12 percent had originated from on-site seeds, 20 percent had colonized the first year from outside the burn, and 22 percent had colonized later. The first decade of postfire succession was thus shaped by species mostly present the first year after fire.

Stickney (1986) also found 63–76 percent of the cover came from only four fire-stimulated plants: fireweed, western brackenfern (*Pteridium aquilinum*),

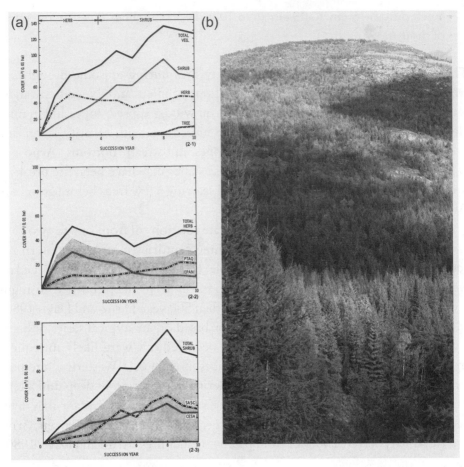

Figure 6.6. Succession after the 1967 Sundance fire in northern Idaho. This was a high-severity fire in moist northwestern mixed-conifer forests. The graphs in part (a) show a composite pattern of development of vegetation in 18 plots during the first decade: (top) total vegetation and life forms, (middle) major herbs (EPAN = fireweed, PTAQ = western brackenfern), and (bottom) major shrubs (SASC = Scouler's willow, CESA = redstem ceanothus); the shaded area indicates the cover contributed by these four major species. The photo in part (b) shows that by year 39 after the fire trees had become quite dense.
Sources: (a) is reproduced from Stickney (1986, p. 8, fig. 2); (b) photo is by the author.

redstem ceanothus, and Scouler's willow. Fireweed and brackenfern dominated the first five years and the two shrubs years 5–10 (fig. 6.6a). However, forbs dominated the first decade on 4 of 18 plots, and shrubs dominated the first decade on 3 of 18 plots, illustrating the diversity possible in postfire succession.

High-severity fires in these forests can leave some surviving trees, particularly fire-resistant western larch on uplands and western red cedar in wet areas (Marshall 1928), that are key sources of postfire tree regeneration. However, seed dispersal over more than a kilometer also led to seedlings in early postfire years (J.A. Larsen 1925b). Trees were quite dense by year 39 after the 1967 Sundance fire (fig. 6.6b), even though they were sparse the first decade after the fire, except on 2 of 18 sites, where lodgepole pine germinated the year after fire (Stickney 1986). However, early tree regeneration can be quite dense. Leiberg (1900b) estimated 2.5 million seedlings per hectare of western hemlock two to three years after fire on a moist site in northern Idaho, and Daubenmire and Daubenmire (1968) emphasized the high density of postfire trees typical of the first one to two decades after high-severity fires in these forests.

Lodgepole pine can dominate large areas after fire, but western larch also can if lodgepole seed is locally rare (H. E. Humphrey and Weaver 1915) or fire recurs before lodgepole reaches seed production (Leiberg 1899b; Antos and Habeck 1981). Western white pine or Douglas-fir may regenerate soon after high-severity fires and eventually overtop lodgepole, which appears to persist for only 100–150 years (Leiberg 1899b; J. A. Larsen 1929; Huberman 1935; McCune and Allen 1985). Initially even-aged western white pine can become uneven aged because of small high-severity fires and mortality of young trees in surface fires (Marshall 1928; Rapraeger 1936), patterns and effects that are also found in western red cedar and western hemlock forests (Arno and Davis 1980).

If fire reburns within a few decades, before lodgepole produces much serotinous seed, a semipersistent brushfield of native shrubs may result (J. A. Larsen 1925b; Antos and Habeck 1981; Shiplett and Neuenschwander 1994). Brushfields favor elk, and wildlife managers have used burns every one to two decades to maintain young shrubs and discourage trees (Leege 1979). Shade-tolerant trees (e.g., grand fir) can invade and eventually overtop the brushfields if fire does not recur (Antos and Habeck 1981). Shade-tolerant western red cedar, western hemlock, and grand fir dominate old-growth forests (J.A. Larsen 1929) 300 or more years after fire (Huberman 1935). Understory diversity may slowly decline into late succession (Habeck 1968).

Land-Use Effects and Restoration

High-severity fires burned extensive areas in the northern Rockies in 1910–36 (Habeck 1970; S. Cohen and Miller 1978; Barrett, Arno, and Key 1991; Barrett 2000) and the late 1800s, particularly in western white pine forests (Leiberg 1899a,b). Leiberg (1897, 61) was concerned about damaging fires set by settlers along railroads and in careless logging in northern Idaho: "The forests of the Coeur d'Alenes in all the accessible portions are becoming mere skeletons of their former state, and soon the last vestiges will be swept away and nothing remain but blackened logs and stumps to mark the former site of the densest forest between the Cascades and the Mississippi." Human-set fires, such as those that concerned Leiberg, may have added substantially to area burned in the late 1800s. Nonetheless, similar large, high-severity fires, including some short-interval reburns, occurred under the HRV (Marshall 1928; Habeck 1970; Antos and Habeck 1981; Barrett, Arno, and Key 1991).

Some studies have suggested that changes in these moist northwestern forests (e.g., increased fuel loads, spruce budworm outbreaks) over the last century were a result of fire exclusion (Habeck and Mutch 1973; McCune 1983). Fires may have been excluded, but the high-severity fire rotation in these forests was likely moderately long (table 6.1), and recent changes are most likely normal fluctuations in interludes between infrequent high-severity fires typical of the HRV (box 3.1).

These forests have suffered from past logging and burning, as well as fragmentation by roads, introduced diseases, and other impacts. Old-growth stands have been depleted by logging, and western white pine has also been decimated by the nonnative disease blister rust (Harvey et al. 2008). Thus, the primary conservation needs, relative to the fire regime, are to manage blister rust, restore additional areas of old-growth forests, and protect the few remaining examples of old-growth forest from high-severity fires until larger areas of old growth are restored and the fire regime can be allowed to function with some relationship to the HRV.

NORTHERN ROCKIES MIXED-CONIFER FORESTS

These mixed-conifer forests in the northern Rockies include Douglas-fir, western larch, lodgepole pine, and lesser amounts of other trees and are confined to places where western larch is found. Western larch is one of the most fire-resistant trees in the Rockies (chap. 2), giving these forests a distinctive fire ecology.

Fire Behavior and Fire History

These mixed-conifer forests in the northern Rockies historically burned in high-severity fires at about 150- to 250-year intervals (table 6.1), but fires could also be mixed in severity (Habeck 1970; Arno, Parsons, and Keane 2000). Sneck (1977) found that most fires in her 2,984-hectare study area over several centuries were small (6–190 hectares) with only portions burning at high intensity, a mixed-severity pattern that may especially occur in small, moist valleys. Small fires of low or moderate severity led to a complex age-class mosaic at times in

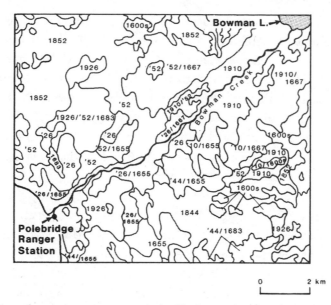

Figure 6.7. Complex age-structure mosaic in Glacier National Park, Montana. Dates indicate single-aged or multiple-aged (underburned) stands reconstructed in 1987.
Source: Reproduced from Barrett et al. (1991, fig. 7a) with permission of the National Research Council, Canada.

Glacier National Park (fig. 6.7; Barrett, Arno, and Key 1991). Surface fires were common but severe enough at times to initiate new age classes (Tesch 1981). A stand studied by Arno et al. (1997) had western larch dating from the early 1500s but also a pulse of Douglas-fir after a fire about 1800 and a pulse of lodgepole pine after fires in about 1844 and 1859, in each case with the old western larch surviving. Arno et al. (1997) interpreted this stand as having had frequent low-severity fires. However, most of the extant trees, other than the scattered old larch, regenerated after three fires between 1800 and 1900, suggesting that moderate-severity fire initiated new age classes, as found by Tesch (1981). Gabriel (1976) found that one part of his study area had an intricate mosaic of age classes, but another had a simpler pattern because of large fires in 1889, 1910, and 1919 that replaced the mosaic with large patches of lodgepole. Similarly, Habeck (1970) observed replacement of a forest of old western larch and understory Douglas-fir by lodgepole pine after a 1929 fire. Fluctuations in pattern and composition typify these landscapes with variable-severity fire and with regeneration contingent on patterns of survival, as discussed next.

Fire Effects and Spatially Contingent Succession

Low-severity fires in these mixed-conifer forests could kill lodgepole pine, leaving surviving western larch and facilitating Douglas-fir regeneration. Ayres

(1900b, 50) noted that in the Swan River Valley in Montana "a mixed stock of larch and lodgepole pine had been run through by light fires, which killed the thin-barked lodgepole pine, but left the thick-barked larch but slightly injured." Ayres (1900a, 314) also observed that on the Flathead River in Montana, "where a fire had run through the red fir [Douglas-fir] and larch about 25 years ago, killing and injuring very few trees, young red fir is found as an undergrowth, very irregular in distribution, but in places forming thickets. These young trees are about 10 feet high and 1 to 2 inches in diameter, and frequently stand 10,000 to the acre [25,000 trees per hectare]."

After high-severity fires, tree regeneration is typically dominated by Douglas-fir, western larch, and lodgepole pine, which may develop into a dense, even-aged, closed forest in about 50 years, possibly followed by renewed regeneration as the canopy opens and lodgepole declines, or after low- to moderate-severity fires that partially open the canopy (Tesch 1981). Further small disturbances and overstory mortality can enhance an uneven age structure by facilitating further regeneration. Lodgepole pine may decline substantially by the time old age is reached (Whitford 1905), but can persist as a component of old-growth forests (Tesch 1981). This predictable successional pattern may be a generalization, however.

Instead, postfire tree regeneration represents *spatially contingent succession*, whose course depends on the composition of the forest around a burn, burn size, the environmental setting of the burn, and time since last fire. This contingency was recognized early in North American ecology by Whitford (1905). He showed photographs of three variants of northern Rockies mixed-conifer forests in Montana's Flathead Valley, with differing early postfire tree composition consistently similar to each nearby unburned forest. He also noted that if a large reburn occurred within 15 years, with only lodgepole pine old enough to produce seed, lodgepole might expand to dominate large areas, as he observed and as was noted in the same area by Ayres (1900a). Similar contingency was observed in these forests after the 1977 Pattee Canyon fire in western Montana (Toth 1992).

Early postfire (i.e., years 1 and 2 after high-severity fire) composition of graminoids, forbs, and shrubs may be less spatially contingent, however; based on study of the 1977 Pattee Canyon fire, it may instead be shaped primarily by resprouting and regrowth of surviving plants, with off-site seed a minor source (Crane, Habeck, and Fischer 1983). Annuals were most abundant, but never common, the first spring after fire. Shrubs began resprouting within weeks and dominated burned ravines by year 5, but uplands stayed forb dominated. The most abundant native postfire plant was fire-stimulated pinegrass, a rhizomatous grass that typically recovers and expands by the first year after fire (appendix C). At Pattee Canyon, it flowered abundantly the first summer after fire, leading to new seedlings in successive years (Crane, Habeck, and Fischer 1983). Fireweed is also common after high-severity fires in these forests (Whitford 1905; Habeck 1970; Crane, Habeck, and Fischer 1983).

Land-Use Effects and Restoration

The high-severity component of the fire regime, at 150- to 250-year intervals, has probably been little affected by fire exclusion, and recent fires (e.g., the 2001 Moose fire in northwestern Montana) probably were similar to fires under the HRV (Barrett 2002). It is unclear whether fire exclusion affected low-severity fires, as the rotation for low-severity fires is unknown. However, surface fires in these forests produced complex responses. They did not simply kill young trees, leading to an open forest, but could encourage dense Douglas-fir regeneration. Prescribed burns to lower understory tree density would thus be misdirected, as they could be followed by renewed tree regeneration. Passive restoration, such as wildland fire use (chap. 11), which may allow these forests to remain within the HRV, may be the most effective restoration approach. Severe fires will occur, as Barrett (2002, 44) suggested: "To me, the question is not how to prevent large fires, but how society will adapt to that dominant, inevitable force."

UPPER-MONTANE MIXED-CONIFER FORESTS

Upper-montane mixed-conifer forests occur from western Montana to central Colorado, generally above the ponderosa pine zone, and are dominated by Douglas-fir, lodgepole pine, limber pine, and quaking aspen, with minor amounts of other trees. As in other mixed-conifer forests, the diversity of trees leads to potentially diverse postfire tree dynamics.

Fire Behavior and Fire History

The fire regime in upper-montane mixed-conifer forests was variable but included large high-severity fires under the HRV. Leiberg (1904a, 27), referring to both subalpine and upper montane forests in the Yellowstone area, said, "The forest fires in this region are remarkable for their destructive force and intensity. Here and there are uneven aged stands, where extremes in age and a mixed composition prove that occasionally the fires did not consume or kill the entire stand. But as a rule most of the older fires made a clean sweep, and in nearly every instance the fires of modern date have done the same." He described

Figure 6.8. Upper-montane mixed-conifer forest likely burned in the 1870s. The location is in the Colorado Front Range, and the photograph illustrates the aftermath of (a) high-severity fire with scattered surviving trees, (b) a surviving island of trees, (c) severe surface fire with surviving trees and tree groups, and (d) a spot fire.
Source: William Henry Jackson, photo number 1232, taken in 1873; courtesy of the U.S. Geological Survey Photographic Library, Denver.

high-severity fires about 1,000–4,000 hectares in area in upper-montane mixed-conifer forests in several townships. In Colorado's Front Range, William Henry Jackson photographed an 1870s mixed-severity fire with mostly high severity but small areas of possible low-severity fire, some moderate-severity fire, and unburned islands (fig. 6.8). Upper-montane mixed-conifer forests often occur in exposed settings, where gusty and variable winds and spotting can lead to erratic, sudden high-severity fire, as killed 15 firefighters in the 1937 Blackwater fire in Wyoming (A. A. Brown 2003).

In a 3,400-hectare study area in southwestern Montana, only two fires burned extensive area over about 150 years in the pre-EuroAmerican era, and a total of 22 high-severity patches varying from 1 to 37 hectares (mean about 30 hectares) in area were found (Korb 2005). High-severity fires left surviving trees, averaging about one per 1.3 hectares. Recent fires in these forests in northwestern Wyoming were similar, if somewhat more severe, and mostly classified as moderate (40–90 percent canopy mortality) to moderately severe (greater than 90 percent mortality), but only 21 percent of prefire mixed-conifer forests burned at high severity (Doyle 1997). In Korb's (2005) study area, low-severity fire was part of the two large mixed-severity fires, and other low-severity fires were small and patchy. Sherriff (2004) found about half of fires in nine plots in northern Colorado mixed-conifer forests were low severity, and about half were moderate or severe, with some canopy replacement. All plots had moderate- or high-severity fire at some point. Similarly, age structure in six of seven stands in Rocky Mountain National Park was associated with high-severity fires in the mid to late 1800s (Hadley and Veblen 1993). Most fires in Montana occurred late in summer (Korb 2005).

In southwestern Montana, about 27 percent of sampled plots in 1876 had a high-severity fire in the previous 26 years, and about 11 percent had a high-severity fire between 1750 and 1850 (Korb 2005). I used these data with equation 5.1 to estimate the high-severity rotation to be about 117 years in 1750–1876, or about 186 years in 1750–1950, the latter date approximately when fire suppression became more effective. This estimate fits the age structure Leiberg (1904a, 22) observed in the Yellowstone area: "Of the entire forest below the subalpine zone 10 per cent is less than 50 years old, 50 per cent more than 50 and less than 120 years, while the remaining 40 per cent comprises veteran stands from 120 to 300 years of age."

Fire Effects and Succession

The first two decades of postfire tree succession in northwestern Wyoming in upper-montane mixed-conifer forests were studied using 14 plots in five fires between 1974 and 1991 (Doyle 1997, 2004). Postfire Douglas-fir seedling density in burned forests was mostly low (less than 0.05 per square meter, or 500 trees per hectare), but about one-third of plots had higher densities (fig. 6.9), particularly where fire severity was low and trees survived nearby. Prefire aspen regenerated to aspen. Prefire stands with some lodgepole had postfire Douglas-fir at low density but were dominated by lodgepole (fig. 6.9), especially in higher-severity areas far from surviving Douglas-firs. These fire effects (severity, distance to surviving trees) were more important in mixed-conifer than subalpine forests and explained the diversity of postfire forests. Postfire tree regeneration may also be complex because of interactions such as competition and facilitation. For example, Douglas-fir survival was increased by shading by limber pine (Baumeister and Callaway 2006).

Forests with Douglas-fir, lodgepole pine, and other conifers in mixtures commonly were replaced by extensive lodgepole pine forests with scattered surviving Douglas-firs after large fires in the 1800s. For example, in 1910 on part of the Targhee National Forest, an early stand exam reported: "Lodgepole

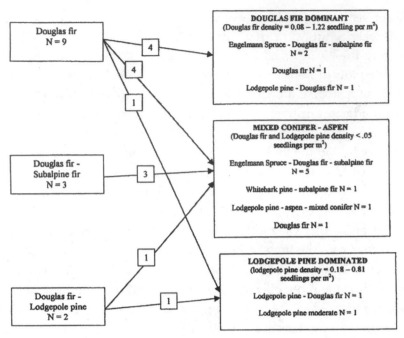

Figure 6.9. Possible postfire tree composition in upper-montane mixed-conifer forests. Shown is possible postfire tree composition (right column) resulting from fires in three types of prefire upper-montane mixed-conifer forests (left column) in northwestern Wyoming. Small boxes and N indicate number of plots.
Source: Reproduced from Doyle (2004, p. 253, fig, 11.3), copyright 2004, with permission of Yale University Press.

occupies the major part of the land occurring usually in immature stands. Scattered through these lodgepole stands . . . are usually a few very large Douglas-firs, the survivors of many forest fires" (Ogle and DuMond 1997, 102). Peirce (1915) observed this pattern in the Bighorn National Forest in northeastern Wyoming, and Bates (1917) noted it throughout Wyoming and Colorado. Lodgepole pine also appeared about 1900 to be taking over Douglas-fir lands in the Payette, Sawtooth, and Targhee National Forests in Idaho, but there was recognition that Douglas-fir would regenerate and eventually overtop the lodgepole (Ogle and DuMond 1997). Leiberg (1904a, 22) observed in the Yellowstone area, "As the lodgepole pine reaches maturity and the stands become more open through natural thinning it will, in course of time, be largely displaced by red fir [Douglas-fir] and spruce." In Rocky Mountain National Park, a stand burned by high-severity fire about 250 years ago initially had lodgepole pine, but Douglas-fir and limber pine were slowly replacing it (Hadley and Veblen 1993).

Lags or failures to regenerate also occurred after fires in upper-montane mixed-conifer forests. Peirce (1915) observed no regeneration of mixed conifers on south-facing, limestone-derived soils about 35 years after high-severity fires in the Bighorn Mountains, although lodgepole was regenerating slowly in places. Henderson (1997) studied 20 stands burned in high-severity fires in a chronosequence in southwestern Montana and found maximum tree density in midseral stands, but herbaceous and shrub cover gradually increased toward mature stands, while graminoid cover declined.

Land-Use Effects and Restoration

Korb (2005) suggested that a 2003 wildfire that burned 5,600 hectares near his southwestern Montana study area was unprecedented relative to historical variability and more severe than past fires because of fire exclusion, as it had about 37 percent high-severity fire, the largest patch about 324 hectares, and no surviving trees within severely burned areas. Korb's conclusion is not warranted, however, as his data document only two large fires prior to EuroAmerican settlement, an insufficient sample for characterizing historical variability in large fires. Moreover, a 1906 survey of the Red Rock Range near Korb's study area described these mixed-conifer forests: "The original forest has been largely destroyed by fire and large areas are now occupied by lodgepole pine and by chaparral [snowbush ceanothus] and popple [quaking aspen]" (Ogle and Du-Mond 1997, 97). Large areas of mixed-conifer forest in this area had burned in high-severity fire. In northern Colorado, Sherriff (2004) found all sampled mixed-conifer forests had high-severity fires and fire-free intervals of 50 to over 100 years in the pre-EuroAmerican era, so that modern fire-free periods are not unprecedented. She suggested that tree thinning would not be needed as part of restoration in these forests and would also likely not lower the risk of high-severity fires. As in other mixed-conifer forests with a substantial component of high-severity fire, the potential effects of fire exclusion are likely mild and do not require active restoration; increased wildland fire use (chap. 11) is likely sufficient to maintain these forests within the HRV.

SOUTHWESTERN MIXED-CONIFER FORESTS

Southwestern mixed-conifer forests are the most southern in the Rockies, occurring from northeastern Utah to southern Colorado and northern New Mexico, where they are characterized by diverse mixtures of white fir, Douglas-fir, ponderosa pine, quaking aspen, blue spruce, and southwestern white pine.

This relatively high tree diversity leads to complex fire-related dynamics that remain relatively poorly understood.

Fire Behavior and Fire History

Fires were most common in southwestern mixed-conifer forests under the HRV between April and early July, before the North American monsoon (Wu 1999). Fires may be less frequent but larger in these forests than in lower-elevation forests, because moist fuels limit fire spread except in droughts (Touchan, Allen, and Swetnam 1996; Wu 1999; Margolis, Swetnam, and Allen 2007). Fires occurred at several locations in 1748, 1851, 1879, 1880, and other known regional drought years (Wu 1999; Grissino-Mayer et al. 2004; Margolis, Swetnam, and Allen 2007), with no detected effect of prior wet years or El Niño–Southern Oscillation (Margolis, Swetnam, and Allen 2007). Fires in some drought years burned in summer, perhaps because the monsoon was delayed or failed (Wu 1999; Grissino-Mayer et al. 2004).

Age structures suggest that the fire regime was variable in severity. In southern Colorado, some stands were even aged, suggesting high-severity fire, others were uneven aged, suggesting lower-severity fire, including patchy high-severity fire with surviving trees or groups (Touchan, Allen, and Swetnam 1996; Wu 1999). However, all-aged stands may have originally been even aged after high-severity fires in 1748, as an all-aged structure may require 200 years or more to fully develop. Both aspen and ponderosa regeneration was limited to the first two decades after fire, also suggesting that canopy-opening high-severity fire occurred (Wu 1999). However, purer aspen patches in these landscapes may have regenerated for as much as 60 years after high-severity fires (Touchan, Allen, and Swetnam 1996), although some regeneration could have followed smaller high-severity or surface fires. Shade-tolerant white fir and blue spruce regenerated after fires and also in extended fire-free periods (Wu 1999). These trees are vulnerable to mortality in surface fires because of dense foliage near the ground; they also are ladders that allow surface fire to become stand-replacement fire, which historically occurred in all sampled forests (Wu 1999).

Fire rotation has not been estimated for these forests, but a first approximation is possible. Wu (1999) found that her nine sites burned in high-severity fires from 1748 to 1880. Use of equation 5.3 indicates that the high-severity rotation would be about 130 years using these two dates, or about 210 years if the period extends to about 1960, when fire control became more effective (chap. 10). Charcoal records from two bogs in northern New Mexico near southwestern mixed-conifer forests (Brunner Jass 1999; Allen et al. 2008) found a fire-episode frequency of 111–222 years over the last 400–500 years (table 8.3), which is compatible with my approximate fire rotation estimates.

Succession from Aspen to Mixed Conifers

Aspen commonly dominates after large high-severity fires or smaller canopy-opening fires if prefire aspen was present (G. A. Pearson 1914). F. S. Baker (1925) thought about 250 years would be required in northern Utah for conifers to regain dominance over aspen after high-severity fires. White fir and aspen were codominant in a stand 82 years after fire (Kleinman 1973), and white fir had two-thirds as many stems as aspen 113 years after fire in the same area (Mueggler 1994). A rephotograph of this area shows, in the mixed-conifer portion of the photo, that conifers increased in the elapsed 83 years but did not fully dominate (fig. 6.4); several more decades (to about 150 years after fire) may be required for conifers to dominate. At six sites in northern New Mexico and southern Colorado, aspen still dominated with just understory conifers, about 125–155 years after high-severity fire (Margolis, Swetnam, and Allen 2007). In Colorado's San Juan Mountains, aspen was absent from two approximately 250-year-old, dry-site mixed-conifer forests but present in three wet mixed-conifer stands burned in high-severity fires in 1748–95 (thus about 200–250 years postfire) and was still 15–20 percent of tree importance in two of the three (Wu 1999). In the montane zone in the San Juans, rephotography showed that about half of 20 scenes of mature aspen-conifer stands (e.g., perhaps a century old at the time), taken near 1900, showed increased conifers a century later, but about half had no change in aspen-conifer proportions (Zier and Baker 2006). The evidence suggests that 150–300 years are needed for conifers to regain dominance after high-severity fires in these mixed-conifer forests, but semistable or perhaps stable mixtures of aspen and conifers can persist for long periods.

Where aspen is uncommon after high-severity fires, Gambel oak or other shrubs of the mixed-mountain shrub zone may dominate, or the site may remain rather bare, with little tree regeneration for some time. Large patches of shrub dominance and lags in tree regeneration may be within the HRV, based on these early observations (G. A. Pearson 1914).

Land-Use Effects and Restoration

Wu (1999) and Grissino-Mayer et al. (2004) both suggested that fire exclusion led to changes, including fuel buildup, in these forests. Grissino-Mayer et al. emphasized that more than 100 years of fuel buildup will lead to higher-intensity fires and cited the 2002 Missionary Ridge fire as a manifestation. Wu suggested that ponderosa pine and aspen had unnaturally low populations in wetter mixed-conifer forests, and that ponderosa was not regenerating and white fir was increasing in dry mixed-conifer forests. However, these authors present no fuels data, and fuel buildup between fires is not a consistent trend in Rocky Mountain forests (chap. 2). Also, both Grissino-Mayer et al. and Wu overlooked that all Wu's plots historically burned in high-severity fires, as well as fires of lower severity, suggesting that a mixed-severity fire with some large areas of high severity, similar to the Missionary Ridge fire of 2002, would be within the HRV in these forests. Recently, Margolis et al. (2007) found greater than 1,000-hectare patches of successional aspen in mixed-conifer landscapes in southern Colorado and northern New Mexico dating to high-severity fires in drought years prior to widespread EuroAmerican settlement. Together the studies of Wu and Margolis et al. show that infrequent, large mixed- or high-severity fires characterized the HRV in southwestern mixed-conifer forests.

Fire exclusion is commonly cited to explain observed changes in forests without sufficient evidence that observed trends are outside the HRV. For example, surface fire decreased in Wu's sampled mixed-conifer forests in the twentieth century, perhaps allowing high initial postfire tree density to persist and some shade-tolerant trees to regenerate, trends commonly thought to have resulted from fire exclusion (Wu 1999; Savage and Mast 2005). Similarly, Savage and Mast (2005) found that about 45 years after the 1956 Ocate fire the density of ponderosa pine in mixed-conifer forests in northern New Mexico was 320 trees per hectare, which these authors thought was exceptionally high due to fire exclusion. However, no evidence (e.g., reconstruction) was presented about young postfire mixed-conifer stands under the HRV, so it is impossible to conclude that current tree density in the Ocate fire area is outside the HRV. Shade-tolerant trees were present in the understory at times under the HRV, since they reached the canopy on some sites and at certain times (Wu 1999). If surface fires under the HRV were periodic and thinned regeneration, the shade-tolerant trees that reached the canopy may have regenerated initially at high density. Observed twentieth-century changes in these forests more likely are typical of normal interludes between infrequent, large high-severity fires, not an effect of fire exclusion (see chap. 10), but more research is needed on forest structure in these forests under the HRV.

Scientific reconstructions of tree density under the HRV are needed before any forest-structure restoration actions are taken, but some actions are clearly not warranted. Lowering fire risk by extensive thinning or fuel reduction is inappropriate, since high- and mixed-severity fires occurred historically in these forests and little is known about tree density or fuel loads under the HRV. Removal of most shade-tolerant trees is not appropriate, as shade-tolerant trees likely at times provided the ladder fuels that facilitated the large high- and mixed-severity fires known to have occurred under the HRV. Large postfire shrub fields and lags in tree regeneration likely also occurred in these forests under the HRV, based on early observations (G. A. Pearson 1914), and should not automatically lead to tree replanting efforts.

FIRE EXCLUSION AND ASPEN DECLINE IN MIXED-CONIFER AND SUBALPINE FORESTS

In studies of recent change in aspen and conifers in mixed-conifer and spruce-fir forests (chap. 8), conifers are often observed to have increased and aspen to have decreased (e.g., Harniss and Harper 1982; Mueggler 1994). The pattern by itself provides no evidence about causes, which may include (1) natural recovery after high-severity fire; (2) conifers favored by early overgrazing, logging, or fuelwood cutting; (3) fire exclusion; or (4) climatic change. Conifers regaining dominance over aspen in a prefire conifer forest is natural recovery (Kulakowski, Veblen, and Drinkwater 2004; Kulakowski, Veblen, and Kurzel 2006). Overgrazing, logging, or fuelwood cutting is documented by conifers established after a change in management (Harniss and Harper 1982) in forest dominated by aspen prior to the last high-severity fire. Fire exclusion or climatic change is documented by conifers gaining dominance in forest dominated by aspen prior to the last high-severity fire. Prefire maps, photographs, historical records, or dating and analysis of burned wood are required to determine whether prefire forests were aspen or conifer dominated. These four competing explanations of change commonly have not all been analyzed, nor has essential information been collected.

Many studies have suggested that aspen decline and conifer increase are unnatural, often citing fire exclusion, but the researchers did not consider competing explanations of change, or did not collect the necessary data to analyze these competing explanations, or used invalid assumptions (Walker, Mann, and McArthur 1996; Bartos and Campbell 1998; Wu 1999; Bartos 2001; Rogers 2002; Kashian, Romme, and Regan 2007). Bartos (2001) reported aspen declines of 49–96 percent across eight western states, based on an assumption (Bartos and Campbell 1998) that, if a conifer stand today has a single aspen, the stand was aspen dominated prior to EuroAmerican settlement. This is an invalid assumption because many of today's aspen or mixed aspen-conifer forests originated after high-severity fires in conifer forests, not in aspen forests, and are now simply recovering prefire conifer dominance (e.g., Kulakowski, Veblen, and Drinkwater 2004; Kulakowski, Veblen, and Kurzel 2006). Rogers (2002, 234) mentioned, but did not analyze, competing explanations, nonetheless concluding that change in 61 percent of plots with aspen present is "transition away from long-term aspen forest sustainability." Gallant et al. (2003) did use maps from 1912 to 1920, which show that aspen had declined. However, they reported a "high degree of stand age uniformity in the area" (p. 389) but did not investigate the likely explanation, which is natural recovery after a large high-severity fire or set of fires in the late 1800s. Late-1800s fires that temporarily increased aspen were found nearby in Yellowstone National Park (Romme et al. 1995). Kashian, Romme, and Regan (2007) suggested that fire exclusion and herbivory explained declining aspen at low elevations in northern Colorado but did not analyze fire history. However, C. S. Crandall (1901) documented extensive high-severity fire in conifer forests in the Kashian, Romme, and Regan study area in 1860–1900, corresponding with the peak in aspen regeneration found by these authors (2007, fig. 4a). Thus, aspen decline in their study area is likely natural recovery from high-severity fires in this period. Walker et al. (1996) attributed observed change in mixed-conifer forests in central Utah to fire exclusion. However, in the same area, Mueggler (1994) emphasized self-thinning and stand development after fire, a recovery explanation. Harniss and Harper (1982), also studying this area, thought increases in conifers might be an effect of heavy grazing, fuelwood cutting, or differences in fire history. Walker, Mann, and McArthur (1996) missed previous studies and also provided no evidence that stands were aspen dominated prior to fire, essential evidence of a fire-exclusion effect. The conclusion of these studies that fire exclusion led to aspen decline is invalid because of no or inadequate analysis of competing explanations of change; moreover, evidence suggests that alternative explanations may better fit the pattern of change.

A more likely explanation of conifer increase is natural recovery of conifers after mid- to late-1800s high-severity fires in mixed-conifer and subalpine forests containing aspen at the time of the fires. These fires led to abundant early-successional aspen forests, where conifers are now reappearing, a natural recovery process. Where prefire vegetation maps (Kulakowski, Veblen, and Drinkwater 2004; Kulakowski, Veblen, and Kurzel 2006) and early historical photographs (Zier and Baker 2006) were actually analyzed in western Colorado, this explanation was supported. Recovery in mixed-conifer forests is also documented to have been slow; mixed-conifer forests more than 250 years old may have aspen codominance (Wu 1999; McKenzie 2001), and these mixtures can persist for centuries (Zier and Baker 2006). There is currently no valid evidence of aspen decline from fire exclusion in mixed-conifer or subalpine forests in the Rockies or nearby, and there is no ecological need for logging or burning conifers to regenerate aspen (e.g., M. Johnson 1994) in mixed-conifer or subalpine forests.

RIPARIAN AND WETLAND VEGETATION

Riparian and wetland vegetation in the Rockies occurs from the lowlands to the alpine in moist settings where fire appears to have been relatively rare, but fire history and fire effects in these ecosystems under the HRV are not well known.

Riparian Vegetation

The riparian setting is generally cooler and moister and has higher relative humidity than adjoining uplands, it is often more shaded, and riparian vegetation has higher foliar moisture content and remains green later in the season (Dwire and Kauffman 2003; Pettit and Naiman 2007). Lightning ignition also is likely less common because of moister vegetation and a low topographic position (Dwire and Kauffman 2003). Along larger, lower-elevation rivers, these effects logically would produce longer fire rotations, lower fire severity, and patchier burns than in adjoining uplands, all which has been found in many studies, but differences with uplands are diminished along higher-elevation, smaller streams (Dwire and Kauffman 2003; Fisk et al. 2004; Pettit and Naiman 2007). In low-order stream reaches in southern Canada, fire dominated disturbance along straight streams lacking bars, with floods dominant only on lateral and point bars (Charron and Johnson 2006). Greater productivity and higher fuel loads in riparian areas might increase fire intensity and severity under dry conditions, relative to adjoining uplands (Pettit and Naiman 2007). Lower-elevation rivers, particularly with extensive unvegetated gravel bars, can function as fire breaks, at least during mild fire weather (Dwire and Kauffman 2003). Under extreme fire weather in the Rockies, fires can burn through nearly all riparian areas, riparian areas can channel wind and fire, and fire can spot across even major rivers.

Riparian vegetation along central and northern Rocky Mountain streams and rivers burned to some extent under the HRV. Early photographs show that major fires of 1910 and 1919 in northern Idaho burned the riparian vegetation down to the water's edge along the Clearwater River (Gruell 1983, plate 1), and late-1800s fires burned conifers along smaller rivers in eastern Montana

(Gruell 1983, plate 46). However, other fires appear to have been patchy in the riparian zone and may not have burned to the water's edge (e.g., Gruell 1980a, plates 10, 37). Although Gruell suggested (1980b, 3) that "periodic wildfire often killed competing conifers and stimulated cottonwood regeneration," I could find clear visual evidence of fire in riparian vegetation in only about 5 of 28 of Gruell's (1980a) early photographs showing riparian vegetation on the Bridger-Teton National Forest, Wyoming, and a few in the northern Rockies (Gruell 1983). I could see no visual evidence of fire in willows. Although the early photographs show that fire occurred in riparian woodlands and forests in the central and northern Rockies, it appeared less common than on adjoining uplands and often appeared to have a patchy pattern.

Fire may have been uncommon in riparian vegetation in the southern Rockies. A focused analysis of riparian vegetation in western Colorado found evidence of fire or insect damage (these could not be distinguished) in only 3 of 143 early photographs of montane narrowleaf cottonwood–blue spruce riparian woodlands and in only 3 of 27 early photographs of subalpine spruce-fir riparian forests (W. L. Baker 1987). Since photographs were taken during or shortly after the late-1800s period of extensive fires in this area (this chap. and chaps. 7–10), this evidence suggests that fire was much less common in riparian woodlands and forests than on uplands and was apparently not a major force structuring riparian vegetation under the HRV.

Native cottonwoods and willows that dominate riparian areas in the montane and subalpine zones of the Rockies commonly are partially or completely topkilled, but most resprout and some may reseed effectively after fire (L. Ellis 2001; Dwire and Kauffman 2003). However, black cottonwood (*Populus balsamifera* ssp. *trichocarpa*) was killed and replaced by bunchgrasses after 1910 and 1929 fires at four locations in Glacier National Park (Singer 1975), even though it resprouted after fire in western Canada (Rood et al. 2007). Plains cottonwood resprouted poorly after fire and was generally replaced by shrubs and grasses (Rood et al. 2007). Conifers common in subalpine riparian areas do not resprout. Rates and patterns of postfire recovery vary with postfire flooding, which, if severe, can erode and damage recovering plants or, if less intense, provide moist conditions that enhance recovery (Pettit and Naiman 2007).

Some have suggested using fire as a disturbance to rejuvenate riparian cottonwood groves (Rood et al. 2007), based on cursory frequency-based fire history at one site. However, fire is not likely to have declined in Rocky Mountain riparian areas since EuroAmerican settlement, as it was likely rare under the HRV. Data are lacking, but it is more likely that fire has increased relative to the HRV. Accidental ignitions have likely increased from roads and housing that often are nearby. Lowering of water tables by water diversions and dams and global warming has allowed fine fuels to increase and fuel moisture to decrease (L. Ellis 2001). Invasion of nonnative tamarisk into lower-elevation riparian areas in the Rockies has increased flammability and fires, as in the Southwest (Busch 1995). Intentionally burning riparian areas is ill advised, given that fire is likely to increase in the future, particularly if invasive weeds are present, and because native trees may fail to regenerate, as illustrated by variation in recovery of black cottonwood and the failure of plains cottonwood to regenerate after fire.

Wetland Vegetation

Nothing is known about how often wetlands dominated by willows, sedges, rushes, and other graminoids burned under the HRV in the Rocky Mountains, but fire likely was rare, given the standing water and high fuel moisture. Under extreme drought in 1988, fire burned wetlands in Glacier National Park (Willard, Wakimoto, and Ryan 1995) and Yellowstone National Park (Kay 1993). In Glacier, the Red Bench fire burned beaked sedge (*Carex rostrata*) meadows with 30–46 centimeters of peat overlying mineral soil (Willard, Wakimoto, and Ryan 1995). The peat was charred but not combusted, except it burned to mineral soil over about 10 percent of the surface where wood, trees, or shrubs occurred. Beaked sedge recovered from rhizomes over much of the area by the end of the first postfire year and was dense by the second postfire summer. Areas of exposed mineral soil had a higher diversity of other plants. In Yellowstone, sedge meadows in some areas burned lightly but in other areas burned to mineral soil (Knight and Wallace 1989), leaving large areas of shallow depressions that filled with water and, surprisingly, young aspen seedlings (Kay 1993).

Studies of prescribed fires in wetland vegetation are few in the West. The L. M. Smith and J. A. Kadlec (1985) and R. P. Young (1987) studies show fire had little effect or slightly increased productivity and shoot density in tule-cattail (*Schoenoplectus-Typha* sp.) marshes, cosmopolitan bulrush (*Schoenoplectus maritimus*) marshes, and wetlands dominated by mountain rush (*Juncus arcticus* ssp. *littoralis*) or common spikerush (*Eleocharis palustris*). However, inland saltgrass was killed when fire was followed by flooding in Utah (L. M. Smith and Kadlec 1985).

SUMMARY

This chapter covers a miscellaneous set of ecosystems, but all the forests were characterized under the HRV by moderate- to long-rotation high-severity (piñon–juniper woodlands) or variable-severity (mixed-conifer forests, riparian) fire. All these forests had diverse and variable structure (tree density, tree composition) under the HRV. After fire, lags in tree regeneration could allow grass or shrub vegetation to dominate for extended periods, or trees could regenerate more quickly. Tree regeneration after fire in mixed-conifer forests appears to have some spatial contingency, in which the severity of the fire, the necessity for seed dispersal, and thus the particular nearby surviving trees determine which trees dominate the postfire environment. In contrast, forest understory plants commonly survive and resprout in place, without spatial contingency, but some forbs and shrubs also germinate profusely from fire-stimulated seed. Fire exclusion since EuroAmerican settlement likely has had little effect on these ecosystems, since fire rotations were long relative to the period of effective fire control, and a general decline in fire is not evident in the Rockies. Available evidence suggests that these ecosystems do not need thinning or burning for restoration and would most benefit instead from weed control and ongoing wildland fire use (chap. 10).

CHAPTER 7

Fire in Ponderosa Pine and Douglas-Fir Forests

Complex ponderosa pine and Douglas-fir forests dominate the montane zone in the Rockies and have been the subject of much scientific research. These forests cover about 12 percent of the land area of the Rockies (table 1.1), they have been major sources of wood products and forage for livestock and wildlife, and recently they have become a focus of ecological restoration. The first section of this chapter covers ponderosa pine–Douglas-fir forests and is followed by sections on variants of these forests that include western larch or are pure Douglas-fir. A final section addresses land-use effects and restoration in all these forests. (See chapter 2 for a discussion of climatic effects on fire in these forests.)

PONDEROSA PINE–DOUGLAS-FIR FORESTS

This section reviews fire behavior, fire history, and fire effects and succession in ponderosa pine–Douglas-fir forests in the Rockies. These are key subjects because recent research has changed our understanding of fire behavior and fire history in these forests under the HRV. This new research shows that fire effects and succession were also often more spatially complex and contingent under the HRV than previously thought, resulting in a richer landscape mosaic of forests varying in age, tree density, and other properties.

Fire Behavior

Fires in ponderosa pine–Douglas-fir forests can burn slowly at low intensity (e.g., flame lengths less than 0.6 m), as in some parts of the 1988 Red Bench fire in Glacier National Park (Willard, Wakimoto, and Ryan 1994), but fires also can spread rapidly at high intensity during drought and high winds. The 2000 Cerro Grande fire near Los Alamos, New Mexico, burned about 1,000 hectares per hour, about one-third of total fire area, for six hours when winds peaked at more than 80 kilometers per hour (Paxon 2000). The 2000 Jasper fire in the Black Hills burned at about 2,400 hectares per hour, over half of total fire area, for seven to eight hours (Lentile 2004). The 2002 Hayman fire in Colorado burned about 3,000 hectares per hour, about one-third of total fire area, for six hours with winds about 30 kilometers per hour (Finney et al. 2003). Earlier fires also burned rapidly. A 1939 Black Hills fire burned about 240 hectares per hour for 34 hours, peaking at 1,175 hectares per hour (A. A. Brown 1940), and a 1936 fire burned about 650 hectares per hour for the first 37 minutes in 40-kilometer-per-hour winds (Cochran 1937). Rapid wind-driven spread led to 13 firefighter deaths at Mann Gulch, Montana, in 1949 (Rothermel 1993). Thus, 6,000–18,000 hectares can burn rapidly at high intensity in six- to eight-hour periods, with up to 30- to 60-meter flame lengths and long-range spotting (0.4–1.6 km; Paxon 2000; Finney et al. 2003).

Recent large fires in these forests had a range of severities, from low to high, but typically a mix (Kotliar, Haire, and Key 2003; Lentile, Smith, and Shepperd 2006). Data for four large fires in 2000 show that fires perceived as severe (e.g., 2000 Cerro Grande) had only 15–30 percent high severity (except for the Hayman, which had about 50 percent), with substantial (about 20–60 percent) low severity (fig. 7.1; Kotliar, Haire, and Key 2003). These fires produced fine-scale patchiness in severity (fig. 7.2). Much, but not all, of the high-severity area was associated with brief, extreme fire weather and rapid spread.

Fire History under the Historical Range of Variability

Available evidence about the fire regime in ponderosa pine–Douglas-fir forests under the HRV includes (1) early historical accounts and records, (2) fire history studies that include forest age-structures, (3) tree ring reconstructions of tree density near EuroAmerican settlement, and (4) paleoecological analyses. Early historical accounts and records of fire include anecdotes from explorers and settlers and reports about forest reserves, which later became national forests (fig. 7.3). These forest-reserve reports, completed by government scientists, include systematic observations in the late 1800s on the pattern and severity of fires, fire sizes, tree density, tree regeneration after fires, and logging and livestock grazing (W. L. Baker, Veblen, and Sherriff 2007).

In the 1990s, combined fire-scar and age-structure analyses, essential to determine severity of fires under the HRV (chap. 5), were first completed in these

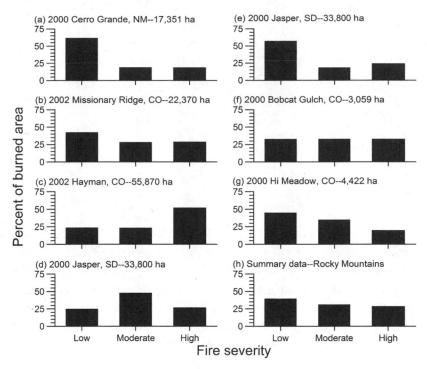

Figure 7.1. Percentage of large fires that burned at low, moderate, and high severity in ponderosa pine–Douglas-fir forests, based on remote sensing analysis.

Sources: Data for (a), (b), (c), and (e) are from differenced normalized burn ratio maps (see Key and Benson 2006) from the National Burn Severity Mapping Project (http://burnseverity.cr.usgs.gov). For (a) and (e), suggested breakpoints (Carl Key, U.S. Geological Survey, pers. comm., July 5, 2006) were used. For (b), breakpoints were used that produced proportions matching the fire-severity map in the BAER report done after the fire; these data are pooled across forest types, not just ponderosa pine–Douglas-fir forests. For (c) breakpoints were used that matched a fire-severity map in the Hayman report (Finney et al. 2003; see fig. 7.2 for actual breakpoints). Data for (f) and (g) are from Kotliar et al. (2003). Data for the Jasper fire in (d) are from Lentile, Smith, and Shepperd (2005). Summary data in (h) are from Robichaud et al. (2000).

forests in the Rockies (e.g., Arno, Scott, and Hartwell 1995). Five Colorado studies have necessary evidence at the landscape scale (table 7.1), and five studies in the Rockies are rated high for age-structure data across landscapes (table 7.2). Many studies have value, but not for fire in landscapes. Some sampled only one to three plots, insufficient for understanding landscapes, and others pooled age data across plots, which can lead to misunderstanding. Individual fires are documented to have burned partly at high severity and partly at low severity (see "Tree Density and Fire Severity" below), a pattern detectable only if age structure and fire history are examined plot by plot across a landscape (W. L. Baker 2006a). Other studies dated only the largest 5–15 trees in a stand to

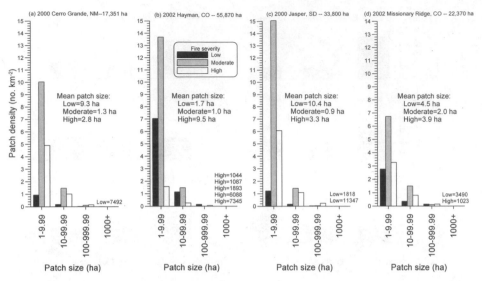

Figure 7.2. Density of discrete patches of low-, moderate-, and high-severity fire. Density is shown by fire severity and by fire-size class for four large fires in ponderosa pine–Douglas-fir forests. The r.le programs for the quantitative analysis of landscape structure (W. L. Baker and Cai 1992) were used to measure patch sizes for maps from classified normalized burn ratio (fig. 7.1). Patches less than 1 hectare in area were dominant but excluded from this analysis. Breakpoints for classifying fire severity are approximations; different breaks could change patch sizes. The lower limit of low-, moderate-, and high-severity fire was defined as follows: Cerro Grande—140, 480, 660; Hayman—120, 250, 400; Jasper—80, 410, 610; Missionary Ridge—205, 440, 660. See Key and Benson (2006) for explanation of severity breakpoints.

estimate stand origins, a sufficient sample where stands clearly originated after high-severity fire but insufficient in general, given that some of these forests lacked high-severity fire (W. L. Baker, Veblen, and Sherriff 2007).

Tree density affects the fire regime in these forests, because as density increases, high-severity rather than low-severity fire becomes more probable (e.g., Lentile, Smith, and Shepperd 2005). Direct estimates made near 1900 and tree ring reconstructions are the best sources of data on tree density under the HRV (table 7.3, fig. 7.4). Reconstructions of tree density near the time of EuroAmerican settlement are made by dating extant trees that were alive near that time; however, this results in an underestimate because some trees that were alive then have since died and decomposed. Direct estimates in early reports overcome this limitation but provide data only in broad ranges.

Tree Density and Fire Severity

Estimates of tree density under the HRV, directly recorded in forest reserve reports and reconstructed using tree rings, are congruent. Both document gener-

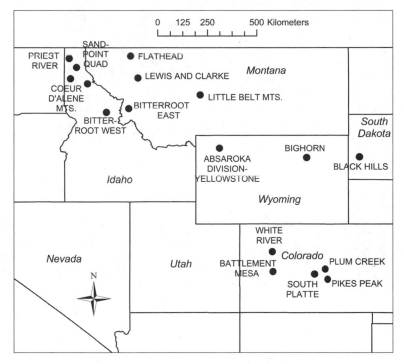

Figure 7.3. Location of 17 forest reserve reports and associated documents written about 1900. See W. L. Baker et al. (2007) for citations to each reserve report.
Source: Figure created by the author.

ally high tree density and large variability, ranging from 17 to 19,700 trees per hectare (table 7.3). Some variation is related to physical setting: Lower-elevation, south-facing slopes had relatively low density, and higher-elevation, north-facing slopes had higher density. Stand development was also important, as young forests after high-severity fire often had 1,000–20,000 trees per hectare (fig. 7.5a,b), maturing forests (less than 100 years after high-severity fire) typically had fewer than 750 trees per hectare, and mature and old-growth forests had 100–400 trees per hectare (table 7.3) and could appear dense (fig. 7.6a). Overall, tree density in these forests in the Rockies was about 5–10 times higher than tree density in the Southwest, which had about 7–60 trees per hectare in mature stands (Covington and Moore 1994). Higher density in the Rockies under the HRV contributed to moderate- and high-severity fire, but during the extreme fire weather also common in the Rockies, low-density forests may also burn at high intensity (e.g., Alexander 2004; Eckert 2004).

Forest reserve reports document that fires from low to high severity occurred in these forests in the mid to late 1800s at many locations in the Rockies and Black Hills (fig. 7.5; W. L. Baker, Veblen, and Sherriff 2007). High-severity fires at times burned large areas (fig. 7.5e) but left some unburned islands or

Table 7.1. Fire History Studies in Ponderosa Pine–Douglas–Fir Forests

Sources & Study Limitations, by Forest Type & Quality of Information	Location	No. Scar Plots	No. Scarred Trees	Age-Structure Est. of Severity	Qual. Est. of Fire Size	No. Stand Origins	Stand-Origin Map	Fire-Year Maps	High-Severity Rotation (yrs)
PONDEROSA PINE–DOUGLAS-FIR									
High									
Ehle & Baker (2003)	N CO	9	154	High, moderate, low	Yes	9	No	No	No est.
Sherriff (2004); Sherriff & Veblen (2006—ponderosa pine sites)	N CO	12	192	High, moderate, low	No	15	No	Yes	No est.
Veblen, Kitzberger, & Donnegan (2000—ponderosa pine sites)[4a,4b]	N CO	~34	~415	No data	No	0	No	Yes	No est.
P. M. Brown Kauffman, & Sheppard (1999); Huckaby et al. (2001)[4a,4b]	C CO	1	153	High, moderate, low	No	224	No	Yes	No est.
Medium									
Barrett (1988)[1b,1c,2b]	C ID	2	102	High, low	Yes	60	No	No	40–200[5]
P. M. Brown & Sieg (1996)[1b,1c,3a,4a,4b]	SW SD	4	57	Insuff. data	Yes	0	No	Yes	No est.
P. M. Brown & Sieg (1999)[1b,1c,3a,4a,4b]	SW SD	3	42	Insuff. data	No	0	No	No	No est.
R. F. Fisher, Jenkins, & Fisher (1987)[2b,3b,4a,4b]	NE WY	10	91	No data	Yes	0	No	No	No est.
Perryman & Laycock (2000)[1b,3b,4a,4b]	NE WY	1	48	No data	No	0	No	No	No est.
Goldblum & Veblen (1992)[1c,3a,4a,4b]	N CO	1	67	High, low	Yes	3	No	Yes	No est.

Wieder & Bower (2004)[3b,4a,4b]	C CO	1	18	No data	Yes	0	No	No	No est.
P.M. Brown & Wu (2005)[1c,3b,4a,4b]	SW CO	1	70	Not used	No	23	No	No	Nc est.
Allen (1989)[1c,3b,4a,4b]	N NM	5	64	No data	Yes	0	No	Yes	No est.
Falk (2004)[1c,2a,3b,4a]	N NM	50	203	No data	Yes	0	No	No	No est.
Foxx & Potter (1984)[3b,4a,4b]	N NM	1	25	No data	Yes	0	No	No	No est.
Low									
Steele, Arno, & Geier–Hayes (1986)[0,1a,1c,2a,2b,3b,4a,4b]	C ID	7	14	No data	No	0	No	No	No est.
Freedman & Habeck (1985)[2b,4a,4b]	NW MT	16	?	Not used	Yes	59	No	No	No est.
Arno (1976—Valley edge)[0,1c,2a,2b,3a,4a,4b]	W MT	3	22	No data	No	0	No	No	No est.
Arno, Scott, & Hartwell (1995—all but F1,F2)[0,1b,2a,2b,4a,4b]	W MT	7	<21	High, moderate, low	Yes	7	No	No	No est.
Tesch (1981)[0]	W MT	2	?	High, low	Yes	2	No	No	No est.
P.M. Brown, Ryan, & Andrews (2000; Upper Pine Cr.)[0,1a,1b,1c,3b,4a,4b]	SW SD	2	19	No data	Yes	0	No	No	No est.
Wienk, Sieg, & McPherson (2004)[0,4a,4b]	SW SD	1	24	High, low	No	12	No	No	No est.
P.M. Brown, Ryan, & Andrews (2000—Ashenfelder)[0,1a,1b,1c,3b,4a,4b]	SE WY	2	22	No data	Yes	0	No	No	No est.
P.M. Brown & Shepperd (2001)[0,3b,4a,4b]	WY, CO	18	272	No data	No	0	No	No	No est.
Boyden, Binkley, & Shepperd (2005)[0,2a?]	N CO	1	?	No data	No	1	No	No	No est.

Table 7.1. Continued

Sources & Study Limitations, by Forest Type & Quality of Information	Location	No. Scar Plots	No. Scarred Trees	Age-Structure Est. of Severity	Qual. Est. of Fire Size	No. Stand Origins	Stand-Origin Map	Fire-Year Maps	High-Severity Rotation (yrs)
P. M. Brown, Ryan, & Andrews (2000; Lone Pine)[0,1a,1b,1c,3b,4a,4b]	N CO	2	19	No data	Yes	0	No	No	No est.
P. M. Brown, Ryan, & Andrews (2000; Hot Creek)[0,1a,1b,1c,3b,4a,4b]	N CO	1	17	No data	Yes	0	No	No	No est.
Laven et al. (1980)[0,4a,4b]	N CO	1	20	Not used	Yes	?	No	No	No est.
Mast, Veblen, & Linhart (1998)[2a;8]	N CO	8	?	High, low	No	20	No	No	No est.
Rowdabaugh (1978)[2b,3b]	N CO	1	22	No data	No	0	No	No	No est.
Skinner & Laven (1982)[2a,3b,4a,4b,7]	N CO	1	17	No data	No	0	No	No	No est.
Donnegan, Veblen, & Sibold (2001)[0,1b,1c,3b,4a,4b]	C CO	11	171	No data	No	0	No	No	No est.
Grissino-Mayer et al. (2004)[0,1b,1c,3b,4a,4b]	S CO	6	133	No data	No	0	No	No	No est.
Allen et al. (2008)[0,1b,3c,4a,4b]	N NM	1	24	No data	Yes	0	No	No	No est.
Savage & Swetnam (1990)[2a,3b,4a,4b,7]	N NM	1	16	No data	No	0	No	No	No est.
Touchan, Swetnam, & Grissino-Mayer (1995)[0,3b,4a,4b,7]	N NM	3	96	No data	No	0	No	No	No est.

PONDEROSA PINE–WESTERN LARCH

Medium

Study	Location								
Arno (1976; Montane slopes)[0,1b,2a,2b,4a,4b]	W MT	10	80	High, moderate, low	Yes	0	No	Yes	No est.
Habeck (1990)[4a,8]	W MT	1	94	Not used	No	10	No	No	No est.

Low

| Arno, Smith, & Krebs (1997; plot L4)[0,1b,2a,2b,4a,4b] | W MT | 1 | 4–8 | High, low | No | 1 | No | No | No est. |

DOUGLAS-FIR

High

| N. T. Korb (2005; xeric)[4a,4b] | SW MT | 1 | 21 | Low | No | 40 | Yes | Yes | No est. |
| Littell (2002)[1b,4a,4b] | MT, WY | 3 | 108 | High, moderate, low | No | 36 | No | Yes | No est. |

Medium

| Arno & Gruell (1986)[1c,2b,4a,4b] | SW MT | 16 | ~32 | Insuff. Data | Yes | 0 | No | No | No est. |
| Heyerdahl, Miller, & Parsons (2006)[3b] | SW MT | 50 | 83 | Not used | No | 50 | No | Yes | No est. |

Low

Singer (1975, 1979)[4a,4b,8]	NW MT	1	<55	High, low	No	?	No	No	No est.
Arno & Gruell (1983)[0,1c,2a,2b,3b,4a,4b]	SW MT	12	<36	No data	No	0	No	No	No est.
Barrett (1994)[1b,1c,2a,2b,3b,4a,4b]	Yellowstone	3	8	No data	No	0	No	No	No est.
Loope & Gruell (1973)[2b,8]	NW WY	1	?	Not used	No	?	No	No	No est.

Notes: Studies were conducted in or near the Rocky Mountains and are ranked in column 1 specifically by the quality of information for understanding fire history across landscapes: *High* = Researchers sampled contiguous landscape areas and (1) included an adequate set of scarred trees with cross-dated fires, providing valid evidence about fire spread across the landscape; and (2) dated tree ages in a series of plots across a landscape, so fire severity and replacement fire rotation could be determined; or only (2). *Medium* = Researchers sampled contiguous landscape areas but (1) did not or could not cross–date fire scars; or (2) did not collect or did not use tree age data, so fire severity is not known; or (3) did not sample contiguous landscape areas but did date tree ages in a series of plots across a multiple-landscape area, so fire severity and replacement fire rotation could be determined. *Low* = Study had multiple limitations, including (1) plots scattered across multiple

Table 7.1. Continued

landscapes or a small number of contiguous plots in a single landscape, so the study provides only small-plot estimates of fire history; (2) biased sampling, low sample size, no age data, and other limitations (see table 5.5); (3) no analysis of fire spread among sampled trees or sample plots; or (4) no presentation of data. Studies are arranged from north to south within each forest type. Footnotes 1–4 use the same codes as in table 5.5, where citations are given explaining each limitation.

[0]Not landscape scale; small plots scattered many km apart or very few (e.g., 1–3) plots within a landscape.

[1a]Biased sample targeting of landscapes with old trees or many fire scars.

[1b]Biased sample targeting of plots with old trees or many fire scars.

[1c]Biased sample targeting of individual trees that are old, have many fire scars, or are thought to preferentially record fire scars.

[2a]Inadequate fire-scar sample size (e.g., <10 trees per plot).

[2b]Fire scars not cross-dated.

[3a]Inadequate age-structure data to allow estimation of fire severity (e.g., a few trees per plot).

[3b]No age-structure data to allow estimation of fire severity, or age data not used.

[4a]Biased fire-interval estimate based on composite fire intervals or individual tree fire intervals.

[4b]Biased fire-interval estimate due to omission of censored fire intervals.

[5]Based on maximum stand ages and longest fire intervals.

[6]Data not fully presented.

[7]No analysis of fire spread.

[8]Data pooled across vegetation types.

Table 7.2. Studies of Tree Age Structure and Stand Origins in Ponderosa Pine–Douglas-Fir Forests

Sources by Forest Type	Location	Rating	Stand Ages (yrs)	No. of Trees	No. of Cores per Stand	Total Stands	Some High Severity	Uneven Aged with Pulses	Uneven aged— no Pulses
STAND AGE									
Ponderosa pine–Douglas-fir									
Arno, Scott, & Hartwell (1995: L1–L3, B1–B4)	W MT	High	Ca. 300–550	591	65–146	7	4	1	2
Ehle & Baker (2003)	N CO	High	Ca. 140–300	686	~75	9	7–8	1	0
Kaufmann, Regan, & Brown (2000); Kaufmann et al. (2003)	N CO	High	Ca. 110–470	575	23	25	25	0	0
Mast (1993: all except NBH1, NBM1); Mast, Veblen, & Linhart 1998	N CO	High	Ca. 60–480	990	7–116	18	No interp.	No interp.	no interp.
Sherriff (2004); Sherriff & Veblen (2006)	N CO	High	Ca. 100–300	1,811	38–144	14	12	2	0
T.A. Morgan, Fiedler, & Woodall (2002)	C MT	Low	Pooled	393	40–60	5	0	5	0
Boyden, Binkley, & Shepperd (2005)	N CO	Low	Ca. 210	100	100	1	1	0	0
P.M. Brown et al. (2001)	N CO	Low	Pooled	1,800	Pooled	Pooled	—	—	—
Hadley & Veblen (1993)	N CO	Low	Ca. 250	207	72–135	2	2	—	—
Knowles & Grant (1983)	N CO	Low	Ca. 125–315	686	99–326	3	2	1	0
Peet (1975)	N CO	Low	Ca. 135–400	115	45–70	2	1	1	0
P.M. Brown & Wu (2005)[a]	SW CO	Low	Pooled	730	Pooled	Pooled	—	—	—
Ponderosa pine–western larch									
Arno, Smith, & Krebs (1997: L4)	W MT	Low	Ca. 385–485	265	93–172	1	1	0	0

Table 7.2. Continued

Sources by Forest Type	Location	Rating	Stand Ages (yrs)	No. of Trees	No. of Cores per Stand	Total Stands	Some High Severity	Uneven Aged with Pulses	Uneven aged— no Pulses
Douglas-fir									
Heyerdahl, Miller, & Parsons (2006)	SW MT	Low	Pooled	1,037	<30	Pooled	—	—	—
Hadley & Veblen (1993)	N CO	Low	Ca. 110	78	78	1	1	—	—
STAND ORIGIN									
Ponderosa pine–Douglas-fir									
T. A. Morgan, Fiedler, & Woodall (2002)	C MT	Low	Pooled	?	40–60	5	0	5	0
Huckaby et al. (2001)	N CO	Low		1,279	5	224	189	0	35?
Douglas-fir									
N. T. Korb (2005)	SW MT	Low	Not shown	?	~15?	40	5	0	27
Littell (2002)	SW MT, NW WY	Low	Not shown	540	15	36	9	0	27?

Notes: Studies were conducted in or near the Rocky Mountains. Stand ages were approximate, as exact dates are seldom provided. A stand-age distribution is a complete sample (usually truncated above a minimum tree size) of tree ages within a plot, with multiple plots across a landscape (areas up to thousands of ha). A stand-origin distribution, in contrast, is a sample of only the suspected oldest trees within a plot, with multiple plots across a landscape. Stand-age distributions allow estimation of within-stand processes affecting tree regeneration (e.g., whether the stand is even aged, uneven aged without visible pulses of regeneration, or uneven aged with regeneration pulses), but this cannot be determined with a stand-origin distribution. Landscape stand-age studies are rated for understanding landscape tree dynamics: (1) *high*—includes multiple plots across a landscape and shows or interprets data for individual stands; (2) *medium*—includes only one to a few plots across a landscape, but data are shown or interpreted for individual stands; (3) *low*—pools data among a set of stands across a landscape, making it impossible to discern landscape structure of patches differing in age.

[a]This study sampled only larger trees (≥20-cm diameter at breast height).

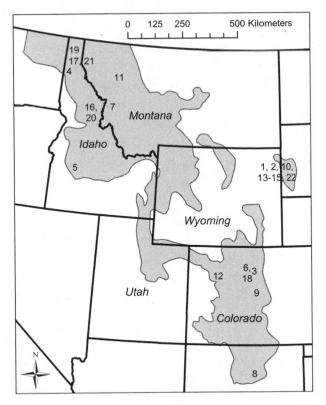

Figure 7.4. Locations of early reconstructions and direct estimates of tree density from near or before 1900. Table 7.3 lists the data for each location.
Source: Modified from W. L. Baker, Veblen, and Sherriff (2007).

groups of trees. Jack (1900, 69) mentions the "few trees and groves which escaped" in a large high-severity fire about 1880 in central Colorado (note the surviving tree and unburned patch in fig. 7.5d). Moderate-severity fires that left many surviving trees and tree groups were also reported (W. L. Baker, Veblen, and Sherriff 2007). Jack (1900, 77), for example, says of fires in Colorado prior to 1900: "Though they have burned over considerable tracts, they have left intervening groups or belts of living timber" (fig. 7.5c,d). Low-severity fires were also noted in the reports (fig. 7.5a). Two early scientists documented a range of patterns from variable-severity fire (table 7.4; fig. 7.5), a topic discussed later in this chapter (see "Variable-Severity Fires and the Shifting Landscape Mosaic").

The five high-rated fire history studies (table 7.1), all in Colorado, also found evidence of high-severity fire in nearly 90 percent of sampled stands. Evidence of high-severity fire is a coherent pulse of tree origins within a few years after a dated fire, sometimes verified by tree deaths (e.g., fig. 7.7a; Ehle and Baker 2003). Only about 10 percent of sampled stands in five high-rated

Table 7.3. Estimates of Tree Density in Ponderosa Pine–Douglas-Fir Forests near or before 1900

Fig. 7.4	Density Range (trees per acre)	Density Range (trees per ha)	Notes	Forest Type	Age of Forest ca. 1900	Sources
1	7–288	17–710	Larger trees only (e.g., >30 cm)	PIPO	Variable	Brown & Cook (2005)[a]
2	10–294	25–725	Mean = 344 trees	PIPO	Unknown	McAdams (1995: <2,000 bf/acre forests)[b]
3	16–1,380	39–3,410	Trees >4 cm	PIPO & PSME	100–250 years	Sherriff (2004)[c]
4	20–30	49–74	Trees >70 cm	PIPO & PSME	Unknown	W. L. Baker et al. (2007: table S4, item 6)
5	22–94	54–232	Trees >90 yrs old	PIPO & PSME	~310 years	Sloan (1998)
6	28–116	68–286	Trees >5 cm	PIPO	100–200 years	Ehle & Baker (2003)[d]
7	47–101	116–249	Pre-1900 trees only	PIPO & PSME	205–445 years	Arno et al. (1995)[e]
8	53–106	130–261	Trees ≥9.1 cm	PIPO	Unknown	Moore et al. (2004)[f]
9	81	200	Trees >1.37 m tall	PIPO	Ca. 90 years	Boyden, Binkley, & Sheppherd (2005)
10	88	217	Trees >12.7 cm	PIPO	Likely >200 years	Pinchot (1908: table 1)
11	93	230	Trees >12.7 cm	PIPO & PSME	Likely >300 years	Pinchot (1908: table 3)
12	100–120	247–296		PSME	Unknown	W. L. Baker, Veblen, & Sherriff (2007: table S4, item 12)
13	107–143	264–353	From ratios in description	PIPO	"Orig. forest" (old growth)	W. L. Baker, Veblen, & Sherriff (2007: table S4, item 10)
14	111–648	275–1,600	Mean = 633 trees	PIPO	Unknown	McAdams (1995: 2–5,000 bf/acre forests)[b]
15	150–200	370–494		PIPO	100 years	W. L. Baker, Veblen, & Sherriff (2007: table S4, item 10)
16	200–300	494–741	Trees >10 cm in "second growth"	PIPO	Likely <100 years	W. L. Baker, Veblen, & Sherriff (2007: table S4, item 7)

17	200–300	494–741		PSME	100–150 years	W. L. Baker, Veblen, & Sherriff (2007: table S4, item 4)
18	402–1,236	992–3,052	Trees > 5 cm	PIPO	20–40 years	Ehle & Baker (2003)[c]
19	800–1,500	1,976–3,705	"In some localities"	PIPO & PSME	Unknown	W. L. Baker, Veblen, & Sherriff (2007: table S4, item 1)
20	800–1,500	1,976–3,705	Trees >10 cm in "second growth"	PSME	Likely <100 years	W. L. Baker, Veblen, & Sherriff (2007: table S4, item 7)
21	1,000–3,000	2,470–7,410		PSME	Young	W. L. Baker, Veblen, & Sherriff (2007: table S4, item 4)
22	7,000–8,000	17,290–19,760		PIPO	Young	W. L. Baker, Veblen, & Sherriff (2007: table S4, item 11)

Notes: Studies were conducted in or near the Rocky Mountains. PIPO = ponderosa pine; PSME = Douglas-fir. This table is modified slightly from W. L. Baker, Veblen, and Sherriff (2007) and includes two additional studies missed by Baker et al. These estimates are either (1) direct reports from near 1900 by scientists or (2) reconstructions based on current trees that were alive near 1900.

[a] This estimate excludes goshawk plots, since some of those were not forested in 1900. Tree density in 1900 is likely underestimated due to loss of small trees that were present in 1874 (P. M. Brown and Cook 2005).

[b] This estimate is of tree density in 1874, not 1900, and is likely underestimated due to loss of small trees that were present in 1874.

[c] This estimate is of tree density just before the year of publication given in the table, not 1900, but stand age was estimated for 1900. These trees were all alive in 1900, but others likely died and have disappeared, so this is an underestimate of 1900 density.

[d] This estimate is of tree density in 1999, not 1900, but stand age was estimated for 1900. These trees were all alive in 1900, but others likely died and have disappeared, so this is an underestimate of 1900 density.

[e] This estimate was derived by adding "number of overstory trees per acre in 1991–93" and "estimated number of overstory trees per acre that died after 1900" from their table 2, excluding flathead stands, which have a mixture of tree species.

[f] This estimate represents the sum of "live trees" and "cut stumps" for two stands, JEMS2A and JEMS3A.

Figure 7.5. Historical photographs of variable-severity fires and tree regeneration, ca. 1900, in ponderosa pine–Douglas-fir forests in the Rocky Mountains: (a) surface fire and regeneration of ponderosa pine in the Black Hills, original photo likely taken by Henry Graves; (b) mixed-severity fire with torching in ponderosa pine in the Black Hills, and dense regeneration, original photo likely taken by Henry Graves; (c) mixed-severity fire in ponderosa pine in the Flathead Valley, Montana; (d) patchy high-severity fire identified by Jack (1900) as in ponderosa pine, and subsequent quaking aspen regeneration, in central Colorado in the Pikes Peak Reserve; original photo taken August 9, 1898, probably by John Jack, labeled "At Halfway House, Beaver Park, timber mostly burned"; (e) high-severity fire identified by Jack (1900) as in ponderosa pine, and subsequent ponderosa pine and quaking aspen regeneration, in central Colorado in the Plum Creek Reserve; original photo, taken August 18, 1989, probably by John Jack, labeled "Looking north at Devils Head (Platte) Mt. from east side"; (f) ponderosa pine forest with understory of Douglas-fir in Lewis and Clarke Reserve, western Montana; original photo likely taken by H. Ayres, labeled "Yellow pine [ponderosa pine] with undergrowth of red fir [Douglas-fir]. Near Woodworth."

Sources: All photos, except (c), were taken by the author of the original photos; the originals are in the National Archives, as follows: (a) FRB no. 103, p. 239, top; (b) FRB no. 101, p. 219, top; (d) FRA no. 003; (e) FRA no. 008, p. 19, no. 8-1; (f) FRD no. 109, p. 69, U10. (c) is reproduced from Whitford (1905, p. 217, fig. 17) with permission of the University of Chicago Press.

Figure 7.6. Logging effects on old-growth ponderosa pine forest in Montana. These photographs were taken of forest southwest of Hamilton, Montana, in the Bitterroot National Forest at Lick Creek: (a) in 1909 before logging (original U.S. Forest Service photo by W. J. Lubken); (b) in 1909 after logging—not at same location as (a); and (c) in 2005 in the Lick Creek area, showing young trees regenerating after thinning.
Sources: Scans of (a) and (b) provided by Keith Hammer, Kalispell, Montana; (c) photograph by the author.

Table 7.4. Early Observations of the Mosaic of Even- and Uneven-Aged
Ponderosa Pine–Douglas-Fir Forests

Black Hills, SD	"Two conditions of stand exist in the yellow pine of this region: even-aged stands, which have resulted very largely from fires and to some extent from logging, and mixed-age stands in which age classes appear in small groups intermingled with each other . . . even-aged stands of mature and over-mature timber occur, which are undoubtedly the result of fires which occurred from 150 to 300 years ago. . . . The mixed–age class stands occur very largely on areas which have never been logged over and on which no fires of very recent date have occurred. They are the result of natural causes, such as insect infestations, fungus attacks, and wind-storms which destroyed small patches of timber throughout the stand at various times. These openings have since been regenerated by seed from the surrounding stands, and new age-classes thereby formed" (P. T. Smith 1915, 297).
Central Colorado	"The large burned tract in the northern part of the reserve [about 24,000 ha] . . . was destroyed by fire about 1880. The few trees and groves which escaped are seeding the ground about them . . ." (Jack 1900, 69, referring to the Pikes Peak reserve). "A great deal of ground shows traces of fire, which must have occurred from thirty to one hundred or more years ago, and upon this is a more or less dense growth of small timber of various ages and sizes, according to the length of time since the fire and the time elapsing before fresh seed stocked the ground. As many of these fires appear to have been comparatively small and local, or to have left living individuals or many intervening strips of living trees which soon produced seed for the burned areas, the ground has become fairly well recovered, much sooner than is possible when many thousands of acres are burned over and no living trees escape" (Jack 1900, 97, referring to the South Platte reserve).

age-structure studies in Colorado and Montana were uneven aged (table 7.2) and had no evidence of extensive crown fire, but most had pulses of tree origins (e.g., fig. 7.7b), possibly from partial canopy fires. Uneven age structure may also have developed after high-severity fire long ago, evident in age structure as a weak postdisturbance cohort (e.g., fig. 7.7c; Arno, Scott, and Hartwell 1995). All five high-rated fire histories (table 7.1) found more low- than high-severity fires. Low-severity fires are documented by a cross-dated fire at more than one location, with intervening trees that predate the fire (chap. 5). Fire history and age structure in northern Colorado suggest a low-severity regime in about 20 percent of the ponderosa pine zone under the HRV, and a variable-severity regime in the other 80 percent (Sherriff 2004; Sherriff and Veblen 2007).

Paleoecological studies (studies of ecology over past millennia) of fire history in these forests are rare because of few sources of long, high-quality sediment records. Along the Payette River, in Idaho, high-severity fires and smaller fires

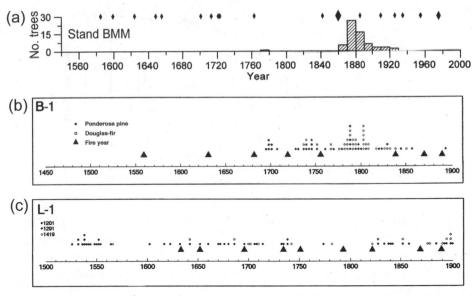

Figure 7.7. Age structure of ponderosa pine–Douglas-fir forests: (a) in Rocky Mountain National Park, illustrating a pulse of regeneration after a high-severity fire about 1860 in Stand BMM; (b) in the Bitterroot National Forest about 125 km south of Missoula, Montana, illustrating pulses of tree regeneration about 1700, 1740, 1790, and 1805; black triangles represent fires; and (c) in the Bitterroot National Forest about 125 km south of Missoula, Montana, illustrating a "canopy-opening" fire about 1525 in a generally uneven-aged stand.

Sources: (a) Reproduced from Ehle and Baker (2003, p. 555, fig. 5), and reprinted with permission of the Ecological Society of America; (b) reproduced from Arno, Scott, and Hartwell (1995, p. 7, fig. 3b, plot B1); (c) reproduced from Arno, Scott, and Hartwell (1995, p. 7, fig. 3a, plot L1).

in these forests were recorded as charcoal-bearing deposits in alluvial fans at the mouth of small, steep watersheds (fig. 7.8; Pierce, Meyer, and Jull 2004; Pierce and Meyer 2008). High-severity fires and erosional events were concentrated in the Medieval Climatic Anomaly (about 950–1350), a period of severe, multi-decadal drought. In the cooler, moister Little Ice Age (LIA; 1350–1900), erosion events and fires were smaller, less severe, and more frequent. However, even in the LIA (i.e., ca. 1600), evidence of severe drought and sedimentation suggest high-severity fire (Pierce, Meyer, and Jull 2004; Pierce and Meyer 2008). Charcoal in Blue Lake, northern Idaho, near the ecotone (boundary) with sagebrush, suggested high-severity fire in these forests about every 175 years from the year 280 to 980, the last about 1280, but low-intensity fire dominated until EuroAmerican settlement (C. S. Smith 1983). In northern New Mexico, charcoal in Alamo Bog near ponderosa pine forests had a mean episode interval of 111 years over the last 500 years, while mean composite fire interval (CFI) from fire

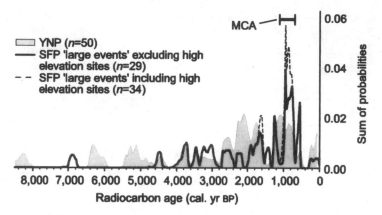

Figure 7.8. Probability of large sedimentation events. These events (e.g., during the MCA, or Medieval Climatic Anomaly) are likely related to severe fires in ponderosa pine–Douglas-fir forests on the South Fork of the Payette River (SFP) in southwest Idaho (graph lines). The corresponding record is shown for lodgepole pine forests in Yellowstone National Park (YNP; shading).
Source: Pierce, Meyer, and Jull (2004, p. 89, fig. 2), copyright 2004, reprinted by permission from Macmillan Publishers Ltd.

scars was about 13 years (Allen et al. 2008). Using multipliers of 3.6 to 16.0 (chap. 5), a CFI of 13 years suggests a point mean fire interval and fire rotation between about 47 and 208 years, which includes the charcoal estimate of 111 years. This is also roughly equivalent to the 50- to 150-year range I estimated for fire rotation for all fires in ponderosa pine forests (see "Fire Rotation" below).

Fire Size

Mixed- and high-severity fires commonly burned thousands of hectares historically, and up to 24,000–52,000 hectares (from forest reserve reports; see W. L. Baker, Veblen, and Sherriff 2007). The 2002 Hayman fire in Colorado had a high-severity patch of about 28,000 hectares (Romme et al. 2003), and forest reserve reports also indicate patches this size (e.g., Jack 1900). Most tree ring studies of fire history were for 1,000–6,000-hectare areas (table 7.5), targeting old-growth forests where high-severity fire was not recently widespread. Fires larger than these study areas could not be detected, but many fires exceeding a few hundred hectares, often 20 percent of detected fires (table 7.5), were found. Only one tree ring study (Veblen, Kitzberger, and Donnegan 2000) covered sufficient area (more than 100,000 ha) to detect large fires, but it did not study fire severity. Based on 41 sample points, 10 fires could have exceeded 10,000 hectares, and 1 might have reached 50,000–75,000 hectares, the scale of the

Table 7.5. Estimates of Fire Rotation in Rocky Mountain Ponderosa Pine–Douglas-Fir Forests

Sources	Location	Study Area (ha)	No. Fires Exceeding a Certain Size	Mean Burned Area (ha)		Estimated Fire Rotation (yrs)
				Low	High	
R. F. Fisher, Jenkins, & Fisher (1987)	NE WY	~2,200	15 of 82 fires in 305 years "area wide"	1,500	2,000	30–40
S. W. Barrett (1988)	C ID	1,215	11 of 59 fires in 213 years were >400 ha	500	750	42–63
		1,619	6 of 18 fires in 205 years were >400 ha	750	1,000	74–98
Allen (1989)	N NM	~6,600	8–14 of 109 fires in 420 years were >1,000 ha	3,000	4,500	59–154
P. M. Brown, Kauffman, & Sheppard (1999); Huckaby et al. (2001)	C CO	~3,090	9 of 77 fires in 350 years were >1,000 ha	1,500	2,500	65–107
P. M. Brown & Sieg (1996)	W SD	~2,500	9 of 18 fires in 309 years were large	1,000	1,750	65–115
Veblen, Kitzberger, & Donnegan (2000)	N CO	>100,000	10 fires in 226 years were >10,000 ha?	20,000	40,000	75–150
Goldblum & Veblen (1992)	N CO	~600	2 of 15 fires in 228 years were >500 ha	600		114
Wieder & Bower (2004)	C CO	~1,000	5 of 20 fires in 256 years were >100 ha	250	500	136–273
Ehle & Baker (2003)	N CO	~3,000	2 of 100 fires in 350 years were >1,000 ha	1,500	2,500	280–467

Note: Table entries are arranged in order by increasing estimated fire rotation. Where authors did not provide estimates, I estimated the size of larger fires in each study from maps of sampling points or other data. Estimated fire rotation is calculated as FR = (period in years)/[(number of fires × mean area × mean burned fraction)/(study area size)]. Mean burned fraction is assumed to be 0.75, which means that on average 75 percent of the area within a fire perimeter actually burned (see W. L. Baker and Ehle 2001).

Hayman fire and the largest fires in forest reserve reports; but many small fires in a year could also explain detection at many sample points (Veblen, Kitzberger, and Donnegan 2000).

FIRE ROTATION AND MEAN FIRE INTERVAL

Fire rotation has not been estimated for these forests, as research in the past focused on small-plot studies. Data on fire sizes (table 7.5) allow a first approximation since most of a fire rotation is from the largest fires (chap. 5). I used fire sizes to make rough estimates of fire-rotation ranges that bracket potential sources of uncertainty (table 7.5). The rotation is for all fires, as they were not partitioned by severity in the original studies. I assumed a mean unburned fraction of one-quarter of fire area (W. L. Baker and Ehle 2001). These rotation estimates, often about 50–150 years, are very imprecise. The high-severity rotation is unknown but is probably long (e.g., 200–400 years); assuming 300 years, partitioning an all-fire rotation of 50–150 years means a low-severity rotation of about 60–300 years (see table 5.3 and chap. 5). This is congruent with rotations estimated by analytical correction of composite fire intervals (W. L. Baker and Ehle 2001). Imprecision can be reduced by more research on fire sizes (chap. 5).

FIRE SEASONALITY

Seasonality of historical fires can be determined by the position of a fire scar in an annual growth ring; positions are often distinguished as early wood, late wood, or dormant season (fall, winter, or spring). The fraction of scars that were early season under the HRV declines from southern Colorado to southeastern Wyoming, a relationship found by P. M. Brown and W. D. Shepperd (2001) and updated here by adding two new studies (fig. 7.9). Fires in the south were most common in the dry period before the North American monsoon begins in early July; those in the north occurred more often after herbaceous plants cured near the end of summer (P. M. Brown and Shepperd 2001).

THE FIRE REGIME IN PONDEROSA PINE–DOUGLAS-FIR FORESTS

Data on variation in fire severity and estimates of fire rotation under the HRV remain sparse or approximate for ponderosa pine–Douglas-fir forests because of a longstanding but incorrect assumption that fires in the Rockies were primarily low severity, as they are thought to have been in the Southwest. However, tree density was generally 5–10 times higher and more variable in the Rockies (e.g., table 7.3) than in the Southwest. In the Rockies, a low-severity fire

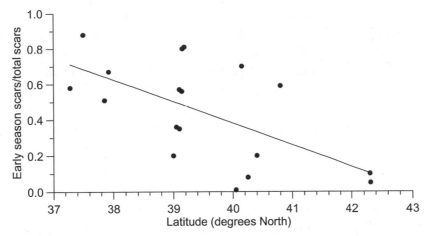

Figure 7.9. Fraction of total fire scars in the early season (about April to end of June) for 20 fire history sites from southern Colorado (about 37 degrees N latitude) to southeastern Wyoming (42.3 degrees N latitude).

Sources: Data for 18 of the sites are interpolated from figs. 3 and 8 in P. M. Brown and Shepperd (2001). The other two points are from data in Donnegan, Veblen, and Sibold (2001) and Grissino-Mayer et al. (2004).

regime may have dominated only in the dry, low-density forests at low elevations and on some xeric sites at higher elevations. In most of these forests, the fire regime was variable severity, with some low, mixed, and high severity (W. L. Baker, Veblen, and Sherriff 2007). Most fires were small, but some were 50,000 hectares or more (W. L. Baker, Veblen, and Sherriff 2007), producing even-aged forests over thousands of hectares or finer mosaics of even- or uneven-aged forests (table 7.4), as has also occurred in recent decades (chap. 5). Fire rotation by severity is unknown under the HRV, but a first approximation is a long rotation for high-severity and a moderate rotation for low-severity fire (e.g., 60–300 years). Fire rotation in low-severity areas at low elevations is unknown.

Fire Effects and Succession

This section reviews research showing that fire under the HRV led to complex changes in fuels, tree mortality and survival, tree regeneration and stand development, and effects on the forest understory. Fire effects varied with fire severity and were spatially heterogeneous, leading to shifting within-stand and landscape mosaics.

CHANGES IN FUELS

Low-severity fires in ponderosa pine–Douglas-fir forests in the Rockies primarily burn surface fuels—including duff; litter; dead wood less than 7.5 centimeters in diameter; and some fine, live plant material—but some canopy trees are typically damaged or killed (see "Tree Mortality and Survival" later in the chapter). Fuel consumption depends on fuel moisture, rate of spread, type of fire (e.g., head fire), and other factors. In 21 prescribed fires in northern Idaho, total fuels (excluding live trees) were reduced by 47 percent on average, about three-quarters of this from reduction in litter and duff (Saveland, Bakken, and Neuenschwander 1990). Total fuels were similarly reduced 38–44 percent by low-intensity fire in western Montana (R. C. Henderson 1967). Duff depth was reduced from 6.6 to 1.3 centimeters by high-intensity fire, which is more than the reduction to 4.0 centimeters by low-intensity fire (Armour, Bunting, and Neuenschwander 1984). However, higher-intensity fires do not always result in lower fuels than low-intensity fires. Although duff was consumed in high-severity parts of the 2000 Jasper fire and litter depth was reduced by 97 percent, the total amount of postfire fuels did not differ with severity (Lentile 2004).

Individual fuels may be initially reduced by fire but at different rates. Litter is flammable and relatively continuous and can be nearly totally consumed in flaming combustion (Bakken 1981). Litter depth remained reduced by 25 percent in low-severity, 36 percent in moderate-severity, and 92 percent in high-severity areas five years after the 2000 Jasper wildfire in western South Dakota (Keyser et al. 2008). Litter was reduced only 29 percent after prescribed fire in this area (Gartner and Thompson 1972). Duff depth was reduced from a mean of 6.6 centimeters in unburned forests to 4.0 centimeters after low-intensity prescribed fire in Montana (Armour, Bunting, and Neuenschwander 1984), similar to a 35 percent mean reduction by prescribed fire in the Black Hills (N. Bosworth 1999). However, five years after the 2000 Jasper wildfire, duff depth remained reduced by 78 percent in low-severity, 84 percent in moderate-severity, and 99 percent in high-severity areas (Keyser et al. 2008). Small, woody fuels were initially reduced, but five years after fire they had recovered to prefire levels in moderate- and high-severity areas in the Jasper fire, though they remained reduced in low-severity areas (Keyser et al. 2008). Prescribed fires initially consume some large, woody fuels (greater than 7.5 cm in diameter)—about 9–22 percent in one case (Bakken 1981), but 50 percent in another during drought (Saab et al. 2006). Consumption in nondrought conditions was mostly limited to rotten logs, but that still increased fire intensity and scorch height on trees (Bakken 1981). By five years after the Jasper fire, large, woody fuels remained reduced in low-severity areas but exceeded prefire levels in moderate- and high-severity areas (Keyser et al. 2008)

Fuel consumption by fire has been a focus of research, but it is only one process affecting net fuel loads after fire. Plants that are killed or damaged may produce substantial quantities of fuel in the first few months and years after fire, as scorched needles and branches are dropped and damaged or killed trees fall and break up. For example, litter and duff depth was initially reduced to 0.5 centimeters (a greater than 90 percent reduction) in moderate-severity areas, but litter and duff depth recovered to 2.2 centimeters, about one-third of the 5.6-centimeter unburned depth, in three years (Lentile 2004). Trees killed in fires (see the next section) initially persist as snags, but after falling they can increase large deadwood to levels higher than in the prefire forest (Graham et al. 1994; Keyser et al. 2008). Tree regeneration and shrub resprouting often are favored after fires, even low-severity ones, and represent ladder fuels that may become as tall as or taller than in prefire forests within a few years after fire (see "Tree Regeneration and Stand Development" below). In a Montana chronosequence, forest floor (duff and litter) thickness increased most rapidly for about 25–50 years after fire, then more slowly (fig. 7.10; MacKenzie, DeLuca, and Sala 2004). During succession, damage and mortality by insects, wind, and diseases affect fuels (W. L. Baker, Veblen, and Sherriff 2007; Lundquist 2007). Large data sets in the Rockies suggest that fuels are spatially heterogeneous and do not build up consistently after fire because multiple processes shape fuel loads in stands of all ages (chap. 2). Fire in these forests does not reduce all fuels (e.g., large deadwood may increase); some fuels that are reduced (e.g., litter) build up

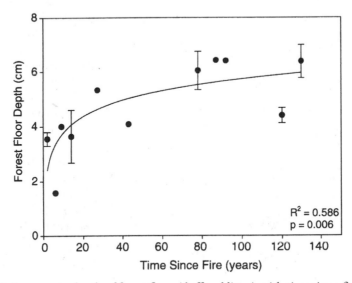

Figure 7.10. Increase in depth of forest floor (duff and litter) with time since fire. The study area for this chronosequence is of ponderosa pine–Douglas-fir forests in western Montana. *Source:* Reproduced from MacKenzie, DeLuca, and Sala (2004, fig. 4), © Elsevier 2004, with permission of Elsevier Limited.

again rapidly, and fire can even stimulate buildup of some fuels (e.g., ladder fuels) above prefire levels. Duff is generally reduced, however, and builds up again slowly.

Tree Mortality and Survival

Prescribed fires, often lower in intensity and severity than wildfires, still kill many or most small trees (e.g., less than 5 centimeters in diameter), as well as some large trees, depending on tree properties, fuel moisture, and other conditions (Gartner and Thompson 1972). Mortality of small trees may be twice as high in summer as in fall fires (table 7.6). Mortality was higher for Douglas-fir than for ponderosa pine of the same size (Wyant, Omi, and Laven 1986) and was highest for trees less than 5 centimeters in diameter, varying from about 25–94 percent, although trees about 15–20 centimeters in diameter were the mean size killed (table 7.6). Mortality was highest in the 3 years after fire but

Table 7.6. Percentage of Trees Killed in Prescribed Fires in
Ponderosa Pine–Douglas-Fir Forests

Sources	Where	When Burned	Years after Fire	Tree Size (cm)	Percent Killed
Ponderosa pine					
Gartner & Thompson (1972)—two separate fires	W SD	Spring	0	"seedlings"	43
			0	"seedlings"	83
Saab et al. (2006)	ID, WA	Spring	1	<7.6 cm dbh	39
				7.6–22.9 cm dbh	30
				>22.9 cm dbh	0
Bakken (1981)	N ID	Fall	1	<5 cm dbh	48–85[a]
				6–15 cm dbh	13–50
				16–50 cm dbh	0–18
Saveland, Bakken, & Neuenschwander (1990)	N ID	Fall	1	>5 cm dbh	42
R. C. Henderson (1967)	W MT	Unclear	1	<5 cm dbh	25–40[a]
				5–10 cm dbh	10–30
				>10 cm dbh	0–10
Wienk, Sieg, & McPherson (2004)	W SD	Spring	1	Not specified	39
Gallup (1998)	N CO	Fall	1	>30 cm tall	13[b]
					59[c]
J. H. Bock & Bock (1984)	W SD	Spring, fall	1	<1.4 m tall	94
				≥1.4 m tall	94
			2	<1.4 m tall	33
				≥1.4 m tall	36

Table 7.6. Continued

Sources	Where	When Burned	Years after Fire	Tree Size (cm)	Percent Killed
Wyant, Omi, & Laven (1986)	N CO	Fall	1	Mean = 18 cm dbh	16
			2	Mean = 18 cm dbh	25
Harrington (1987)	SW CO	Fall	1	>5 cm dbh	5
			5	>5 cm dbh	12
		Spring	1	>5 cm dbh	17
			5	>5 cm dbh	26
		Summer	1	>5 cm dbh	21
			5	>5 cm dbh	29
Ryan & Frandsen (1991)	NW MT	Summer	6	>50 cm dbh	21
Harrington (1993)	SW CO	Fall	10	4–11 cm dbh	28
		Spring	10	4–11 cm dbh	57
		Summer	10	4–11 cm dbh	64
		Fall	10	11–19 cm dbh	19
		Spring	10	11–19 cm dbh	27
		Summer	10	11–19 cm dbh	38
		Fall	10	19–24 cm dbh	0
		Spring	10	19–24 cm dbh	10
		Summer	10	19–24 cm dbh	12
		Fall	10	>24 cm dbh	7
		Spring	10	>24 cm dbh	7
		Summer	10	>24 cm dbh	8
Douglas-fir					
Bevins (1980)	C MT	Various	1	Mean = 19 cm dbh	43
Wyant, Omi, & Laven (1986)	N CO	Fall	1	Mean = 14 cm dbh	36
			2	Mean = 14 cm dbh	62
Ryan, Peterson, & Reinhardt (1988)	W MT	Fall, spring	1	Mean = 18 cm dbh	22
			2	Mean = 18 cm dbh	42
			8	Mean = 18 cm dbh	50

Note: Studies were conducted in or near the Rocky Mountains and are arranged by years after fire within each forest type; dbh = diameter at breast height.

[a]Estimate was interpolated from a graph.

[b]Low-density forest with 135 trees per hectare^{-1}.

[c]High-density forest with 2,579 trees per hectare^{-1}.

continued for 10 years (Harrington 1993). Ten-year mortality for large trees (more than 24 centimeters in diameter) was 7–8 percent from one fire (table 7.6). If similar fires occurred every 10 years, canopy trees could be half killed and not replaced in about 100 years (100×0.93^{10}). In contrast, burning at the shortest end of the estimated rotation under the HRV (50 years; see table 7.5) would allow half the canopy trees to survive for about 500 years, even if no regeneration occurred. Rates of mortality thus also suggest that fire rotations were more likely in the 50- to 150-year range than near 10 years.

Old trees with thicker bark and tall, limb-free boles survive fire better than small, thin-barked trees with low branches. Equations are available to predict whether a tree will survive or die after fire, based on tree diameter, bark thickness, stem char, scorch height, percent crown scorch, and other variables (appendix D). Some variables predict mortality before fire, but others predict survival, given damage (e.g., stem char, scorch height, percent crown scorch) after fire. The best predictor of postfire mortality is usually percent crown scorch, often along with bark thickness or tree diameter (D. L. Peterson 1985; D. L. Peterson and Arbaugh 1986; Wyant, Omi, and Laven 1986; Ryan and Reinhardt 1988). J. F. Fowler and C. H. Sieg (2004) have reviewed mortality models, and equations are available (see the sources listed in appendix D). Nomograms are also available from studying 2,356 trees in 43 prescribed fires in Idaho, Montana, Oregon, and Washington (fig. 7.11).

Prediction equations and nomograms do not identify mortality causes, which may include crown scorch, cambium heating to lethal temperatures, and burning of duff mounds leading to disruption of root system function (Ryan and Frandsen 1991; Gallup 1998). Ponderosa pine survive high levels of crown scorch, and may suffer only 30 percent mortality even with 90 percent crown scorch (Harrington 1993). Gallup (1998) observed that 15 percent of trees in a prescribed fire had some *torching*, in which generally low-severity fire climbs into and burns individual tree canopies, and 9 percent torched to their tops. However, if flames remained close to the trunk, the tree could survive; moreover, torching had not occurred in half of killed trees, so crown scorch was only one cause of mortality (Gallup 1998).

Wildfires commonly kill trees at higher rates than prescribed fires do. The 1988 Red Bench fire in Montana burned through old-growth ponderosa pine–Douglas-fir. In 7 of 16 sample sites, all trees were killed, and only low–intensity fires with flame lengths of less than 0.6 meter allowed survival of mature pines (Willard, Wakimoto, and Ryan 1994). By two years after the 1977 La Mesa fire in northern New Mexico, 60 percent of 897 trees in nine plots were dead (L. D. Potter and Foxx 1984). One year after a 1931 wildfire in Idaho, 74 percent of trees greater than 25 centimeters in diameter in an unlogged area were dead or unlikely to survive (Connaughton 1936). Five years after the 2000 Jasper fire in South Dakota, 64 percent of small trees (5–15 cm in diameter) in low-severity areas and 100 percent in moderate-severity areas were dead (Keyser et al. 2008). Of course, high-severity fire could kill nearly all trees over large areas, as documented in forest reserve reports.

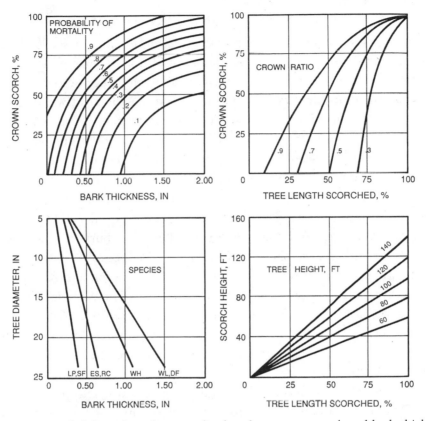

Figure 7.11. Probability of postfire mortality based on crown scorch and bark thickness. This is a nomogram for estimating the probability of mortality after fire based on crown-scorch percentage and bark thickness by measuring tree diameter, scorch height, and tree height. Use the lower left figure to estimate bark thickness from tree diameter for lodgepole pine (LP), subalpine fir (SF), Engelmann spruce (ES), western hemlock (WH), western larch (WL), and Douglas-fir (DF); then follow the line for bark thickness up to the upper left figure. Use the lower right panel to estimate the percentage of tree length scorched, using scorch height and tree height; then follow the vertical line up to the upper right figure to estimate crown-scorch percentage from crown ratio. Probability of mortality is obtained from the intersection of bark thickness and crown-scorch percentage.
Source: Reproduced from Reinhardt and Ryan (1989, fig. 1).

Dead trees may be viewed as a lost resource, but resulting snags are centers for biological diversity, supporting fungi, insects, birds, and small mammals (H. Y. Smith 1999; Saab et al. 2006). The most persistent snags were from old trees (H. Y. Smith 1999), with relatively low crown scorch, that survived for two to three years before dying (Harrington 1996). Snags ultimately become coarse woody debris (chap. 2), which is a resource that is also valuable for wildlife (Saab et al. 2006).

TREE REGENERATION AND STAND DEVELOPMENT

Ponderosa pines have episodic seed production, are sensitive to seedbed conditions, and suffer high postgermination mortality. Over 22 years, ponderosa produced about one fair or better cone crop to almost three poor crops in western Montana and a one-fair to two-poor ratio in eastern Montana (Boe 1954), only three good crops in 23 years in Idaho (Foiles and Curtis 1973), and only one good crop every 4–6 years in Colorado (Shepperd, Edminster, and Mata 2006). Most seeds are lost to seed predators, such as squirrels (Foiles and Curtis 1973). Seed germination and seedling survival are favored on mineral soils (Harrington and Kelsey 1979; R. B. Boyce 1985), but most seedlings die from drought, competition, herbivores, frost heaving, and other events (Foiles and Curtis 1973; Shepperd, Edminster, and Mata 2006). Seedlings on burned seedbeds had higher shoot and root weights than seedlings on seedbeds that had undergone other treatments (e.g., mechanical scarification; Harrington and Kelsey 1979). Some native bunchgrasses have roots that strongly compete with ponderosa seedlings for moisture, particularly in dry years or dry settings (Kolb and Robberecht 1996), and seeds that land on grasses and forbs germinate at low rates compared to seeds that land on mineral soil or burned seedbeds (Harrington and Kelsey 1979).

Ponderosa thus exhibits highly irregular regeneration, except in the Black Hills (Shearer and Schmidt 1970). In Idaho, the summer of 1963 was cool and wet, and abundant seedlings appeared, the first abundance since 1941 (Foiles and Curtis 1973). Successful seedling regeneration in northern Colorado is also favored by above-average precipitation (Shepperd, Edminster, and Mata 2006). In forest-grassland ecotones in northern Colorado, regeneration was concentrated in 4 years out of 40, and these were years with adequate seed production and high spring and fall moisture linked with El Niños (League and Veblen 2006).

Although ponderosa regeneration is episodic, it may follow fire, but at differing rates depending on fire severity. After the 2000 Jasper fire in South Dakota, low-severity burned areas, relative to unburned forest, had 15 percent higher seed production, much lower seed predation, 612 seedlings per hectare by year 3 (Lentile 2004; Lentile, Smith, and Shepperd 2005) and even higher seedling density by year 5 in both low- and moderate-severity areas (Keyser et al. 2008). Even denser regeneration occurred after low-severity fire in this area about 1900 (see fig. 7.5a). Long-term survival of seedlings after low-severity fire may be limited by competition with canopy trees (Lentile, Smith, and Shepperd 2005). Seed production was 41 percent lower in moderate-severity burned areas than in unburned forest with high seed predation but still led to an average of 450 seedlings per hectare 3 years after fire (Lentile 2004; Lentile, Smith, and Shepperd 2005). These seedlings likely have a higher probability of long-term survival because canopy mortality reduced competition and allowed light to reach the forest floor (Lentile, Smith, and Shepperd 2005). Figure 7.5b shows that regeneration was abundant after a moderate-severity burn about 1900. Regeneration may be concentrated in the first 10 years after low-severity fire and may cease during intervals without fire (Ehle and Baker 2003; Sherriff 2004). Forest reserve reports document that after low-severity fires in these forests, trees sometimes regenerated in dense thickets (W. L. Baker, Veblen, and Sherriff 2007), as in the photos in figure 7.5a,b.

After high-severity fire in these forests under the HRV, even-aged, postfire, regenerating trees could be quite dense (table 7.3) but also sparse (fig. 7.5c) or delayed for decades (W. L. Baker, Veblen, and Sherriff 2007). Dense postfire stands are documented under the HRV in several locations in the Rockies (W. L. Baker, Veblen, and Sherriff 2007). Age-structure analysis has showed that, after high-severity fire, tree regeneration was concentrated within 20–25 years (Ehle and Baker 2003), 15–30 years (Huckaby et al. 2001), or 19–60 years (Sherriff 2004). However, initial lags could occur on dry sites. After an 1879 high-severity fire in the Black Hills, regeneration was delayed about 5 years (Wienk, Sieg, and McPherson 2004). A fire south of Pikes Peak, Colorado, that likely burned in 1851 included high-severity fire in ponderosa pine forests; the area was described in 1905, about 50 years after the fire: "The reproduction of forests after the fire has been slow and irregular in the extreme. In some localities a new growth almost as dense as a field of grain now occupies the ground. In others there is not a vestige of new growth among the bleached trunks of dead pines and spruces—not even aspen occurring. Between these two extremes may be found every degree of reforesting" (Gardner 1905, 106–107).

Delays in regeneration averaged 21–46 years and ranged up to 119 years in ponderosa pine forests north of Pikes Peak (Kaufmann et al. 2003). However, a site in northern New Mexico had 218–318 young trees per hectare 16 years after high-severity fire (Foxx 1996). Regeneration after high-severity fire under the HRV thus could be rapid or substantially delayed, more commonly on dry sites, allowing openings to persist for 150 years or more (Kaufmann, Regan, and Brown 2000).

In retrospective age-structure studies, episodes of tree regeneration often followed fires, with more regeneration after larger, more severe fires (Kaufmann, Regan, and Brown 2000; Ehle and Baker 2003; Sherriff 2004). However, other causes of regeneration (insect outbreaks, favorable climate) must be distinguished from fire in retrospective studies, and criteria have been developed to do this (Ehle and Baker 2003). Climatically related cohorts have been documented in recent decades in some settings (Foiles and Curtis 1973; League and Veblen 2006). Some retrospective studies, however, have concluded that climate, rather than fire, explains past tree regeneration, even though fire preceded distinct pulses of regeneration visible in age structures (e.g., 1684 in P. M. Brown and Wu 2005; Boyden, Binkley, and Shepperd 2005). Where confounded like this, valid conclusions cannot be drawn.

Tree regeneration may be spatially heterogeneous after fire, particularly high-severity fire. After the 2000 Jasper fire, about 45 percent of the high-severity area was more than 40 meters from unburned forest—beyond normal dispersal distance for ponderosa pine seed—and had low seedling density (less than 100 seedlings per hectare) three years after fire (fig. 7.12; Bonnet, Schoettle, and Shepperd 2005; Lentile, Smith, and Shepperd 2005). However, the 55 percent of the high-severity area within normal dispersal range had higher seedling density (400–1,100 seedlings per hectare) in year 2 after fire. Seedlings were associated with burned conifer needles on mineral soil—common in the scorch zone bordering high-severity fire—close to seed sources (Bonnet, Schoettle, and Shepperd 2005). Similarly, two years after high-severity fire in northern New Mexico, about 95 percent of ponderosa seedlings were associated with conifer litter, particularly if it was less than 2 centimeters deep (L. D. Potter and Foxx 1984). Burned needles and litter act as mulch that retains moisture, reduces temperature fluctuations, and retains seeds (Harrington and Kelsey 1979; Bonnet, Schoettle, and Shepperd 2005).

Shade may also favor ponderosa pine seedlings in postfire settings (L. D. Potter and Foxx 1984; Bonnet, Schoettle, and Shepperd 2005). Postfire ponderosa seedlings in New Mexico and South Dakota were found where live ground cover was low but some overstory shade from surviving trees or scorched canopies existed (L. D. Potter and Foxx 1984; Bonnet, Schoettle, and Shepperd 2005). In New Mexico, 30 percent of postfire seedlings were found with little overstory shade but greater than 20 percent cover of bunchgrasses and low shrubs (L. D. Potter and Foxx 1984). Some early-postfire shrubs, such as snowbush ceanothus (Wahlenberg 1930) and alderleaf mountain mahogany (Liang 2005), may act as nurse plants for ponderosa pine seedlings, increasing their rate of survival. Others, such as Gambel oak, may hamper ponderosa regeneration (F. S. Baker and Korstian 1931). Some foresters sought to kill Gambel oak with herbicides or burning, with limited success (e.g., Harrington 1985), which is not surprising given its roots (fig. 9.8). Recently, the ecological value of Gambel oak in southwestern forests has been better recognized (Abella and Fulé 2008).

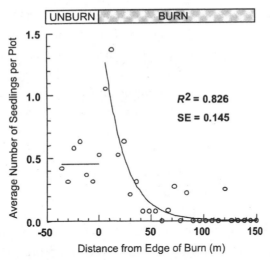

Figure 7.12. Mean number of seedlings vs. distance from unburned forest edge. Shown is the mean number of ponderosa pine seedlings and their distance from a forest edge containing unburned crowns after the Jasper fire in the Black Hills.
Source: Reproduced from Bonnet, Schoettle, and Shepperd (2005, fig. 2) with permission of the National Research Council, Canada.

After high-severity fire, an initially even-aged cohort of trees may gradually become uneven aged because of self-thinning, low- or moderate-severity fires, disease, or insects that kill overstory trees, followed by small pulses of regeneration, spreading out tree ages (table 7.4; Lundquist and Negron 2000; Ehle and Baker 2003). Old-growth age structures reflect low-level regeneration that ascended to the canopy over a 300- to 500-year period (Knowles and Grant 1983; Arno, Scott, and Hartwell 1995). More distinct, large pulses of regeneration are also evident at different times in different plots within a region, suggesting local events. Ponderosa pine and Douglas-fir also may regenerate at distinct times and each experience long periods (e.g., 100 years) without successful regeneration in a particular stand (Arno, Scott, and Hartwell 1995).

Tree density or basal area may approach prefire levels or become relatively stable within about 60–100 years after high-severity fire in Montana (Klebenow 1965; MacKenzie, DeLuca, and Sala 2004). Basal-area increase slowed to a low level after about 100 years (MacKenzie, DeLuca, and Sala 2004).

Douglas-fir may naturally increase in mid- to late succession after high-severity fire in these forests. Oldest ponderosa pines often are a century older than oldest Douglas-fir where found together on moister sites (Mast 1993; Arno, Scott, and Hartwell 1995; P. M. Brown and Wu 2005). This pattern may reflect greater drought tolerance of ponderosa pine on open sites found after moderate- and high-severity fire, and shade tolerance of Douglas-fir, allowing regeneration beneath a developing ponderosa pine canopy. Douglas-fir accounted for much of an increasing trend in basal area over a 130-year chronosequence (MacKenzie, DeLuca, and Sala 2004). Young Douglas-fir beneath ponderosa pine canopies thus may not mean unnatural fire exclusion but instead may indicate natural recovery and succession after high-severity fire.

FIRE EFFECTS AND RECOVERY OF THE FOREST UNDERSTORY

Low-severity fires have modest effects on understory plants, which recover in a few years, according to many studies (table 7.7). Regrowth of grasses may be evident a few weeks after low-severity fires but can be delayed by competition for water (Defossé and Robberecht 1996). In the ecotone with plains grasslands in South Dakota, several grasses (e.g., big bluestem, needle and thread) were reduced, but others increased (e.g., wheatgrasses, blue grama) in year 1 after fire (Gartner and Thompson 1972; Forde 1983). Most recovered to prefire levels by year 2 (Forde 1983). Cover and number of forbs increased in year 1 after fire (Gartner and Thompson 1972). In the Black Hills, after low-severity fire, herbaceous cover was reduced but nearly reached prefire levels in two years (Lentile 2004). Resprouting shrubs were little different one year after fire than before (Gartner and Thompson 1972), but the shrub leadplant (*Amorpha canescens*) increased substantially in years 1–2 after fire (J. H. Bock and Bock 1984).

In Idaho ponderosa pine, cover of native perennial graminoids (bluebunch wheatgrass, Idaho fescue) was reduced for at least three years after low-severity fire (Merrill, Mayland, and Peek 1980). Shrub biomass declined 50 percent the first year after fire but exceeded prefire levels by year 4 (Merrill and Mayland 1982). Snowberry cover was not reduced in the first three years after another fire in Idaho (Armour, Bunting, and Neuenschwander 1984). Herbaceous production was 1.3–2.2 times prefire production for four years after fire, primarily from native annual forbs and nonnative cheatgrass (Merrill, Mayland, and Peek 1980). After low-severity fire in northern New Mexico, prefire native perennial grass cover remained reduced by about half and native shrub cover by about three-quarters, but forb cover was little changed by year 16 after fire (Foxx 1996).

High-severity fire that killed canopy trees reduced perennial grasses and some forbs for up to four years but favored other forbs. Herbaceous cover was reduced the first year but reached prefire levels three years after high-severity fire in South Dakota (Lentile 2004). High-intensity fire reduced cover of native perennial graminoids for at least three years postfire, because of mortality from smoldering combustion in dry duff, but did not change shrub cover (Armour,

Table 7.7. Studies of Postfire Responses of Plants in Ponderosa Pine–Douglas-Fir Forests

Sources and Type of Observation by Vegetation Type	Location	Time Since Fire (yrs)	Severity L	M	H	Variable Studied I	G	R	D	P
PONDEROSA PINE										
Observations/plots after fire										
Defossé & Robberecht (1996)	N ID	6 mo.	X			X				
Gartner & Thompson (1972)	W SD	1	X			X				
Wienk, Sieg, & McPherson (2004)	W SD	1	X				X	X	X	X
Petterson (1999)	N CO	1	X				X		X	
Guthrie (1984)	N NM	1		X	X	X				
Forde (1983)	W SD	1–2	X			X				
J. H. Bock & Bock (1984)	W SD	1–2, 5	X				X	X	X	
Foxx (1996)	N NM	1–2, 8, 16	X				X	X	X	X
Armour, Bunting, & Neuen-schwander (1984)	N ID	1–3	X				X	X	X	
Lentile (2004)	W SD	1–3	X	X	X			X	X	
Merrill, Mayland, & Peek (1980)	N ID	1–4	X				X	X		X
Merrill and Mayland (1982)	N ID	1–4	X				X			X
Keyser et al. 2008	W SD	1–5	X	X	X			X		
Eichhorn & Watts (1984)	E MT	1–14		X		X				
Bonnet, Schoettle, & Shepperd (2005)	W SD	2		X				X		
Lentile, Smith, & Shepperd (2005)	W SD	2	X	X	X			X		
L. D. Potter & Foxx (1984)	N NM	2				X	X	X	X	
Willard, Wakimoto, & Ryan (1994)	NW MT	4	X					X		
Newland & DeLuca (2000)	W MT	4, 10?	X	X		X				
Klebenow (1965)	W MT	42				X	X		X	
Chronosequence studies										
MacKenzie, DeLuca, & Sala (2004)	W MT	2–132				X		X	X	
DOUGLAS-FIR										
Observations/plots after fire										
Lyon (1966, 1971)	SC ID	1–7, 20				X	X	X	X	
Eichhorn & Watts (1984)	E MT	1–19, 28				X	X	X		

Note: Studies were conducted in or near the Rocky Mountains. Entries are ordered by time since fire and whether observations were of one or more fires or a chronosequence of fires. L = low severity, M = moderate severity, H = high severity, I = individual species, G = species groups (i.e., forbs, graminoids, shrubs), R = tree regeneration or density (or other tree population variables), D = diversity, P = production or growth (e.g., annual basal area increment).

Bunting, and Neuenschwander 1984). Native graminoids were reduced to less than 5 percent cover for four years in one case in eastern Montana but reached nearly 45 percent cover by year 12 after fire; in another case, graminoids were low one year after fire and then rose to high levels (fig. 7.13). Several forbs were reduced, but others greatly increased after high-intensity fire (Armour, Bunting, and Neuenschwander 1984). In eastern Montana, forb cover was low in one case in the first four years after high-severity fire but increased to about 15 percent by year 9; in another case, forb cover was highest in years 1–2 after fire. Shrubs resprouted at lower rates by year 2 after severe fire in northern New Mexico, in forests not burned for more than 50 years, with higher resprouting in a stand burned 17 years previously (L. D. Potter and Foxx 1984). Forb and grass cover increased for 8–16 years after high-severity fire in this area (Foxx 1996). Shrubs increased after high-severity fires in 1974 and an explorer observed that grayleaf red raspberry (*Rubus idaeus* ssp. *strigosus*) covered mountainsides in the Black Hills after high-severity fires in the late 1800s (J. H. Bock and C. E. Bock 1984).

Only two studies included more than the first few years after high-severity fires. Figure 7.14 shows that shrub cover increased for more than a century after fire in western Montana, while graminoids and forbs had less cover and either were stable or declined slightly for a century after fire. Also, Klebenow (1965) reported that 42 years after severe fire in western Montana, shrubs dominated, particularly snowbush ceanothus, while graminoids and forbs were common but less abundant. Together, the evidence suggests that although shrubs, graminoids, and forbs can be temporarily reduced or increased, they generally recover within a few years after either low- or high-severity fire in ponderosa pine–Douglas-fir forests.

LOW- AND MODERATE-SEVERITY FIRES AND THE WITHIN-STAND SHIFTING MOSAIC

Low- and moderate-severity fires that kill individual trees or larger groups may contribute to a shifting mosaic of canopy disturbance patches within a stand (e.g., 100 hectares of forest), a concept pioneered by Watt (1947). The mosaic is often difficult to detect visually in the forest overstory and may become evident only if trees are dated and mapped (Arno, Scott, and Hartwell 1995). Fire is only one of several disturbance agents, including root diseases, lightning, bark beetles and other insects, ice and snow, and wind (P. C. Johnson 1966; Lundquist and Negron 2000). These agents create canopy or subcanopy disturbed areas within which trees or other plants may find conditions favorable for regeneration.

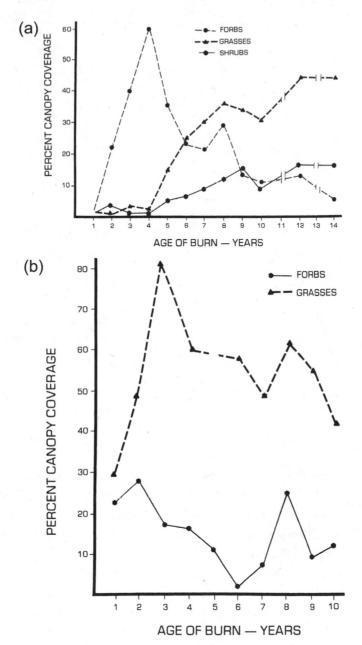

Figure 7.13. Percent cover changes in the first two decades after high-intensity fire. These are chronosequences showing changes in percent cover of forbs, grasses, and shrubs with time since high-intensity fire in (a) ponderosa pine–juniper, and (b) ponderosa pine–wheatgrasses in central Montana.

Sources: Part (a) Reproduced from Eichhorn and Watts (1984, p. 28, fig. 3); part (b) reproduced from Eichhorn and Watts (1984, p. 30, fig. 4)—both with permission of the Montana Academy of Sciences.

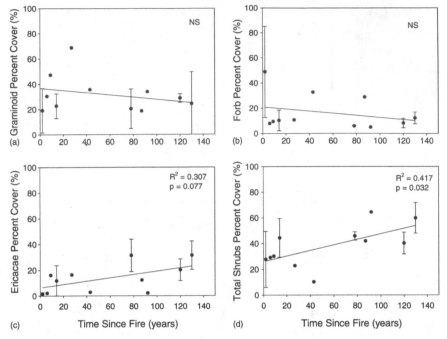

Figure 7.14. Percent cover changes in the first 140 years after fire. Chronosequence showing changes in percent cover of graminoids, forbs, Ericaceae, and shrubs with time since fire in ponderosa pine–Douglas-fir forests of western Montana.
Source: Reproduced from MacKenzie, DeLuca, and Sala (2004, p. 337, fig. 3), © Elsevier 2004, with permission of Elsevier Limited.

Where tree regeneration is favored in these disturbed patches, stands may contain a mosaic of patches in various stages of development, from small seedlings to groups of trees just reaching the canopy (Lundquist and Negron 2000).

A shifting within-stand mosaic has been found in ponderosa pine–Douglas-fir forests in the Rockies, but the mosaic is poorly understood. Lundquist and Negron (2000) mapped and analyzed a patchy mosaic pattern in ponderosa pine forests in the Black Hills and documented a role for fire and other agents in shifting the mosaic. Similarly, regenerating trees on the Salmon National Forest, Idaho, in 1911 were described as follows:"Where the forest cover has been broken here, dense reproduction of yellow pine is found" (Ogle and DuMond 1997, 76). Arno, Scott, and Hartwell (1995) documented a mosaic in canopy trees and correspondence between mortality caused by overstory fire and tree regeneration in groups. On the Payette National Forest, Idaho, in 1912, this observation was made: "Reproduction in the yellow pine type was patchy with seedlings occurring in groups of even-aged classes" (Ogle and DuMond 1997, 66); however, it is not known what triggered regeneration or whether regeneration corresponded with canopy patches. P. M. Brown and R. Wu (2005) said understory tree cohorts were not related to overstory mortality, but they did not present evidence. Boyden, Binkley, and Shepperd (2005) formally analyzed this in central Colorado and found no relationship between recent seedling cohorts and overstory mortality. In Rocky Mountain National Park, ponderosa pines were only weakly clumped, suggesting a weak or absent mosaic of patchy regeneration (Ehle and Baker 2003). Ehle and Baker suggested that overstory mortality alone may not trigger tree regeneration, as small trees would still face competition from understory grasses and forbs. Trees did regenerate after low-severity fires, suggesting that low- and moderate-severity fires may trigger regeneration cohorts, as in early photos (fig. 7.5a,b), although not always in patches. However, in the pine-grassland ecotone in the park and near the prairie-forest border, evidence was found of tree cohorts in patches (Mast and Veblen 1999). Later work in these areas found that cohorts corresponded with wet periods (League and Veblen 2006). Further research is needed on how fire and other agents shape stands, particularly where a low-severity fire regime occurs.

VARIABLE-SEVERITY FIRES AND THE SHIFTING LANDSCAPE MOSAIC

A spatially and temporally shifting mosaic of patches at the landscape scale is a natural consequence of the variable-severity fire regime that likely prevailed in most ponderosa pine–Douglas-fir forests in the Rockies (box 3.1). The landscape patch mosaic in variable-severity fire regimes may be fine grained and spatially complex if fires burn short distances at low-severity, then flare up and run in the crowns or shift direction, leaving unburned islands and strips (fig. 7.15). Four fires in 2000–2002 illustrate this patchiness (fig. 7.2). Patches in these four fires were relatively small (see also Lentile, Smith, and Shepperd 2005), about three-quarters less than 1 hectare, and mean patch sizes were often 1–10 hectares. Patches greater than 100 hectares were rare, but low- and high-severity patches of more than 1,000 hectares occurred (e.g., an 11,347-hectare low-severity patch in the Jasper fire, a 7,345-hectare high-severity patch in the Hayman fire). Moderate-severity patches were smallest and densest, perhaps because surface fire torched, but was not sustained, as crown fire. Fine patterns of severity across all four fires, even where forests were relatively homogeneous, suggest that fire severity is not fully explained by simple factors such as tree density.

At an instant in time, landscapes in a variable-severity fire regime may contain several patterns described by two early scientists (Jack 1900; P. T. Smith

Figure 7.15. Complex mosaic pattern of burning in the 2002 Missionary Ridge fire. Shown is a part of the fire in ponderosa pine forests east of Durango in southwestern Colorado.

Source: Photo by the author.

1915): (1) large patches of even-aged forests from extensive high-severity fire; (2) a finer mosaic of even-aged patches, varying in age, from smaller high- or moderate-severity fires; and (3) patches having a fine, uneven-aged, within-stand mosaic from low- or moderate-severity fire (table 7.4). Temporary grassland or shrubland openings after high-severity fire (Kaufmann et al. 2001) are a fourth type.

Temporal fluctuation in the landscape mosaic is the hallmark of a variable-severity fire regime. Episodic large fires, such as the 2000–2002 fires, are followed by interludes with widespread forest recovery and smaller high-severity fires or other disturbances that increase heterogeneity in the mosaic (W. L. Baker, Veblen, and Sherriff 2007). In the interludes, patches become older and more uneven aged as self-thinning, surface fires, and other agents kill individual trees or tree groups, or possibly from regeneration in wet periods. The next large fire may replace this complex mosaic with coarser patches of even-aged forest or semipersistent openings, leaving some remnants of the prefire mosaic. Thus, landscapes in variable-severity fire regimes have times of low spatial heterogeneity and extensive young, dense forests and times of high spatial heterogeneity, diverse age structure, and more extensive old, uneven-aged forests. At times of high spatial heterogeneity, adjacent stands may differ in structure (e.g., in tree density), because a single fire burned through them with distinct severity

(Arno, Scott, and Hartwell 1995; Ehle and Baker 2003; Sherriff 2004) or because local, within-stand processes varied. Ponderosa pine landscapes fluctuate substantially in forest age, tree density, fuel loads, and pattern (W. L. Baker, Veblen, and Sherriff 2007).

This fluctuation is documented in the Pikes Peak area, Colorado. Here, landscapes described by Jack (1900) and Gardner (1905) about 1900 reflect a period about 20–50 years after large, high-severity fires (fig. 7.5e), including extensive patches of forest probably burned about 1851 and 1880, patches that escaped these fires (fig. 7.5d), and patches from smaller preceding fires (table 7.4). By about 2000, in part of this area near Cheesman Lake, some of the high-severity patches had regenerated to dense, middle-aged forest; some openings were reforesting; and remnant old-growth forests that had not burned in several hundred years remained (Huckaby et al. 2001). The 2002 Hayman fire replaced this patchy, recovering mosaic with extensive burned forests (Romme et al. 2003), producing a landscape similar to that described a century earlier by Jack (1900) and Gardner (1905). In the late 1800s and early 2000s, both landscapes contained large patches of high-severity fire (about 28,000 hectares in the 2002 Hayman fire and 24,000 hectares in the 1851 and 1880 fire areas), along with finer mosaics of even-aged and uneven-aged forests that escaped these fires (table 7.4).

Fire historians who reconstructed fire history and forest structure in ponderosa pine forests in the past typically sampled only in part of this mosaic. Specifically, they sampled in old-growth patches that to become old had to have endured a long period without high-severity fire, and they omitted sampling in parts of landscapes with younger, even-aged forests and other patterns (e.g., Grissino-Mayer et al. 2004). This sampling focused on old growth provides a biased history of landscapes by inherently omitting moderate- and high-severity fire areas (W. L. Baker and Ehle 2001). Similarly, in the Black Hills, fire history in old stands suggested a low-severity fire regime in ponderosa pine forests (P. M. Brown and Sieg 1996, 1999), but historical evidence across larger landscapes documented that moderate- and high-severity fire was common (Shinneman and Baker 1997).

PONDEROSA PINE–WESTERN LARCH FORESTS

Fire behavior in ponderosa pine–western larch forests has been little studied separately from studies of ponderosa pine–Douglas-fir forests. Evidence of fire history and age structure in a stand in western Montana (stand L4) that burned in high-severity fire about 1663 indicates that some larch survived (table 7.1; Arno, Smith, and Krebs 1997). Subsequent moderate-severity underburns are indicated by small regeneration pulses of larch, ponderosa pine, and lodgepole pine (Arno, Smith, and Krebs 1997). In three areas in western Montana, fire history and tree regeneration data suggest that fires were often low severity, but many were followed by some tree regeneration, suggesting higher severity (Arno 1976). In Glacier National Park, some old ponderosa pine stands occurred near 50-year-old lodgepole pine stands, suggesting that lodgepole had replaced ponderosa after high-severity fire (Lunan 1972). Fifty-one fires burned from 1557 to 1918 in ponderosa pine–western larch forests in Pattee Canyon near Missoula, Montana, but fire severity was not studied (Habeck 1990). Tree density near 1900 in these forests was estimated from extant live trees to have been about 93 trees per hectare on south-facing and 172 trees per hectare on north-facing slopes in a forest about 250 years old (Habeck 1990), which is comparable to historical tree densities in ponderosa pine–Douglas-fir forests near that age (table 7.3).

Fire sizes are poorly known, but several fires burned through 80- to 300-hectare study areas in Montana between 1734 and 1900 (Arno 1976). Arno et al. (1997) considered ponderosa pine–western larch forests to be cooler and moister than ponderosa pine–Douglas-fir forests, and believed the fire regime to be variable in severity but perhaps with more high-severity fire. Fire rotation is not known. The interval between charcoal peaks at Foy Lake in northwestern Montana has averaged about 25–30 years since ca. 1700 (Power et al. 2006), but these peaks likely represent fires of varying severity and size, and several peaks are probably needed to equal a fire rotation (chap. 5).

DOUGLAS-FIR FORESTS

Pure or nearly pure Douglas-fir forests cover nearly 5 percent of the area of the Rockies (table 1.1) and often occur in moister settings (e.g., north-facing slopes) near ponderosa pine forests. Douglas-fir forests have received less research attention than ponderosa pine forests. This section reviews what is known about fire behavior and fire history, along with fire effects and succession, in Douglas-fir forests in the Rockies.

Fire Behavior and Fire History

The importance of high- and low-severity fire likely varied from place to place in Douglas-fir forests, leaving a complex diversity of evidence. Singer (1975) thought fires in these forests in northwestern Montana were generally low-severity surface fires. At Soda Butte Creek in Yellowstone, fire history and stand-origin evidence identified high-severity fire in one-quarter of plots, spanning a 286-year sample period up to 1870 (Littell 2002). Several fires with some high severity were low severity in other areas, indicating mixed-severity fire.

In southwestern Montana's Red Rock Range, a stand exam in 1906 reported: "Originally all the mountain slopes between 6,000 or 6,500 and 7,800 feet were probably occupied by Douglas-fir forest. Of this, only remnants are left. . . . There is little doubt that most of this damage was due to crown fires which swept over large areas killing whole forests outright. Ground fires have probably occurred in greater number, but their effects have been rather indirect" (Ogle and DuMond 1997, 100). This stand exam, which covered about 8,000 hectares, found about one-third lodgepole pine forests, one-third brush land (mostly snowbush ceanothus), and about one-third Douglas-fir forests. The brush land was identified as former Douglas-fir forest burned in high-severity fires. In the Douglas-fir forests about 13 percent was described as dense and even aged; 35 percent as broken and many aged; and 47 percent as open and many aged, with trees in clumps, suggesting variable-severity fires.

In the Centennial Valley, a few kilometers west of the area of the 1906 stand exam, Korb (2005) studied a xeric forest probably representing the most open Douglas-fir forests in this area (Korb also conducted a separate study in mixed-conifer forests; see chap. 6). In Korb's study area, based on fire-scar and stand-origin evidence, only a few small high-severity patches (all less than 10 hectares) occurred between about 1750 and 1900, suggesting a long high-severity rotation. Low-severity fires also occurred but were commonly small, also suggesting a long rotation. However, large areas of brush land in the nearby 1906 stand exam suggest that high-severity fire was much more important in the broader Douglas-fir landscape.

The forest reserve reports emphasize high-severity fires, documenting that low-severity fires were not commonly observed in pure Douglas-fir forests or mixed Douglas-fir–ponderosa pine with Douglas-fir dominant, and that fires were predominantly moderate- to high-severity with some large areas of high-severity (W. L. Baker, Veblen, and Sherriff 2007). A quote from Leiberg (1899a, 276) about these forests in the Bitterroot Valley, Montana, provides an example: "The destruction has been greatest in the pure, or nearly pure, red-fir [Douglas-fir] forest, where we estimate that upon more than 50,000 acres the forest has been destroyed from 60 per cent to total." Moderate- to high-severity fires are also described in many early stand exams (1910–1915) in Idaho (Ogle and

DuMond 1997, 50–54). Age-structure data are too limited to provide much evidence (table 7.2). Available evidence suggests that the fire regime may generally have been variable severity in these forests, with driest sites having less high-severity and more low-severity fire, but most forests having substantial moderate- to high-severity and less low-severity fire.

Fire-size estimates were limited by size of study area, all of which were less than 3,600 hectares. Of 20 fires documented by scars, 15 fires were found on only one or two scarred trees in Korb's (2005) 3,600-hectare study area, with only two extensive fires (in 1800 and 1855) between 1750 and 1900. In Heyerdahl et al.'s (2006) 1,030-hectare study area, only three fires between 1700 and 1855 exceeded 200 hectares. In Littell's (2002) Soda Butte Creek study area of about 1,100 hectares, only about four fires exceeded 100 hectares between 1584 and 1870. Thus, fires were often small in these forests, scarring one or two trees, exceeding 100 hectares only about once every 50–100 years in a 1,000–3,500-hectare watershed, suggesting long fire rotations. The 2,628 hectares of "brushland" in 1906 in the Red Rock Range suggest that high-severity fires could reach this size. As in other fire regimes, much of the fire rotation is from the few fires greater than 100 hectares. The 1800 and 1855 fires found in Korb's study area were also at least present if not extensive in Littell's (2002) study area in Yellowstone (about 150 kilometers away) and in Heyerdahl et al.'s (2006) study area in southwestern Montana (about 140 kilometers away), suggesting that individual large fires and much of the fire rotation can occur with some regional synchrony.

Fire rotation has not been estimated in these forests; available data allow approximation but only for xeric, low-density forests. I used Littell's (2002) fire-year maps for Soda Butte Creek. I summed the areas in mapped fire perimeters to be about three-quarters of the study area in 286 years, which is a 380-year rotation for fires, regardless of severity. However, Littell's perimeters minimally enclose contiguous evidence and omit disjunct points; a broader interpretation of each fire year suggests that the rotation could be nearer 200 years. If high-severity fire were one-quarter of total fire area, partitioning a 200-year rotation (table 5.3) suggests a high-severity rotation of 800 years and low-severity rotation of 267 years. Littell targeted concentrations of fire scars in choosing the study area, and thus the extent of high-severity fire in these forests may be substantially underestimated (W. L. Baker and Ehle 2001).

In southwestern Montana, area burned was estimated for each of 11 fires between 1700 and 1860 in a mosaic of low-density Douglas-fir and mountain big sagebrush in a 1,030-hectare study area (Heyerdahl, Miller, and Parsons 2006). Using their data and equation 5.1 (chap. 5), I estimated fire rotation to be about 160 years (W. L. Baker, in press). This is a pooled estimate for Douglas-fir and mountain big sagebrush. Heyerdahl, Miller, and Parsons presented only pooled age structure and did not correlate fire dates with regeneration pulses, so there is no basis for calling the fires "surface fires." However, high-severity fires were uncommon in Korb's (2005) and Littell's (2002) study areas, so it is plausible that much of the rotation was low-severity fire. Assuming an 800-year high-severity rotation, for example, would lead to about a 200-year low-severity rotation, near the estimate from Littell's area. Twelve to 17 fire events per 1,000 years (a 59- to 83-year mean event interval) occurred in the last 2,000 years, according to charcoal analysis in a Yellowstone lake near a mosaic of Douglas-fir and sagebrush-grassland (Millspaugh, Whitlock, and Bartlein 2004). To be compatible with tree ring evidence, about three of these events would have to equal a fire rotation.

Tree density under the HRV varied over a large range from xeric to moist Douglas-fir forests. Reconstructed tree density in 1855 in southwestern Montana averaged 45 trees per hectare, varying across 20 plots from about 40- to 260-trees per hectare (Heyerdahl, Miller, and Parsons 2006). Also in this area, reconstructed tree density averaged about 130 trees per hectare in all but two plots, which had 712 and 1,155 trees per hectare in 1876 (Korb 2005). Both estimates, however, were for immediately after large fires, when tree density was probably at a low point in the HRV. Sudworth (1900b, p. 128) estimated that tree density in the densest stands in northwestern Colorado near 1900 was about 250–300 trees per hectare. Leiberg (1900b) estimated that 100- to 150-year-old forests in northern Idaho near 1900 had about 500–750 trees per hectare, and younger forests had 2,500–7,500 trees per hectare. Forests a little south, likely less than 100 years old, had 2,000–3,700 trees per hectare (Leiberg 1900b).

The evidence together suggests that Douglas-fir forests were quite variable in tree density and fire history under the HRV. Low-density forests (e.g., probably with more than 50–150 trees per hectare) near mountain big sagebrush likely had the most low-severity fire, at rotations of about 200–270 years, but with some high-severity fire at long rotations (e.g., 800 years). More typical mature Douglas-fir forests likely had more than 250 trees per hectare and were sometimes very dense (e.g., more than 1,000 trees per hectare), particularly if young. In these typical Douglas-fir forests, high-severity fire likely dominated, with unknown rotation.

Fire Effects and Succession

Douglas-fir had a ratio of one fair or better cone crop to seven-tenths of a poor crop in 22 years in Montana, so good cone crops were more frequent than in ponderosa pine (Boe 1954). Douglas-fir germination and survival are favored

by burned seedbeds, more so as duff depth decreases (R. B. Boyce and Neuenschwander 1989). Shade is also beneficial to seedlings (Ryker 1975), and some
shrubs, particularly snowbush ceanothus, can enhance regeneration (Wahlenberg 1930; M. H. Jones 1995). Other shrubs may reduce regeneration after fire
(Cholewa 1977).

As in ponderosa pine forests, trees at times regenerated densely after high-
severity fires under the HRV, but substantial lags also occurred. Regarding the
present Gallatin National Forest south of Big Timber, Montana, Leiberg (1904,
31) said: "On tracts burned over thirty or forty years ago close-set stands of red
fir [Douglas-fir] are coming in abundantly." In contrast, in some areas burned in
high-severity fire in 1879 in the Bighorn Mountains, Wyoming, Douglas-fir regeneration was sparse on north-facing slopes and lacking on south-facing
slopes about 35 years later (Peirce 1915). In northern Idaho, Douglas-fir regeneration lagged until 5–10 years after high-severity fire, in part because of competition with resprouting shrubs (Cholewa 1977).

Regeneration lags may have created some semipersistent mountain big
sagebrush areas after fires. Some sagebrush areas in the Bighorn Mountains,
Wyoming, may have resulted from successive high-severity fires in Douglas-fir
ca. 1705 and 1805, and sagebrush about 35 years after an 1879 fire was only
slowly being reinvaded by Douglas-fir (Peirce 1915). A stand exam in Idaho in
1906 reported, "On the drier and steeper slopes there is often nothing but a little grass and sagebrush. Most of the land was originally occupied by a Douglas
fir forest, as is shown by numerous charred stubs and fallen logs, and most of it
will in time be occupied by Douglas fir again, but the process will at best be a
slow one" (Ogle and DuMond 1997, 99). Little is known about how much
time is required for trees to recover after high-severity fires in areas with lags.
Planted and naturally seeded Douglas-fir were expected to reach 10 centimeters in diameter and crown closure after about 30 years and reach maturity 60
years after fire in Montana (Lyon 1971), perhaps the fastest recovery possible after high-severity fire in these forests.

The best available evidence about recovery of the understory after fire in
these forests is from Lyon (1966, 1971), who studied the first 7 years after a prescribed high-severity fire in a logged Douglas-fir forest at Neal Canyon, south-
central Idaho, and a nearby 20-year-old wildfire. Live ground cover reached 27
percent in year 1 and 69 percent in year 2 after fire, more than double prefire
cover, then declined before increasing to about 60 percent in year 7 after fire
(Lyon 1971). American dragonhead (*Dracocephalum parviflorum*) was uncommon
before fire but by year 2 covered 37 percent of the surface, then declined rapidly (Lyon 1971). Streambank wild hollyhock (*Iliamna rivularis*) and fireweed
(*Chamerion angustifolium*) then dominated until year 6, when snowbush cean-

othus became prominent. This shrub did not resprout but reseeded at high density and by year 1 had nearly 500,000 seedlings per hectare (Lyon 1971). All these plants, except fireweed, have fire-stimulated seed banks. Fireweed survives fire well, resprouts, then reseeds rapidly, so a delayed increase occurs. By 2 years after fire, density of resprouting shrubs and shrub seedlings had nearly doubled, but shrubs were short (less than 0.5 meter), with crown volume only about two-thirds of prefire volume. Scouler's willow was much more prominent than before fire, but Rocky Mountain maple was reduced (Lyon 1966). By year 7 after fire, shrub density was 10 times higher than before fire, and Rocky Mountain maple again dominated. By year 9, snowbush ceanothus was likely to exceed the combined volumes of other shrubs and remain dominant for 30–50 or more years, until tree cover was reestablished (Lyon 1971). The general sequence of postfire recovery at Neal Canyon is shown in fig. 7.16a.

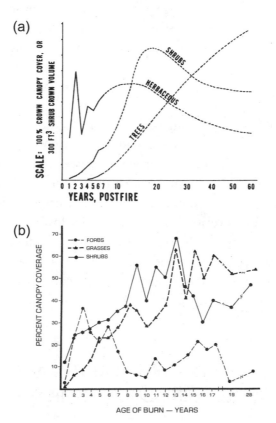

Figure 7.16. Postfire succession in Douglas-fir forests. Succession is shown by species group (e.g., forbs) (a) after prescribed high-severity fire at Neal Canyon, central Idaho; and (b) after wildfire in the Missouri River breaks in central Montana.
Sources: (a) Reproduced from Lyon (1971, p. 26, fig. 15); (b) reproduced from Eichhorn and Watts (1984, p. 26, fig. 1) with permission of the Montana Academy of Sciences.

Early postfire succession, described in 1910 on the present Targhee National Forest in Idaho, supports the role of snowbush ceanothus:

> Brush cover, as a rule, occurs on Douglas fir land where repeated fires have finally destroyed all or almost all of the fir. *Ceanothus velutinus* probably occurs more extensively than any other brush. In the gulches and on the slopes on the south side of the Great Bend Ridge, chaparral [Ceanothus] grows in dense large patches over many thousand acres. Very frequently excellent seedling growth of Douglas fir exists under chaparral in this part of the forest. (Ogle and DuMond 1997, 101–102)

In eastern Montana, high-severity fire reduced native graminoid cover for three years, but cover generally exceeded prefire levels in years 8–28 after fire (Eichhorn and Watts 1984). Bluebunch wheatgrass was initially reduced but was the main graminoid by year 12 after fire. Resprouting shrubs (e.g., chokecherry, snowberry [*Symphoricarpos* spp.], and rose) exceeded prefire levels more than one year after fire, and shrub cover was no different from that in an unburned control area the first four years after fire, exceeding that level in years 5–28. The sequence of postfire recovery in eastern Montana is shown in fig. 7.16b.

The evidence suggests that high-severity fires in these forests led to initially high bare areas the first year after fire, with bare ground declining to relatively low levels by year 2. In both cases, burned forests became dominated by shrubs by about the first decade after fire, and shrubs dominated for several decades while graminoids were second in importance and forbs least. Seed bank species were favored by high-severity fire, resprouters by lower-severity fire in these forests in southern Canada (Stark, Arsenault, and Bradfield 2006). As Lyon (1971, 25) emphasized, "The major constituents of the postburn community were represented in the preburn forest stand." Generalized successional patterns and more detail about individual species' response to fire in Idaho can be found in Steele and Geier-Hayes (1995).

LAND-USE EFFECTS AND RESTORATION IN PONDEROSA PINE–DOUGLAS-FIR FORESTS

Land-use effects on ponderosa pine–Douglas-fir forests can be difficult to discern decades after they occur, and most of these forests have a history of multiple land uses. This section reviews the effects of fire exclusion, tree-density increases and regeneration lags after fire, logging and livestock grazing, and invasive species; it also considers whether land-use changes led to more severe fires in these forests in recent years and looks at the options for ecological restoration.

Effects of Fire Exclusion

Small-plot studies in ponderosa pine–Douglas-fir forests suggest that fire declined after EuroAmerican settlement, and possible causes include overgrazing and barriers to fire spread, direct fire control, and unfavorable climatic periods. These effects are partly temporally confounded, and separating causes is difficult (chap. 10). Two isolated mesas in Utah (Madany and West 1983) and Washington (Rummel 1951) that lack human-created barriers to fire spread or direct control still had little fire in the twentieth century, lending weight to climate as the explanation of fire decline. However, in New Mexico the timing of fire decline is correlated with that of sheep grazing (Savage and Swetnam 1990). Overall, data from small-plot studies are insufficient to determine whether fire declined across larger landscapes (chap. 10).

Fire decline, regardless of cause, likely leads to net reduction in regeneration. Ponderosa regeneration is enhanced on burned seedbeds and so is reduced when fire declines. Where regeneration is naturally triggered by other disturbances or moist episodes, decline in fire increases survival of regeneration. However, with estimated surface fire rotations of 60–300 years (see earlier in chapter) and rare or small regeneration events from favorable climate or small disturbances, the net effect of fire decline is likely to be reduced regeneration, not increased regeneration, as commonly assumed. Regeneration of ponderosa pine has been low since EuroAmerican settlement where fires were rare and grazing, logging, or other land-use triggers were absent, as in Rocky Mountain National Park (Ehle and Baker 2003), and on isolated, ungrazed, unlogged mesas elsewhere in the West (box 7.1).

Shade-tolerant trees, particularly Douglas-fir, in the understory of ponderosa pine forests have been incorrectly thought to be solely a symptom of fire exclusion or decline (e.g., Arno, Scott, and Hartwell 1995; Kaufmann et al. 2001), and some authors have even recommended removing them in restoration (e.g., Kaufmann et al. 2003). Research in an Idaho wilderness showed that stands unburned for the last 80–120 years had a higher density of Douglas-fir than stands burned two to four times in this period (Keeling, Sala, and DeLuca 2006), suggesting that fire-free periods could allow Douglas-fir to increase. However, both tree ring studies (Sherriff and Veblen 2006) and forest reserve reports show that Douglas-fir was commonly found in the understory and

BOX 7.1

Control of Tree Regeneration by Fire and Competition

There is little doubt that surface fires in forests often kill small trees, limiting successful tree regeneration, but it has long been known that competition from understory grasses and other plants or from overstory trees also can limit regeneration (Weidman 1921; G. A. Pearson 1934). Disruption of this competitive control of regeneration, which is most apparent in ponderosa pine forests, can come from both livestock grazing that reduces understory plants and logging that opens the canopy. Subsequent surface fires, if they occur, could offset the loss of competition.

Available evidence from combinations of grazing and surface fire, while limited, suggests the hypothesis that the restrictive effects on tree regeneration from competition and surface fire are similar (table B7.1.1): (1) Competition from a healthy understory of native plants, particularly native bunchgrasses, can prevent or dampen tree regeneration, even without surface fires, so periodic fires may not be needed to control tree density in forests. (2) Periodic surface fires, if they occur often enough, can also alone prevent or reduce regeneration, even if competitive control by native understory plants has been reduced by livestock grazing. Competition and surface fire thus offer overlapping and similar constraints to tree regeneration. If only one is reduced, regeneration may not be abundant, but if both are reduced, regeneration is likely to be common. It also appears that the ability of the understory to restrict regeneration can be restored, if native plants were not destroyed, by excluding or reducing grazing. The forest canopy or root competition from overstory trees may exert a similar control over tree regeneration, judging by a common regeneration increase after logging.

canopy of these kinds of forests under the HRV (e.g., fig. 7.5f). Douglas-fir regeneration was often observed in the understory around 1900 in stand exams in the central and northern Rockies (Ogle and DuMond 1997). The Douglas-fir also was favored by low-severity fires, sometimes more so than ponderosa pine. For example, Leiberg (1900a, 378) said of northern Idaho, "Here, likewise, the grass assists in spreading the fires, which in this type of forest kill the yellow-pine saplings, but appear to promote the spread of the red fir [Douglas-fir], a tree that everywhere in this zone crowds the growth of the yellow pine on the

BOX 7.1

Continued

Table B7.1.1. Tree Regeneration in Ponderosa Pine Forests, Given Combinations of Fire and Livestock Grazing

		Livestock Grazing Effect on Competition			
		No grazing—isolated mesas	Past grazing that did not destroy bunchgrasses, followed by reduction or exclusion of herbivores	Past grazing that destroyed bunchgrasses, followed by reduction or exclusion of herbivores	Ongoing grazing, not heavy enough to prevent tree regeneration
Current Fire	Ongoing periodic surface fire	*Little tree regeneration?* No case studies	*Little tree regeneration?* No case studies	*Little tree regeneration* N AZ, north rim, Fulé et al. (2002)	*Little tree regeneration?* No case studies
	Periodic surface fire reduced	*Little tree regeneration* E WA, Rummell (1951); SW UT, Madany & West (1983)	*Little tree regeneration* N CO, Ehle & Baker (2003); N NM, early sheep grazing, Savage (1991)	*Abundant tree regeneration* N AZ, south rim? Fulé et al. (2002); N NM, later reduction in grazing a contributor, Savage (1991)	*Abundant tree regeneration* E WA, Rummell (1951); SW UT, Madany & West (1983)

Note: These two factors have potential interactions in controlling tree regeneration in bunchgrass-dominated ponderosa pine forests. Empirical case studies that included particular combinations of these two factors are listed. Hypotheses that lack empirical studies are followed by a question mark. Empirical studies are placed where I suspect they belong, but authors did not always identify the impacts of grazing or the current fire regime; thus, placements are tentative.

fire-swept areas." Also, after high-severity fire in these forests, ponderosa pine may dominate early succession and Douglas-fir become prominent in mid to late succession (see "Tree Regeneration and Stand Development" above). Thus, the situation for shade-tolerant trees is complex. Douglas-fir was common in these forests under the HRV and could be favored by fire, but it also may regenerate during extended fire-free periods. Shade-tolerant trees should not be systematically removed in restoration unless their abundance can be clearly linked to a land-use effect (W. L. Baker, Veblen, and Sherriff 2007), which is difficult to do (chap. 10).

Fuels were commonly thought to build up with fire decline in these forests, but evidence instead shows that some fuels may decline, others build up, and clear trends in total fuels are lacking. Large deadwood may diminish with fire decline, as wood stays locked up in live trees (Harmon 2002). In Rocky Mountain National Park, there was no trend in buildup of large deadwood on the forest floor in these forests after EuroAmerican settlement, and forest floor wood was often ancient, dating to the 1700s or earlier (W. L. Baker, Veblen, and Sherriff 2007). Ladder fuels, such as shrubs and regenerating trees, that may allow fire to climb into canopies, could build up in some cases, but these often are reduced by fire decline. Shrubs have variable response to fire, but shrub resprouting and density are often increased by fire, as mentioned previously, so fire decline may lead to reductions. Regeneration of ponderosa pine is generally reduced by fire decline, and Douglas-fir has variable responses to fire and fire decline, as explained earlier. Duff is one fuel component that may build up during postfire succession (MacKenzie, DeLuca, and Sala 2004). Total fuels, however, may not build up with fire decline because of the multiple fuel-producing processes that affect stands of all ages (J. K. Brown and See 1981). Lack of fuel buildup since EuroAmerican settlement has been documented by large fuel studies in these forests around 2000; they showed that total fuel loads were rather low, were possibly deficient for long-term forest health, and did not present a high fire hazard (Robertson and Bowser 1999; J. K. Brown, Reinhardt, and Kramer 2003). Reducing fuel loads is not generally ecologically warranted in ponderosa pine and Douglas-fir forests in the Rockies.

Tree Density Increases and Regeneration Lags after Fire

Temporally varying and spatially heterogeneous tree density is characteristic of ponderosa pine and Douglas-fir forests in the Rocky Mountains. Open, low-density forests with low crown closure were thought characteristic of these forests by some (Kaufmann et al. 2001, 2003). Kaufmann et al. also thought tree density increases over the last century were a consequence of twentieth-century fire exclusion. However, the recovery process after moderate- to high-severity fire in variable-severity fire regimes is naturally characterized by variable and often increasing tree density. Where initially low, tree density may increase until the next extensive moderate- to high-severity fire or until self-thinning and other within-stand processes lead to uneven-aged stands with lower density (W. L. Baker, Veblen, and Sherriff 2007). Lowering tree density to the particular levels found after nineteenth-century fires (Kaufmann et al. 2001,

2003) is inappropriate, since tree density fluctuated under the HRV. If trees are removed, regeneration will likely recur (e.g., fig. 7.6c) until canopy trees have again become large and dense enough to discourage extensive regeneration (Arno, Scott, and Hartwell 1995).

One study suggested that high-density tree regeneration, or a lag in regeneration, after high-severity fires in the 1940s–70s, in the part of the study area in the southern tip of the Rockies, was outside the HRV (Savage and Mast 2005). Savage and Mast found that the 1977 La Mesa fire in northern New Mexico had 97 larger trees per hectare (greater than 6 centimeters in diameter) and 380 smaller trees per hectare about 25 years after fire, and they considered this to be unnaturally high tree density. However, they presented no data on the density of postfire regeneration under the HRV. The density of larger trees is actually at the low end of the range for early postfire forests in the Rockies under the HRV (table 7.3). A postfire density of 97 trees per hectare is also a little below that of two mature stands near Santa Fe, which in 1911 had 130–261 live trees per hectare (greater than 9.1 cm in diameter; M. M. Moore et al. 2004). Tree density after the La Mesa fire does not appear to be outside the HRV, based on these data, but better data are needed on postfire tree density in this area.

Lags in regeneration after fire have also occurred recently in the Rockies, as they did under the HRV. In Glacier National Park, seedlings were absent the first three years after fire, and rare in year 4 (Willard, Wakimoto, and Ryan 1994). T. A. Morgan, Fiedler, and Woodall (2002, 18) described a 70,000-hectare fire in eastern Montana after about 15 years as "largely unregenerated," but similar lags occurred under the HRV in other parts of the Rockies (e.g., Kaufmann et al. 2003). Tree regeneration reported after fires in the twentieth century appears to be within the HRV, which included areas with high tree density, low tree density, and lags in regeneration.

Logging and Livestock Grazing

These forests have been extensively logged, including railroad logging and high-grade logging that removed the largest and best trees (Paulson and Baker 2006). Often large canopy trees were logged, and this was followed by a pulse of tree regeneration that became a dense understory and later a moderately dense middle-aged forest (Gruell et al. 1982). The pre-logging forest at Lick Creek in Montana, for example, was old and dense, with little understory tree regeneration (fig. 7.6a), most likely because high canopy density or cover, and understory cover, competitively excluded tree regeneration (Arno, Scott, and Hartwell 1995), and moderate-rotation low-severity fires may also have killed young trees.

The U.S. Forest Service circulated a poster promoting the idea that the forest at Lick Creek had originally been very open and that fire exclusion, not logging, had led to dramatic increases in tree density (Hammer 2000). However, they did not show a photograph of the pre-logging forest (fig. 7.6a) but instead showed the forest immediately after selective logging (fig. 7.6b), implying that that was the original condition of the forest (Hammer 2000). The pre-logging forest, however, was much denser, with large pines.

The density of 125 trees per hectare that preceded logging in 1907–11 became more than 1,500 trees per hectare by 1948 (H.Y. Smith and Arno 1999). Gruell et al. (1982, 39) described the logging, the postlogging condition the Forest Service implied was the original condition (fig. 7.6b), and the response to logging:

> Large quantities of overstory pines were felled, creating sizable openings. Logs were skidded and slash was burned in piles, locally scraping off or consuming surface vegetation, pine needle litter, and humus, and exposing mineral soil. The photo sequences covering the next 40 years show that tall shrubs (especially Scouler willow) and tree regeneration became established in direct proportion to the amount of stand opening and ground disturbance . . . dense pole stands developed on much of the area 30 to 40 years after logging.

Tree regeneration after logging depends on pattern and extent of tree removal and understory damage. Expansive removal of canopy trees and slash fires led to slow recovery in Colorado (Gary and Currie 1977). However, most forests in northern Colorado logged around 1900 had dense young trees by the mid-1980s (Veblen and Lorenz 1986). Logging typically removes large, fire-resistant trees, leaving behind and breaking up smaller trees, branch wood, needles and other fine fuels that substantially increase flammability and fire severity if they are not burned in prescribed fires (W. L. Baker, Veblen, and Sherriff 2007; chap. 10). Logged forests are often also deficient in large wood, since tree boles are removed. Logging can substantially increase Douglas-fir, relative to ponderosa pine, in the understory (Boe 1947–48).

Livestock grazing removes or decreases native grass cover, reducing competition and allowing more trees to regenerate (G. A. Pearson 1942; Madany and West 1983; see also chap. 10). Grazing also may reduce the spread of low-severity fires, as in northern New Mexico, where fire decline was associated with the onset of sheep grazing (Savage and Swetnam 1990). Grazing in ponderosa pine and Douglas-fir forests may favor Douglas-fir and lead to dense, young trees that provide ladder fuels, potentially increasing the risk of high-severity fires (Zimmerman and Neuenschwander 1984).

Invasive Species and Fire

Fires may increase nonnative weeds. After prescribed fire in the Black Hills, total understory biomass increased in year 1 by a factor of about seven, mostly from increases in nonnative forbs (Wienk, Sieg, and McPherson 2004). In Montana, thinning and prescribed burning, more than thinning or burning alone, increased cover and richness of about 16 nonnative species (Dodson 2004). Five years after low-severity prescribed fire and high-severity wildfire in Idaho, nonnative richness and cover were much higher in the high-severity area, and the low-severity area had lower nonnative richness and cover than unburned controls (Acton 2002). Low-severity fire may be preferred from this standpoint (Dodson and Fiedler 2006), but fires of any severity can cause expansion of invasives.

One of the most serious invasives in ponderosa pine forests in the Rockies is cheatgrass, which is favored by fire. Cheatgrass increased from 9 percent cover before burning to 24 percent one year after an Idaho wildfire (Merrill, Mayland, and Peek 1980). Cheatgrass is most common at low elevations but is becoming a threat throughout the ponderosa pine zone in the Rockies. Even low-intensity fire can cause cheatgrass expansion in ponderosa pine forests, increasing rather than reducing fire risk (Keeley and McGinnis 2007). Pine needle litter inhibits cheatgrass, whereas needle combustion, particularly beneath trees (fig. 7.17), favors cheatgrass (Gundale, Sutherland, and DeLuca 2008). Reduced burning is advised if cheatgrass is present (Keeley and McGinnis 2007).

Land-Use Changes and More Severe Fires

Many scientists have suggested that fires are becoming more severe in ponderosa pine and Douglas-fir landscapes because trees have become unnaturally dense (Arno et al. 1995; Allen et al. 2002; Kaufmann et al. 2004). However, evidence does not support the idea that fires are now more severe than they were under the HRV or that trees are more dense in most of these forests (W. L. Baker, Veblen, and Sherriff 2007). Systematic analysis showed that there is scant evidence in fire histories about fire severity in these forests under the HRV and little scientific basis to conclude that fires have increased in severity (W. L. Baker and Ehle 2003). However, other historical evidence, reviewed previously in this chapter, is more conclusive, showing that these forests burned extensively in the late 1800s and early 1900s in drought years (e.g., 1851, 1880, 1889, 1910), with

Figure 7.17. Cheatgrass beneath ponderosa pine trees scorched in a prescribed fire. The location is at an elevation of about 2,700 m in Rocky Mountain National Park, Colorado. *Source:* Photo by the author.

high-severity fire over tens of thousands of hectares. Some recent large fires (2000 Cerro Grande, 2002 Hayman) were set by humans, but do not appear outside the HRV in the mix of fire severities or size of high-severity patches, based on this historical evidence. Many of these fires initially appeared to have been high severity but often were actually dominated by low-severity, with smaller amounts of high-severity fire (fig. 7.1).

Analysis of 24 years of data on fires from 1980 to 2003 in these forests in the Rockies showed that the fire rotation for moderate- to high-severity fire was 244–278 years, and for high-severity fire was 500–714 years (Rhodes and Baker 2008). These are long fire rotations for western landscapes, suggesting that high-severity fire is not excessive in these forests. Available evidence suggests that fires have not increased in severity in ponderosa pine–Douglas-fir landscapes in the Rockies since EuroAmerican settlement.

Beyond Thinning in Restoration

It has been common practice to focus restoration on thinning to lower tree density and fire risk in ponderosa pine and Douglas-fir forests (e.g., Arno et al. 1995; P. M. Brown et al. 2001). However, this is a symptom-based approach that is unlikely to succeed and is also inappropriate in most of these forests in the Rockies (W. L. Baker, Veblen, and Sherriff 2007).

Effective and appropriate forest restoration first identifies the historical fire regime for a site, then determines land-use effects and how to reverse those effects while reforming the land uses, before undertaking restoration of all affected components of forest structure (chap. 11). The first step is to determine whether the fire regime was variable severity or low severity, which requires collection of age-structure and fire-scar evidence, historical evidence, or use of a GIS-based predictive model (W. L. Baker, Veblen, and Sherriff 2007). Next, it is essential to determine how the forest was altered by the major land uses in its history. If the fire regime was variable severity, as in much of the Rockies, distinguishing land-use effects on tree density is difficult, because of large natural variability in density (table 7.3). Moreover, it is essential to compare the density and structure of the stand being restored to that of a stand that was approximately the same age near or before 1900 (table 7.3), not with old-growth forests (W. L. Baker, Veblen, and Sherriff 2007).

Present tree densities in most ponderosa pine–Douglas-fir forests in the Rocky Mountains are likely to be within the HRV for stands of their age (W. L. Baker, Veblen, and Sherriff 2007). Thinning is thus not warranted for restoration unless it is part of a landscape-scale restoration to hasten development of patches of old-growth structure (see W. L. Baker, Veblen, and Sherriff 2007), a legitimate goal given the loss of old forests to logging. Most ponderosa pine–Douglas-fir landscapes in the Rockies were subject to variable-severity fire under the HRV and contained a diversity of patches that varied in age and tree density, and this landscape variation remains an appropriate restoration framework. At low elevations, if the low-severity model applies, thinning to lower tree density may be more appropriate (W. L. Baker, Veblen, and Sherriff 2007).

However, restoration thinning may be ineffective unless the land uses that led to the need for thinning (logging, grazing) are reformed; native understory plants are restored; invasive species are controlled; and fuels, including wood and snags removed by logging, are restored. Thinning alone may simply lead to renewed tree regeneration (fig. 7.6c) and a need for further thinning in a potentially endless, futile, and costly cycle (W. L. Baker, Veblen, and Sherriff 2007).

SUMMARY

Fires in ponderosa pine–Douglas-fir landscapes in the Rockies under the HRV were typically variable in severity, with a mixture of low- and high-severity fire. A first approximation of fire rotation and point mean fire interval for low-severity fire is 60–300 years in ponderosa pine forests and 200 years in Douglas-fir forests, with the high-severity rotation unknown but likely centuries. Much of the burned area in these forests is from infrequent large fires, up to 50,000 hectares or more, including some large high-severity patches (e.g., 20,000 ha). Spatial and temporal variability in fire size and severity under the HRV led to fluctuating and spatially variable landscapes, including substantial variability in tree density and age. Fire generally favored tree regeneration, sometimes leading to high-density patches of regenerating trees, other times to substantial lags before forest recovery. Forest understory plants generally recover within a few years after either low- or high-severity fire. Fires may have declined since EuroAmerican settlement, but fuels have not built up to abnormal levels, nor are trees abnormally dense or fires more severe. Where these forests are unhealthy in the Rocky Mountains, it is because of past unsustainable land uses, particularly high-grade logging and overgrazing by livestock. Health can be restored only by remedying the specific effects of each land use and reforming each land use with a new goal of sustaining both the structure and processes, including fire, that shaped these forests under the HRV.

CHAPTER 8

Fire in Subalpine Forests

Subalpine forests are extensive in the Rockies, covering 18 percent of the land area (table 1.1) and extending from the top of the montane zone to the tree line. The subalpine zone includes forests on moister sites dominated by spruce-fir, lodgepole pine, and quaking aspen, and on drier sites dominated by five-needle pines, including bristlecone pine, limber pine, and whitebark pine. The subalpine zone is moister, has higher fuel loadings, and has more lightning than the montane zone, leading to less frequent but potentially larger and more intense fires. High-severity fires may be followed by a diversity of successional vegetation, including dense to sparse conifer regeneration, resprouting quaking aspen patches, and semipersistent meadows.

SPRUCE-FIR AND LODGEPOLE PINE FORESTS

Although lodgepole pine may often dominate lower elevations and spruce-fir upper elevations of the subalpine zone, these two forest types also commonly overlap in elevation. In areas of overlap, lodgepole can be either successional to spruce-fir or form stable forests. The interaction of these trees is affected by the unique capability of lodgepole pine to store seeds in cones that may be opened by the heat of a fire. This section covers recent fire behavior, fire history and fire behavior under the HRV, fire effects and succession, and land-use effects in spruce-fir and lodgepole pine forests.

Recent Fire Behavior

Recent fires in spruce-fir and lodgepole pine forests have often included a range of fire intensities, from low-intensity surface fire to high-intensity crown fire (e.g., Beighley and Bishop 1990). In Yellowstone, a 1988 fire perimeter of about 230,000 hectares contained 28 percent unburned forest, 16 percent burned in light surface fire that did not kill overstory trees, 25 percent burned in severe surface fire that killed overstory trees, and 31 percent burned in crown fire that killed overstory trees. Severe surface fire was predominantly on crown fire margins (M. G. Turner, Hargrove et al. 1994). Small fires are common in these forests—95 percent of fires in one compilation were less than 4 hectares (J. K. Brown 1975).

Spread that leads to large fires is uncommon but is favored by curing of vegetation and by the passage of cold fronts, low-level jets, or downsloping winds that increase in late summer (W. L. Baker 2003). Large central and northern Rocky Mountain fires commonly have a southwest-to-northeast orientation, reflecting rapid spread during passage of cold fronts (fig. 8.1). High-intensity fires with large burned area often are wind driven but can be plume dominated (chap. 2). The 1985 Butte fire in Idaho overran a fire-fighting crew during intense, plume-dominated spread at more than 800 hectares per hour with flames 60–90 meters high (Rothermel and Mutch 1986). Exceptional wind-driven fires can burn 1,000–4,500 hectares per hour for 16–24-hour periods, with flames to 90 meters high and spotting 0.5–1.5 kilometers in front of the fire (Rothermel 1991a; Rothermel, Hartford, and Chase 1994). The 1988 Canyon Creek fire, in Montana, ignited in June, but in September a low-level jet led to wind gusts of more than 80 kilometers per hour, and the fire burned about 4,500 hectares per hour for 16 hours (Goens 1990). In Yellowstone, the North Fork fire burned about 156,000 hectares in nine days, averaging 725 hectares per hour during an unusually extended spread (fig. 8.1; D. A. Thomas 1991).

Severity of recent fires was inconsistently related to prefire forest structure, perhaps reflecting a threshold effect. A fire in a mosaic of spruce-fir and lodgepole burned much hotter in the lodgepole (Barth 1970). Pre-1988 Yellowstone high-severity fires occurred more in old-growth spruce-fir and mature (not old-growth) lodgepole pine (Renkin and Despain 1992). After the 1988 fires, sites with larger trees were weakly associated with lower fire severity, and larger trees (more than 15 cm in diameter) had less damage than smaller trees (M. G. Turner, Romme, and Gardner 1999). Severity of the 1988 Yellowstone fires was unrelated to prefire tree density or environment (slope, aspect), but late-

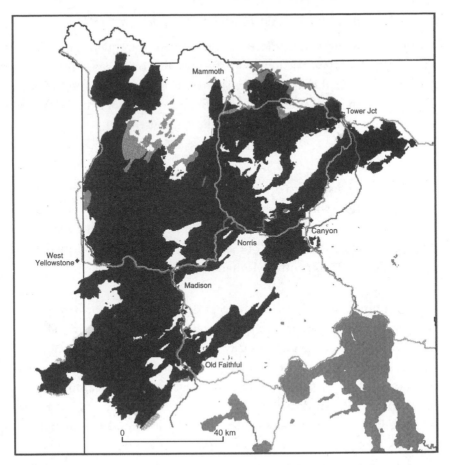

Figure 8.1. Southwest-to-northeast orientation associated with wind-driven fire spread. Shown are the 1988 North Fork and Fan fires in Yellowstone National Park, Wyoming. Each day of fire spread has a different shading.
Source: Reproduced from Rothermel, Hartford, and Chase (1994, p. 29).

successional forests and forests with severe mountain pine beetle or mistletoe had more crown fire than expected if fire had burned randomly (M. G. Turner, Romme, and Gardner 1999). In general, prior mortality from bark beetles in northwestern Colorado and Yellowstone did not increase fire severity significantly (box 4.2), but prior blowdown did (Kulakowski and Veblen 2007). The effect of prefire fuels and forest structure may be a threshold phenomenon, more important in shaping fire when fire weather is less severe and less important under severe fire weather (M. G. Turner and Romme 1994).

Fire Behavior and Fire History under the HRV

Fires in spruce-fir and lodgepole pine forests were infrequent and often high severity under the HRV, but low-severity fires also occurred. This section reviews what is known about fire severity, fire size, fire intervals and the potential for reburns after short intervals, estimated fire rotation under the HRV, fire seasonality, and fire-climate relationships in these forests.

FIRE SEVERITY

As in recent fires, fire severity under the historical range of variability may have been relatively little shaped by fuel abundance, which was generally high enough for high-severity fire if fuels dried (Bessie and Johnson 1995). Fuels are poorly known under the HRV, but Leiberg (1904a, 28) described them in lodgepole pine forests in central Montana about 1900:

> Duff, or humus, is nearly lacking, except on a few of the northern slopes, and plays no particular part in the spread and intensity of the fires. The general ground cover consists of moss, usually a thin layer two or three inches in depth, a slight sprinkling of pine needles, low shrubs, mostly species of huckleberry, and more or less of a grassy turf or sward. During the dry season all this material burns readily, but does not make a hot or high flaming fire. It is different with the litter. The great mass of dry or partly dry wood of which it is composed makes hot and flaming fires, consuming or killing all live timber.

Leiberg (1904a, 27–28) also described the fire regime north of Yellowstone, mostly in lodgepole pine forests: "The forest fires in this region are remarkable for their destructive force and intensity. Here and there are uneven-aged stands, where extremes in age and a mixed composition prove that occasionally the fires did not consume or kill the entire stand. But as a rule most of the older fires made a clean sweep, and in nearly every instance the fires of modern date have done the same."

C. S. Crandall (1901) found that about 70 percent (some 360,000 hectares) of about a 520,000-hectare area of forest in northern Colorado extending to southern Wyoming, much of it spruce-fir and lodgepole pine, had burned at high severity since about 1860, leading to large areas of young lodgepole pine forest. Crandall reported some low-severity surface fires, high-severity surface fires, and crown fires, the latter two killing nearly all trees. Forest reserve reports from about 1900 in Idaho and Montana (Leiberg 1899a, 1900a; Ayres 1900a,b;

Leiberg 1904a), Wyoming (Town 1899), and Colorado (Sudworth 1900a) also reported high-severity fire and large areas burned in the preceding few decades. Tree ring reconstructions similarly show that the fire regime under the HRV was dominated by high-severity fire (tables 8.1, 8.2). These fires left some un-burned tree groups in "belts and tongues along the edge of the fire" (F. E. Clements 1910, 11).

Surface fires have been documented in Colorado, Wyoming, and Montana by direct observation (C. S. Crandall 1901; F. E. Clements 1910) and tree ring re-constructions (Kipfmueller and Baker 2000; E. Howe and Baker 2003; Kipf-mueller 2003; Buechling and Baker 2004; Sibold, Veblen, and González 2006). Necessary tree ring evidence includes cross-dated fire scars or growth releases separated by older forest, documented by adequate tree ages, that predates the fire. Cross-dating is needed because small fires may occur every year (Loope and Gruell 1973; Kipfmueller and Baker 2000), so raw ring counts that differ by a few years may not mean a single fire year. Surface fires were reported in Montana (Singer 1975; Arno 1976; Gabriel 1976; Arno, Reinhardt, and Scott 1993; J. K. Brown et al. 1994) and Colorado (Zimmerman and Laven 1984), but lack of cross-dating precludes assessment of whether they were common in these areas.

Well-documented low-severity surface fires commonly occurred on the margin of high-severity fires where the fire was going out, in wet areas where fuel moisture reduced severity, and in dry, rocky areas and open stands where tree density was low (F. E. Clements 1910; Donnegan, Veblen, and Sibold 2001; Sherriff, Veblen, and Sibold 2001; Kulakowski, Veblen, and Bebi 2003; Buech-ling and Baker 2004). Discrete surface fires have also been documented (fig. 8.2; F. E. Clements 1910; E. Howe and Baker 2003; Kulakowski, Veblen, and Bebi 2003; Buechling and Baker 2004). Surface fires in lodgepole pine, indicated by scars through a stand, led to no detectable mortality or postfire tree regenera-tion (Sibold et al. 2007); thus these rare fires have little or no effect on tree pop-ulations. A few moderate-intensity fires could open the canopy and allow new regeneration in small areas (Singer 1975; T. L. Franklin and Laven 1991; Barrett 1994; Kulakowski, Veblen, and Bebi 2003). The rotation for surface fire was 1,480 years in the Park Range in northern Colorado (E. Howe and Baker 2003) and nearly 7,600 years in one part of Rocky Mountain National Park (Buechling and Baker 2004); in another part, less than 3 percent of the forest had seen surface fire in the past few centuries (Sibold, Veblen, and González 2006), a fire rotation exceeding 10,000 years. Low- to moderate-severity fire was generally a minor component of the fire regime in subalpine lodgepole pine and spruce-fir forests in the Rockies.

Table 8.1. Fire History Studies in Subalpine Forests

Sources & Study Limitations, by Forest Type & Quality of Information	Location	Study Area (ha)	No. Scar Plots	No. Scarred Trees	No. Stand-Origin Plots	No. Dated Trees	Stand-Origin Map	Fire-Year Maps	High-Severity Fire Rotation (mean and/or range, in yrs)[1]	Low-Severity Fire Rotation (mean, in yrs)[1]
STABLE LODGEPOLE & SPRUCE-FIR										
High										
Kipfmueller (2003); Kipfmueller & Baker (2000); W. L. Baker & Kipfmueller (2001)	ID, MT	3,133	101	96	101	1,904	No	Yes	191; 139–341 (L)[2]	No est.
Buechling & Baker (2004)[3]	SE WY	3,241	54	65	54	731	Yes	Yes	127 (H)	No est.
	N CO	9,200	231	113	231	3,461	Yes	Yes	346 (H)	7,587 (H)
E. Howe & Baker (2003)	N CO	3,700	77	63	77	1,504	Yes	Yes	175 or 281 (H),	1,480 (H)[4]
Kulakowski & Veblen (2002)	N CO	4,400	52	32	52	726	Yes	No	no est.	No est.
Sibold, Veblen, & González (2006); Sibold & Veblen (2006)	N CO	28,040	487	676	487	6,152	Yes	Yes	249; 145–326 (H)[2]	No est.
Kulakowski, Veblen, & Bebi (2003)	NW CO	4,600	54	46	54	825	Yes	No	No est.	No est.
Medium										
Gabriel (1976)[3,5,6,7]	NW MT	33,000	—	222	>22	?	No	Yes	150–200 (M)	No est.
Singer (1975, 1979)[7]	NW MT	7,130	—	55	—	1,000	No	No	250–370 (M)	No est.
Barrett (1994)[5,7,8]	Yellowstone	36,210	0	0	98	431	Yes	Yes	~200 (L)	No est.
Romme (1982)[5,7]	Yellowstone	7,300	36	44	36	?	Yes	No	300–400 (L)	No est.
Romme & Despain (1989a,b);[9] Schoennagel, Turner, & Romme (2003)	Yellowstone	129,600	?	?	?	?	No	No	~350 (M)[10]	No est.
M. P. Murray (1996); M. P. Murray, Bunting, & Morgan (1998)[3,5,6,7]	C ID	~16,500	<232	<130	<232	<900	Yes	No	184; 173–201 (M)	No est.

Reference	Location										
Veblen et al. (1994)[11]	NW CO	594	18	?	18	>900	Yes	No	521 (H)	No est.
Alington (1998)[8,5]	S CO	9,500	11	19	61	≥305	No	No	No est.	No est.
Low										
Barrett, Arno, & Key (1991)[5,7,8,11]	Glacier, MT	?	0	0	13	≥65	Yes	No	202 (L)	No est.
Barrett (2000)[7,9]	NW MT	~8,500	44	?	44	?	No	No	150 (L)	No est.
J. K. Brown et al. (1994)[5,7,8,9,12]	ID, MT	<32,840	20	?	20	?	No	No	112 (L)	No est.
Arno (1976)[5,7,9,11]	W MT	<1,000	12	69	?	?	No	Yes	No est.	No est.
Mathews (1980)[7,8]	W MT	3,650	71	?	71	~570	No	No	No est.	No est.
Arno, Reinhardt, & Scott (1993)[7,11]	W MT	202	35	?	35	>175	No	Yes	No est.	No est.
Loope & Gruell (1973)[9]	NW WY	?	?	?	?	?	No	No	No est.	No est.
Romme & Knight (1981)[5,7,8,9]	SE WY	4,500	.	0	26	?	No	No	300 (L)	No est.
Wadleigh & Jenkins (1996)[5,7,9,11]	N UT	1,036	14	62	?	?	No	Yes	No est.	No est.
Alexander (1981)[5,8,11,12,13]	N CO	—	4	4	0	0	No	No	No est.	No est.
F. E. Clements (1910)[7,9,12]	N CO	?	?	?	?	?	No	No	No est.	No est.
Huckaby & Moir (1995)[5,8]	N CO	9,400	0	0	70	≥210	Yes	No	No est.	No est.
Sherriff, Veblen, & Sibold (2001)[5,12,13]	N CO	—	2	33	0	0	No	No	No est.	No est.
Skinner & Laven (1982)[5,8,12]	N CO	—	3	10	0	0	No	No	No est.	No est.
Skinner & Laven (1983)[5,8,12]	N CO	9,238	9	?	0	0	No	No	No est.	No est.
Donnegan, Veblen, & Sibold (2001)[5,12,13]	C CO	—	1	15	0	0	No	No	No est.	No est.
Zimmerman & Laven (1984)[5,7,9]	C CO	?	15	≥30	15	?	No	No	No est.	No est.
Margolis, Swetnam, & Allen (2007)[12]	S CO, N NM	~4,500	182	74	12	398	No	No	No est.	No est.

Table 8.1. Continued

Sources & Study Limitations, by Forest Type & Quality of Information	Location	Study Area (ha)	No. Scar Plots	No. Scarred Trees	No. Stand-Origin Plots	No. Dated Trees	Stand-Origin Map	Fire-Year Maps	High-Severity Fire Rotation (mean and/or range, in yrs)[1]	Low-Severity Fire Rotation (mean, in yrs)[1]
STABLE BRISTLECONE PINE										
Medium										
W. L. Baker (1992)[8,12]	S CO	—	0	0	41	~500	No	No	~300 (M)	No est.
Low										
Sherriff, Veblen, & Sibold (2001)[5,12,13]	C CO	—	4	49	0	0	No	No	No est.	No est.
Donnegan, Veblen, & Sibold (2001)[5,12,13]	C CO	—	3	35	0	0	No	No	No est.	No est.
STABLE LIMBER PINE										
Medium										
Buechling & Baker (2004)[3]	N CO	9,200	231	113	231	3,461	Yes	Yes	346 (H).	7,587 (H)
Low										
Sherriff, Veblen, & Sibold (2001)[5,12,13]	N CO	—	4	42	0	0	No	No	No est.	No est.
Donnegan, Veblen, & Sibold (2001)[5,12,13]	C CO	—	1	11	0	0	No	No	No est.	No est.
STABLE WHITEBARK PINE										
Medium										
Gabriel (1976)[3,5,6,7]	NW MT	33,000	1	222	>22	?	No	Yes	150–200 (M)	No est.
Keane, Morgan, & Menakis (1994)[8,12]	NW MT	600,000	0	0	111	>2,500	No	No	228 (M)	No est.
M. P. Murray (1996); M. P. Murray, Bunting, & Morgan (1998)[3,5,6,7]	C ID	~6,350	<232	<130	<232	<900	No	No	184; 173–201(M)	No est.

Study	Location									
Mattson & Reinhart (1990)[8,3,9]	Yellowstone	?	0	0	31	?	No	No	300 (M)	No est.
Low										
J. K. Brown et al. (1994)[5,7,8,9,12]	ID, MT	<32,840	12	?	12	?	No	No	180 (L)	No est.
Larson (2005)[11,12]	W MT	300–500	3	111	3	862	No	No	No est.	No est.
Walsh (2005)[8,11,12]	MT, NW WY	~560	7	15	7	202	No	No	No est.	No est.
P. Morgan & Bunting (1990)[8,11]	NW WY	750	3	14	10	~400	No	No	200–300 (L)	No est.
Barrett (1994)[8,5,9]	Yellowstone	?	0	0	11	?	Yes	Yes	~340 (L)	No est.

Notes: Studies were conducted in or near the Rocky Mountains and are ranked by quality of information for understanding fire history across landscapes: *High* = Researchers sampled contiguous, sufficiently large (i.e., at least a few thousand ha) landscape areas, included an adequate set of scarred trees with cross-dated fires, and dated sufficient tree ages (i.e., >10 per plot) in a sufficient series of plots across a landscape, usually leading to a stand-origin map and possibly also fire-year maps. *Medium* = Researchers sampled sufficiently large, contiguous landscape areas, with a series of plots including fire scars and tree ages, but collected few or no fire scars or did not or could not cross-date fire scars or did not collect sufficient tree age data per plot. *Low* = Study had multiple limitations, including (a) not being conducted at landscape scale or plots being scattered across multiple landscapes or a small number of contiguous plots in a single landscape, so they provide limited or no information about fire history at the landscape scale; (b) insufficient sample size (few sample locations, few scarred trees, small number of trees dated per sample location, or small study area); and (c) insufficient detail about sample size or method of collection. Where data were collected and pooled across more than one vegetation type or across both montane and subalpine forests, the study is listed in each category, and the pooling is indicated. Studies are arranged from north to south within each forest type.

Table 8.1. Continued

[1]Fire rotation estimates based on high, medium, or low quantity and quality of evidence: (H) = fire areas from reconstructed fire-year maps; (M) = estimated mean or median from a histogram of stand-origin dates or from fire sizes estimated from a stand-origin map; (L) = estimated mean from a small set of dated plots or based on the time required for fuel buildup and development of mature forests.

[2]Weighted mean, weighted by the area of separate subareas.

[3]Data were collected and pooled across more than one of the subalpine forest types listed in the table.

[4]Fire rotation for surface fire was not reported but was calculated from data in table 2 in Howe and Baker (2003) from 280 years/(700 ha/3,700 ha).

[5]Insufficient sample size: small number of trees dated per sample location or no age data.

[6]Data were collected and pooled across both montane and subalpine forests.

[7]Fire scars were not cross-dated.

[8]Insufficient sample size: few or no scarred trees.

[9]Insufficient detail about sample size or method of collection.

[10]Based on data in figure 4 in Romme and Despain (1989a) and the H method in note #1, I estimated that about 0.72 of the study area burned in a 250-year period for a fire rotation (250/0.72) of about 350 years.

[11]Insufficient sample size: small study area.

[12]Study included plots scattered across multiple, disjunct landscapes or a small number of contiguous plots in a single landscape, so it provides limited or no information about fire history at the landscape scale.

[13]Insufficient sample size: few sample locations.

Table 8.2. Studies of Tree Age Structure in Subalpine Forests

| Sources | Location | Stable Aspen | Stable Lodge-Pole | Spruce-Fir & Successional | | | | Stable Limber | Stable Whitbark | FTE Spruce-Fir | FTE Limber | FTE Bristle-cone |
				Aspen	Lodge-Pole	Limber	Whitebark					
Bigler (1976)	NW MT						5					
Daniels et al. (2005)	W MT						1					
Larson (2005)	W MT						3	.				
Muir (1993)	W MT		48									
Tomback, Sund, & Hoffman (1993)	W MT						2					
Kipfmueller & Kupfer (2005)	C ID				21							
P. G. Anderson (1994)	NW WY				5				2			
Morgan & Bunting (1990)	NW WY						10					
Kashian, Turner, & Romme (2005)	Yellowstone		48									
D. L. Taylor (1969)	Yellowstone		5									
Kulakowski, Veblen, & Drinkwater (2004)	NW CO	2		2								
Kurzel, Veblen, & Kulakowski (2007)	NW CO	8		6								
P. C. Miller (1970)	NW CO				2							
Veblen, Hadley, & Reid (1991)	NW CO				9							
Veblen et al. (1994)	NW CO				7							
Aplet, Laven, & Smith (1988)	N CO				5							
P. M. Brown et al. (1995)	N CO				1							

Table 8.2. Continued

Sources	Location	Stable Aspen	Stable Lodge-Pole	Spruce-Fir & Successional Aspen	Spruce-Fir & Successional Lodge-Pole	Spruce-Fir & Successional Limber	Spruce-Fir & Successional Whitebark	Stable Limber	Stable Whitebark	FTE Spruce-Fir	FTE Limber	FTE Bristle-cone
Huckaby (1991)	N CO									3	2	1
Kashian, Romme, & Regan (2007)	N CO	1										
Knowles & Grant (1983)	N CO			2								
A. J. Parker & Parker (1983)	N CO		1									
A. J. Parker & Peet (1984)	N CO				1							
Peet (1975)	N CO		6		17							
Rebertus, Burns, & Veblen (1991)	N CO					9		2		2	3	
Roovers & Rebertus (1993)	N CO				3							
Shankman & Daly (1988)	N CO									3	3	
Shea (1985)	N CO				2							
Veblen (1986)	N CO				4	2						
Veblen, Hadley, & Reid (1991)	N CO				3							
Veblen & Lorenz (1986)	N CO				6							
Whipple & Dix (1979)	N CO		3		10							

Note: Studies were conducted in or near the Rocky Mountains and are arranged from north to south by area and alphabetically within an area. Stand age is based on the oldest trees. Many stands contain other, subdominant trees. Table entry is number of sampled stands. Authors did not always identify whether the forest was stable or successional, and thus the assignment of stands is mine and is tentative. I have omitted studies that used stand-origin dating, which does not include a full age structure. FTE = forest-tundra ecotone.

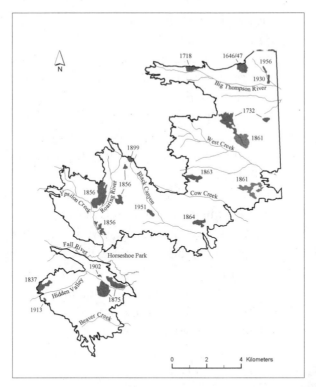

Figure 8.2. Reconstructed extent of subalpine surface fires, including surface components of mixed-severity fires, between 1533 and 2002 in a portion of Rocky Mountain National Park, Colorado.

Source: Reprinted from Buechling and Baker (2004, p. 1267, fig. 5), with permission of the National Research Council, Canada.

Fire Size

Forest reserve reports and tree ring reconstructions are primary sources of data about fire size under the HRV. Reserve reports indicate that very large areas burned in these forests in the late nineteenth century in the Rockies, but individual fires were seldom mapped or dated to year, so fire size is unknown. However, Town (1899) observed that about 28,340 hectares of lodgepole pine burned in 1898 in the Bighorn Mountains, including a fire of about 12,000 hectares and several of 4,000–5,000 hectares. Tree ring reconstructions of fire size are limited by the area it is possible to study, given the laborious nature of this work. Most studies sampled less than 10,000 hectares, so very large fires could not be fully mapped, but a few covered 25,000–50,000 hectares, and the largest, in Yellowstone, covered about 130,000 hectares (table 8.1). It is also difficult to sample all small areas, so a minimum sampling unit (e.g., patches of more than 5 hectares [Romme and Despain 1989a]; of more than 8 hectares

[Buechling and Baker 2004]) or a minimum grid-cell size (e.g., 100 hectares [M. P. Murray, Bunting, and Morgan 1998]) is used. Thus, fire-size data are often doubly truncated, missing numerically dominant small fires and the full extent of the largest fires. Fire-size distributions are based on few fires, as fire extent has not commonly been reconstructed.

Reconstructed fire-size data are available for only 54 fires in 13 subalpine basins in northern Colorado and southern Wyoming (fig. 8.3a). These small basins each had about 1,000–6,000 hectares of forest, and many fires were confined to a basin, though some burned more than one (Buechling and Baker 2004). Maximum proportion of a basin's forested area burned was about 80 percent (W. L. Baker and Kipfmueller 2001), and fires up to about 50 percent of a basin's forested area were not uncommon. A drop occurs above the 2,000 to 2,999-hectare size class (fig. 8.3a), likely because fires encountered barriers to further spread. The size distribution of 40 fires in 5 basins, from stand-origin maps (which may underestimate fire size), also drops above the 2,000 to 2,999-hectare class (fig. 8.3b); but in the largest study area (Barrett 1994), it extends up to the class of 5,000+ hectares. Fire size may commonly be limited by basin size

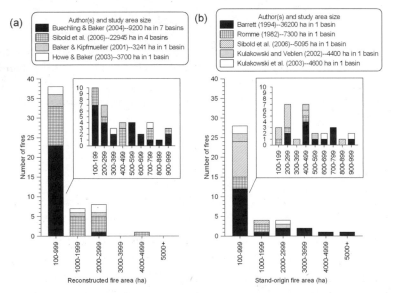

Figure 8.3. Fire-size distribution for spruce-fir and lodgepole pine forests. Data are for the Rocky Mountains under the HRV, based on tree ring reconstructions, including (a) reconstructions of original fire extent and (b) extant area from a stand-origin map. Data used extend to the period of active fire suppression identified by the authors, typically in the early 1900s.

Sources: Buechling and Baker (2004, table 1 to 1915), Sibold et al. (2006, table 4 to 1915), W. L. Baker and Kipfmueller (2001, table 1), Howe and Baker (2003, table 1 to 1910), Barrett (1994, table 1), Romme (1982; interpolated from fig. 2c), Kulakowski and Veblen (2002, table 2), and Kulakowski et al. (2003, table 2).

and natural fire breaks (W. L. Baker 2003). On the Yellowstone Plateau, a large burned area from the 1700s suggests that large fires can occur where lodgepole is continuous and fire breaks are few (Romme and Despain 1989a).

FIRE INTERVALS AND REBURNS

Fires vary in size, so a fire-interval distribution based on a count of fires is not meaningful. Data are needed on land area burned at various intervals. Also, two 100-hectare fires 10 years apart do not represent a fire interval of 10 years unless the fires overlap. Fire-interval distributions thus require spatial overlays of reconstructed or mapped fire boundaries, found in only two studies (fig. 8.4). Data sets are small, but they suggest that increasing area is burned as intervals increase up to about 200 years; more area was burned at very long intervals in one case (E. Howe and Baker 2003).

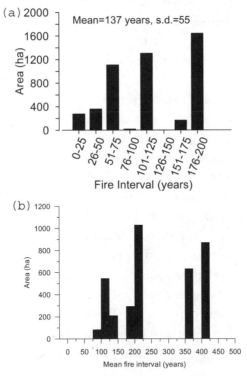

Figure 8.4. Area burned vs. time since last fire in subalpine forests. The graphs are based on reconstructed and overlaid fires (a) between 1680 and 1868 in a subalpine forest in southeastern Wyoming; and (b) between 1580 and 2000 in a subalpine forest in the Park Range, northern Colorado.

Sources: Part (a) is reprinted from W. L. Baker and Kipfmueller (2001, p. 253, fig. 3) with permission of Wiley-Blackwell Publishing Ltd.; part (b) is reprinted from E. Howe and Baker (2003, p. 803, fig. 5A) with permission of Wiley-Blackwell Publishing Ltd.

Some short-interval reburning is known. Ayres (1900a) mentions reburning of subalpine forests in northwestern Montana that had burned about 40 years earlier. Gabriel (1976) found much of the forest in this area that was killed by fire in 1889 reburned in 1895 (6 years later), 1904 (15 years later), and 1919 (30 years later). North of Yellowstone, Leiberg (1904a) observed that forests that reburned after 15–20 years were dominated by snowbush ceanothus or other shrubs, delaying tree regeneration for one or two decades. A reburn 18 years after a 1901 fire favored spruce-fir over lodgepole, which was too young to seed (Kipfmueller and Kupfer 2005). Low serotiny in young lodgepole pine (box 8.1) also suggests that short-interval reburns were rare under the HRV (Perry and Lotan 1979). In the Selway-Bitterroot wilderness, fire atlas data show that only 7.1 percent of the subalpine forest burned twice between 1880 and 1996, and another 0.6 percent burned three times (Rollins, Morgan, and Swetnam 2002). Short-interval reburning in subalpine forests over the last century has been low, as it likely was under the HRV (fig. 8.4).

BOX 8.1

Serotiny in Lodgepole Pine

Lodgepole pine has cones that open when mature, and it also has closed or serotinous cones that require heat to melt the resin that holds the cone scales together, preventing seed release. Serotinous cones may remain attached to trees for decades, with seed still viable (Lotan 1975). Serotiny level increases from the dominantly open-coned Sierra Nevada–Cascade Mountains to the Rockies (Critchfield 1957). Within the Rockies, there is considerable variability in serotiny level, but individual trees generally are either serotinous or open coned (Lotan 1976). Variability in fire severity and timing selects for mixed open and serotinous cones (Perry and Lotan 1979). However, pine squirrels select against serotiny by harvesting nearly all serotinous cones and seeds; whereas open cones allow some seed release, avoiding complete predation (Benkman and Siepielski 2004). Five isolated mountain ranges that lack pine squirrels had a median serotiny of 92 percent versus 34 percent in the Rockies in general (Benkman and Siepielski 2004).

Fires that favor serotinous cones are hot enough to melt resin but not to ignite cones. High-intensity surface fire, with torching that did not ignite all the cones, was strongly linked to high seedling density after the 1988 Yellowstone fires (Ellis et al. 1994). Modeling and analysis show the rates of fire spread and intensity sufficient to open serotinous cones but

BOX 8.1
Continued

not ignite them (fig. B8.1.1; E. A. Johnson and Gutsell 1993). These authors show that "over the 10–20-meter range of heights, if the cones are opening . . . then the canopy will also be killed," leading to high light and a mineral seedbed favorable for lodgepole regeneration (1993, 750). Despain et al. (1996) showed that short-duration (10- to 20-second) crown fire stimulated seed germination, and videos taken during the 1988 Yellowstone fires showed that the mean duration of flames in lodgepole crown fires was 24.5 seconds, insufficient to ignite most cones.

In Yellowstone, serotiny levels declined toward more mesic sites and higher elevation, with a threshold near 2,300–2,400 meters in elevation

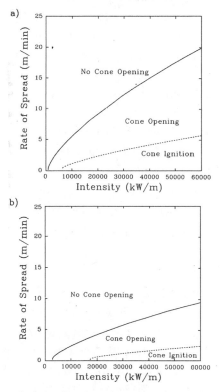

Figure B8.1.1. Serotiny in lodgepole pine. The conditions of rate of fire spread and intensity under which serotinous lodgepole pine cones will not open, will open, or will ignite for (a) a tree 10 m tall, and (b) a tree 20 m tall.
Source: Reproduced from E. A. Johnson and Gutsell (1993, p. 749, fig. 3) with permission of Opulus Press.

BOX 8.1

Continued

(Tinker et al. 1994), possibly because of longer fire intervals at higher elevations (Schoennagel et al. 2003). Serotiny levels did not consistently decline with increasing elevation in other areas (Lotan 1975; Muir and Lotan 1985; Doyle 2004) or in relation to other single environmental variables (Muir and Lotan 1985; Doyle 2004). In Yellowstone, below about 2,400 meters in elevation, serotiny level is lowest in stands less than 100 years old and highest in stands about 70–200 years old (Schoennagel et al. 2003). Old stands and long fire intervals might lead to high-intensity fires favoring serotiny (Muir and Lotan 1985), but in Yellowstone serotiny was low in stands more than 200 years old (Schoennagel et al. 2003).

Serotiny varies across landscapes. Areas within 1 kilometer of each other tend to be relatively homogenous in serotiny level, perhaps corresponding with the modal size of individual patches created by fires, whereas variability in serotiny was high at scales of 1–10 kilometers (Tinker et al. 1994). However, high variability in sapling density was not uncommon among adjacent 50×50-meter areas 11–13 years after the 1988 Yellowstone fires; this scale is finer than that from fire severity and may at least partly be associated with the scale of variation in serotiny (Kashian et al. 2004). J. K. Brown (1975, 441) suggested that "the turbulent pulsating nature of high-intensity fires also could cause wide variation in the amount of viable seed released over short distances." Further research is needed on variation in serotiny across landscapes (Kashian et al. 2004).

FIRE ROTATION

Fire rotation under the HRV can be estimated using age structures (table 8.2), stand-origin data in forest reserve reports, stand-origin and fire-scar dating, and paleocharcoal studies (fig. 8.5). A few forest reserve reports have sufficient data. Leiberg reported in 1904 (1904a, p. 27) that more than 70 percent of the forest (mostly lodgepole) north of Yellowstone had burned in the preceding 120 years, a fire rotation of about 170 years (120/0.7); he also reported that 24 percent of the forest had burned in the previous 22 years and 45 percent in the 130 years before that, a rotation of about 220 years (152/0.69). Fire rotation in the Bitterroots in Idaho and Montana can be estimated (Leiberg 1900a, 387, fig. 2) as about 285 years (179/0.629) before 1859. This reserve is only partly spruce-fir and lodgepole, but much of the fire was in these forests. In the Little Belt

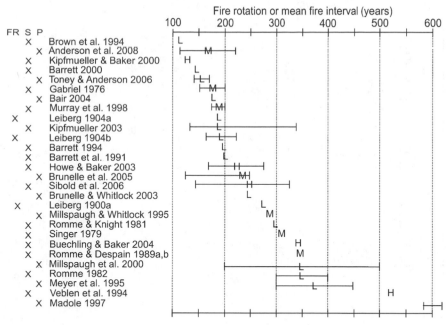

Figure 8.5. Estimated fire rotation for spruce-fir and lodgepole pine forests. Shown are available data for the Rocky Mountains, based on forest reserve (FR), stand-origin and scar-dating (S), and paleocharcoal (P) studies. Brackets indicate the low and high estimates for a particular study or for individual watersheds within a larger study area. A capital letter indicates the estimated value or the mean when a range is given, and the rating of the estimate of fire rotation (L = low, M = medium, H = high). Ratings for forest reserve estimates are all considered low. Contemporary stand-origin and fire-scar ratings are based on table 8.1. Paleocharcoal estimates are considered low if they are based on a single lake, alluvial fan, or other data source or they lack calibration with a contemporary stand-origin and fire-scar study in the same area; they are considered medium if they include a calibration and more than one lake.

Mountains, in central Montana, about half the forests are spruce-fir and lodge-pole pine; Leiberg (1904b, 23) estimated that 22 percent had burned in the previous 40 years and 58 percent in the 110 years before that, a rotation of about 188 years (150/0.8). Stand-origin dating with fire-scar analysis is the largest data source (table 8.1, fig. 8.5), potentially providing estimates of fire rotation from fire-year maps (chap. 5), but is limited to the last few centuries. Paleocharcoal studies cover the Holocene, but intervals represent fire episodes that may not equal a fire rotation (table 8.3, chap. 5).

Overall, most estimates of rotation for high-severity fire in spruce-fir and lodgepole pine forests are about 175–350 years, but some possible trends occur (fig. 8.5). First, Montana and Idaho forests appear to have somewhat shorter rotations, mostly 150–300 years, compared to 225–350 years in Wyoming and

Table 8.3. Charcoal-Based Paleoecological Studies of Fire History

Sources	Location	Species/Ecosystem						Charcoal		Mean Fire-Episode Interval
		SF	LP	WP	DF	PP	MC	Sm	Lg	
Lake, pond, wetland studies										
Power et al. (2006)	Foy Lake, NW MT				X	X			X	25–30 years since 300 BP
Brunelle & Whitlock (2003)	Burnt Knob Lake, N ID	X	X	X					X	250 years since 750 BP
Brunelle et al. (2005)	Burnt Knob Lake, N ID	X	X	X					X	250 years
	Hoodoo Lake, MT	X	X		X					250 years
	Baker Lake, MT	X		X						250 years
	Pintlar Lake, ID	X	X							125 years
C. S. Smith (1983)	Blue Lake, N ID	X	X		X	X		X		No est.
Karsian (1995)	Mary's Frog Pond, W MT	X	X		X			X		No est.
Hemphill (1983)	Sheep Mountain Bog, W MT	X	X	X				X		No est.
Mehringer, Arno, & Petersen (1977)	Lost Trail Pass Bog, SW MT	X	X	X	X			X		No est.
Millspaugh & Whitlock (1995)	Grizzly, Dryad, Mallard, Yellowstone & Duck Lakes, Yellowstone		X						X	290–300 years since ca. 1700; 110–140 years before 1700
Millspaugh, Whitlock, & Bartlein (2000)	Cygnet Lake, Yellowstone		X						X	200–500 years since 2000 BP
Millspaugh et al. (2004)	Cygnet Lake, Yellowstone		X						X	See above
	Slough Creek Lake, Yellowstone				X				X	59–83 years since 2000 BP
Whitlock et al. (2008)	Trail Lake, Yellowstone	X	X							No est.
Benedict (2000)	Bob Lake, N CO	X	X							1,320–1,760 years (krummholz)
Fall (1997)	Keystone Ironbog, C CO	X	X						X	No est.
Petersen (1988)	Beef Pasture, SW CO	X						X		No est.
Toney & Anderson (2006)	Little Molas Lake, SW CO	X							X	143–167 years since 1000 BP

Study	Location			Fire interval / notes
Anderson et al. (2008)	Hunters Lake, SW CO	X		110 years (treeline)
	De Herrera Lake, SW CO	X		100–200 years
	Brazos Ridge Marsh, N NM	X		165–225 years
Bair (2004)	Jicarita Bog, N NM	X		~180 years since 550 BP
Brunner Jass (1999); Allen et al. (2008)	Alamo Bog, N NM		X X	111 years since 500 BP
	Chihuahueños Bog, N NM		X	222 years since 400 BP
Alluvial fan studies				
Pierce, Meyer, & Jull (2004); Pierce & Meyer (2008)	S. Fk. Payette River, SW ID		X X	No est.; high-severity fire possibly at 350 years BP & 650–1050 BP
G. A. Meyer, Wells, & Jull (1995)	Yellowstone Nat. Park, WY	X		300–450 years since 4000 BP
G. A. Meyer & Pierce (2003)	Yellowstone Nat. Park, WY	X		See above
	S. Fk. Payette River, SW ID		X X	See above
Madole (1997)	Rocky Mt. Nat. Park, N CO	X	X X	580–620 years in past several thousand years
Elliott & Parker (2001)	Buffalo Creek, C CO		X X	900–1,000 years for high-severity fire + flooding; also multidecadal episodes

Note: Studies were conducted in or near the Rocky Mountains and are arranged from north to south. BP = before present, SF = spruce–fir, LP = lodgepole pine, WP = whitebark pine, DF = Douglas-fir, PP = ponderosa pine, MC = southwestern mixed conifer, Sm = small charcoal, Lg = large charcoal.

Colorado. The trend may be real, but most Montana and Idaho studies used weaker methods of estimating fire rotation and are rated low. Second, fire rotation is shorter at lower elevations, particularly in lodgepole, and longer in higher-elevation spruce-fir. Schoennagel et al. (2003) found mean fire interval was 135–185 years in the lower subalpine of Yellowstone and 280–310 years at slightly higher elevations. Sibold et al. (2006) found 162- to 216-year rotations in lower-elevation lodgepole and 401- to 713-year rotations in higher-elevation spruce-fir. Buechling and Baker (2004) found rotations of 275–281 years at lower elevations (below 3,035 m) and 344–350 years at higher elevations (above 3,035 m). Third, summer-dry settings on the west side of the Bitterroots and Yellowstone had fewer fire episodes per 1,000 years as July insolation declined over the Holocene; also, recent mean fire-episode intervals were longer relative to summer-wet locations to the east, where intervals also became shorter rather than longer over the Holocene (fig. 8.6). This moisture trend is repeated in northern Colorado, where the wetter western slope (E. Howe and Baker 2003; Sibold, Veblen, and González 2006) had shorter rotations than the drier eastern slope (Buechling and Baker 2004).

A caveat about these rotation estimates is that individual watersheds in a larger landscape may have rotations that are 100 years, or more, shorter or longer than in the larger landscape (Sibold, Veblen, and González 2006). Subalpine watersheds, particularly small ones, often had only a few large fires in the past few centuries, and the addition or subtraction of one fire can substantially change the rotation (E. Howe and Baker 2003). Some short or long estimates (fig. 8.5) thus may arise from stochasticity or limited spatial and temporal samples. Accurate estimates require observation over more than one rotation in a large land area, which was often not possible.

FIRE SEASONALITY AND FIRE-CLIMATE RELATIONSHIPS

The forest-reserve reports, done about 1900, mention subalpine fires that ignited in midsummer and burned until extinguished by rain or snow in early fall (e.g., Leiberg 1900a; Sudworth 1900a). The cumulative probability of fire-stopping precipitation during a fire started in mid-July reaches about 80 percent by mid-September (Latham and Rothermel 1993).

Regional drought during the year of the fire was strongly correlated with large fires in these forests in Montana (Kipfmueller 2003) and Colorado (Buechling and Baker 2004; Schoennagel et al. 2005; Sibold and Veblen 2006), but drought the preceding year may also contribute in Montana (Kipfmueller 2003). Large fire years in the northern Rockies were explained not by low annual precipitation but by low winter and spring precipitation and lack of peri-

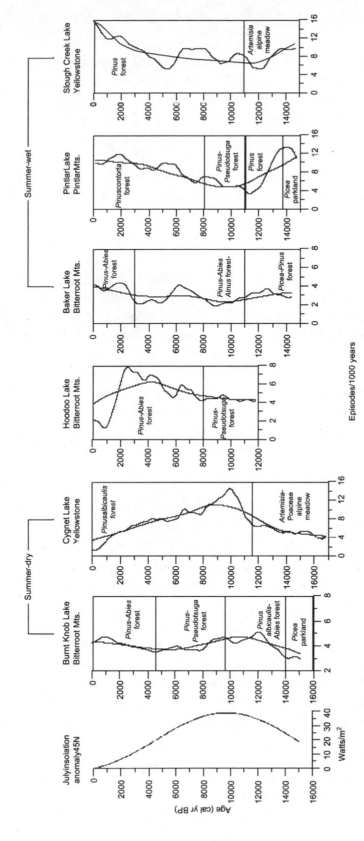

Figure 8.6. Trends over the Holocene in insolation, vegetation, and fire. Shown are trends in July insolation, vegetation types, and number of fire episodes per 1,000 years, contrasting summer-dry and summer-wet subalpine forest sites in the central and northern Rocky Mountains.

Source: Reprinted from Brunelle et al. (2005, p. 2294, fig. 8) © Elsevier 2005, with permission of Elsevier Limited.

odic summer rain (Larsen and Delavan 1922). Several synoptic climate patterns may lead to drought (W. L. Baker 2003), and teleconnections with oceanic and atmospheric conditions in both the Atlantic and Pacific are strongly linked to subalpine forest fires (chap. 2).

Fire Effects and Succession

This section covers changes in fuels, tree regeneration and stand development, recovery of the forest understory, and contingency and the shifting landscape mosaic in spruce-fir and lodgepole pine forests.

CHANGES IN FUELS

Three-quarters of ignitions in subalpine forests are in snags, duff, and down wood (Barrows 1951a). Fuel loadings are generally high in subalpine forests (table 2.3). Largest loadings are duff and large deadwood, with small amounts of crown foliage, litter, smaller deadwood, and live and dead understory vegetation (Habeck 1976; Alexander 1979; table 2.3). Low-intensity surface fires may have little effect, killing few or no lodgepole pines (Sibold et al. 2007) and reducing surface fuels, which are rapidly replaced, by less than 50 percent (Lawson 1972). High-severity fires, in contrast, consume most fine, dead fuels and duff, as well as some large wood. Biomass increases as subalpine forests recover from fire, but fuel, the biomass that is flammable, does not necessarily build up predictably (chap. 2).

TREE REGENERATION AND STAND DEVELOPMENT

Trees regenerate after fires in lodgepole pine and spruce-fir forests, but the extent and pattern of regeneration and the tree species composition during post-fire succession and stand development depend on the type of prefire forest, including stable lodgepole pine forests, stable and successional quaking aspen forests, and stable spruce-fir forests.

Stable lodgepole pine forests. Lodgepole pine forests can be stable, rather than successional to spruce-fir, on drier sites, on lower-elevation sites, or on infertile soils, as few other trees are potential successors (Whipple and Dix 1979; Peet 1981; Despain 1983). Lodgepole was successional to spruce-fir in central Colorado 6,400–4,400 years BP (before present) but has been stable during the past 2,600 years (P. L. Fall 1997).

These forests have very high variability in postfire tree density, based on observations after high-severity fires. Regeneration lags of 6–10 years were noted in northern Colorado (C. S. Crandall 1897a,b) and up to 30 years at some sites in the northern Rockies (Trimble and Tripp 1949). After the 1988 Yellowstone fires, seedling density increased through year 2, then stabilized (M. G. Turner et al. 1997). Density varied from 80 to 1.8 million trees per hectare across 12 sites at year 2 (M. Ellis et al. 1994) and from 0 to 598,500 trees per hectare among 149 plots in years 11–13 (mean = 22,718, median = 2,813 trees per hectare; Kashian et al. 2004). Fourteen other studies of the first 20 years after fires (table 8.4) found densities in these same ranges.

Estimates of tree density in these forests around 1900 (table 8.5) are consistent with observations after recent fires. Lodgepole pine forests dominated by saplings and small trees up to about 35 years of age were at times very dense (i.e., 25,000–100,000 trees per hectare), just dense (i.e., less than 7,000 trees per hectare), or even somewhat sparse. Leiberg (1899a, 270) described young lodgepole pine after high-severity fire in Montana's Bitterroot Mountains: "The closeness of growth is very striking. The trees frequently stand so close that it is difficult for a man even on foot to force his way between them."

Variation in postfire tree density is strongly shaped by fire severity. Crown fires can kill or reduce the viability of canopy seeds, even in serotinous cones, whereas moderate-severity fire (intense surface fire with torching) allows more survival and release of seed from serotinous cones and still produces a favorable mineral seedbed (J. K. Brown 1975; J. E. Anderson and Romme 1991; M. Ellis et al. 1994; M. G. Turner et al. 1997; J. E. Anderson et al. 2004). Moderate-severity fires in Yellowstone in 1988 led to second-year seedling density 4–24 times as high as in crown-fire areas (fig. 8.7), but light surface fires also had high seedling density (M. G. Turner, Romme, and Gardner 1999), also observed in Colorado (T. L. Franklin and Laven 1991). Seedlings were sufficient to replace dead overstory trees in all severities (M. Ellis et al. 1994; M. G. Turner, Romme, and Gardner 1999).

Lodgepole seed-dispersal distance is generally less than 120 meters, one of the shortest of Rocky Mountain conifers (fig. 3.13). In crown fire areas where regeneration can be dependent on seed dispersal, seedling density was negatively correlated with distance to a seed source (M. G. Turner, Romme, and Gardner 1999), and sites more than 100 meters from a moderately burned or unburned site had 2–7 times fewer seedlings (M. Ellis et al. 1994; fig. 8.7). About 50 percent of crown-fire area after the 1988 Yellowstone fires was within 50 meters of live forest, and 75 percent was within 200 meters, where eventual regeneration is likely (M. G. Turner, Hargrove et al. 1994). However, a large (about 3,700-ha) crown fire patch, five years postfire, had less than 10 seedlings per hectare and may become nonforested (M. G. Turner et al. 1997).

Postfire lodgepole seedling density is also shaped by the level of serotiny in the prefire forest (box 8.1). Prefire percent serotiny was linearly related to the log of second-year seedling density after the 1988 Yellowstone fires, and

Table 8.4. Studies of Postfire Responses of Plants in Subalpine Forests

Sources and Type of Observation by Vegetation Type	Location	Time since Fire (yrs)	Severity			Variable Studied				
			L	M	H	I	G	R	D	P
STABLE LODGEPOLE										
Observations/plots after fire										
J. E. Anderson & Romme (1991)	Yellowstone	1		X	X	X		X	X	
Ellis et al. (1994)	Yellowstone	1–2		X	X			X		
S. L. Miller et al. (1998)	NW WY	1–2			X			X		
Barth (1970)	N CO	1, 3			X	X	X	X	X	
M. G. Turner, Romme, & Gardner (1999)	Yellowstone	1–4	X	X	X			X	X	
Mason (1915)	CO, MT	1, 8, 22			X			X		
M. G. Turner et al. (1997)	Yellowstone	2–5	X	X	X	X	X	X	X	
Nyland (1998)	Yellowstone	3, 8			X			X		
J. E. Anderson et al. (2004)	Yellowstone	3–9	X	X	X			X		
M. A. Wood (1981)	Yellowstone	4–5				X	X	X		
Crandall (1897a, 1901)	N CO	4, 9, 13			X			X		
Ament (1995)	Yellowstone	5			X	X	X	X	X	
Beidleman (1957, 1967)	N CO	7, 9, 20			X			X		
R. A. Reed et al. (1999)	Yellowstone	9			X					X
Kashian et al. (2004)	Yellowstone	11–13		X	X			X		
M. G. Turner et al. (2004)	Yellowstone	11–12		X	X			X		X
Schoennagel, Turner, & Romme (2003)	Yellowstone	12			X			X		
Schoennagel, Veblen, & Romme (2004)	Yellowstone	12			X	X	X	X	X	
Crandall (1897b)	N CO	22+			X			X		
A. J. Parker & Parker (1994)	C CO	120–140			X			X		
Chronosequence studies										
D. L. Taylor (1969, 1971, 1973)	Yellowstone	1–300				X	X	X	X	
Trimble & Tripp (1949)	N WY, W MT	1–300+				X		X		
Clagg (1975)	N CO	1–370				X	X	X	X	
Clements (1910)	N CO	4–202	X			X	X	X		
Kashian, Turner, & Romme (2005)	Yellowstone	12–350				X		X		
Smith & Resh (1999)	SE WY	15–260				X				X
Peet (1975, 1978, 1981)	N CO	20–350				X		X		X
Kashian et al. (2005)	Yellowstone	50–350				X				X
Kipfmueller & Baker (1998b)	SE WY	50–500				X	X			
Moir (1969)	N CO	67–107				X		X		
Pearson, Knight, & Fahey (1987)	SE WY	75–240				X				X

Table 8.4. Continued

Sources and Type of Observation by Vegetation Type	Location	Time since Fire (yrs)	Severity			Variable Studied				
			L	M	H	I	G	R	D	P
QUAKING ASPEN SEEDLINGS										
Observations/plots after fire										
Kay (1993)	Yellowstone	1, 3			X		X			
Romme et al. (1997)	Yellowstone	3–5	X	X	X		X			
Kay (1993)	Grand Teton	4, 6			X		X	X		
M. G. Turner, Romme, & Reed (2003)	Yellowstone	8	X	X	X		X			
Romme et al. (2005)	Yellowstone	8–12			X		X			
STABLE QUAKING ASPEN[a]										
Observations/plots after fire										
Doyle (1997, 2004)	NW WY	1–25	X	X	X	X	X			
Zier & Baker (2006)	SW CO	~55–130			X		X			
Manier & Laven (2002)	SW CO	~85–100			X		X			
Kulakowski, Veblen, & Drinkwater (2004)	NW CO	100			X		X			
Kulakowski, Veblen, & Kurzel (2006)	NW CO	100			X		X			
STABLE SPRUCE-FIR AND SUCCESSIONAL LODGEPOLE, ASPEN, LIMBER, WHITEBARK										
Observations/plots after fire										
Ashton (1930), R. A. Nelson (1934, 1954)	N CO	0, 4, 7, 24			X	X	X			
Farris, Neuenschwander, & Boudreau (1998)	C ID	1			X		X			
Barth (1970)	N CO	1, 3			X	X	X	X	X	
Doyle (1994); Doyle et al. (1998)	NW WY	1–3, 9, 17	X	X	X		X	X		
Patton & Avant (1970)	N NM	1–5			X		X			
Lyon (1976, 1984)	W MT	1–21			X	X	X	X	X	
M. A. Wood (1981)	Yellowstone	3–4			X	X	X			
Lowdermilk (1925)	W MT	10			X		X			
Beidleman (1957, 1967)	N CO	31, 41			X		X			
Zier & Baker (2006)	SW CO	~55–130			X		X			
Jenkins, Dicus, & Hebertson 1998	NE UT	~93–150			X		X			
Kulakowski, Veblen, & Drinkwater (2004)	NW CO	100			X		X			
Kulakowski, Veblen, & Kurzel (2006)	NW CO	100			X		X			
Franklin & Laven (1991)	N CO	100	X		X		X			
W. L. Baker (1991)	C CO	120			X		X			
Chronosequence studies										
Doyle (1997, 2004)	NW WY	1–25		X	X		X			

Table 8.4. Continued

Sources and Type of Observation by Vegetation Type	Location	Time since Fire (yrs)	Severity			Variable Studied				
			L	M	H	I	G	R	D	P
Hodson & Foster (1910)	CO	1–30			X		X			
Clagg (1975)	N CO	1–400			X	X	X	X		
Arno, Simmermann, & Keane (1985)	W MT	3–300+			X	X				
Ives (1941)	N CO	10–78			X			X		
Stahelin (1943)	CO, WY	50–82			X			X		
Kipfmueller & Kupfer (2005)	C ID	80–800			X			X		
Peet (1975, 1978, 1981)	N CO	73–500+			X			X		X
Aplet, Smith, & Laven (1989)	N CO	125–700			X			X		X
Aplet, Laven, & Smith (1988)	N CO	175–575			X			X		
Whipple & Dix (1979)	N CO	220–500			X			X		
Veblen (1986)	N CO	240–550			X			X		
Romme & Knight (1981)	SE WY	260–640			X			X		
STABLE BRISTLECONE PINE										
Chronosequence studies										
W. L. Baker (1992b)	S CO	50–750			X			X		
STABLE LIMBER PINE										
Chronosequence studies										
Peet (1981)	N CO	50–250?			X			X		
Rebertus, Burns, & Veblen (1991)	N CO	90 to >1,000			X			X		
STABLE WHITEBARK PINE										
Observations/plots after fire										
Ash & Lasko (1990)	W MT	1–3		X	X			X		X
Sund (1988); Tomback, Hoffman, & Sund (1990); Tomback, Sund, & Hoffman (1993)	W MT—Sleeping Child fire	26		X	X			X		
Tomback, Hoffman, & Sund (1990); Tomback, Sund, & Hoffman (1993); Tomback (1994)	W MT—Saddle Mt. fire	28			X			X		
Doyle (2004)	NW WY	6–8			X			X		
Tomback et al. (1995)	N ID	25			X			X		
Kendall & Arno (1990)	N ID, W MT	80			X			X		
Chronosequence studies										
Murray, Bunting, & Morgan (2000)	C ID, W MT	0–240						X		
P. Morgan & Bunting (1990)	NW WY	25–320	X	X				X		

Table 8.4. Continued

Sources and Type of Observation by Vegetation Type	Location	Time since Fire (yrs)	Severity L	M	H	Variable Studied I	G	R	D	P
T. Weaver, Forcella, & Dale (1990)	ID, MT, WY	29–650			X			X		X
Forcella & Weaver (1977)	W MT	29–650			X			X		X
Weaver & Dale (1974)	W MT	40–420			X	X		X	X	
FOREST-TUNDRA ECOTONE										
Observations/plots after fire										
Bollinger (1973)	N CO	10–85			X	X		X		
Stahelin (1943)	CO, WY	50–82			X			X		
Shankman & Daly (1988)	N CO	66–87			X			X		
Huckaby (1991)	N CO	68–122			X			X		
Shankman (1984)	N CO	78			X			X		
Billings (1969)	SE WY	160			X	X		X		

Notes: Studies were conducted in or near the Rocky Mountains. Table entries are ordered by time since fire and whether observations were of one or more fires or a chronosequence of fires. L = low severity, M = moderate severity, H = high severity, I = individual species, G = species groups (i.e., forbs, graminoids, shrubs), R = tree regeneration or density (or other tree population structural variables), D = diversity, P = production or growth (e.g., annual basal area increment).

[a]Authors often did not identify whether burns were in stable or successional aspen, so this is tentative.

serotiny alone explained 45–50 percent of variation in postfire seedling density (J. E. Anderson et al. 2004). Another study found prefire percent serotiny was the best predictor of year 2 postfire seedling density in three areas in Yellowstone (M. G. Turner et al. 1997). However, across 50 stands, prefire percent serotiny explained only about 13 percent of variation in year 12 postfire seedling density (Schoennagel, Turner, and Romme 2003). Effect of serotiny, moreover, was limited to lower-elevation sites (less than 2,400 m), as serotiny declines with elevation. At lower elevations, fires occurring at shorter intervals (less than 100 years) had the lowest postfire seedling density and lowest prefire serotiny, while fires after intermediate intervals (70–200 years) had the highest prefire serotiny and postfire seedling density (Schoennagel, Turner, and Romme 2003).

Postfire seedling density is also affected by competition, seedbed character, environment, seed predation, and mycorrhizae. Competition by resprouting postfire plants, particularly blueberries (*Vaccinium* sp.), reduced seedling density by half (F. E. Clements 1910). Other plants that reduced seedlings include kinnikinnick (*Arctostaphylos uva-ursi*; F. E. Clements 1910; Mason 1915) and pinegrass (*Calamagrostis rubescens*; J. E. Anderson et al. 2004). Third-year postfire

Table 8.5. Estimates of Tree Density in Subalpine Forests near or before 1900

Location	Elevation (m)	Density ca. 1900 (trees per acre)	Density ca. 1900 (trees per ha)	Notes	Forest Type	Age of Forest ca. 1900 (yrs)	Sources
NW MT	?	40,000	98,800	Total trees	Lodgepole	?	Ayres (1900a, 294)
W MT	?	20,000–30,000	49,400–74,100	Total trees	Lodgepole	Sapling	Leiberg (1900a, 347)
N CO	2,900	17–68	42–168	Includes seedlings	Lodgepole	9	Crandall (1901, 63–72; 1890 fire)
N CO	2,500–2,900	200–1,200	494–2,964	Includes seedlings	Lodgepole	~20–25	Crandall (1901, 46–47; scattered type)
SE WY	2,650	2,708	6,689	Includes seedlings	Lodgepole	~30	Crandall (1901, 56–58; scattered type)
N CO	2,500–2,900	11,612–17,420	28,682–43,027	Includes seedlings	Lodgepole	~30–35	Crandall (1901, 46–47; dense type)
W MT	?	1,000–2,000	2,470–4,940	>10 cm	Lodgepole	60–80	Leiberg (1900a, 363)
SE WY	2,500–2,900	267	659	>10 cm	Lodgepole	66	Crandall (1901, 61–63)
		161	398	<10 cm			
		274	677	Small seedlings			
		702	1734	Total trees			
W MT	?	500–800	1,235–1,976	>10 cm	Spruce-fir	75–175	Leiberg (1900a, 363)
SE WY	2,500–2,900	597	1,475	>10 cm	Lodgepole	87	Crandall (1901, 61–63)
		440	1,087	<10 cm			
		184	454	Small seedlings			
		1,221	3,016	Total trees			
NW CO	?	148–316	366–781	>25 cm	Spruce-fir	110–220	Sudworth (1900b, 154–162; 14 plots)
NW CO	?	160	395	>25 cm	Lodgepole	145	Sudworth (1900b, 158; 1 plot)
N CO	2,500–2,900	267–773	659–1,909	>10 cm	Lodgepole	154–235	Crandall (1901, 79)
N CO	>2,900	1,082	2,673	>5 cm	Lodgepole–spruce-fir	~180	Crandall (1901, 96–98)

Note: Studies were conducted in or near the Rocky Mountains. These estimates are direct reports from scientists ca. 1900. Entries are arranged by age of forest.

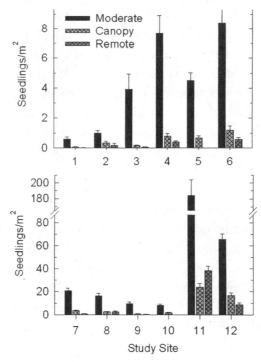

Figure 8.7. Seedlings of lodgepole pine two years after the 1988 Yellowstone fires. Shown is the mean density of seedlings in paired areas of moderate and severe (canopy) fire within 60 m of each other and in a remote canopy-fire area more than 100 m from an unburned or moderately burned forest that might be a seed source. Numbers along x-axis represent the 12 study sites.

Source: Reproduced from J. E. Anderson et al. (2004, p. 69, fig. 4.2), copyright 2004, with permission of Yale University Press, New Haven.

seedling density was significantly higher where postfire herbaceous density was lower (Barth 1970). A general effect of competition by other plants on postfire seedling density was lacking after the 1988 Yellowstone fires, and postfire seedlings were favored on burned litter or duff, rather than mineral soil (J. E. Anderson et al. 2004). Third-year seedling density was higher on south-facing slopes and in lower slope positions (Barth 1970), and twelfth-year seedling density was also higher on gentler slopes and less fertile soils. Soil fertility was the strongest predictor of year 12 seedling density in Yellowstone (Schoennagel, Turner, and Romme 2003). Lodgepole seedlings that survived into year 2 after fire in Wyoming were all mycorrhizal, perhaps because mycorrhizae increase water-use efficiency (S. L. Miller et al. 1998).

In spite of large differences in tree density early in succession, forest structure generally converges during stand development, based on chronosequence and age-structure studies (tables 8.2, 8.4). Peet (1981) studied 23 stands and Clagg (1975) 17 stands in Rocky Mountain National Park and found that density increased until 70–125 years after fire; then it declined, as did density variation (fig. 8.8a,b,e). Kashian et al. (2005) studied 48 stands in Yellowstone and found that tree density and variation in density declined until stands were about 200 years old (fig. 8.8c,d). Stands with high initial density self-thinned, and sparse stands filled in with new regeneration. Dense stands could persist, however; a 70-year-old Montana stand had about 250,000 trees per hectare (Mason 1915), and a "mature" stand appeared still quite dense near 1900 (fig. 8.8f). In Colorado, most stands from about 125–250 years after fire were mature, with little regeneration and slow growth (F. E. Clements 1910; Clagg 1975; Peet 1981). About 1900, maturing lodgepole pine forests (i.e., about 60–100 years after fire) typically had about 1,700–5,000 trees per hectare, while stands about 100–250 years after fire had about 350–2700 trees per hectare (table 8.5). Canopy mortality allowed renewed regeneration from about 250–325 years after fire (Whipple and Dix 1979; Peet 1981; Romme 1982), but rare low-severity fires with torching also could open the canopy and allow regeneration (J. K. Brown 1975).

Stable and successional quaking aspen forests. Stable quaking aspen forests occur in both the montane (chap. 6) and subalpine zones; they are rarer above 3,000 meters in elevation from central Colorado north (Kulakowski, Veblen, and Drinkwater 2004; Kulakowski, Veblen, and Kurzel 2006; Kashian, Romme, and Regan 2007), although common in southern Colorado (Zier and Baker 2006). Less than half of the aspen forest above 3,000 meters elevation in northwestern Colorado in the late nineteenth century was stable or persisted until the late twentieth century; the rest became conifer dominated (Kulakowski, Veblen, and Drinkwater 2004; Kulakowski, Veblen, and Kurzel 2006).

Aspen is successional in many subalpine conifer forests. It increased in western and southern Colorado following nineteenth-century fires and other disturbances in subalpine conifer forests (Manier and Laven 2002; Zier and Baker 2006; Margolis et al. 2007). Conifers increased relative to aspen over the last century in subalpine forests in northern New Mexico, based on 24 rephotographs (Sallach 1986). After 1860s subalpine fires in northern Colorado, conifers regained dominance within about 75 years (Ives 1941), but in central and southern Colorado, spruce-fir forests that had burned in high-severity fires remained aspen dominated 62–82 years later (Stahelin 1943) and even about 100–145 years later (Margolis, Swetnam, and Allen 2007). In northwestern Colorado, about 41 percent of spruce-fir forest in the Flat Tops (Kulakowski et

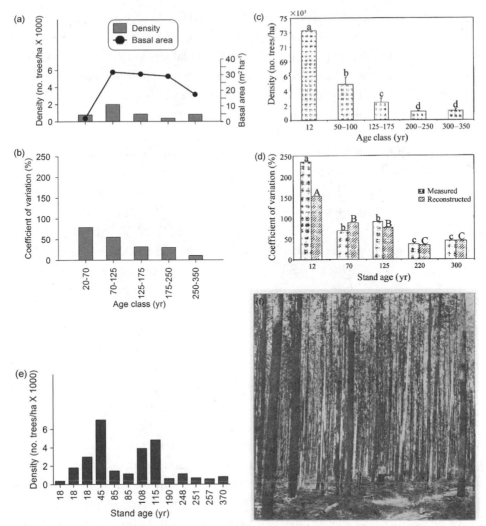

Figure 8.8. Changes in forest structure during stand development after high-severity fire. Changes are based on chronosequence studies: in Rocky Mountain National Park, of (a) tree density (left axis) and basal area (right axis), and (b) coefficient of variation in tree density; in Yellowstone National Park, of (c) tree density, and (d) coefficient of variation in tree density; (e) of lodgepole pine more than 1 meter tall in Rocky Mountain National Park; and (f) of "mature lodgepole pine forest in Swan Valley, showing a dense stand in which the trunks are 15 to 20 centimeters in diameter."

Sources: Parts (a–b) based on data in Peet (1981); parts (c–d) reproduced from Kashian et al. (2005, p. 647, fig. 1) with permission of the Ecological Society of America; part (e) based on data in Clagg (1975); (f) reproduced from Whitford (1905, p. 293, fig. 23) with permission of the University of Chicago Press.

al. 2006) and about 62 percent on Grand Mesa (Kulakowski, Veblen, and Drinkwater 2004) that was burned in the late nineteenth century was still aspen dominated in the late twentieth century (Kulakowski et al. 2006). In Utah, F. S. Baker (1925) thought subalpine conifers might regain dominance by about 150 years after fire, but overall it appears that from 75 to more than 150 years may elapse before subalpine conifers regain dominance after fire where aspen is successional.

Stable aspen-conifer mixes may also be common. In the San Juan Mountains in southern Colorado, more than half of early photographs containing subalpine aspen showed stability or increase in aspen a century later (Zier and Baker 2006). At high elevations in Utah, conifers might never fully dominate, and conifer-aspen mixes may be stable (F. S. Baker 1925). A mix of aspen and conifers was found in stands more than 250 years old in central western Colorado (McKenzie 2001). Conifer-aspen mixes were common in the subalpine in the San Juans a century ago, and many lacked more conifers or aspen a century later (Zier and Baker 2006).

Aspen even dominated after fire in some lodgepole pine forests (R. A. Nelson 1934, 1954). A mosaic of aspen-lodgepole was found 45 years after fire in central Colorado (Langenheim 1962) and 75–80 years after fire in northern Colorado (Parker and Parker 1983; Kashian, Romme, and Regan 2007). These trees alternately dominated after fires over the past 8,000 years in central Colorado (P. L. Fall 1997).

Although seedlings of aspen are rare in the western United States, they appeared in burned lodgepole pine forests, aspen forests, and even wetlands after the 1988 Yellowstone fires; their fate remains uncertain, however (table 8.4; Kay 1993; Romme et al. 2005). Seedlings were most abundant in severe burns at lower elevations, near live adult aspen trees, and downwind of large geographical concentrations of unburned aspen (Turner et al. 2003). Seedlings originated most abundantly in year 1 after fire and at declining levels through year 8 (Romme et al. 2005). However, they were still less than a meter tall 12 years after fire, as a result of browsing pressure and unfavorable environments that may not allow persistence (Romme et al. 2005).

Stable spruce–fir forests. Above the elevation in which lodgepole is common, spruce-fir forests appear stable—generally lacking other trees that can succeed them—and they can progress through several stages of stand development after high-severity fire. First, a colonization stage of postfire tree regeneration lasts 20–100 years (Clagg 1975; Peet 1981; Veblen 1986) or even 200 years (Aplet, Laven, and Smith 1988). Fir, spruce, or both may predominate (Hodson and Foster 1910; Lowdermilk 1925; Beidleman 1957; Doyle 2004). In a second stage—spruce exclusion—regeneration of spruce declines, while older spruce overtop and dominate fir, which is shade tolerant and may continue regenerating (Clagg 1975; Whipple and Dix 1979; Peet 1981; Aplet, Laven, and Smith 1988). In the third stage—spruce reinitiation—fir reaches its roughly 300-year life span, and mortality of fir and some spruce reduces basal area and opens the canopy, allowing renewed spruce regeneration (Peet 1981; Aplet, Laven, and Smith 1988). A fourth stage—second-generation spruce-fir forest—may begin about 500 years after fire if renewed regeneration reaches the canopy, with spruce dominance favored by small-scale canopy disturbance and fir favored by its shade tolerance (Peet 1981; Aplet, Laven, and Smith 1988). This sequence is not often realized, as it is altered by insect outbreaks, wind, and other disturbances (W. L. Baker and Veblen 1990; E. Howe and Baker 2003), and even by tree-regeneration failures.

Postfire spruce-fir regeneration is most favored on north-facing slopes at moderate elevations in the subalpine zone, particularly if trees survive and provide seed, soils are granitic or fertile (Hodson and Foster 1910), and postfire vegetation is dominated by *Vaccinium* (Stahelin 1943). Under these conditions, thousands of seedlings per hectare are common after fire (Lowdermilk 1925; Doyle 2004).

However, postfire tree regeneration may be slow, deficient, or even absent on dry sites, such as on south-facing slopes and at high elevations (box 8.2). Hodson and Foster (1910, 13) noted in Colorado: "Large burned tracts are often wholly without spruce seedlings, and on many such tracts there is little prospect of reproduction" because of lack of seed or adverse soil conditions after fire. A meadow created by fire in spruce-fir in Colorado about 1915 generally lacked seedlings 51 years later, except at the margin (Beidleman 1967), and large areas of burned spruce-fir remained grasslands 55–65 years after fire in central Colorado (Langenheim 1962). Stahelin (1943) studied eight fires burned 50–82 years earlier, in Colorado and southern Wyoming, and found that conifer regeneration was poor where there were few surviving trees to provide seed, on drier south- or east-facing slopes, on nongranitic soils, and with postfire graminoid cover. Doyle (2004) also found that long distance from surviving trees and dense herbaceous cover impeded conifer regeneration. In northwestern Colorado, about 10–13 percent (on Grand Mesa; Kulakowski, Veblen, and Drinkwater 2004) and 31 percent (in the Flat Tops; Kulakowski et al. 2006) of spruce-fir forest present in the late 1800s was "grassland/shrubland" in the late 1900s, presumably a result of fire. About 60–69 percent of the grassland/shrubland present in the late-1800s was aspen or spruce-fir forest a

BOX 8.2

Fire in the Forest-Tundra Ecotone

The forest-tundra ecotone (FTE) often consists of a lower zone of closed subalpine forest; a zone of patch forest with intervening subalpine meadows; and a krummholz zone of shorter, wind-trimmed trees (Weisberg and Baker 1995). Fires did occur, even in krummholz (fig. B8.2.1) or in ribbon forest common in the central and northern Rockies (Billings 1969). The recurrence interval for fire in the FTE is 1,320–1,760 years, based on a charcoal study in northern Colorado (Benedict 2000). W. L. Baker, Honaker, and Weisberg (1995) mapped all fires visible in aerial photos over a 36,750-hectare area of FTE in Rocky Mountain National Park and found 25 separate fires covering 1,604 hectares in total, a mean of 64 hectares per fire. Fires that burned in the late 1800s were still visible. Thus, using equation 5.1 (chap. 5), fire rotation for the FTE is estimated as 100 years/(1,604/36,750) = about 2,300 years, which is longer than Benedict's estimate. Burned area was greater and fire rotations thus shorter in closed forest (985 hectares, about 1,750 years) than in patch forest (569 hectares, about 2,175 years) or krummholz (50 hectares, about 14,000 years), supporting Huckaby's (1991) observation that fires often burn up only through the closed forest. The Baker, Honaker, and Weisberg estimate is not very reliable, as only about 100 years of observation were used to es-

Figure B8.2.1. Fire in the forest-tundra ecotone. Winter scene after a high-severity fire in limber pine krummholz on Long's Peak, northern Colorado, in 1927, by "Skinner."

Source: Courtesy of Rocky Mountain National Park archives, negative no. 2496.

BOX 8.2
Continued

timate long rotations, but it is likely, based on these studies that fire rotation in the forest-tundra ecotone exceeds 1,000 years.

Given this long rotation, the position of the FTE is likely not controlled by fire, as suggested by some (Peet 1981; Noble 1993), or by other disturbances, at least in northern Colorado (W. L. Baker, Honaker, and Weisberg 1995). However, postfire recovery can be slow, and some postfire meadows have persisted for one to two centuries after forest fires (Stahelin 1943; Billings 1969; Habeck 1969; Bollinger 1973; Shankman and Daly 1988). Other FTE areas burned in the last century (table 8.4) have regenerating trees, but often with slow growth and low density, particularly at higher elevations of the FTE (Bollinger 1973; Peet 1981; Shankman 1984; Shankman and Daly 1988; Huckaby 1991) or where dense postfire graminoid turf dominates (Stahelin 1943). Burns in the FTE generally regenerated slowly until a pulse of regeneration occurred between the mid-1940s and the 1970s (Huckaby 1991; Moir and Huckaby 1994); this was a cool, snowy period during a negative Pacific Decadal Oscillation (Alftine, Malanson, and Fagre 2003). Moreover, regeneration in the FTE is heterogeneous, favored in patch forests more than krummholz, and in snowbeds, willow wetlands, and areas of *Vaccinium* (Weisberg and Baker 1995). Time required for full recovery of burned FTE forests is probably several centuries (Shankman and Daly 1988; Rebertus, Burns, and Veblen 1991). Recovery may be shorter in moister settings and if cool, snowy decades occur, particularly if surviving trees with seed are nearby; but recovery may be very long on xeric sites or in dense meadows, particularly if warm, dry decades occur.

century later (Kulakowski, Veblen, and Drinkwater 2004; Kulakowski, Veblen, and Kurzel 2006), illustrating slow but dynamic fluctuations in amounts of grassland/shrubland and subalpine forest.

At lower elevations of the spruce-fir zone and on drier sites, spruce and lodgepole or just lodgepole can dominate postfire succession; but at higher elevations, spruce may colonize first, followed by lodgepole (Whipple and Dix 1979; Veblen 1986; Doyle et al. 1998). Where prefire lodgepole serotiny is low and seed dispersal controls postfire regeneration, spruce seedlings may dominate, with highest seedling density in moderately burned areas that have surviving seed trees (Doyle et al. 1998). Very short (i.e., less than 20 years) or long (more than 250 years) fire-free intervals may favor spruce and fir in early postfire succession, as they are better postfire dispersers than lodgepole (Kipf-mueller and Kupfer 2005). If spruce and fir colonize and dominate early, pro-gression to late-successional spruce-fir dominance may proceed more quickly than if lodgepole dominates (J. K. Brown 1975; Kipfmueller and Kupfer 2005).

Where serotiny is higher and lodgepole dominates before fire, it is likely to also dominate after fire, with postfire density strongly shaped by prefire serotiny level. In Doyle's (2004) Grand Teton study area, lodgepole seedling density was higher in severely burned than moderately burned areas, unlike in Yellowstone's stable lodgepole forests (see "Stable Lodgepole Pine Forests" section above); however, these study areas have different environments. Lodgepole seedling density was also high after high-severity fire in successional lodgepole in Mon-tana, averaging about 84,000 seedlings per hectare in the first few postfire years (Lyon 1984). Where lodgepole dominates after fire, a peak in fir regeneration can be delayed 60–120 years (Whipple and Dix 1979; Veblen 1986; P. G. Ander-son 1994). Eventual decline in lodgepole regeneration and breakup of the lodgepole canopy allow increasing spruce or fir dominance by about 300 years after fire in Colorado (Peet 1981; Veblen 1986) and more than 250–350 years in the northern Rockies (Kipfmueller and Kupfer 2005). Rare surface fire with some torching can occasionally create small gaps that allow renewed lodgepole regeneration and decrease fir regeneration, delaying replacement of lodgepole (T. L. Franklin and Laven 1991). With initial lodgepole dominance, full recovery of spruce-fir forest may take 200–300 years in drainage bottoms, 400 years on mesic uplands, and centuries longer on xeric uplands, because of the 350- to 400-year life span of lodgepole in this area. On xeric sites and at lower elevations, where spruce and fir regeneration are less favored, mature forests may contain a mixture of fir, spruce, and lodgepole (Romme and Knight 1981). Data on spruce-fir tree density about 1900 are meager but suggest that these forests may have been less dense than comparably aged lodgepole pine forests (table 8.5).

On xeric sites in the northern Rockies, whitebark pine (Doyle 2004; Kipf-mueller and Kupfer 2005), and in the southern Rockies, limber pine (Rebertus, Burns, and Veblen 1991), can expand after fire at high elevations, even if not present in the prefire forest. Expansion is favored by long-distance dispersal by Clark's nutcracker (fig. 8.9) and tolerance for xeric postfire conditions. On xeric sites, limber can precede establishment of lodgepole, spruce, and fir by 10–130 years and may facilitate the establishment of these trees (Veblen 1986; Re-bertus, Burns, and Veblen 1991; Donnegan and Rebertus 1999). Similarly,

Figure 8.9. Clark's nutcracker.
Source: Photo by the author.

whitebark pine can precede and facilitate the establishment of subalpine fir in the northern Rockies (Callaway 1998; Tomback et al. 2001). As the spruce and fir mature, limber and whitebark regeneration declines and mortality increases (e.g., Donnegan and Rebertus 1999). Full replacement of early-successional limber by spruce and fir may take several hundred years (Peet 1981), and a high-severity fire is typical, given the fire rotation (table 8.1) before this occurs (Rebertus, Burns, and Veblen 1991). Thus, the spruce-fir and successional limber or whitebark landscape is one of fluctuating dominance by these trees, shaped by contingencies: seed dispersal by Clark's nutcracker and wind, severity of the site, time since the last fire, the fire's extent, and shifting competition and facilitation among trees.

Recovery of the Forest Understory

In both lodgepole pine and spruce-fir forests (table 8.4), surviving plants resprouted and dominated the first year or two after fire, persisting into later stages of succession (Clagg 1975; J. E. Anderson and Romme 1991; Doyle et al. 1998), suggesting the initial-floristics succession model (chap. 3). Resprouting was high after the 1988 Yellowstone fires, probably because in the highest-severity burns, mean depth of soil char was only about 1.4 centimeters, shallow relative to typical rhizome and root depths (M. G. Turner, Romme, and Gardner 1999). Seed dispersal was the source of few plants after subalpine forest fires (J. E. Anderson and Romme 1991; M. G. Turner et al. 1997; Doyle et al. 1998). And only a few of those, including American dragonhead (*Dracocephalum parviflorum*) and streambank wild hollyhock (*Iliamna rivularis*), commonly originate from a seed bank (J. E. Anderson and Romme 1991; Doyle et al. 1998). Seed banks in subalpine forests have low seed density, are only about 20–35 percent similar to aboveground vegetation, and are reduced about 77 percent by high-severity fires, which suggests that they are not very important sources of postfire plants (Whipple 1978; D. L. Clark 1991). Geographic variation in prefire vegetation was most important in explaining vegetation variation after the 1988 Yellowstone fires (M. G. Turner et al. 1997). It is the local mix of prefire plants, since most resprout, that largely shapes postfire vegetation.

Dominant resprouting perennials appear resilient to successive severe fires regardless of how soon another fire occurs. Time since the last fire before 1988 had little influence on postfire plant composition and abundance five years after the 1988 Yellowstone fires (Ament 1995). At year 12, total cover, species richness, and cover of resprouting perennials did not differ between sites burned after short (7–100 years) compared to long (100–395 years) intervals; short intervals did lead to higher cover of annuals and lower cover of shrubs and fire-intolerant perennials, but these were relatively minor components of the prefire community (Schoennagel et al. 2004).

Higher fire severity, which is often also associated with larger fire patches, led to generally lower total plant cover in the first five years after fire in lodgepole pine in Yellowstone (J. E. Anderson and Romme 1991; M. G. Turner et al. 1997). Cover of forbs, graminoids, and shrubs all decreased from light surface burns to severe surface burns to crown fires (M. G. Turner et al. 1997). By years 4 and 5 after fire, light and severe surface burns did not differ in total plant cover, but crown fires still had less cover (M. G. Turner et al. 1997). Lyon (1984, 16) said that vegetation right after fire often is sparse: "If live vegetation is the only criterion, there is reason for concern and little that can be done to alleviate it." However, five years after the 1988 Yellowstone fires, bare ground had less than 20 percent cover in all burned forests, except at tree line (Ament 1995).

Cover of perennial forbs that dominated the prefire forest understory increased from year 1 to year 3 after high-severity fire in lodgepole in Colorado (Barth 1970). Forbs recovered to prefire levels three to four years after a spruce-fir fire and four to five years after a lodgepole fire in Yellowstone (M. A. Wood

1981). By three years after the 1988 Yellowstone fires, forb cover was higher in the most severely burned areas than in unburned or less severe burns (M. G. Turner, Romme, and Gardner 1999). However, after a Montana lodgepole fire, perennial forbs increased rapidly the first three to four years and continued to increase into the second decade after fire (Lyon 1984).

Native perennial graminoids recovered to prefire levels three to four years after a spruce-fir fire and four to five years after a lodgepole fire in the 1970s in Yellowstone (M. A. Wood 1981). Perennial graminoid cover recovered to prefire levels on light and severe surface burns in the first two years after the 1988 Yellowstone fires, but graminoid cover remained reduced for five years on crown fire sites (M. G. Turner, Romme, and Gardner 1999). Graminoids continued a gradual increase over a period of two decades after high-severity fire in Montana (Lyon 1984).

Shrub cover remained reduced in crown fire areas five years after the 1988 Yellowstone fires (M. G. Turner et al. 1997), as well as three to five years after 1970s crown fires in Yellowstone (M. A. Wood 1981). Grouse whortleberry (*Vaccinium scoparium*), one of the dominant forest shrubs, reseeds poorly and has rather shallow rhizomes that make it vulnerable to significant reduction and slow recovery at higher fire severity (M. G. Turner et al. 1997). However, it may recover quickly in some cases, particularly compared to nonsprouting shrubs such as common juniper (*Juniperus communis*), which reached a peak in abundance 100–300 years after fire (Clagg 1975). Annuals were initially uncommon after the 1988 Yellowstone fires but increased to a peak about three years after fire before declining. One annual, maiden blue-eyed Mary (*Collinsia parviflora*) can disperse long distances and increased on large, severe burns (M. G. Turner et al. 1997). A few postfire annuals, including *Collinsia*, are common in a variety of burned forests (Ament 1995).

Details of individual species' response to fire are documented in studies in western Montana, Yellowstone National Park, Grand Teton National Park, and northern Colorado (table 8.4). Species diversity and productivity during postfire succession are reviewed in chapter 3.

CONTINGENCY AND THE SHIFTING LANDSCAPE MOSAIC

A shifting mosaic of disturbance patches characterizes most landscapes subject to fire (box 3.1). In subalpine forests, spatial variation in the prefire landscape, the fire itself, and the immediate postfire environment shapes the mosaic (table 8.6). Prefire subalpine forest landscapes are spatially heterogeneous, in part because of the underlying topographically complex mountain environment and its diversity of physical settings and natural fire breaks (fig. 2.8). Added to this is variation

Table 8.6. Effects on Postfire Responses of Vegetation in Subalpine Forest Landscapes

Conditions and Characteristics	Examples of Some Effects
Prefire conditions	
Topographic setting	Slower tree recovery occurs on south-facing slopes.
Natural fire breaks	These limit, slow, and shape the pattern of fire.
Time since last replacement disturbance	This affects fuel characteristics, level of serotiny.
Types of disturbances and resulting stand structure	These affect fuel characteristics.
Level of serotiny, if lodgepole pine is present	This affects the potential for postfire lodgepole regeneration.
Characteristics of the fire	
Surface fire or crown fire and fire intensity	This affects tree mortality and survival.
Consumption of seed, if lodgepole pine is present	High consumption lowers postfire lodgepole.
Glowing/smoldering combustion and soil heating	These affect the survival of shrubs, forbs, graminoids, and microbes and potential germination from the seed bank.
Pattern of unburned islands, surviving trees	These affect the sources of postfire seed.
Area	Large fires mean longer distance to surviving seed.
Postfire conditions	
Distance to unburned forest or surviving seed trees	This affects the sources of postfire seed.
Competition between trees, shrubs, and other plants	Graminoid cover can slow tree regeneration.
Facilitation	This affects spruce-fir regeneration within limber pine forest.

in the prefire vegetation state, a function of time since the last high-severity fire, fuel legacies from other disturbances, level of serotiny, and other factors. During the fire, variation in the intensity of flaming combustion affects the consumption of serotinous lodgepole cones, while duration and intensity of glowing or smoldering combustion affect seed banks and resprouting plants. However, infrequent high-severity fire dominates, and the postfire legacy of unburned vegetation (e.g., unburned trees) and fire size shape postfire succession.

As discussed in the last section, after fire there is considerable contingency (e.g., were serotinous cones opened or consumed; what was the total heat flux reaching roots?), and thus a diversity of potential postfire successional situations, which may coexist in a mosaic across a landscape. Successional stages from fires in previous centuries remain evident and ecologically important today, across many subalpine forest landscapes, due to the short growing season and long re-

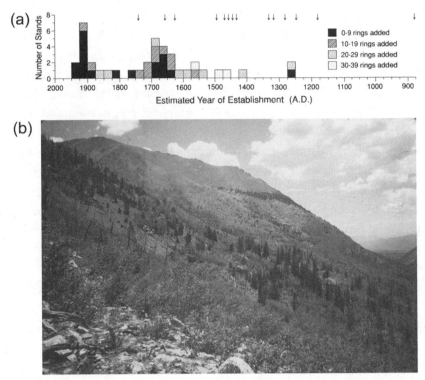

Figure 8.10. Stand origins and postfire succession in bristlecone pine forests: (a) Stand-origin dates for bristlecone pine forests in southern Colorado. The number of tree rings added to estimate the age of the oldest tree is indicated by the shading. Arrows indicate the oldest date obtained in 14 stands where the oldest trees had rotten centers. (b) Recovery following a stand-replacement fire ca. 1900, typical of bristlecone pine forests, on Mount Taylor in the Sawatch Range, Colorado.

Sources: Part (a) is reproduced from W. L. Baker (1992b, p. 23, fig. 8), copyright 1992, reprinted with permission of the Regents of the University of Colorado; (b) photo by the author.

covery time (fig. 8.10b). Subalpine forests are landscapes of recovery, because brief large fires are followed by centuries of slow recovery from a diversity of postfire states.

The landscape mosaic shifts over time as large fires occur, followed by interludes with small fires, leading to temporal variation in the proportions of the landscape in various successional states (box 3.1), and thus fluctuations in landscape diversity (Romme 1982). The proportions of states in the mosaic do not reach stability at some larger spatial extent (Romme 1982), in part because fires are large relative to underlying topographic heterogeneity (W. L. Baker and Kipfmueller 2001). In a sense, subalpine landscapes become more diverse when fires are smaller, allowing more states within a given land area, while landscape diversity may be reduced by large fires that sweep across and homogenize the mosaic (W. L. Baker 1995), at least in terms of age. However, even large severe fires are patchy, and one to two years after the 1988 Yellowstone fires, tens of thousands of discrete patches of burned forest, differing in severity, were visible (M. G. Turner, Hargrove et al. 1994). By years 11–13, fine-scale patchiness in tree regeneration was high (Kashian et al. 2004). Simulation shows that more frequent fire leads to younger stands more variable in tree density, thus increasing heterogeneity (Schoennagel et al. 2006).

Fire is not alone in shaping these landscapes. Occasional large windstorms occur (W. L. Baker et al. 2002). Outbreaks of spruce beetle (*Dendroctonus rufipennis*), mountain pine beetle, and other insects may be as important as fire (W. L. Baker and Veblen 1990) and may also influence fire (chap. 4). Spatial and temporal variability are inherent in subalpine forest landscapes.

Land-Use Effects

Fire control began shortly after EuroAmerican settlement but was generally ineffective until the late 1950s (chap. 10; Kipfmueller and Baker 2000). A half century of effective control is unlikely to have had much effect in forests with fire rotations of 175–350 years (Romme and Despain 1989a). Moreover, fire declines for multiple reasons (chap. 10). Buechling and Baker (2004) found that decreased twentieth-century subalpine fire in Rocky Mountain National Park was related to less drought. Sibold and Veblen (2006) found that the twentieth century in the park included teleconnections favorable to extensive subalpine fires but in relatively brief periods. These authors suggested that low-flammability young forests regenerating after nineteenth-century fires may also have limited large fires in the twentieth century.

Changes in fire rotations in spruce-fir and lodgepole pine forests since EuroAmerican settlement cannot yet be detected, even if they have occurred, because fire rotations were generally 175–350 years (fig. 8.5). At best, only a century of data is available, and at least a full rotation of data is needed to detect change (W. L. Baker 1995). Much of the rotation occurs from episodes of extensive fire during drought, followed by long interludes during which fire rotation appears very long (box 3.1). Thus, analysis based on short datasets (e.g., 12 years [J. K. Brown et al. 1994]) provides unreliable evidence about change (chap. 10).

Neither tree densities nor fuel loads are likely outside the HRV from fire exclusion. Low-severity fire was rare, with very long rotations, and had little effect on trees (Sibold et al. 2007); thus, exclusion of surface fire could not explain tree-density increases. It would be difficult to detect a fire-exclusion effect on fuels, even if it had occurred, as fuels cannot be reconstructed, early historical data are few, and variation in fuels among stands of the same age is very high. Moreover, fuels do not consistently build up as subalpine forests age (chap. 2). Also, quaking aspen has likely not declined in these forests as a result of fire exclusion (see "Fire Exclusion and Aspen Decline in Mixed-Conifer and Subalpine Forests" in chapter 6).

Although these forests have likely been little affected by fire exclusion (see also chap. 10), they have been modified by logging, livestock grazing, and postfire seeding. Clear-cutting of lodgepole pine forests led to a net loss of about 80 megagrams per hectare of coarse woody debris; whereas fire led to a net gain of 95 megagrams per hectare (Tinker and Knight 2000). This wood provides important wildlife habitat, maintains soil fertility and structure, and plays a key role in streams. If timber harvesting is to more closely mimic the impacts of fire, more wood must be left after harvest (Tinker and Knight 2004). Livestock grazing after fire can alter postfire succession, in one case discouraging reestablishment of spruce and fir, which led to dominance by more drought-tolerant bristlecone pine (W. L. Baker 1991). See chapter 10 for information on the effects of seeding and logging after subalpine forest fires.

FIVE-NEEDLE PINE FORESTS

Five-needle pine forests, dominated by a set of pines with five needles per bundle, cover relatively little land area in the Rockies (table 1.1) but provide important landscape diversity in otherwise large areas of lodgepole pine and spruce-fir forests. This section covers bristlecone pine forests, limber pine forests, and whitebark pine forests, which can occur in similar environments—often on ridges, rocky sites, and southerly-facing slopes—along a geographical gradient in the Rockies from northern New Mexico to the Canadian border.

Stable Bristlecone Pine Forests

Fire history in stable bristlecone pine forests is known primarily from stand-origin dating. Stand origins were estimated in 65 stands across the range of bristlecone pine in central and southern Colorado, revealing two peaks in origin dates (fig. 8.10a), one from 1900–1925 and another from 1625–1700. These peaks corresponded with periods of warmer than normal temperatures and drought that may have promoted fires (W. L. Baker 1992b). Mature stands typically had a unimodal tree-size distribution, few or no seedlings, and no evidence of disturbance agents other than fire. In contrast, seedlings and small trees were common in stands burned a century ago (fig. 8.10b). Bristlecone pine thus behaves as a very long-lived pioneer tree that regenerates primarily after high-severity fires (W. L. Baker 1992b).

Fire rotation for the high-severity fires that are the primary structuring disturbance in these forests can be estimated from fig. 8.10a as approximately 300 years. Donnegan, Veblen, and Sibold (2001) found three stands, Sherriff et al. (2001) found four stands, and P. M. Brown and Schoettle (2008) found a single stand containing more than 10 fire-scarred trees, from which they estimated fire frequency and the relationship of fires to climate. However, sites with abundant fire scars and low-severity fires are atypical of bristlecone pine forests, based on rangewide sampling (W. L. Baker 1992b), and the sampled low-severity fires and their relationship with climate are unlikely to represent the fires that structure most of these forests.

High-severity fires in these forests appear to be followed by regeneration of bristlecone pine, along with common associated trees, including quaking aspen, Engelmann spruce, and subalpine fir (fig. 8.10b). Engelmann spruce and subalpine fir are enduring codominants of bristlecone pine on some sites in these forests. Engelmann spruce might eventually overtop and dominate bristlecone pine on these sites, since it is often actively regenerating inside these forests, while bristlecone is not; but the time required may be as much as 1,000 years (W. L. Baker 1992b). High-severity fire is likely to reset succession before this occurs. Unfortunately, the introduced pathogen, white pine blister rust, was discovered on bristlecone pine in the Sangre de Cristo and Wet Mountains, Colorado, in 2003 (Blodgett and Sullivan 2004). This pathogen caused widespread mortality of five-needle pines in the northern Rockies (see "Stable Whitebark Pine Forests" below) and could similarly devastate bristlecone pine forests.

Stable Limber Pine Forests

As with bristlecone and whitebark pine, limber pine occurs in very old, low-density stands that can be found on rocky sites near the tree line; it also occurs in some denser forests. Limber pine can dominate after fires or be mixed with other trees at somewhat lower elevations. Donnegan, Veblen, and Sibold (2001) found one site and Sherriff et al. (2001) four sites on dry, rocky ridges and exposed areas dominated by limber pine, where fire scars were common. These authors did not determine the severity or spatial extent of the fires represented by scars; however, although three of four sites sampled by Sherriff et al. were within a 300-hectare search area, only one of 13 fires between 1298 and 1988 had burned more than one of the three sites, suggesting that fires were generally small and likely had a long rotation. Buechling and Baker (2004) found that limber pine stands near spruce-fir forests often burned in the same fires. Limber pine made up only 3 percent of their study area but likely had roughly the same

350-year high-severity and 7,500-year low-severity rotation. Some limber pine trees can survive high-severity fires, suggesting that fires can be patchy because of discontinuous fuels (Peet 1981).

Limber pine can regenerate after fire in burned limber pine forests (Peet 1981) and can also expand and temporarily dominate burned spruce-fir forests (see spruce-fir section above) due to its tolerance of xeric sites and seed dispersal by Clark's nutcrackers (Rebertus, Burns, and Veblen 1991). Recovery after fires can be very slow on high-elevation sites with limber pine (fig. 8.11). Two old-growth stands on xeric sites in Colorado were about 450 and 700 years old, uneven aged, and seedlings were common, suggesting stability (Rebertus,

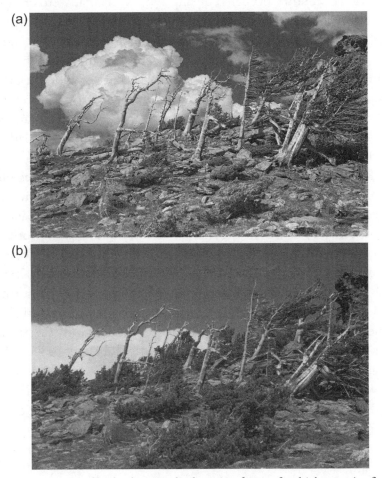

Figure 8.11. Recovery of high-elevation limber pine forest after high-severity fire. Shown is a forest on Trail Ridge, Rocky Mountain National Park, after high-severity fire in 1860 (dated by Buechling and Baker 2004; this is the edge of the fire area, which is mostly to the left of the photos): (a) in 1962, and (b) in 1996.

Sources: Photo (a) by R. Cantor, courtesy of Rocky Mountain National Park archives, negative number 4273; photo (b) by the author.

Burns, and Veblen 1991). White pine blister rust has spread into southeastern Wyoming and northern Colorado, where it is killing mature limber pine (Kearns and Jacobi 2007) and may prevent successful regeneration.

Stable Whitebark Pine Forests

Whitebark pine, an important food source for bears, squirrels, birds, and other animals (Tomback, Arno, and Keane 2001), dominates stands on rocky sites at high elevations and can dominate after fires at lower elevations or occur in mixtures with other trees (Arno 1986). Most recent fires in these forests were small (i.e., less than 4 hectares). Extensive fires that account for most of the burned area over the last century occurred during a few years that were drier than average (P. Morgan and Murray 2001; Rollins, Swetnam, and Morgan 2001). High-severity fire burned more than 2,000 hectares between 1910 and 1926 in Gabriel's (1976) study area in northwestern Montana. On the Idaho–Montana border, fires from 1979 to 1990 included 2,222 hectares of whitebark pine, with 14 percent of the area inside fire perimeters in crown fire, 47 percent in lethal surface fire, 12 percent in nonlethal surface fire, and 27 percent in unburned islands (J. K. Brown et al. 1994). Thus, high-severity fire, particularly intense lethal surface fires, has dominated since EuroAmerican settlement.

Fire history is poorly known across whitebark pine landscapes under the HRV. Only three fire histories that included whitebark pine are landscape scale and have adequate fire-scar and age-structure or stand-origin data (Gabriel 1976; Mattson and Reinhart 1990; M. P. Murray, Bunting, and Morgan 1998). However, data from these studies are unfortunately not specific to whitebark pine but are pooled across forest types (table 8.1). In another study, Keane, Morgan, and Menakis (1994) sampled 111 whitebark pine stands in northwestern Montana, found few fire scars, and suggested that large, high-severity fires dominated the fire regime. Fire rotation estimates for high-severity fire are 150–340 years, with the most reliable estimate specific to whitebark pine at 228 years from Keane, Morgan, and Menakis (table 8.1). Two paleoecological sites (Burnt Knob Lake, Idaho, and Baker Lake, Montana) included high-elevation cirques with whitebark pine forest and spruce-fir forest (table 8.3), where the mean episode interval for high-severity fire over the past few thousand years was 250 years (Brunelle et al. 2005), consistent with the Keane, Morgan, and Menakis rotation estimate.

Fire scars are common on some sites (P. Morgan and Bunting 1990; J. K. Brown et al. 1994; Larson 2005), but fires on those sites were small under the HRV and low-severity fire was thus of limited importance. Arno (1986) suggested that surface fires likely killed small subalpine fir and spruce and prevented them from becoming dominant but allowed larger whitebark pine to survive. This mechanism could operate in small areas, but to affect much of the whitebark range it would require very large surface fires. However, evidence suggests that nearly all surface fires were quite small. Arno and Hoff (1989, 8) said, "Tiny spot fires are most common because fuels are generally sparse and conditions moist and cool," and other authors agree that surface fires most often scarred few trees (P. Morgan and Bunting 1990; Barrett 1994). Fuels were insufficient to allow fire spread at seven sites in northwest Wyoming and southwest Montana (Walsh 2005). Based simply on the presence of fire scars in small areas, the fire regime was described as mixed severity (J. K. Brown et al. 1994; Arno 2001; Larson 2005). Other studies found that low-severity fires were absent (Kipfmueller and Kupfer 2005) or prominent at only one of seven sites (Walsh 2005). At three of seven sites, Walsh (2005) found no evidence of past fire at all. Studies covering large areas found fire scars rare and low-severity fire of little importance (T. Weaver and Dale 1974; Gabriel 1976; Keane, Morgan, and Menakis 1994). Where fire had a major role in stable whitebark pine forests, high-severity fire (either crown fire or intense, patchy surface fire) likely dominated, given discontinuous fuels and varying tree density, with low-severity fire of limited extent in some areas (Arno 2001).

Lodgepole pine can dominate or codominate after fire in some whitebark pine forests (Kipfmueller and Kupfer 2005). Where lodgepole is lacking, whitebark regeneration can be abundant within large burns, as Clark's nutcrackers harvest and transport the seed and store it in soil caches, from which groups of trees regenerate, often over a period of years (Sund 1988; Tomback, Hoffman, and Sund 1990; Tomback, Sund, and Hoffman 1993). Regeneration is typically near trees, rocks, or other objects and with grouse whortleberry nearby (Tomback, Sund, and Hoffman 1993). Regeneration declined exponentially with increasing distance from parent trees, with low densities beyond about 2 kilometers but extending to more than 8 kilometers. Regeneration can be sparse up to seven years after fire (Ash and Lasko 1990; Tomback, Sund, and Hoffman 1993; Doyle 2004), increasing during wetter periods (Tomback, Sund, and Hoffman 1993) and possibly peaking 50–75 years after fire (Larson 2005). Some sites had little regeneration even 80 years after fire (Kendall and Arno 1990). Subalpine fir may also regenerate in these burns, particularly smaller ones with an upwind seed source (Tomback, Hoffman, and Sund 1990). After high-severity fire, most postfire plants were shrubs and forbs that resprouted the year after fire, but shrub and forb production continued to increase during years 1–3 after fire (Ash and Lasko 1990).

Some analyses of changes in fire in these forests are based on unsupported assumptions or insufficient data. J. K. Brown et al. (1994) thought fire had declined but assumed that the mean composite fire interval from fire scars is equal to the fire rotation, which is not supported (see chap. 10). K. T. Peterson (1999) thought that low rates of burning over 20 years (1978–1998) in Glacier National Park meant that whitebark pine forests were outside the HRV for fire. Rollins et al. (2001) estimated from fire atlases that 29 percent of upper subalpine forests with whitebark pine in the Selway-Bitterroot Wilderness burned between 1880 and 1996, a rotation of about 400 years, which is longer than rotations under the HRV (table 8.1). However, fire rotations more than 200 years long cannot be accurately estimated with two decades or even a century of data, since nearly all the rotation can occur from episodes of fire any time during the rotation (box 3.1). Short-term rates of burning are of interest, but it is not possible to determine yet whether fire rotation has declined in these forests since EuroAmerican settlement.

Changes in whitebark pine abundance over the last century were also attributed to fire exclusion (Arno, Reinhardt, and Scott 1993; Keane, Morgan, and Menakis 1994; P. Morgan et al. 1994; M. P. Murray, Bunting, and Morgan 2000), but the episodic nature of fire and recovery was not considered. The proportion of the landscape occupied by late-seral forests increased in the 1900s, while midseral forests, containing more whitebark pine, declined (Arno, Reinhardt, and Scott 1993; M. P. Murray, Bunting, and Morgan 2000). These authors attributed the change to fire exclusion, thinking that abundant fires in the mid-1800s were the norm and that the post-1900 decline in fire was from fire exclusion. Exclusion of surface fires was also thought to have allowed subalpine fir to increase (Arno 1986; P. Morgan and Bunting 1990).

However, long-rotation fire regimes are typified by episodes of large fires followed by long periods with little fire (box 3.1). The mid- to late 1800s had exceptional fire in the Rockies (e.g., C. S. Crandall 1901; Barrett, Arno, and Menakis 1997), including in whitebark pine forests (Larson 2005), which was likely atypical of a long-term mean. Abundant whitebark pine in the early 1900s and prominent late-successional stages in the late 1900s likely are not from fire exclusion. More likely is recovery and succession after the exceptional nineteenth-century high-severity fires and, in some cases, insect outbreaks. Whitebark pine can expand after high-severity fires (Doyle 2004), particularly on severe sites and where fires are large. Subalpine fir may eventually recolonize through seed dispersal, and early establishing whitebark can facilitate fir establishment (Callaway 1998; Tomback, Arno, and Keane 2001). Increasing subalpine fir dominance late in succession may also reflect its greater shade tolerance and ability to overtop whitebark. Mountain pine beetle outbreaks that kill overstory whitebark can also allow subalpine fir to increase (Kipfmueller and Kupfer 2005).

The evidence reviewed earlier in this section suggests that surface fires, which might have preferentially killed subalpine fir under the HRV, were not widespread in these forests. Moreover, pooled age structure in 10 stands in Wyoming (P. Morgan and Bunting 1990) shows that subalpine fir was common in these forests under the HRV; it regenerated abundantly in whitebark pine stands during a period 130–180 years before Morgan and Bunting's 1990 study, before extensive EuroAmerican land uses, and with no trend in subalpine fir regeneration after EuroAmerican settlement (fig. 8.12). In western Montana age structures, subalpine fir was also younger than co-occurring whitebark, but subalpine fir regeneration in one stand began about 1700, and in two others fir expanded after nineteenth-century fires, with no trend in regeneration or even a decline in the 1900s (Larson 2005). At one of the locations sampled by Larson, another study found relatively continuous subalpine fir regeneration since about 1700 (Daniels et al. 2005). These studies show that subalpine fir has not increased because of exclusion of surface fires that killed small trees under the HRV.

Whitebark pine forests that lack competing conifers, especially those with an understory of grouse whortleberry, may remain dominated by whitebark for hundreds of years (T. Weaver and Dale 1974). Tree density is quite variable but generally high (about 1,500–4,500 trees per hectare) the first century after high-severity fire, declining to about 1,000–2,000 trees per hectare by about 200 years after fire (Forcella and Weaver 1977; T. Weaver, Forcella, and Dale

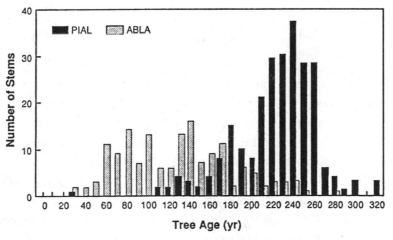

Figure 8.12. Age structure for subalpine fir forests with successional whitebark pine. This is pooled age structure (*n* = 10 stands) for subalpine fir (ABLA) forests with successional whitebark pine (PIAL) on the Shoshone National Forest, Wyoming.
Source: Reproduced from P. Morgan and Bunting (1990, p. 168, fig. 1).

1990). Tree productivity increased for about the first century after fire, then continued to increase but more gradually (T. Weaver, Forcella, and Dale 1990), while plant species richness was highest in a 40-year-old stand, then remained rather low in stands about 100–420 years old (T. Weaver and Dale 1974).

The introduced pathogen, white pine blister rust, has caused large-scale mortality of mature whitebark pine (Keane, Morgan, and Menakis 1994; Tomback et al. 2001). Fire exclusion was thought to have reduced early-successional sites favorable to whitebark, adding to the decline (Keane and Morgan 1994; Keane and Arno 2001). However, this is more likely normal landscape fluctuation than fire exclusion (see earlier in this section). Attention can thus shift to strategic use of fire in restoration. Unfortunately, after large fires, such as the 1967 Sundance fire, whitebark regeneration can be poor. In the Sundance case, prefire loss of mature trees to blister rust had reduced seed density, and postfire seedlings became infected even at long distances from unburned forest (Tomback et al. 1995). Thus, a more complex, contingent, and directed use of fire is now envisioned to produce rust-resistant whitebark in the long run. In some cases, this may require fire control or fuel reduction to protect valuable rust-resistant trees, or small burns where whitebark regeneration is favored and subalpine fir is not, as well as perhaps mechanical removal of competing conifers or fuels and other methods for eventual whitebark pine restoration (Hoff et al. 2001; Keane and Arno 2001). Unfortunately, these approaches may soon be needed to address the decline of limber pine and bristlecone pine in the southern Rockies from blister rust, as well.

SUMMARY

Spruce-fir and lodgepole pine forests have fire rotations of 175–350 years, shortest in the lower subalpine and northern Rockies and longest in the upper subalpine and southern Rockies. Fuel loadings are generally high, and fire is most limited by rare dry conditions. Large fractions of burned area are from infrequent, large, often severe fires during droughts and high winds, when unstoppable fires with flames of up to 90 meters can burn thousands of hectares per hour. High-severity fire dominates; low-severity surface fire is rare, with rotations of millennia. The early postfire landscape can be very heterogeneous in tree composition and structure due to variability in fire severity and lodgepole-pine serotiny, seed dispersal contingent on distance to surviving trees, and competition with resprouting herbaceous vegetation. Most shrubs, forbs, and graminoids that were present before fire resprout readily and recover within a few years or at most a decade or two; seed banks play a minor role after fire. Landscape mosaics fluctuate temporally, as episodes of large fires punctuate long interludes with little fire. These are landscapes of recovery. Variable and often slow recovery of trees after fires, except where lodgepole dominates, lead to spatially heterogeneous mixtures of meadows, shrublands, and recovering forests. Five-needle pine forests on drier subalpine sites are also subject to multicentury fire rotations, also generally dominated by infrequent, high-severity fire, with low-severity fire rare or localized. Fire suppression likely has had little effect in subalpine forests. The primary human effects on fire have instead been from logging, livestock grazing, and postfire treatments such as seeding and logging.

CHAPTER 9

Fire in Shrublands
and Grasslands

Shrublands and grasslands cover about one-third of the land area of the Rocky Mountains (table 1.1). These ecosystems may occur as open areas in a matrix of forests or dominate broad, open landscapes—particularly on dry, low-elevation slopes—extending to the tree line. They may also form a shrubland-grassland mosaic within an opening. These ecosystems are major sources of forage, provide important habitat for wildlife, and are prominent features of many Rocky Mountain landscapes. Fire research has expanded, but much is still unknown about fire history and fire effects in these ecosystems under the HRV.

SALT DESERT SHRUBLANDS

Fire was likely uncommon in salt desert shrublands under the HRV (West 1983), but little actual evidence exists. Crawford et al. (2004) suggested a mean fire return interval greater than 500 years. Fires do occur in a variety of salt desert shrublands in the West; wildfires burned recently in greasewood (Sheeter 1968; Boltz 1994), winterfat (Pellant and Reichert 1984; Yensen et al. 1992; Boltz 1994; West 1994), shadscale saltbush (Groves and Steenhof 1988; Yensen et al. 1992; West 1994), and Gardner's saltbush (West 1994). Some shrubs of these communities (i.e., greasewood, Gardner's saltbush) resprout after fire, but others (shadscale saltbush) do not or vary in resprouting ability (winterfat). Lack of fine fuels may limit fire in these communities on arid sites or during droughts, but some (e.g., winterfat, greasewood, shadscale saltbush) had native perennial grass under the HRV and could have burned, particularly after wet years when grass production may have increased.

Salt desert shrublands may be threatened by too much fire. During and following the exceptionally wet 1982–83 El Niño, cheatgrass and other annuals increased dramatically in these communities, followed by extensive fires in parts of Utah and Nevada (West 1994) and in Idaho (Groves and Steenhof 1988; Yensen et al. 1992). These fires further expanded cheatgrass and other nonnative annuals (fig. 9.1). Burned shadscale saltbush regenerated poorly after 1980s fires (Yensen et al. 1992; West 1994). If native shrubs do not resprout after fire, expensive replanting is needed to avoid dominance by nonnatives (Pellant and Reichert 1984). Thus, it is logical to seek control of nonnative annuals before fires occur, avoid prescribed fires in these areas, and actively suppress fires ignited in them.

SAGEBRUSH SHRUBLANDS

Sagebrush shrublands alone cover about 12 percent of the area of the Rockies (table 1.1), and these shrublands provide key summer and winter habitat for

Figure 9.1. Cheatgrass replacement of natural vegetation. Cheatgrass has replaced shadscale saltbush and Wyoming big sagebrush over extensive areas of the Snake River Plains, Idaho.
Source: Photo by the author.

native ungulates, birds, and other species. This section reviews the link between sagebrush fire behavior and the landscape mosaic under the HRV, new estimates of fire rotation that suggest fire was not frequent in these ecosystems, fire effects on plants, and important human effects on sagebrush fire regimes.

Fire Behavior and the Landscape Patch Mosaic under the HRV

Fireline intensity in modern sagebrush fires has varied over a large range because of variation in fuel loads, moisture, and weather (Sapsis and Kauffman 1991); but fires are virtually all high severity, topkilling the shrubs. Britton and Clark (1985, 23) reported, "It is relatively unimportant how fast the fire moves, how hot the fire is, or what the fire intensity is . . . if a fire front passes through an area, the sagebrush will be killed." Fires thus do not thin sagebrush by burning beneath the shrubs and thinning stems, as implied by Winward (1991). Large mixed- or low-severity fires are unknown. Instead, fire kills sagebrush in patches or extensively, creating and controlling a patchy landscape mosaic of burned and unburned areas. Tens of thousands of hectares can burn, and multiple ignitions can merge into a large complex (Komarek 1966).

The landscape patch mosaic (figs. 4.4, 9.2) is shaped by variable fuels and weather. Fuels may be discontinuous, however, and insufficient to carry fire. About 20 percent cover of sagebrush and 300 kilograms per hectare of fine fuels are needed for fire to carry (Britton and Clark 1985), although strong

Figure 9.2. Yellow rabbitbrush and spineless horsebrush 11 years after sagebrush fire. The fire is the 1994 Butte City fire in Wyoming big sagebrush in southern Idaho described by B. W. Butler and Reynolds (1997).
Source: Photo by the author in 2005.

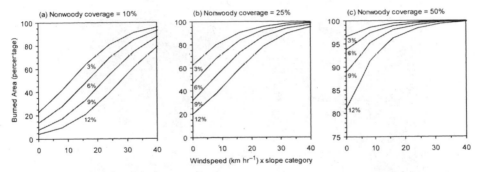

Figure 9.3. Modeled effect of three factors on burned area in sagebrush. The three factors are nonwoody coverage (fine fuels), windspeed-slope interaction, and 10-hour fuel moisture content (percentage) shown as lines. Slope categories are <5% = 1; 5–15% = 2; 16–25% = 3; 26–35% = 4; >35% = 5.
Source: Reproduced from C. S. Wright and Prichard (2006, p. 497, fig. 2).

winds may lower this need (J. K. Brown 1982). Patchy burns (fig. 4.4) may be common in moister parts of landscapes, cool periods, or droughts, when fine fuels are reduced (Clifton 1981; Kuntz 1982; Boltz 1994; C. S. Wright and Prichard 2006). Extensive fires with little unburned area (fig. 9.2) may occur under dry, windy conditions if fine fuels are sufficient (Kuntz 1982; B. W. Butler and Reynolds 1997). Empirical models of area burned in sagebrush, developed from 17 sites in California, Oregon, Nevada, and Wyoming, summarize these trends, showing how the burned area percentage declines as fine fuels and wind speed decline and fuel moisture increases (fig. 9.3). A logical hypothesis is that individual pre-EuroAmerican fires were more extensive and less patchy than those of today, as fine fuels have been reduced by overgrazing by livestock (Wrobleski and Kauffman 2003).

Fire Rotation under the HRV

Fire-rotation estimates are from four sources: (1) fire scars on trees near sagebrush, (2) fire rotation in adjoining forests, (3) macroscopic charcoal in sediments near sagebrush, and (4) rate of recovery of sagebrush after fire. Fire-scar studies include Burkhardt and Tisdale (1976) near mountain big sagebrush, and Houston (1973) and Arno and Gruell (1983, 1986) near grasslands with scattered sagebrush (discussed with grasslands later in chapter). Fire-rotation estimates in adjoining forests include Shinneman and Baker (in press) and Heyerdahl et al. (2006) near mountain big sagebrush, and Floyd et al. (2004) and Shinneman and Baker (in press) near Wyoming big sagebrush. Macroscopic charcoal from a spring in central Nevada near Wyoming big sagebrush indicated that fire

Figure 9.4. Mountain big sagebrush recovery 17 years after the 1988 Yellowstone fires. Mountain big sagebrush has partly reseeded and recovered.
Source: Photo by the author in 2005.

episodes occurred at 200- to 500-year intervals under the HRV (Mensing, Livingston, and Barker 2006). More than one fire episode may be needed to equal a fire rotation, as episodes could represent fires that burned only part of a landscape (chap. 5).

Sagebrush recovery after fire (fig. 9.4) might also indicate the fire rotation, since fires likely did not occur more often than the time required for sagebrush to regain prefire density and cover. Density only measures shrub presence, not fully mature plants; thus, cover is the best measure. Rate of recovery is affected by fire intensity, fire size, herbivory, environment, and climate (Pechanec and Stewart 1944; Blaisdell 1953; Lesica, Cooper, and Kudray 2005), but data are insufficient to estimate these effects. Recovery time can be estimated from chronosequences, but data are few, except for mountain and Wyoming big sagebrush (fig. 9.5).

Mountain big sagebrush in the Rockies (fig. 9.5a) recovered to more than 85 percent of prefire cover in 25–35 years in some cases (fast track), but it more often reached less than 40 percent of prefire cover in that time, likely requiring 75 years or more for full recovery (W. L. Baker, in press). Moreover, more than 70 years would be required for mountain big sagebrush to return to the interior of a large burn in Idaho, and a few decades more (about 100 years total) to regain former cover (Welch and Criddle 2003). Thus, the range of recovery may be about 25–35 years for fast-track and 75–100 years for slow-track mountain big sagebrush, the tracks possibly reflecting variation in precipitation, fire size, or other factors (W. L. Baker, in press).

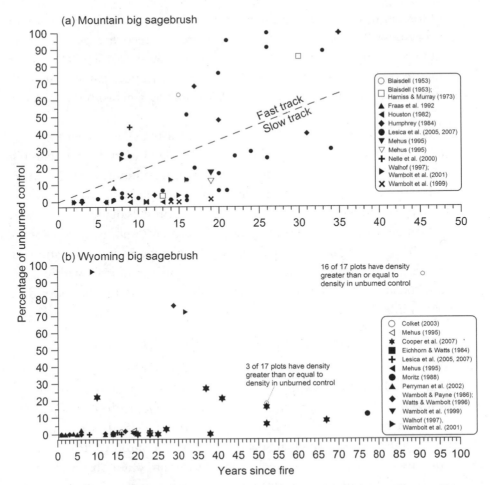

Figure 9.5. Recovery of sagebrush after fire relative to unburned control areas. Data are shown for (a) mountain big sagebrush, and (b) Wyoming big sagebrush. Sagebrush density is shown as open symbols, sagebrush cover as closed symbols. Points represent a single sample plot, and similar symbols represent a single study. Possible fast-track and slow-track recovery trajectories in (a) are separated by a dashed line and explained in the text. Data in S. V. Cooper, Lesica, and Kudray (2007) were interpolated from their graph. I could not generally determine years for points less than 20 years since fire; thus, all but one of those points have been omitted. All others have 0% recovery.

Recovery of Wyoming big sagebrush appears surprisingly slow (fig. 9.5b), with almost no recovery in the first 25 years, and, except for three points, no more than 30 percent recovery in the first 80 years. Full recovery no doubt occurred under the HRV, but scientific observation of full recovery after fire is lacking. A slight upward curve to the points (fig. 9.5b) might suggest 150–200 years for full recovery if extrapolated, but this is highly speculative. The time required for full recovery of Wyoming big sagebrush is currently unknown. In other ecosystems, fire rotation is twice or more the recovery period (W. L. Baker 2006b). Using a factor of 2.0, fire rotations would be 150–200 years or more in typical slow-track mountain big sagebrush, uncertain in fast-track mountain big sagebrush, and unknown in Wyoming big sagebrush (W. L. Baker, in press).

Using the four sources together, W. L. Baker (in press) estimated fire rotations (table 9.2) in expanses of sagebrush to be 150–300 years in mountain big sagebrush, 200–350 years in Wyoming big sagebrush, and more than 200 years in little (low) sagebrush. Where sagebrush is intermixed with woodlands and forests, the fire regime is likely similar to that of the forest. Rotation is estimated in mountain big sagebrush to be 160 years in intermix with Douglas-fir forests and 400–600 years in intermix with piñon-juniper woodlands. Rotation in Wyoming big sagebrush is estimated at 400–600 years in intermix with piñon-juniper woodlands. Rotation in little sagebrush is estimated to be more than 425 years in intermixes.

Fire Effects on Plants in Sagebrush

After fires in sagebrush, understory plants most commonly resprout or germinate from a seed bank, but some also reseed from off-site, according to a diversity of studies (table 9.1). After Idaho fires in threetip sagebrush (Akinsoji 1988) and mountain big sagebrush (L. D. Humphrey 1984; Seefeldt, Germino, and DiCristina 2007), most perennial grasses and forbs resprouted from underground stems and roots. Postfire sagebrush seedlings are often observed within a few years after fire, likely from surviving seed banks in the centers of burns, with added seed from surviving plants on the margin (Mueggler 1956). Survival of other plants in the seed bank is common after sagebrush fires, but fire generally reduces seed banks. The native seed bank in unburned Wyoming big sagebrush in Utah was sparse before fire (i.e., about 45 native seeds per square meter), and fire reduced the seed bank by more than half; most native plants present before fire had some representation in the postfire seed bank (Hassan and West 1986). Weedy nonnative annuals dominated the seed bank before and after fire in threetip sagebrush in Idaho, but several native annuals and perennials, especially six-weeks fescue, survived in the seed bank (Akinsoji 1988).

Extent of mortality to a particular species and rate of recovery of that species depend on the season of fire, its intensity and duration, and plant morphology. A June fire killed most needle and thread, but an August fire killed none (H. A. Wright and Klemmedson 1965). Similarly, this study found that mortality may be strongly dependent on fire intensity; 70 percent of Thurber's

Table 9.1. Studies of Postfire Responses of Plants in Sagebrush and Sagebrush–Antelope Bitterbrush

Sources	Location	Time since Fire (yrs)	Taxon							Variable Studied				
			ARTRT	ARTRV	ARTRW	Low	Silver	Three-tip	Bitter-brush	Indiv. Spp.	Spp. Gps.	Regen.	Divers.	Prod.
Observations/plots after fire														
Tracy (2004)	Yellowstone	0		X										X
Blaisdell & Mueggler (1956)	SE ID	0–1		X				X		X				
Rennick (1981)	SE MT	0–1					X			X	X	X		X
H. A. Wright & Klemmedson (1965)	SW ID	0–1			X					X				
Akinsoji (1988)	SE ID	1		X				X		X			X	
Seefeldt & McCoy (2003)	SE ID	1						X			X		X	
Jirik & Bunting (1994)	SW ID	1		X				X		X				X
Nimir & Payne (1978)	SW MT	1		X			X			X	X			X
Petterson (1999)	N CO	1			X						X		X	
Raper et al. (1985)	SW WY	1		X					X		X			X
White & Currie (1983)	C MT	1					X					X		
Clifton (1981)	SC ID	1–2			X	X				X		X		X
Ratzlaff & Anderson (1995)	SE ID	1–2						X		X	X		X	
Schwecke & Hann (1989)	W MT	1–2		X						X			X	X

Table 9.1. Continued

Sources	Location	Time since Fire (yrs)	Taxon							Variable Studied				
			ARTRT	ARTRV	ARTRW	Low	Silver	Three-tip	Bitter-brush	Indiv. Spp.	Spp. Gps.	Regen.	Divers.	Prod.
Norland, Singer, & Mack (1996)	Yellowstone	1–2		X							X			X
Singer, Coughenour, & Norland (2004)	Yellowstone	1–2		X							X			X
West & Hassan (1985)	N UT	1–2			X						X	X		X
Cook, Hershey, & Irwin (1994)	SE WY	1–3		X							X			X
Kuntz (1982)	SE ID	1–3		X					X	X	X	X		
Seefeldt, Germino, & DiCristina (2007)	SE ID	1–3		X						X	X			
McGee (1982)	NW WY	1–3		X							X			
Mueggler & Blaisdell (1958)	SE ID	1–3		X						X	X	X		X
Mangan & Autenrieth (1985)	SE ID	1–4		X						X				
Peek, Riggs, & Lauer (1979)	C ID	1–4			X						X			
Shariff (1988)	SE WY	1–4		X			X				X	X		X
Tracy & McNaughton (1997)	Yellowstone	1–5		X							X			
Colket (2003)	SE ID	1–7	X		X						X			X

Study	Location	
Van Dyke & Darragh (2006)	SC MT	1–9
Perryman et al. (2002)	NW CO	1–12
Blaisdell (1953)	SE ID	1–15
Wambolt & Payne (1986)	SW MT	1–17
West & Yorks (2002)	N UT	1–19
Watts & Wambolt (1996)	SW MT	1–29
Harniss & Murray (1973)	SE ID	1–30
R. B. Murray (1983)	SE ID	1–43
Omphile (1986)	SE WY	2
Singer & Harter (1996)	Yellowstone	2
Pechanec & Stewart (1944)	SE ID	3–9
S. V. Cooper, Lesica, & Kudray (2007)	SE MT	4–67
Fraas, Wambolt, & Frisina (1992)	SW MT	8
Hoffman (1996)	SW MT, Yellowstone	9–13
Houston (1982)	Yellowstone	9–13
Rens (2001)	Yellowstone	10–11
Mehus (1995)	Yellowstone	19
Chronosequence studies		
Nelle, Reese, & Connelly (2000)	SE ID	1–36

Table 9.1. Continued

Sources	Location	Time since Fire (yrs)	ARTRT	ARTRW	ARTRV	Low	Silver	Three-tip	Bitter-brush	Indiv. Spp.	Spp. Gps.	Regen.	Divers.	Prod.
Colket (2003)	SE ID	1–92	X									X[a]	X	
Wambolt, Walhof, & Frisina (2001)	SW MT	2–32		X							X	X		X
Walhof (1997)	SW MT	2–32		X							X	X		X
Moritz (1988)	SE ID	2–>35		X						X	X	X		
Humphrey (1984)	SE ID	2–36	X		X					X	X	X		
Lesica, Cooper, & Kudray (2005, 2007)	SW MT	3–34	X					X		X	X	X	X	
Eichhorn & Watts (1984)	C MT	5–14		X							X	X		
Wambolt, Hoffman, & Mehus (1999)	SW MT	9–19	X							X		X		X

Notes: Studies were conducted in and near the Rocky Mountains. The dominant taxon on each site is identified. ARTRT = *Artemisia tridentata* ssp. *tridentata* (basin big sagebrush); ARTRV = *Artemisia tridentata* ssp. *vaseyena* (mountain big sagebrush); ARTRW = *Artemisia tridentata* ssp. *wyomingensis* (Wyoming big sagebrush); Low = little (low) sagebrush, black sagebrush, and alkali sagebrush. Variables studied are individual species, species groups (i.e., forbs, graminoids, shrubs), sagebrush regeneration (regen.), species diversity (divers.), and production (prod.). Table entries are ordered by time since fire (range of years studied) and whether observations or permanent plots were in one or more fires or a chronosequence of fires.

[a] Studied only for years 1–7.

needlegrass survived heating to 200° centigrade, but most large plants died in a hotter wildfire. Fire intensity may be lower in prescribed fires, causing less mortality and explaining why wildfires led to a lower postfire increase in native perennial grasses (Lesica, Cooper, and Kudray 2005). Plant morphology shapes the intensity and duration of fire at the plant. Squirreltail, for example, has a solid stem with less vegetative material and combusts less near or below the soil surface, which is common among fine-leaved fescues and needlegrasses (H. A. Wright and Klemmedson 1965; H. A. Wright 1971). Mat-forming forbs and fibrous-rooted forbs were most damaged by fire in mountain big sagebrush in Idaho, but taprooted and rhizomatous forbs did not appear damaged, and most resprouted (Blaisdell 1953; Kuntz 1982).

Resprouting shrubs are common after fire in sagebrush (fig. 9.2), but these shrubs may be initially damaged if fire intensity is high or from fire in particular seasons. Threetip sagebrush can resprout, but only a small fraction commonly do, and the shrub often is initially reduced by fire (Pechanec and Stewart 1944; L. D. Humphrey 1984; Akinsoji 1988; Lesica, Cooper, and Kudray 2005). Silver sagebrush resprouted vigorously the summer after April fires in Montana (Rennick 1981; White and Currie 1983). Higher fire intensity led to 40 percent mortality of silver sagebrush after spring burns but 75 percent after fall burns (White and Currie 1983). Fire reduced the sprouting shrubs yellow rabbitbrush and spineless horsebrush (fig. 9.2) the first few years after fires, especially if hot, in mountain big sagebrush (Harniss and Murray 1973; Kuntz 1982). These shrubs expanded dramatically in the next two to three decades in mountain big sagebrush (Harniss and Murray 1973; L. D. Humphrey 1984; Mehus 1995) and in the next three to eight decades in Wyoming big sagebrush in Idaho (W. E. Moritz 1988). Antelope bitterbrush survival and resprouting vary with geography or ecotype, plant age, and competition with other plants, but bitterbrush is often damaged by fire, particularly summer fire, or when carbohydrate reserves are low (R. E. Martin and Driver 1983; C. L. Rice 1983; J. G. Cook, Hershey, and Irwin 1994). Bitterbrush may resprout or regenerate from seed rapidly at times, but it can require up to 30 years to fully recover from fire (R. B. Murray 1983).

Native perennial grasses, especially Idaho fescue, often are reduced a year after fire in sagebrush (Pechanec and Stewart 1944; Mueggler and Blaisdell 1958; Nimir and Payne 1978; West and Hassan 1985; West and Yorks 2002; Seefeldt et al. 2007). Damage to Idaho fescue can persist for more than 12 years (Blaisdell 1953; Harniss and Murray 1973) but was lacking in a set of Montana sites (Lesica, Cooper, and Kudray 2005). In another set, Idaho fescue cover commonly was reduced ($n = 4$) or no different ($n = 8$) 2 to 32 years after burning, with inconsistent recovery time (Wambolt, Walhof, and Frisina 2001).

Effects of fire on native perennial grasses vary somewhat among sage-brushes, perhaps reflecting an underlying moisture gradient, but trends after fire are also inconsistent, suggesting that other factors (e.g., fire intensity, season) affect responses. On the driest end of the gradient, on a semidesert Wyoming big sagebrush site in Utah, native perennial grasses were reduced for 4 years after fire (West and Yorks 2002). On a Montana site with Wyoming big sagebrush, western wheatgrass cover remained below unburned cover 5–14 years after fire (Eichhorn and Watts 1984), but reanalysis of this site and others in the area showed a 77 percent increase for a half century or more (S. V. Cooper, Lesica, and Kudray 2007). A few years after fire, native perennial grasses, including bluebunch wheatgrass, were unchanged in Wyoming big sagebrush in Idaho (Peek, Riggs, and Lauer 1979; Colket 2003) and Montana (Lesica, Cooper, and Kudray 2005). However, in southwestern Montana, bluebunch wheatgrass had higher production in years 1–6 and 17 after fire in Wyoming big sagebrush than in unburned sagebrush (Wambolt and Payne 1986). At three other sites, no difference occurred in bluebunch wheatgrass cover 9 and 32 years after fire (Wambolt, Walhof, and Frisina 2001). Bluebunch wheatgrass also lacked change after fire in southeastern Montana (S. V. Cooper, Lesica, and Kudray 2007).

On moister sites in mountain big sagebrush, native grasses, especially blue-bunch and rhizomatous grasses, may respond positively to fire, but with many exceptions. Increased perennial grass lasted more than 12 years on some Idaho sites (Harniss and Murray 1973) but was lacking 8 years after fire on a site in southwestern Montana (Fraas, Wambolt, and Frisina 1992) and on seven other sites in this area 1–16 years after fire (Wambolt, Hoffman, and Mehus 1999). Increased native perennial grass lasted up to about 30 years on a set of Montana sites (Lesica, Cooper, and Kudray 2005) but was only about 13 percent higher after prescribed fires and 4 percent higher after wildfires (Lesica, Cooper, and Kudray 2005). Western wheatgrass increased for a few years after fire in silver sagebrush (Rennick 1981; Shariff 1988) and mountain big sagebrush (Omphile 1986). Streambank wheatgrass production did not change on a light or moderate burn but increased two to three times on a hotter burn a year after fire in mountain big sagebrush in Idaho and stayed elevated for 15 years (Blaisdell 1953). Graminoid cover, however, was still reduced 2 years after spring prescribed fire and especially after fall fire on a mountain big sagebrush site in Wyoming (McGee 1982).

Native annual forbs vary from dramatic increase to no change after fire in sagebrush. They increased dramatically in year 1 after fire in mountain big sagebrush in Idaho (Blaisdell 1953; Kuntz 1982) and southeastern Wyoming (J. G. Cook, Hershey, and Irwin 1994) and in year 2 in Wyoming big sagebrush in

Idaho (Peek, Riggs, and Lauer 1979), but they declined in year 3 in both cases. Native annual and biennial forbs did not increase after fire in other cases in mountain big sagebrush in Idaho (Harniss and Murray 1973; Seefeldt, Germino, and DiCristina 2007) or on other Wyoming big sagebrush sites in Idaho (Colket 2003) and Montana (Wambolt and Payne 1986).

Native perennial forbs generally show little change after fire in sagebrush, but they can decline or increase. Little change in perennial forb cover occurred after fire in Wyoming big sagebrush by years 1–4 in Idaho (Peek, Riggs, and Lauer 1979), years 1–17 in southwest Montana (Wambolt and Payne 1986), years 4–67 in southeastern Montana (S. V. Cooper, Lesica, and Kudray 2007), and years 1–12 in Colorado (Perryman et al. 2002). Native forbs were reduced the first 2 years after fire in mountain big sagebrush in Idaho (Pechanec and Stewart 1944; Harniss and Murray 1973) and Wyoming (McGee 1982). Cover of major forbs did not differ among burns 1–14 years after fire on other mountain big sagebrush sites in Idaho (Nelle, Reese, and Connelly 2000; Seefeldt, Germino, and DiCristina 2007) nor did a standing crop of forbs generally change 1–3 years after fire on two mountain big sagebrush sites in Wyoming (J. G. Cook, Hershey, and Irwin 1994). Forb production increased 20–129 percent (depending on fire intensity) 3 years after fire in mountain big sagebrush in Idaho and remained higher 15 years after fire (Blaisdell 1953). Forb production was elevated for up to 9 years after fire in mountain big sagebrush in Montana (Van Dyke and Darragh 2006).

Production typically shifts from shrubs to herbaceous plants after fire in sagebrush, but total production of postfire herbaceous vegetation usually does not exceed prefire production of the combined shrub–herbaceous complex (Blaisdell 1953; Mueggler and Blaisdell 1958; Peek, Riggs, and Lauer 1979). However, total plant production increased by about 20 percent in year 2 after 1988 fires in mountain big sagebrush in Yellowstone (Singer and Harter 1996). Other studies focused on herbaceous production, which was reduced in year 1 but higher in year 2 after fire (Omphile 1986) and in years 2 and 3 after fire (J. G. Cook, Hershey, and Irwin 1994) in mountain big sagebrush in Wyoming. Graminoid production increased the year after fire in Yellowstone mountain big sagebrush but returned to prefire levels by year 5 after fire (Tracy and Mc-Naughton 1997). Up to 3 years after fire in Idaho mountain big sagebrush, graminoid production was 1.3 to 2.3 times higher, and remained 6–16 percent higher 15 years after fire (Blaisdell 1953). A spring burn increased herbaceous production 2 years after fire by a factor of 1.3 to 4.9 on a mountain big sagebrush site, but a fall burn decreased production by a factor of 0.8 to 0.9 (Schwecke and Hann 1989). In Wyoming big sagebrush in Utah, herbaceous production returned to prefire levels 1 year after fire, but production was mostly from cheatgrass (West and Hassan 1985). In Idaho on a Wyoming big sagebrush site without significant cheatgrass, herbaceous production did not change after fire (Peek, Riggs, and Lauer 1979), but on a Montana site, production (excluding sagebrush) was significantly elevated in years 2–13 and 17 after fire (Wambolt and Payne 1986).

Diversity at the alpha (local) level may be little affected by fire in sagebrush. One year after fire in threetip sagebrush (Akinsoji 1988) and in (most likely) mountain big sagebrush (Petterson 1999), plant species richness was unchanged. Humphrey (1984) found that plant species richness and Shannon diversity, a combined measure of richness and evenness, were a little higher in year 2 after fire in mountain big sagebrush in Idaho but had declined by year 3 and did not change over the next three to four decades. Plant species richness and Shannon diversity were unchanged at most of six sites 1–12 years after fire in Wyoming big sagebrush in Colorado (Perryman et al. 2002). Plant species richness increased 1.2–2.1 times on two mountain big sagebrush sites in Montana (Schwecke and Hann 1989) and even more on another site (Van Dyke and Darragh 2006). No differences in species richness occurred between burned and unburned sites for basin, Wyoming, and mountain big sagebrush in a larger sample in Montana (Lesica, Cooper, and Kudray 2005). Mean richness per plot was 32 species in unburned Wyoming big sagebrush and 26 in burned plots 4– 67 years after fire in southeastern Montana (S. V. Cooper, Lesica, and Kudray 2007). Alpha diversity may be little affected, because most species survive fire and resprout (L. D. Humphrey 1984).

Human Effects on Sagebrush Fire Regimes and Recovery after Fire

Recent fire rotation in sagebrush (undifferentiated by sagebrush taxa), from government fire data for 1980–2007, is 70 years on the Snake River Plains in southern Idaho and southwestern Montana, 197 years in the silver sagebrush region in eastern Montana and Wyoming, 249 years on the Colorado Plateau in southwestern Colorado, and 345 years in the Wyoming Basin in western Wyoming and northwestern Colorado (W. L. Baker, in press). These estimates are based on periods of data too short for accurate estimation, but comparing these estimates to rotation under the HRV (table 9.2) suggests that fire exclusion may have had little net impact and is not likely a cause of invasions of trees into sagebrush (box 9.1). Fire rotation in sagebrush is likely not outside the HRV in the Rockies, except on the Snake River Plains and other areas with abundant cheatgrass, where rotation is likely short relative to the HRV.

Burning is not needed to restore sagebrush ecosystems, and it is especially inadvisable where native plants are depleted and cheatgrass is present (Pechanec

Table 9.2. Estimated Fire Rotations for Rocky Mountain Shrublands and Grasslands

Vegetation Type	Estimated Recovery Period (yrs)	Estimated Fire Rotation (yrs)
Salt desert shrublands	Unknown	More than 500
Mountain big sagebrush	75–100 in slow track, 25–35 in fast track (fig. 9.5)	150–300 in large expanses, 160 in intermix with Douglas-fir, 400–600 in intermix with piñon-juniper
Wyoming big sagebrush	Unknown	200–350 in large expanses, 400–600 in intermix with piñon-juniper
Little (low)	Unknown	More than 200 in large expanses, More than 425 in intermixes
Curlleaf mountain mahogany	Unknown	Unknown
Mixed mountain shrublands	20	100
Shortgrass prairies	3–5	More than 200?
Dry mixed-grass prairies	1–5	140–160?
Montane grasslands	3–5	50–115 in large expanses, 90–200 in intermixes with forests
Subalpine grasslands	Unknown	50–170 in large expanses, 90–300 in intermixes with forests

Sources: See corresponding sections of text in chapter 9.

and Stewart 1944; W. L. Baker 2006b). Burning does not restore sagebrush but removes it for decades to centuries (fig. 9.5). Sagebrush has been cleared and degraded and would benefit from rest, not more disturbance.

A cheatgrass-fire cycle (box 9.2) can end in replacement of sagebrush by cheatgrass, as on the Snake River Plains (Pickford 1932; Whisenant 1990; Knick 1999). The 70-year rotation there in the last two decades is too short to allow Wyoming big sagebrush to fully recover after fire (fig. 9.5b). Cheatgrass can increase dramatically the first one to two years after fire in Wyoming big sagebrush (West and Hassan 1985; W. E. Moritz 1988) or threetip sagebrush (Seefeldt and McCoy 2003). About one-third of the cheatgrass seed bank survived high-intensity fire in threetip sagebrush in Idaho (Akinsoji 1988), and about half the cheatgrass seed bank survived wildfire in Wyoming big sagebrush in Utah (Hassan and West 1986). The seed bank rebounded to twice the prefire seed bank by 14 months after fire (fig. 9.6), especially in "hot spots" where shrubs burned (Hassan and West 1986). Cheatgrass may not always increase after fire (Clifton 1981; L. D. Humphrey 1984; Colket 2003) and may decline as native perennials recover, especially on ungrazed sites (West and Yorks 2002; Seefeldt and McCoy 2003). However, many sagebrush stands are deficient in native grasses and forbs from overgrazing and are vulnerable to cheatgrass dominance after fire (Pechanec and Stewart 1944). Fires are occurring at shorter rotations in sagebrush with cheatgrass, leading to lower species richness, loss of

BOX 9.1
Tree Invasion and Fire

Some authors (Houston 1973; Arno and Gruell 1983) believed that fire was frequent under the HRV and prevented shrub and tree invasion into grasslands and shrublands, and that fire exclusion thus could explain such invasions since EuroAmerican settlement. These authors incorrectly interpreted mean composite fire intervals to represent mean fire interval and fire rotation (chap. 5). Mean fire intervals and fire rotations in shrublands and grasslands in the Rockies under the HRV were long enough to allow trees to invade (table 9.2). However, fire may have killed invading trees at times and could create some temporary openings.

Forest openings in the Rockies may be (1) temporary after high-severity forest fires where tree regeneration represents the natural recovery of the prefire forest, and (2) more persistent where tree regeneration may be a temporary response to climatic fluctuations or a potentially more persistent effect of land uses. Temporary openings after high-severity forest fire may have dead or charred wood; a weak or lacking mollic epipedon (a soil layer indicating long-term grass dominance); and normally incipient or abundant tree regeneration, indicating recovery of the forest. More persistent openings have little dead or charred wood; may have a mollic epipedon; and may generally lack much tree regeneration, unless temporarily favored by climatic episodes or triggered by land uses. Small trees in forest openings in the Rockies can thus represent natural forest recovery in temporary openings, natural fluctuations in tree abundance related to climatic episodes, or unnatural invasion in persistent openings. If persistent openings are controlled by fire, trees might invade in episodes without fire and then be swept away by fire.

Both temporary and persistent openings occurred in Rocky Mountain landscapes under the HRV. After high-severity fires in spruce-fir forests, meadows with slowly recovering trees may exist for a century or more because of limited tree seed, or because seeds that arrive and germinate face adverse climate and competition from grass-sedge turf (Stahelin 1943; Billings 1969). In montane forests in Colorado, temporary openings from high-severity fire could persist for more than 150 years (chap. 7), gradually becoming reforested. Persistent openings were evident in green fescue parks in northern Idaho, where charcoal from burned wood was

BOX 9.1

Continued

scattered, suggesting that parks were not formerly well forested (Daubenmire 1981). Similarly, meadows in western Montana (Sindelar 1971), central Idaho (D. R. Butler 1986), southeastern Wyoming (Doering and Reider 1992), and northern New Mexico (Allen 1984, 1989) also had little or no charcoal, suggesting that grasslands did not arise from burning of forest. Lack of extensive charcoal and a deep mollic epipedon in grassland soils are indicators that grasslands were present at least for hundreds of years (Allen 1984, 1989).

Lack of invasion in periods lacking fire, and stability of a forest-opening boundary location after fire, also suggest persistent openings not controlled by fire. Little tree invasion occurred when fire was lacking for decades in northern Idaho parks (Daubenmire 1981). Many forest-grassland boundaries in Montana lacked evidence of fire for the previous 50–90 years but also lacked tree invasion (Sindelar 1971). After fire across forest-grassland boundaries in northern Wyoming (Despain 1973) and western Montana (Sindelar 1971), trees returned in the burned forest, but not the meadow, leaving the boundary position unchanged. If fire controlled the location of these boundaries, they should have retreated into the forest after fire. Trees invaded *after* some fires in grasslands in Glacier National Park (Singer 1975) and in Idaho (D. R. Butler 1986), suggesting that fire itself can reduce competition from grasses and forbs, allowing tree regeneration.

It is essential to determine, if possible, whether openings are temporary or persistent. For example, in part of northern New Mexico (Coop and Givnish 2007), ponderosa pine invasion on a mountain led to a 62.5 percent grassland loss in 61 years. This rate is consistent with the rate of natural forest recovery where high-severity fire created a temporary opening in ponderosa pine forests (chap. 7). The authors interpreted the invasion as linked to cessation of fires in a grassland, but evidence of prior long-term grassland persistence was lacking, and natural recovery after forest fire remains a possible explanation.

Where invasions have occurred in persistent openings, their timing has correlated not with fire exclusion but with climatic fluctuations or livestock grazing. Tree invasion did not intensify until decades after the onset of fire exclusion in northern New Mexico (Allen 1989; Coop and

(continued on next page)

BOX 9.1
Continued

Givnish 2007) and not until the 1940s in central Wyoming (Dunwiddie 1977) and parts of northern New Mexico (Coop and Givnish 2007). This lag could be consistent with a spatial lag expected from fire exclusion (see box 10.1) where fire rotations are long (table 9.2). However, invasions were better correlated with favorable climatic episodes (D. R. Butler 1986; Jakubos and Romme 1993; Coop and Givnish 2007) or with livestock grazing (e.g., Coop and Givnish 2007). Sindelar (1971) showed that a lightly grazed area in Montana had few invading Douglas-fir and only 2.5 percent cover of mountain big sagebrush, while across a fence on a site with a history of livestock overgrazing, Douglas-fir reached 25,000 trees per hectare and sagebrush averaged 14.4 percent cover. Tree invasion in central Wyoming began after a decline in livestock grazing that allowed tree regeneration at a time when competition from grasses and forbs had been reduced (Dunwiddie 1977). Tree invasion in northern New Mexico also coincided with a shift from sheep to cattle grazing (Coop and Givnish 2007).

Tree invasions can be substantial in local areas (e.g., Hansen, Wyckoff, and Banfield 1995), but analysis over larger land areas showed only minor loss of opening area over the last century. At the scale of mountain ranges, opening loss was small in montane and subalpine grasslands and mountain big sagebrush (Andersen and Baker 2006; Zier and Baker 2006) and also in Wyoming big sagebrush (Manier et al. 2005; Shinneman 2006). Replacement of openings by trees, at rates over the past century, would require more than 1,400 years in Wyoming (Andersen and Baker 2006). In 10,449 hectares of grassland in northern New Mexico, a 12.4 percent decrease from 1935 to 1996 suggests nearly 500 years (61/0.124) for full grassland loss. Moreover, these rates of invasion may have been increased by livestock grazing; thus, natural rates could be slower.

Even these loss rates, perhaps elevated by livestock grazing, would not have required frequent fire to maintain openings. Some tree invasion could naturally occur during interludes between large fires, which could then erase invasions, since large, infrequent fires do most of the work in Rocky Mountain ecosystems (box 3.1). Given the fire rotations I estimated (table 9.2), one- to two-century periods without fire, punctuated

BOX 9.1
Continued

by large fires, would be within the HRV. Given that full invasions may require 500–1,400 years, fires every one to two centuries should be sufficient to remove invading trees, even where invasion has been elevated by grazing.

Trees could, of course, be intentionally removed from forest openings using fire, but frequent fire is inappropriate, given the one- to three-century fire rotations that likely characterized grasslands and shrublands under the HRV (table 9.2). Frequent fire is also unnecessary, given documented rates of invasion. A single burn once every century or two would be sufficient. However, burning temporary meadows where trees are naturally recovering after high-severity forest fire or are naturally invading during interludes between infrequent, large fires, would not be ecological restoration and would simply interrupt natural fluctuations in ecosystems. If climatic episodes caused tree invasion, tree removal is also not ecological restoration.

To choose ecologically appropriate action, it is essential to first determine whether an opening is persistent or temporary, then identify the causes of a particular tree invasion. If the opening is temporary or persistent and the cause is natural, it is reasonable to do nothing. If land uses (e.g., livestock grazing) caused tree invasion in a persistent opening, tree removal may be warranted. However, before removal, the land use needs to be reformed and the vegetation restored. Otherwise, removal of trees may be futile, as renewed regeneration may occur.

In any case, frequent fire did not occur under the HRV in Rocky Mountain grasslands and shrublands. Evidence suggests little or no decline in fire relative to the HRV (chap. 10), and the limited data for large land areas also suggest little tree invasion is occurring overall. Fire likely infrequently created temporary openings that may have persisted for a century or more, and fire could certainly infrequently remove some trees that invaded persistent openings during favorable climatic episodes. A fire every century or two may suffice and seems to be occurring.

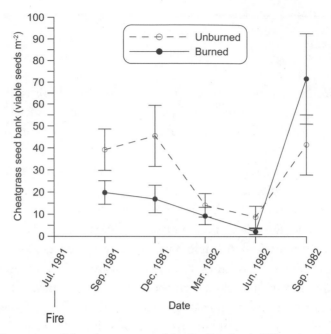

Figure 9.6. Recovery of the cheatgrass seed bank after fire in Wyoming big sagebrush in northern Utah.
Source: Data are from Hassan and West (1986).

biological soil crust, and failure of sagebrush and other shrubs to regenerate (Whisenant 1990; Boltz 1994). Full fire control is essential if sagebrush is to be maintained where cheatgrass is common.

Reseeding sagebrush sites after fire has seldom been successful at reducing cheatgrass (Shinneman 2006) and may delay recovery of native plants (Ratzlaff and Anderson 1995) or increase cheatgrass if seed mixes are contaminated (Getz and Baker 2008; chap. 11). Moreover, physical disturbance from some re-seeding (e.g., drill seeding) may favor cheatgrass (Ratzlaff and Anderson 1995). If restoration of biological diversity is the goal, postfire reseeding of burned sagebrush should be suspended until development of native seed mixes proven to resist cheatgrass while restoring native diversity (Shinneman 2006; Getz and Baker 2008).

Postfire grazing may alter recovery after fire. Immediate postfire livestock grazing of burned mountain big sagebrush in Idaho reduced native grasses by 23 percent, and sagebrush recovery occurred within nine years; while a similar site rested for a year, then, conservatively grazed, had 106 percent more native perennial grasses and less than 10 percent as much sagebrush (Pechanec and Stewart 1944). Postfire livestock grazing reduced recovery of native peren-nials and increased persistence of cheatgrass on a semidesert Wyoming big

BOX 9.2

Cheatgrass Fire Cycle

Cheatgrass is a nonnative annual grass, of Eurasian origin, that had spread across the western United States by the early 1900s, facilitated by over-grazing, agriculture, and other disturbances (Mack 1981). By the 1920s, it occurred in dense stands after fires in sagebrush (Pickford 1932), and by 1940 reports of large fires in cheatgrass were common (Billings 1990). Although the cheatgrass seed bank is reduced the year after fire, it rebounds to the same or a higher level in one to two years (Hassan and West 1986; Akinsoji 1988; L. D. Humphrey and Schupp 2001). Smoke stimulates production and elongation of leaves and increases overall growth and seed production (J. A. Young and Evans 1978; Blank and Young 1998). Cheatgrass greens up and depletes soil moisture early in the spring, increasing drought stress and lowering productivity of native plants (Melgoza, Nowak, and Tausch 1990). Awned seed allows cheat to be readily dispersed by animals, wind, and vehicles (Mack 1981). These traits allow cheat to expand after fire in salt desert shrubs, Wyoming big sagebrush, piñon-juniper (West and Hassan 1985; W. E. Moritz 1988; Shinneman 2006) and increasingly in other vegetation (e.g., ponderosa pine and aspen forests). However, cheatgrass can also expand unexpectedly, even in relatively undisturbed vegetation (Floyd et al. 2006).

Grasses, particularly annual grasses (e.g., cheatgrass) are especially flammable for several reasons (D'Antonio and Vitousek 1992; Brooks et al. 2004): (1) their life form includes elevated, standing dead material with a high surface-volume ratio that dries easily; (2) the packing ratio (see table 2.1) of grass crowns is favorable for fire spread; (3) grass crowns are situated near the ground, where air temperature and drying effects are high; and (4) grasses, particularly annuals, can create spatially continuous fuels in favorable years. Nonnative annual grasses can also effectively lengthen the fire season and change the timing of fires, since they use spring moisture and dry up before native grasses, promoting earlier fires.

Annual grasslands in the Great Basin are characterized by large fires, favored the year after a wet year from a carryover of built-up fine fuel (Knapp 1998). On the Snake River Plains, Idaho (fig. 9.1) where more than half the land is cheatgrass dominated, fires burned about 36 percent

(continued on next page)

BOX 9.2
Continued

of a 290,000-hectare area in 18 years, a fire rotation of about 50 years, much shorter than the rotation prior to EuroAmerican settlement (W. L. Baker 2006b). Thus, a positive feedback loop or grass-fire cycle has developed, which leads to near cheatgrass monoculture (Pickford 1932; Whisenant 1990; Knick 1999), as well as lower native-species richness, and slows or prevents regeneration of sagebrush and other nonsprouting shrubs (Whisenant 1990; Boltz 1994).

Vulnerability to a cheatgrass fire cycle is partly dependent on prefire vegetation. Cheatgrass may not always increase after fire (Clifton 1981; L. D. Humphrey 1984; Colket 2003) and can decline as native perennials recover, especially on ungrazed sites (West and Yorks 2002). Burned sagebrush had significantly higher postfire cheatgrass cover than did burned piñon-juniper woodlands in western Colorado (Shinneman 2006). Sagebrush stands deficient in native grasses and forbs are especially vulnerable to cheatgrass after fire (Pechanec and Stewart 1944). In a semiarid area of western Colorado, the highest postfire cheatgrass cover was where fires burned livestock-degraded sagebrush that had low cover of biological soil crust, high cheatgrass cover, and low forb diversity and cover (Shinneman 2006).

---

sagebrush site in Utah (West and Yorks 2002). Heavy herbivory by elk and deer on Yellowstone's northern winter range significantly reduced the rate of recovery of sagebrush after fires (Wambolt 1998).

CURLLEAF MOUNTAIN MAHOGANY

Stands of curlleaf mountain mahogany, a tall shrub or small tree (fig. 9.7), called mahogany for simplicity, burned under the HRV in crown fires, surface fires, and mixed-severity fires. Scheldt (1969) dated a high-severity fire, with some survivors, from the late 1890s. A few fires in the mid-1800s, visible in early photos, included partial canopy replacement (Gruell, Bunting, and Neuenschwander 1985). Fires can be hot, consuming some stems and leaving little charcoal (Scheldt 1969). Recent fires have sometimes been high severity over

Figure 9.7. Curlleaf mountain mahogany in the Wasatch Mountains, Utah.
Source: Photo by the author.

large areas, but scars and charcoal at the base of stems in some stands suggest that surface fire also occurred (Arno and Wilson 1986). Mahogany has thick bark and can survive light surface fires (Gruell, Bunting, and Neuenschwander 1985). However, the proportion of severities under the HRV cannot be estimated from these limited observations.

Fire history from mahogany stands is also insufficient to estimate fire rotation or variation in fire severity. Scars can be dated in some cases from surviving fire-scarred stems on the edge of high-severity fires (Scheldt 1969; Arno and Wilson 1986). However, annual rings are unclear (scars from a known 1940 fire dated from 1941 to about 1949), and scars from other agents have made identification of fire difficult (Arno and Wilson 1986). Pith dates from mahogany in four stands in Idaho corresponded with some fires on nearby ponderosa pines, hinting at mixed-severity fire (Arno and Wilson 1986). However, most pine scar dates did not correspond with either mahogany pith dates or mahogany scar dates. The sample also was targeted at pines with multiple scars, was not cross-dated, and included only two to three scarred trees per site, leaving the extent of the fires and the rotation in question (chap. 5). Further research is needed.

Mahogany is considered a weak resprouter (Gruell, Bunting, and Neuenschwander 1985), often dependent on regeneration from seed after fire. Only 2–4 percent of burned stems resprouted after high-severity fires in two stands (Gruell, Bunting, and Neuenschwander 1985). Seedlings are favored on exposed mineral soil, but lags in postfire regeneration are common, with no regeneration 3, 17, and even 30 years after fires (Scheldt 1969; Gruell, Bunting, and Neuenschwander 1985). Stands of mahogany can be stable or successional to conifers (Gruell, Bunting, and Neuenschwander 1985). Stands in western Utah were predominantly stable (J. N. Davis 1976). Stable stands appear common on rocky sites, but some overgrazed grasslands on deeper soils may have been invaded by mahogany (Scheldt 1969; Gruell, Bunting, and Neuenschwander 1985).

Gruell et al. (1985) thought that reduction of fire had allowed an increase in stem density in many stands and a decline in regeneration in older stands. However, their figure 2 shows that a low-density stand known to have burned in a partial crown fire before 1868 had maturing mahogany and higher stem density by 1982. This likely represents recovery from severe fire, not an abnormal increase in stem density from fire exclusion. Moreover, population structure of mahogany stands may typically include a suppressed understory subpopulation capable of persisting for a century or more until an overstory individual dies (Schulz, Tueller, and Tausch 1990). Also, young postfire stands are typically dense, thinning as the stand matures (Schulz, Tausch, and Tueller 1991). Thus, an abundance of young stems today, when stem ages are pooled across stands (Gruell, Bunting, and Neuenschwander 1985), may simply reflect normally low-density, mature mahogany stands, combined with normally high-density, postdisturbance stands. Mahogany stem density and stand structure today, where mahogany occurred in the pre-EuroAmerican era, are likely not outside the HRV, but more systematic analysis of age structure and fire across landscapes is warranted.

MIXED MOUNTAIN SHRUBLANDS

Under the HRV in mixed mountain shrublands, fires were likely high severity, topkilling aboveground vegetation, as also observed recently (H. E. Brown 1958; Wadleigh, Parker, and Smith 1998; Floyd, Romme, and Hanna 2000).

High severity is expected, given dense canopies and continuous surface fuels. Occasional lighter burns left some surviving stems, even within high-severity fires (Floyd, Romme, and Hanna 2000). Some fire scars were found (H. E. Brown 1958) where high-severity fires were going out (Floyd, Romme, and Hanna 2000) or in other lightly burned areas. Floyd et al. (2000) estimated that fire rotation in this vegetation in southwestern Colorado was about 100 years, the only estimate for the Rockies. Fire rotation was estimated by stand-origin dating, found reliable up to 150 years after fire. Fire size under the HRV may have reached as high as 15,000 hectares. Sudworth (1900a) reported that fires in 1898 burned about 13,000–17,000 hectares of "brush and grass land" on Battlement Mesa in northwestern Colorado. He also said, "Early fires have swept through thousands of acres of this growth" on the White River Plateau (1900b, 176). Fires become large during high winds and drought (Poreda 1992), but spread occurs in these continuous fuels even without strong winds, particularly given the often steep slopes (H. E. Brown 1958) and few fire breaks (Floyd, Romme, and Hanna 2000).

Many of the dominant shrubs resprout from root crowns or rhizomes (chap. 3; appendix B; Harper, Wagstaff, and Kunzler 1985). Gambel oak also has a *lignotuber*, an enlarged stem base with dormant buds that can sprout after fire, an adaptation common in chaparral but rare in the Rockies (fig. 9.8). Mountain big sagebrush and black sagebrush are killed by fire (Kufeld 1983; Floyd, Romme, and Hanna 2000) and recover slowly. Alderleaf mountain mahogany and antelope bitterbrush can be killed and reinvade slowly in some areas if severity is high (Floyd, Romme, and Hanna 2000). Mountain mahogany sprouted well even after high-severity fire in the Colorado Front Range (Liang 2005). Mountain snowberry and Oregon boxleaf are among the slowest resprouting shrubs to recover; whereas chokecherry, Utah serviceberry, and Gambel oak often resprout profusely but regain prefire cover more slowly (McKell 1950; Kufeld 1983). These three shrubs may exceed prefire stem density by two to four times in years 1 and 2 after fire (McKell 1950; Poreda 1992; Poreda and Wullstein 1994), perhaps stimulated by postfire nutrients (W. L. Baker 1949). In northwestern Colorado, Sudworth (1900b, 176–77) noted, "The severity of these fires is attested now only in the larger, blackened, dead stems surrounded by impenetrable thickets of new root sprouts," which is also the case today (fig. 9.9). Stem density thins gradually as the stand matures (McKell 1950; H. E. Brown 1958). In Utah, the rate of recovery of Gambel oak declines with elevation (fig. 9.10).

Postfire succession varies, depending on fire intensity and density of shrubs and openings, based on studies at a dozen sites (table 9.3). Year 1 postfire may have either increased or reduced cover and production of native perennial

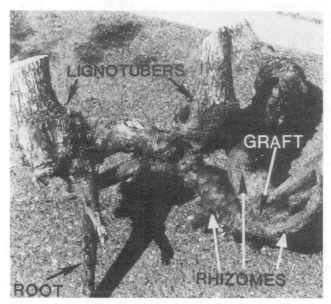

Figure 9.8. Lignotubers and rhizomes of Gambel oak.
Source: Reproduced from Tiedemann, Clary, and Barbour (1987, p. 1066, fig. 1), with permission of the Botanical Society of America.

Figure 9.9. Recovery of mixed mountain shrubs three years after a 2002 fire in Mesa Verde National Park, Colorado.
Source: Photo by the author.

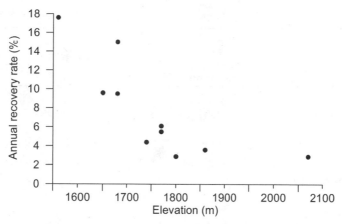

Figure 9.10. Rate of burned Gambel oak height recovery after fire vs. elevation. The study area was in the Uinta National Forest, Utah, and vicinity. The rate is calculated as percent recovery, compared to height in paired unburned stands, divided by age of burn for burns 3–15 years prior to sampling.
Source: Based on data in Kunzler and Harper (1980).

grasses and forbs, but increased annual forbs are common (McKell 1950; Kunzler, Harper, and Kunzler 1981; J. G. Cook, Hershey, and Irwin 1994; Poreda and Wullstein 1994). Perennial and annual forb production was elevated in years 2 and 3 postfire (Kufeld 1983; J. G. Cook, Hershey, and Irwin 1994), lasting up to 10 years postfire (McKell 1950; Kufeld 1983). After year 1, graminoid cover and production were elevated for up to a decade in Utah (McKell 1950) but declined for 5 or more years in Colorado (Kufeld 1983). Postfire composition and cover of forbs and graminoids were substantially different from those of unburned stands in Utah for up to 4 years after fire but were similar to unburned stands by years 9–19 after fire in one case (Kunzler, Harper, and Kunzler 1981) and year 18 in another (McKell 1950). Plant species diversity may increase or not change in the first year after fire, depending on the density of shrubs and openings (Poreda 1992), but diversity may decline during the first two decades of succession (Steinhoff 1978).

The mixed mountain shrub community appears to be a stable vegetation type over large areas, but it is also successional to forests on large areas. Experimental planting of ponderosa pine failed in most mixed mountain shrub areas, so fire is not the sole factor excluding trees (F. S. Baker and Korstian 1931). The shrubs are more drought tolerant than the trees, the trees cannot endure dense shade, and the shrubs appear favored by a clay layer in the soil (F. S. Baker and Korstian 1931; Steinhoff 1978). Nonetheless, the mixed mountain shrub community is also successional to forests on many sites; shrubs may be nurse plants

Table 9.3. Studies of Postfire Responses of Plants in Mixed Mountain Shrublands

Sources	Location	Time since Fire (yrs)	Severity			Variable Studied				
			L	M	H	I	G	S	D	P
Observations/plots after fire										
W. L. Baker (1949)	N UT	0–3		X			X			
Poreda (1992); Poreda & Wullstein (1994)	N UT	1		X		X	X	X	X	
D. L. Young & Bailey (1975)	N CO	1–2	X	X				X		X
Floyd, Romme, & Hanna (2000)	SW CO	1–2		X				X		
J. G. Cook, Hershey, & Irwin (1994)	SE WY	1–3		X			X	X		X
Frischknecht & Plummer (1955)	N UT	2		X	X			X		X
Kufeld (1983)	W CO	2–10		X			X	X		X
H. E. Brown (1958)	W CO	3		X				X		
Chronosequence studies										
McKell (1950)	N UT	1–18		X		X	X	X		
Kunzler & Harper (1980)	N UT	3–15		X				X		
Kunzler, Harper, & Kunzler (1981)	N UT	3–30		X	X			X		
Steinhoff (1978)	SW CO	5–100+	X			X	X	X	X	

Note: Studies were conducted in and near the Rocky Mountains. Fires were low (L), moderate (M), or high (H) in severity. Variables studied are individual species (I), groups of species (G; i.e., forbs, graminoids, shrubs), shrub regeneration or growth (S), species diversity (D), and production or growth (P). Table entries are ordered by time since fire (range of years studied) and whether observations were of one or more fires or a chronosequence of fires.

for twoneedle piñon (M. E. Floyd 1982) at low elevations and for Douglas-fir or white fir at higher elevations (F. S. Baker and Korstian 1931; Christensen 1964). Many mixed mountain shrub stands in southwestern Colorado may be temporary after fires or logging of Douglas-fir and ponderosa pine forests (H. E. Brown 1958; Harper, Wagstaff, and Kunzler 1985). Stable and successional mixed mountain shrub communities are difficult to distinguish, and on intermediate sites it could be that the abundance of trees naturally fluctuates, depending on the timing of fires and climatic fluctuations.

Both fire control and human ignitions affect some areas. For example, in a 64,000-hectare mixed mountain shrub area adjacent to Ogden, Utah, only 5 hectares were burned by lightning fires from 1973 to 1997, because of fire control, but 3,465 hectares were burned by human-ignited fires (Wadleigh et al. 1998), an unusually human-controlled fire regime for the Rockies (chap. 10).

MONTANE AND SUBALPINE GRASSLANDS

Fire behavior in mountain grasslands (montane and subalpine grasslands) is poorly known. Even in extreme years, such as 1988 in Yellowstone, fires in mountain grasslands can leave unburned area within perimeters, due to erratic winds (Singer and Harter 1996). A mosaic was also observed in recent prescribed fires (C. A. Johnson 1989). Rate of spread may be modest—Yellowstone grassland fires spread at 2.4 kilometers per hour (Singer and Harter 1996).

Fire rotation can be only roughly estimated but was likely moderate under the HRV, based on observations of trees and charcoal in meadows. Recovery is so fast (see later in this section) that its rate probably does not control fire rotation through fuel buildup. Trees in meadows in some cases suggest that fire was infrequent under the HRV. Subalpine green fescue parks in northern Idaho may have burned at long intervals (e.g., 300 years), as scattered old trees in parks are thin barked and would likely have been killed if fire had been frequent (Daubenmire 1981). Lack of charcoal in mountain grasslands has been reported in northern Idaho (Daubenmire 1981), western Montana (Sindelar 1971), northern Utah (Schimpf, Henderson, and MacMahon 1980), and northern New Mexico (Allen 1984). Schimpf et al. (1980, 19) suggested that "the absence of charcoal in the upper few centimeters leads us to conclude that fire has not been a major force in the meadow ecosystems during the past few centuries."

Another measure of fire rotation in mountain grasslands is the rotation in adjacent forests, which might control the grassland fire regime where grassland patches are small. High-severity fires have the greatest probability of spreading into grasslands; thus, fire rotations might range up to about 200 years in montane grasslands and to about 300 years in subalpine grasslands, based on high-severity rotations in nearby forests (chaps. 6–8). Some surface fires in forests may also spread into grasslands; thus, another source of fire history includes fire-scarred trees adjacent to grasslands. This evidence is limited to three studies in northwestern Wyoming and southwestern Montana (Arno and Gruell 1983, 1986; Houston 1973). These studies targeted sites with old trees and trees with long scar records, while areas with no scars or few scars—representing long intervals without fire—were not sampled. Thus, the fire intervals reported in these studies likely represent the shortest intervals experienced in the grasslands. These authors summarized the mean composite fire interval (CFI) in the pre-EuroAmerican era as less than 35–40 years (Arno and Gruell 1983), as 25 years (Arno and Gruell 1986), and as 20–25 years (Houston 1973).

CFI values require correction to estimate fire rotation or mean fire interval. Using a middle CFI estimate, 25 years, and the lowest multiplier of 3.6 (chap. 5) to estimate fire rotation from CFI yields an estimated rotation of 90 years at the low end, which applies to grasslands in intermixes with forests. Thus, estimates for intermix grasslands are 90–200 years in montane grasslands and 90–300 years in subalpine grasslands. In large expanses of grassland, fire rotation is likely shorter than in forests; a correction for this could be estimated using modern fire data, but necessary data are not available. Using an adjacency correction for sagebrush of 0.57, the only one available (W. L. Baker, in press), the low end of the rotation would be about 50 years (90 × 0.57) in large expanses of grassland, and the high end 115 years in the montane (200 × 0.57) and 170 years in the subalpine (300 × 0.57). Thus, a rough approximation for fire rotation is 50–115 years in expanses of montane grasslands and 50–170 years in subalpine grasslands. These estimates can be improved by more research on adjacency correction and fire history.

Plant response to fire in grasslands depends on fire intensity, physiological status of the vegetation, and season of the burn, based on a half dozen studies (table 9.4). Bunchgrasses may be especially damaged if mosses or forbs grow near their bases, causing greater heating (C. G. Johnson 1998). Fescues that dominate many mountain grasslands are often initially damaged and require from a few years to more than five years to recover (Humes 1960; Antos, McCune, and Bara 1983; C. A. Johnson 1989, 1998; Menke and Muir 2004). Rough fescue recovered only to 65 percent of unburned cover by three years after fire in western Montana (Antos, McCune, and Bara 1983), but it recovered by year 2 after fire in eastern Montana (Jourdonnais and Bedunah 1990). Large

Table 9.4. Studies of Postfire Responses of Plants in Mountain Grasslands

Sources	Location	Time since Fire (yrs)	Individual Species	Species Groups	Production
Humes (1960)	W MT	0–2	X	X	X
Antos, McCune, & Bara (1983)	W MT	0–3	X	X	X
Jourdonnais & Bedunah (1990)	E MT	0–3	X	X	X
Menke & Muir (2004)	SW ID	1	X	X	—
Fuller (2007)	N NM	1	—	—	X
Singer & Harter (1996)	Yellowstone	1–3	X	X	X
Merrill, Mayland, & Peek (1980)	C ID	1–4	X	X	X
C. G. Johnson (1998)	NE OR	1–5	X	—	—
Tracy (1997)	Yellowstone	2–3	—	—	X

Note: Studies were conducted in and near the Rocky Mountains. Variables studied are individual species, groups of species (i.e., forbs, graminoids, shrubs), and production. Table entries are ordered by time since fire (range of years studied) and represent observations or plots completed after fire.

clumps (i.e., more than 20 cm in basal diameter) of this fescue were vulnerable to mortality, as litter accumulation led to hot fire that burned the plant to the roots (Antos, McCune, and Bara 1983; Jourdonnais and Bedunah 1990). On the other hand, Humes (1960) found that Idaho fescue was more consistently damaged than rough fescue, in the same general area as Antos et al. studied. Idaho fescue recovered by year 2 after fire in eastern Montana (Jourdonnais and Bedunah 1990) and Yellowstone (Singer and Harter 1996). Mountain grasslands dominated by fescues may suffer less initial damage from late-season burns after plants are dormant (Jourdonnais and Bedunah 1990).

Native grasses, other than fescues, may be less damaged by fire in mountain grasslands. Bluebunch wheatgrass, for example, has smaller clumps and buds 1 centimeter below the soil surface and did not change (Merrill, Mayland, and Peek 1980; Antos, McCune, and Bara 1983; Menke and Muir 2004) or increased (Humes 1960; Singer and Harter 1996) after fire; however, it can be reduced the first year after fire (C. A. Johnson 1989). Sandberg bluegrass, which has small clumps and completes growth early, increased after fire in some cases (Humes 1960; Antos, McCune, and Bara 1983; Singer and Harter 1996), but not others (C. G. Johnson 1998). Grasses and forbs may generally produce more flowers and seed stalks after fire (Humes 1960; Singer and Harter 1996). Slow recovery of litter may enhance spring greenup and green forage availability for herbivores in mountain grasslands (Humes 1960; C. A. Johnson 1989; Jourdonnais and Bedunah 1990).

Native perennial forbs with buds near the surface declined after fire but partly recovered by year 3 after fire; deeper rooting forbs increased or did not change (Antos, McCune, and Bara 1983). Native annual forbs increased the first year after fire but declined to prefire levels by three (Antos, McCune, and Bara 1983) or four years (Merrill, Mayland, and Peek 1980) after fire. Mosses and lichens were reduced year 1 after fire, although total cover—but not composition—reached prefire levels by year 3 after fire (Antos, McCune, and Bara 1983).

Total plant and grass production was modestly higher in years 2–3 after the 1988 fires in Yellowstone grasslands, but few differences were found in plant nutrient content (Singer and Harter 1996; Tracy 1997). In Idaho, production averaged about 1.6 times higher on burned than on unburned sites, for four years after fire, peaking in year 1, but some increase was cheatgrass (Merrill, Mayland, and Peek 1980). Standing crop biomass was reduced about 50 percent in upland montane grasslands in northern New Mexico the summer after a fall prescribed fire, but grass had higher protein content (Fuller 2007). Mountain grasslands may recover composition and production by about three years after fire, but recovery is enhanced by high precipitation (Jourdonnais and Bedunah 1990) or low-intensity fire (Linn 1980; Singer and Harter 1996). Higher-intensity fire that kills bunchgrasses may delay recovery for several years (Linn 1980).

Fires can increase invasive species in mountain grasslands. Cheatgrass was generally unchanged or reduced by fire in Montana (Humes 1960; Antos, Mc-Cune, and Bara 1983), but number of tillers and spikelets per plant doubled (Humes 1960). In Wyoming and Idaho, cheatgrass greatly increased after fires (C. S. Williams 1963; Merrill, Mayland, and Peek 1980) and was 40 percent taller for two years (Merrill, Mayland, and Peek 1980).

Estimated fire rotation for the northern Rockies, from fire atlas data for 1900–2003, is about 170 years (see table 10.8). This compares with my esti-mates under the HRV of 50–115 years for expanses of montane grasslands and 50–170 years for expanses of subalpine grasslands. My estimates are too approx-imate to warrant strong conclusions, but they may indicate fire decline. How-ever, that decline, if real, could be remedied by one fire every century or two.

MIXED-GRASS AND SHORTGRASS PRAIRIES

Mixed-grass and shortgrass prairies represent large expanses of fine fuel in a windy environment interrupted by few natural firebreaks, and historical records document some very large fires in these prairies. Perhaps the largest, a fire in the Canadian prairies in the 1850s, may have exceeded 75 million hectares (J. G. Nelson and England 1978). Fires of 0.4–1.0 million hectares occurred in the southern plains in the late 1800s (Haley 1929; Jackson 1965) and also in 2006 (C. B. Clements et al. 2007). Nearly 500 kilometers were burned along the Platte River about 1850 (Hart and Hart 1997), and a 1905 fire in Nebraska burned about 260 kilometers until stopped by rivers (Komarek 1966).

Fire spread in grasslands may be limited below about 10 kilometers per hour wind speed if grasses are not cured or are discontinuous because of aridity or grazing, but spread can reach 20–25 kilometers per hour with continuous cured grasses and winds more than 50 kilometers per hour (Cheney, Gould, and Catchpole 1998). Observed spread in late-1800s fires was up to 13–16 kilome-ters per hour (Haley 1929; Jackson 1965), and fires appeared to burn in narrow lines of wind-driven wedges, sometimes spotting a few meters ahead (Haley 1929). Maximum flame height was a surprisingly high 18–24 meters with 9–19 kilometer-per-hour winds in April, and about 6 meters in July prescribed fires in shortgrass prairie in New Mexico (Ford and Johnson 2006).

Lightning fires in the mid-1900s in the northern plains peaked in July to August (Komarek, 1964; K. F. Higgins 1984). Historical accounts differ (C. T. Moore 1972; J. G. Nelson and England 1978; K. F. Higgins 1986; Wendtland and Dodd 1992), but spring and fall fires may have been most extensive because of cured fuels and windier conditions (Sieg 1997). Most historical accounts are of small fires; only about 10 percent burned for more than a day (K. F. Higgins 1986). Recent fires were typically less than 10 hectares, limited by postfire rain from lightning storms (K. F. Higgins 1984).

Lightning is a common source of ignition in plains grasslands (Komarek 1966; Rowe 1969; K. F. Higgins 1984), but some have suggested that these grasslands were largely the result of burning by Indians (O. C. Stewart 1955; Pyne 1986). However, Indians were likely not a primary ignition source. Early explorers provide few reliable accounts of ignition sources, as they attributed nearly all fires to people (C. T. Moore 1972; K. F. Higgins 1986), generally not understanding that lightning can ignite fires (W. L. Baker 2002). Of 418 fires reported in the plains prior to 1890, only 20 (about 5 percent) were reliably observed to be Indian-set fires (C. T. Moore 1972). Nonetheless, these accounts suggest that Indians set fires for communication (28 percent of accounts), warfare (25 percent), and hunting (19 percent), although hunting was not prominent near the Rockies. Accidental fires also occurred and caused some hardship, including undesirable game movements, loss of firewood near camping areas, and even starvation (C. T. Moore 1972; K. F. Higgins 1986). These accounts suggest that Indians did set fires, but that large conflagrations were not regularly or intentionally set, instead occurring from accidents or lightning.

Fires varied temporally in mixed-grass prairies. Charcoal in a lake in the northern plains suggests that fires were rare during the drier (than today) Little Ice Age (about 1550–1900) and alternated between periods with high fire and with little or no fire over the past 4,500 years (K. J. Brown et al. 2005). Widespread fire may have peaked in wet periods with abundant fuels (Rowe 1969; Umbanhowar 1996; K. J. Brown et al. 2005). Fires can ignite within a few days after rain (K. F. Higgins 1986), but increased fire in wet periods did not always occur (Umbanhowar 2004). Regional coherence in periodicity of high fire occurred at about 140- to 160-year intervals, correlated with a cycle of solar variability (Umbanhowar 2004; K. J. Brown et al. 2005). Fire was also more common in the moister eastern plains, based on charcoal abundance (Unbanhowar 1996) and observations of fire behavior. Fires burning in mixed-grass prairies commonly burned less rapidly or even went out when they encountered shortgrass prairies, and early settlers escaped fires in mixed-grass prairies by moving to shortgrass areas (Haley 1929; Ross 1947). A fire history based on fire-scarred trees at Capulin Volcano National Monument in northeastern New Mexico (Guyette et al. 2006) is not relevant or reliable for adjoining shortgrass prairies, as it targeted scarred trees in atypical and limited areas of forest; did not correct for adjacency; and used composite fire intervals, which do not accurately estimate fire rotation (chap. 5).

Together the evidence, while limited, suggests that the fire regime on the plains was similar to that in other vegetation types, with small fires totaling little burned area and infrequent large fires accounting for most of the burned area (chap. 5). Fire rotation in the mixed-grass prairie may have occurred near the regionally coherent charcoal peaks at 140- to 160-year intervals, with fire in the shortgrass prairie at even longer rotations. However, paleoecological fire research is relatively new. It may be too early to characterize the historical fire rotation of these prairies, but fires likely burned much less often than suggested in the past, and Indians likely played a less important role. If fire rotation under the HRV was in the 140- to 160-year range, fire exclusion likely had little effect. Limited evidence is available about recent fire rotation, but in Scotts Bluff National Monument, in western Nebraska, fire rotation was about 85 years between 1935 and 1990, using equation 5.1 and data in Wendtland and Dodd (1992, table 1).

A useful resource on fire effects on soil, plants, and animals in shortgrass and mixed-grass prairies is K. F. Higgins, Kruse, and Piehl (1989). Mixed-grass prairies recover from fire within a few years, even within one year in moist areas, but fires during droughts and in drier areas (e.g., shortgrass) require longer recovery (H. A. Wright and Bailey 1980). Spring fires may increase warm-season (e.g., blue grama) relative to cool-season (e.g., needle and thread, western wheatgrass) grasses in moister mixed-grass prairies (Steuter 1987). Dormant-season (i.e., fall) fires may have less impact on shortgrass prairies (Brockway, Gatewood, and Paris 2002). Blue grama and buffalograss respond in a neutral or positive way to dormant-season fire but are reduced for up to two years by growing-season fire (Ford 1999; Ford and Johnson 2006). H. A. Wright and Bailey (1980) saw little ecological value in burning shortgrass or dry mixed-grass prairies, the main prairies within and near the Rocky Mountains. The limited available evidence also suggests that fire may not be outside the HRV.

SUMMARY

Fire in Rocky Mountain shrublands and grasslands likely burned at longer intervals (table 9.2) than suggested in the past. Shortest fire rotations may have been in mixed mountain shrublands, where lightning and drought are common and fuels are continuous, and in montane and subalpine grasslands. Longest rotations may have been in little (low) and Wyoming big sagebrush, salt desert shrublands, and shortgrass prairies, where fine fuels are sparse and ignition is limited. All these vegetation types, except possibly curlleaf mountain mahogany, burn primarily in high-severity fires that topkill plants. Landscape mosaics of burned and unburned areas are known in sagebrush and prairies, due to wind shifts and variability in fuels, but may have occurred in other shrublands and grasslands.

Recovery after fire is comparatively rapid in grasslands, likely an adaptation to drought and grazing—not just fire—and also in mixed mountain shrublands, where shrub resprouting is common and may be an adaptation to fire. Nonsprouting sagebrush represents the slowest recovery, requiring a century or more on the most arid sites. Response to fire in all of these vegetation types often includes an initial pulse of native annuals; little change in native forbs; and a shift in production from shrubs, where they occur, to herbaceous plants, with little change in total production. Some plants may be more susceptible to damage (e.g., fescues, antelope bitterbrush), which may delay their recovery by one to three decades. Even the common native perennial grasses show inconsistent responses to fire. Response is so variable because it depends on varying moisture conditions, fire intensity, and plant morphology. Variability in response to fire reflects the difficulty in predicting short-term (i.e., 1–10 years) responses, but over the longer term all these vegetation types recover after fire. An exception may be Wyoming big sagebrush, not yet observed over a full cycle of recovery after fire.

Fire rotations under the HRV are imprecisely known (table 9.2), but recent fire rotations are not likely outside the HRV in most Rocky Mountain shrubland and grassland ecosystems. Deficiency of fire, if any, could be remedied in most cases by one fire every century or two, given the long rotations under the HRV. Fire is not advisable at all where cheatgrass or other invasives might expand. The most significant impacts on Rocky Mountain shrubland and grassland ecosystems are replacement by cheatgrass, knapweed, and other invasives; conversion to other land uses; fragmentation; and overgrazing by livestock. Where it is possible to remedy these legacies, this is certainly needed more than intentional burning and would allow Rocky Mountain shrublands and grasslands to better recover after future wildfire.

CHAPTER 10

People and Fire: Land–Use Legacies across Landscapes

Land uses differ in magnitude, pattern, and duration of direct effects on fire but also interact and have delayed, indirect effects. To attribute change in fire to a land use requires detecting the change (e.g., fire decline), then analyzing and excluding competing potential causes. Logical arguments and observations provide some evidence about land-use effects but do not fully resolve how much change is attributable to each land use. As with twentieth-century climatic change (Stott et al. 2000), quantitative analysis is needed to estimate potential forcing by each cause (table 10.1). Unfortunately, changes in fire are poorly known, hampering such analysis. Fire history research (table 10.2) has focused on pre-EuroAmerican fires and subsequent broad changes, rather than the pattern, timing, and magnitude of changes since EuroAmerican settlement. Details of causes and interactions in particular places have been of less interest, and few studies have analyzed all potential causes. In this chapter, I review the logic and evidence concerning the potential spatial pattern, characteristics, and magnitude of land-use effects on fire. I also review the evidence of change in fire since EuroAmerican settlement and qualitatively analyze how well potential causes explain observed changes. Generalizations about the human effects on fire regimes do not hold across the Rockies, as land uses have not been uniform in time or space.

LAND-USE EFFECTS ON FIRE REGIMES

This section reviews land uses that may have altered fire regimes after EuroAmerican settlement, including the decline in ignitions by Indians,

Table 10.1. Human and Natural Factors Affecting Fire Regimes since EuroAmerican Settlement

Land-Use and Natural Factors	Temporal Pattern	Spatial Pattern
INCREASED FIRE SINCE SETTLEMENT		
Miscellaneous human-caused fires (e.g., smoking, arson)	Consistent since about 1920	Near roads, settlements, logging, recreation
Mining: intentional and accidental ignitions	Esp. before about 1920	Near mining areas
Livestock grazing: tree regeneration and increased fuels	Esp. since 1930s, with reductions in stocking rate	Esp. in ponderosa pine, Douglas-fir, mixed-conifer forests
Livestock grazing: fires ignited by sheep herders	From settlement to 1920	Sheep-grazing areas
Logging: initial, direct increase in activity fuels	Esp. 1950s–1970s	Logging areas, esp. N. Rockies
Logging: drier, windier microclimate	Esp. 1950s–1970s	Logging areas, esp. N. Rockies
Roads: increased human-caused fires and burned area	Increasing as road system expands	Near roads
Landscape fragmentation: edge effect increased flammability in adjacent vegetation	Since settlement	Near logging, roads, settlements, agriculture
Invasive-plant fire cycles	Esp. since 1940s	Areas with cheatgrass, tamarisk
Fire control: lagged effect on fuel buildup	Mostly since ~1960?	Only part of Rockies, mostly low elevations
19th-c. fires: increased flammable early-succ. vegetation	Noted 1910–1930s	Not noted outside N. Rockies
19th-c. fires: increased middle-aged dense forests	Noted late 1900s–present	Noted in southern Rockies
Climatic teleconnections and drought favorable to fire	Varies across Rockies	Within broad regions of Rockies
Global warming: earlier springs	Detected since about 1980	Esp. N. Rockies mid-elevations
DECREASED FIRE SINCE SETTLEMENT		
Reduced ignitions by Indians	Since settlement	Low-elevation camping and travel areas
Livestock grazing: seasonal reduction in fine fuels	Esp. settlement to 1930s	Grazing areas, esp. low elevations
Livestock grazing: persistent reduction in productivity	Ongoing	Grazing areas, esp. low elevations
Logging: reduced area of old, flammable forests	Esp. since 1950s	Logging areas, esp. N. Rockies
Fire control: direct reduction in burned area	Ongoing	Only part of Rockies, mostly low elevations

Table 10.1. Continued

Land-Use and Natural Factors	Temporal Pattern	Spatial Pattern
Landscape fragmentation: reduced fire spread	Ongoing	Esp. near roads, settlements, agriculture
Wet periods, decreased drought	In 20th century	Detected in subalpine forests
19th-c. fires: young, less flammable forests	Esp. in early 1900s	Areas burned in 19th-c. fires

Sources: Based on studies conducted in and near the Rocky Mountains. See the chapter text for citations and explanation of each land-use effect.

human-caused fires since EuroAmerican settlement, livestock grazing, logging, roads, nonnative plants, and intentional fire control. I first review only the effects, spatial pattern, and timing of each land use. A later section analyzes how each land use may have contributed to observed changes in fire.

Indians and Fire: The Myth of the Humanized Landscape

If Indians burned large areas, the removal of Indians could have decreased fire. Indians lived in the Rockies for thousands of years. By the mid-1800s, an estimated 19,300 remained (Mooney 1928), which was perhaps only 60 percent of the population that existed before decimation by disease, genocide, forced removal, and relocation after Spanish settlement (Thornton 1987). Given this loss, about 32,000 Indians may have lived in the Rockies, mostly in Montana, near the year 1500 (W. L. Baker 2002). Even if this estimate is low, the Indian population in the Rockies was small relative to other parts of the United States (Denevan 1992). Further disruption and change accompanied the Indians' acquisition of horses and guns, in 1650–1750, from Spanish settlements, increasing warfare and spawning a plains culture centered on bison (Shimkin 1986). Evidence about Indian use of fire comes from (1) accounts by early EuroAmericans; (2) Indian oral histories; (3) fire history data; and (4) pollen and charcoal data, a minor source (see W. L. Baker 2002).

A recurring theme in the literature is that the Indians burned everything and managed the inhabited world like a garden, leaving no real wilderness. Since vegetation everywhere was historically highly managed, the argument follows, we should today abandon protecting wild nature and restoring ecosystems and their fire regimes, and create whatever kind of world we want. For example, Charles Mann, in the book *1491*, says: "Native Americans ran the continent as they saw fit. Modern nations must do the same. If they want to return as much of the landscape as possible to its state in 1491, they will have to create

Table 10.2. Change in Fire since EuroAmerican Settlement, relative to the HRV

Sources by Vegetation Type	Location	Adequacy of Fire History Data for Detecting Change	Change +	Change −	Loss of Ignition by Indians	Livestock Grazing	Fire Control	Fire Exclusion Syndrome	Recovery from 19th-c. Fires	Climatic Change
						Land Uses			Natural	
PIÑON-JUNIPER										
Eisenhart (2004)	WC CO	No data								
Shinneman (2006)	WC CO	No data								
M. L. Floyd, Romme, & Hanna (2000)	SW CO	Short	n	Y	*	C	C		*	M
M. L. Floyd, Hanna, & Romme (2004)	SW CO	Short	n	n						
PONDEROSA PINE–DOUGLAS-FIR										
Barrett (1988)	C ID	Not adequate[a]								
P. M. Brown & Sieg (1996)	SW SD	Not adequate[b]								
P. M. Brown & Sieg (1999)	SW SD	Not adequate[b]								
P. M. Brown (2006)	SW SD	Scattered,[c] short	n	Y				S	*	*
R. F. Fisher, Jenkins, & Fisher (1987)	NE WY	Not adequate[a,d]								
Perryman & Laycock (2000)	NE WY	Not adequate[e]								
Ehle & Baker (2003)	N CO	No data								
Goldblum & Veblen (1992)	N CO	Not adequate[d]								
Mast, Veblen, & Linhart (1998)	N CO	Not adequate[e]								
Sherriff & Veblen (2006)	N CO	Scattered,[c] short	n	Y	*	C	C		c	C
Veblen, Kitzberger, & Donnegan (2000)	N CO	Not adequate[f]								
P. M. Brown, Kauffman, & Sheppard (1999)	C CO	Scattered,[c] short	n	Y	*	C	M		*	*

Table 10.2. Continued

Sources by Vegetation Type	Location	Adequacy of Fire History Data for Detecting Change	Change +	Change –	Loss of Ignition by Indians	Land Uses: Livestock Grazing	Fire Control	Fire Exclusion Syndrome	Natural: Recovery from 19th-c. Fires	Climatic Change
Huckaby et al. (2001)	C CO	Short	n	Y	*	C	M		*	*
Wieder & Bower (2004)	C CO	No data								
P. M. Brown & Wu (2005)	SW CO	Not adequate[b,d]								
Allen (1989)	N NM	Scattered,[c] short	N	Y	*	M	C		*	C
Falk (2004)	N NM	No data								
Foxx & Potter (1984)	N NM	Not adequate[e,g]								
PONDEROSA PINE–W. LARCH										
Arno (1976: montane slopes)	W MT	Not adequate[a]								
Habeck (1990)	W MT	Not adequate[a,d]								
DOUGLAS-FIR										
Arno & Gruell (1986)	SW MT	Not adequate[a,e]								
Heyerdahl, Miller, & Parsons (2006)	SW MT	Not adequate[d]								
N. T. Korb (2005: xeric)	SW MT	Scattered,[c] short	n	Y	*	M	C		*	*
Littell (2002)	MT, WY	Not adequate[f]								
QUAKING ASPEN (STABLE)										
Romme et al. (2001)	SW CO	No data								
MOIST NW MIXED CONIFER										
Barrett, Arno, & Key (1991: maritime sites)	NW MT	Not adequate[h]								
McCune (1983)	W MT	Not adequate[i,j]								
Barrett (2000)	C ID	Unknown								

Citation	Location	Evidence							
N. ROCKIES MIXED CONIFER									
Barrett, Arno, & Key (1991: dry sites)	NW MT	Not adequate[a]							
Gabriel (1976)	NW MT	Not adequate[a,h]							
Sneck (1977)	NW MT	Not adequate[a,e]							
Barrett (2000)	C ID	Unknown							
UPPER–MONTANE MIXED CONIF.									
N.T. Korb (2005: mesic)	SW MT	Scattered,[c] short	n	Y	*	M	C	*	*
Sherriff & Veblen (2006)	N CO	Scattered,[c] short	n	N					
SOUTHWESTERN MIXED CONIFER									
Wu (1999)	S CO	Scattered,[c] short	N	Y	*	C	M	*	C
STABLE LODGEPOLE–SPRUCE-FIR									
Gabriel (1976)	NW MT	Not adequate[a,h]							
Singer (1975, 1979)	NW MT	Not adequate[a,j]							
Kipfmueller (2003)	ID, MT	Not adequate[b,i,j]							
Barrett (1994)	Yellowstone	Not adequate[a]							
Romme (1982)	Yellowstone	Not adequate[i]							
Romme & Despain (1989a,b)	Yellowstone	Short	n	N					
M.P. Murray, Bunting, & Morgan (1998)	C ID	Short	n	Y	*	M	M	C	*
Kipfmueller & Baker (2000)	SE WY	Short	Y	Y	n	n	M	*	N
Kulakowski, Veblen, & Bebi (2003)	NW CO	No data							
Veblen et al. (1994)	NW CO	No data							
Buechling & Baker (2004)	N CO	Short	Y	Y	n	n	N	C	M
E. Howe & Baker (2003)	N CO	Short	N	Y	n	C	C	*	*
Kulakowski & Veblen (2002)	N CO	No data	N	Y	n				
Sibold & Veblen (2006)	N CO	Short	Y	Y		N	C	C	N
Alington (1998)	S CO	Not adequate[e]							

Table 10.2. Continued

Sources by Vegetation Type	Location	Adequacy of Fire History Data for Detecting Change	Change +	Change −	Loss of Ignition by Indians	Land Uses Livestock Grazing	Fire Control	Fire Exclusion Syndrome	Natural Recovery from 19th-c. Fires	Climatic Change
STABLE BRISTLECONE PINE										
W. L. Baker (1992b)	S CO	No data								
STABLE LIMBER PINE										
Buechling & Baker (2004)	N CO	Short	Y	Y	n	n	N		C	M
STABLE WHITEBARK PINE										
Gabriel (1976)	NW MT	Not adequate[a]								
Keane, Morgan, & Menakis (1994)	NW MT	No data								
M. P. Murray, Bunting, & Morgan (1998)	C ID	Short	n	Y	*	M	M		C	*
Mattson & Reinhart (1990)	Yellowstone	No data								
SAGEBRUSH										
W. L. Baker (2006b)	Rockies	No data								
CURLLEAF MTN. MAHOGANY										
None available	—									
MIXED MOUNTAIN SHRUBLANDS										
M. L. Floyd, Romme, & Hanna (2000)	SW CO	Best	n	Y	n	C	C		*	M
GRASSLANDS AND PRAIRIES										
Estimates in this book	Rockies	No data								

Notes: Shown are the adequacy of data and identified causes for each study. The table includes all medium- and high-quality fire history studies in the Rockies (see chapters 5–9). Studies are arranged by vegetation type and from north to south within a type. N = no, identified by author; n = no, identified from author's data; Y = yes, identified by author; M = identified by author as a major cause; C = identified by author as a contributing cause; ★ = of unknown importance, not analyzed; S = part of a fire-exclusion syndrome, without analysis of specific data; SHORT = fire rotation (FR) for a big area (bigger than the largest fires in the area), based on stand-origin maps and fire-year maps, from an adequate sample of sites and tree ages per site (≥5), but a modern period of data shorter than the FR under the HRV; BEST = FR for a big area (bigger than the largest fires in the area), based on stand-origin maps and fire-year maps, from an adequate sample of sites and tree ages per site (≥10), but a modern period of data nearly equal to the FR under the HRV.

[a] CFI with a lack of cross-dated scars.

[b] CFI with a smaller sample size after EuroAmerican settlement than before.

[c] CFI for a large area (bigger than the largest fires in the area), using many scattered sample sites, with an adequate sample size (≥10 fire-scarred trees per site), an approximately equal sample size in the pre- and post-EuroAmerican settlement periods, and cross-dated fire scars.

[d] CFI with a small study area relative to the size of fires in the region.

[e] CFI with an inadequate fire-scar sample size (<10 fire-scarred trees per site).

[f] CFI with a targeted sampling of stands or a study area containing older trees.

[g] CFI with a small set of sample sites for the size of the study area.

[h] FR with a small sample of tree ages (<5) per sample site.

[i] FR with a small study area relative to the size of fires in the region.

[j] FR with a modern period of data substantially less than one fire rotation.

the world's largest gardens" (Mann 2005, 326). This idea has been presented by historians (Pyne 1982; G. W. Williams 2004), anthropologists (O. C. Stewart 1954), and geographers (Denevan 1992). However, their narratives appear founded on social ideology and often ignore ecological evidence, overstating human influence and substituting a "myth of the humanized landscape" for an earlier "myth of pristine nature" (Vale 1998, 231). Because available ecological evidence about the Indian use of fire was seldom a focus of such works, Thomas Vale (2002) compiled a critical review of the evidence, to which I contributed (W. L. Baker 2002). The following paragraphs summarize that contribution.

Early EuroAmericans commonly thought Indians were the cause of most fires. A USDA correspondent in northern New Mexico reported, "The Indians have been mostly removed to reservations, except a few stragglers who come back to hunt, but still these fires will rage and we cannot tell how they originate" (Hough 1882, 199). John Wesley Powell, encountering and mapping large burned areas in Utah in 1878, said "In the main these fires are set by Indians" (1879, 17). Other government scientists, sent out to inspect future national forests, attributed nearly all fires to Indians or settlers. By the 1920s, evidence showed that wildland fires were often ignited by lightning (fig. 10.1a), but early EuroAmericans were commonly unaware of this and incorrectly assigned nearly all ignitions to Indians or other people (W. L. Baker 2002). Early Euro-American accounts, the chief source of the idea that Indians pervasively

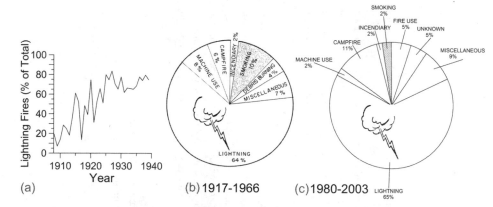

Figure 10.1. Causes of fires in the Rockies: (a) the rise, between 1908 and 1939, in the percentage of fires attributed to lightning; (b–c) the percentage of total fires in the Rocky Mountains by cause.

Sources: The graph in (a) is based on data in Wellner (1970). The 1917–66 pie chart in (b) is reproduced from USDA Forest Service (1966, p. 48); data for the 1980–2003 pie chart in (c) are from the USDI Bureau of Land Management (2004), which covers fires only on federal and Indian lands, ~75 percent of land area in the Rockies.

burned, are thus generally unreliable in assessing the extent of burning by Indians.

Oral histories from Indians are more reliable, as Indians apparently understood lightning and considered fires unremarkable (Barrett and Arno 1982). A large sample is needed of elders who remember their ancestors' practices near or before EuroAmerican settlement, along with a standard research protocol to avoid anecdotes and bias. The only study from the Rockies meeting these criteria is by Barrett (1981; also Barrett 1980a,b; Barrett and Arno 1982, 1999). This research makes it clear that Indians used fire for many reasons: for hunting; to stimulate fruiting in desired plants; to facilitate travel and improve grazing for horses; to protect the forest from dangerous crown fires; to clear campsites of tall vegetation that might conceal enemies; and in communication and warfare, sometimes as an expression of hostility toward EuroAmericans. Of course, Indian campfires at times escaped, and other accidental fires occurred.

Fire history research, using fire-scar records, has found that lower-elevation valleys and hunting or travel zones, known to have been heavily used by Indians, had shorter mean fire intervals than places more remote from Indian use (Barrett 1981). However, heavily forested canyons near low-elevation valleys lacked evidence of shorter fire intervals (McCune 1983), which suggests that burning was confined to heavy use areas. At high elevations, where Indian use was low and lightning was sufficient to explain the fire history, as in Jackson Hole and Yellowstone, burning by Indians is considered minor (Loope and Gruell 1973; Romme and Despain 1989).

Evidence about burning by Indians, using the three main sources (early EuroAmerican accounts, Indian oral histories, and fire history data), is much less abundant than previously thought, because early historical observations are unreliable. However, the evidence is sufficient to reject the notion (Denevan 1992; Mann 2005) that Indians burned everywhere and significantly modified Rocky Mountain vegetation. The evidence instead suggests that some Indians did burn at times and had logical reasons for doing so. The impacts of their burning are not easily detected but appear to be most common in low-elevation valleys, along travel routes, and in hunting areas, where use was concentrated (Barrett 1981). At higher elevations and away from heavy use areas and travel routes, evidence suggests little burning by Indians, partly because these areas are typically too moist to burn except during drought (chap. 2).

Some narratives after the publication of Vale (2002) continued to exaggerate Indian use of fire and ignored or misrepresented the evidence (see G. W. Williams [2004] and response by Barrett, Swetnam, and Baker [2005]). Charles Mann said, "Most scientists have changed their minds about Indian fire," and

"Vale is in the minority now," as "using clever laboratory techniques" and the arguments of historians, scientists have become convinced that Indians were "routinely enshrouding miles of countryside in smoke and ash" (Mann 2005, 252). Mann cited no evidence from laboratory techniques and no evidence showing scientists have changed their minds.

In contrast, a welcome response to Vale (2002) was a further discussion of the evidence. Kay (2007) estimated the mean density of Indians and suggested that accidental ignitions from their escaped campfires were likely much greater than ignitions by lightning. He used lightning-ignition data over the last few decades from the National Interagency Fire Center (NIFC). However, three problems with his analysis and conclusions suggest that they are not valid. First, NIFC data do not record the numerous ignitions that burn small areas; thus, these data are not valid for the comparison Kay made. Second, there is no basis for estimating the rate at which unattended campfires would escape and lead to significant burned area. Citing no evidence, Kay speculated one ignition from an escaped campfire per adult Indian per year, but the actual number is crucial. If the number was a more plausible 1 per 20 Indians, for example, accidental ignitions would likely have been insignificant relative to lightning. If escapes from Indian campfires were like those from EuroAmerican campfires today, they would have led to small fires, as today's campfires often are started during unfavorable weather and in locations unfavorable for spread. Third, if escaped Indian campfires were a significant source of ignition, they should have been evident to early EuroAmerican observers and should have shown up in oral interviews with Indians, who instead attributed their fires mostly to intentional ignitions. Indians viewed escaped fires as damaging, potentially removing forage for horses and other animals (Barrett 1981). Neither Mann's expansively gardening Indian nor Kay's routinely careless Indian fits the evidence.

Human-Caused Fires after EuroAmerican Settlement

In the Colorado Front Range, early photographs and forest age structures show that much montane and subalpine forest was burned (and logged) during the early period of EuroAmerican settlement (Veblen and Lorenz 1986, 1991). Miners apparently set fires to expose rocks, but fires also escaped, as suggested by the spatial association of fire with mining areas and, in some cases, the occurrence of fires shortly after mining onset (Sudworth 1900a; Veblen and Lorenz 1991; Kipfmueller and Baker 2000; Veblen, Kitzberger, and Donnegan 2000; E. Howe and Baker 2003). Sheep herders burned subalpine ranges in the fall to improve forage and used campfires, many of which reportedly escaped, to keep coyotes away (Dubois 1903; Hatton 1920). Dubois (1903) attributed 75 percent of burned area in Colorado's San Juan Mountains in the late1800s to herders, although it is not clear whether Dubois understood that lightning could ignite fires (W. L. Baker 2002). By about 1920, laws and regulations reduced burning by herders (Hatton 1920). Other early fires were the result of escaped campfires and other accidents, railroads, and logging, as well as retribution by Indians and settlers against each other (e.g., Sudworth 1900b) and the government (Leiberg 1899a).

In the Rockies, lightning still starts the majority of fires and accounts for the majority of burned area (fig. 10.1b,c), in contrast to much of the United States, which has higher proportions of human-caused ignitions. Human-caused fires and burned areas are concentrated near people and their infrastructure (fig. 10.2). Arson and accidental fires are often ignited from roads and railroads and decrease in density with distance (Show et al. 1941; C. C. Wilson 1979). Arson-caused fires remained 2 percent of total fires and burned area over the last century. Smoking-caused fires declined in number (fig. 10.1b,c) but are often large, accounting for 11 percent of total burned area. Railroads were the most common cause of fires in national forests in Idaho and Montana in the early 1900s (Larsen and Delavan 1922). However, by the 1930s, railroad fires totaled only about 3 percent of all fires (Barrows 1951b) and, from 1980 to 2003, they totaled less than 1 percent of all fires and 0.1 percent of burned area.

Fires are also ignited in and escape from campgrounds and other recreational locations, towns, exurban developments, and other concentrated human activity. Recreational fires and those near towns and cities are numerous but small (Pew and Larsen 2001), probably because fuels are discontinuous and access is available for control (Prestemon et al. 2002). In the Rockies, escaped campfires between 1980 and 2003 totaled 11 percent of all fires but only 3.5 percent of burned area. In the Missoula Valley, Montana, human-caused fires increased with population density but leveled off above about 2,000 people per square mile (fig. 10.2d). Lower elevations in Southern California represent a human-dominated endpoint, as more than 90 percent of burned area is from human-set fires (Keeley and Fotheringham 2003). That endpoint could be reached near towns and cities in the Rockies, judging from fires in the Missoula Valley (fig. 10.2).

Livestock Grazing

Domestic livestock (particularly cows, sheep, and goats) can have direct effects on fuels and flammability during the grazing season and indirect effects lasting years (table 10.3). In general, livestock have most effect where vegetation lacks a long evolutionary history of large, native herbivores, where bunchgrasses

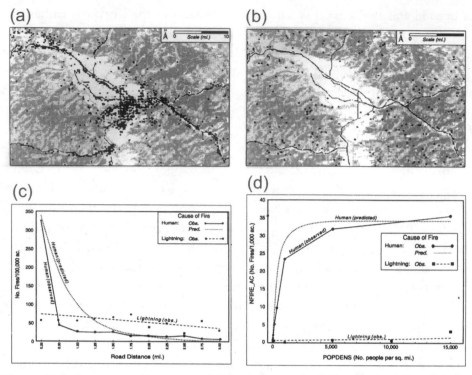

Figure 10.2. Locations of human- and lightning-caused wildfires in the Missoula Valley, 1981–90: (a) human-caused fires; (b) lightning-caused fires; (c) the distribution of human-caused and lightning-caused fires in 0.25-mile classes of distance from roads, showing a predicted trend from a regression analysis; and (d) human-caused and lightning-caused fire density in relation to human population density classes, along with a regression prediction equation.

Source: Reproduced from Close and Wakimoto (1998, pp. 244–45, figs. 6a, 6b, 7, and 8), with permission of Kelly Close, Fort Collins, Colorado.

rather than sod-forming grasses were dominant, and where stocking rates were long high (Milchunas and Lauenroth 1993). Vegetation was less susceptible to alteration on the eastern slope of the Rockies near the plains and in parts of the northern Rockies, where bison and sod grasses were prominent. The western slope and the central and southern Rockies were more susceptible, as bison were sparser and bunchgrasses dominated.

The direct effect of grazing in the growing season is to decrease fire (table 10.3). Livestock consume forage, reducing the fine, live fuels and litter that carry fire (Ellison 1960; Zimmerman and Neuenschwander 1984; Hobbs 1996; Belsky and Blumenthal 1997; A. Jones 2000). Livestock also trample unconsumed vegetation, break it up, and work it partly into the soil, reducing its amount and its flammability (Ingram 1931). In tallgrass prairies, heavily grazed

Table 10.3. Direct and Indirect Effects of Livestock Grazing on Ignition, Spread, and Intensity of Fires

Ignition	Spread	Intensity	Effect
			DIRECT EFFECTS DURING A GRAZING SEASON
+			Increased accidental and intentional ignitions by herders
−	−	−	Loading reduced by consumption of fine fuels
			• Fine, live fuels reduced
			• Litter reduced
−	−	−	Trampling effects
			• Breakdown and decomposition of fine fuels enhanced
			• Breakage of small woody fuels, leading to increased loadings and flammability but also increased decomposition
−	−	−	Fire reduced by stock driveways acting as fuel breaks
			INDIRECT EFFECTS BEYOND A GRAZING SEASON
−	−	−	Reduced annual net primary productivity and plant cover, as well as increased bare ground
	+	+	Increased tree regeneration
			• Increased ladder fuels
			• Increased tree density
			• Increased shade, favoring shade-tolerant trees (e.g., Douglas-fir, white fir, grand fir)
+	+	+	Increased fuel loadings accompanying tree regeneration
±	±	±	Changed vegetation composition, possibly decreasing or increasing flammability
±	±	±	Increased invasive weeds, some flammable (e.g., cheatgrass)

Sources: See text for citations and explanations of effects. − = decrease, + = increase, ± = either an increase or a decrease.

patches have lower fire spread and intensity (Hobbs et al. 1991). Early stock driveways for moving livestock could slow or stop fire spread (e.g., in Gallatin National Forest, Montana; Hatton 1920). Until about 1920, land managers encouraged grazing to reduce fires (Murray, Bunting, and Morgan 1998). The net direct effect of cattle grazing on fire may have been to lower ignition, spread, and intensity in the grazing season. However, sheep herders likely set fires, and the net direct effect of sheep grazing may have been to increase, not decrease, fire. Livestock grazing had lasting deleterious impacts on biological diversity in the Rockies (H. A. Wallace 1936; Shinneman, Baker, and Lyon 2008) and would likely still have if it was used to reduce vegetation to a level that would limit fire spread.

The indirect effects of livestock grazing on fire are from changes in fuels, tree regeneration, invasive weeds, and compositional changes in vegetation (table 10.3). Livestock grazing can cause lasting reductions in native plant cover and productivity and increase bare ground (Zimmerman and Neuenschwander

1984; Shinneman, Baker, and Lyon 2008). By the 1930s, grazing capacity was reduced across the West by two-thirds in sagebrush and more than half in piñon-juniper (H. A. Wallace 1936). Nearly 1,200 plots in the Rockies showed that grazing reduced the density of plant cover in ponderosa pine forests by 21 percent and in aspen forests and mixed-mountain shrublands by 45 percent (H. A. Wallace 1936). Experimental and reference-area studies in the Rockies, from low to high elevations and across vegetation types, also showed these trends (Evanko and Peterson 1955; W. M. Johnson 1956; Laycock 1967; R. F. Miller, Svejcar, and West 1994; Shinneman, Baker, and Lyon 2008). Lower productivity and cover and more bare ground reduce potential fire intensity and spread.

Grazing at high intensity can nearly prevent tree regeneration through browsing or trampling (Ingram 1931; Currie, Edminster, and Knott 1978), but increased regeneration often follows grazing reduction after a high-intensity period (Heerwagen 1954). High-intensity, long-duration grazing reduces native plant cover and the competition that these plants, particularly bunchgrasses, exert on tree seedlings for water and nutrients (Kolb and Robberecht 1996). At lower grazing intensity, tree seedlings may be grazed less than other plants, which favors the tree seedlings. Increased tree density from livestock grazing was found in comparisons of isolated mesas lacking livestock with paired mesas having continued grazing (box 7.1). As tree regeneration increases, litterfall and shading increase, raising annual fuel inputs and reducing decomposition, so fuel loadings rise. Higher loadings occurred in grazed ponderosa pine in Idaho than in an exclosure (fig. 10.3). Livestock grazing may increase invasive weeds, some of which are more flammable than the natives they replace. Livestock also select against unpalatable plants, changing vegetation composition, and some of the plants that increase may be more flammable (Leopold 1924; Ingram 1931; Ellison 1960).

In general, the indirect effects of livestock grazing eventually increase potential ignition, spread, and intensity of fire and the potential for crown fire (table 10.3). Livestock grazing effects on fire ignition, spread, and intensity are inherently lagged across landscapes (box 10.1), whereas the effects of the removal of competition on tree regeneration may be rapid and nearly synchronous. Thus, rapid and rather uniform ecological changes, such as tree invasion shortly after livestock introduction (e.g., R. F. Miller and Rose 1999), cannot have been caused by reduction in fire by livestock and more likely were caused by reduction or removal of competition exerted by native plants.

Logging

Logging has long been a primary source of fuel buildup, which leads to increased fire risk in Rocky Mountain and other forests (Plummer 1912; Lyman

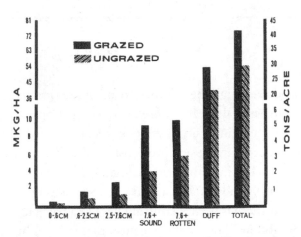

Figure 10.3. Effect of livestock grazing on fuels. Mean fuel loadings by size class are shown for an "ungrazed" exclosure in a ponderosa pine–Douglas-fir forest fenced since the early 1940s and an adjoining area subject to livestock grazing, both in central Idaho. MKG/HA is fuel loading in units of 1,000 kilograms per hectare.
Source: Reproduced from Zimmerman and Neuenschwander (1984, p. 108, fig. 4), copyright 1984, with permission of the Society for Range Management.

1947; C. C. Wilson and Dell 1971; Pew and Larsen 2001). Removing canopy trees for wood increases flammability in the harvested area and adjoining forests in several ways, increasing the probability of more and larger fires, as well as crown fires (table 10.4). Effects vary with the type and intensity of logging. Logging focused on restoration or lowering fire risk, rather than wood products, is covered in chapter 11.

Logging commonly replaces large, fire-resistant trees with smaller, often more numerous and more flammable trees, as noted by many authors (e.g., Weatherspoon 1996). In the Rockies, density of small trees typically, but not always, increases after logging (e.g., W. L. Baker, Veblen, and Sherriff 2007). Since resistance to heat damage is related to the square of bark thickness (chap. 3), smaller trees after logging are much more vulnerable to mortality from surface fires than were the large, thick-barked trees that typically are logged. Also, small postlogging trees have shorter crown base heights and are usually more dense (W. L. Baker, Veblen, and Sherriff 2007), and thus are more vulnerable to crown fire initiation and spread. This is offset a little by more shade and lowered wind velocity from denser trees (Raymond and Peterson 2005). Removing overstory trees also typically increases shrubs, forbs, and graminoids, increasing live fine fuels that become flammable in droughts (Keyes and Varner 2006). Logging often removes shade-intolerant canopy trees, so that Douglas-fir and shade-tolerant trees, such as white fir or grand fir, increase (Boe 1947–48; Arno et al.

BOX 10.1
Spatial Lags in Land-Use Effects

Widespread land uses might be assumed to affect fire simultaneously across landscapes, but the spatial nature of fire leads to lags. Some land-use effects that alter fire spread or intensity (livestock grazing, global warming, large human-set fires) occur nearly synchronously and widely across large landscapes. Others (smaller human-set fires, logging, fire control) have lagged effects that effectively move across a landscape over a longer period.

For example, there is no fire control effect in a particular area until an ignition occurs, followed by a control action that reduces burned area (fig. B10.1.1). The net affected area is only the area that would have burned without control minus the area burned by the time of control (W. L. Baker 1993). The affected area tracks fires across the landscape, as there is no effect in a particular area until a fire ignites there and is controlled. Even if fire is suddenly completely controlled in a landscape, the effect of

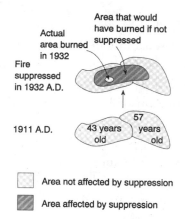

Figure B10.1.1. Spatial lags in land-use effects. The diagram illustrates how a spatial lag effect on fire occurs from a land use applied uniformly across a landscape. In this hypothetical example, fire suppression started in 1911 in a landscape with two patches that had been burned 43 and 57 years before. The first suppressed fire was in 1932, when part of it (unshaded) burned before it was suppressed; the area that would have burned without suppression is indicated by dark shading. There are two key points: First, no effect of fire suppression occurred in the landscape until 1932, since that was the year of the first fire that was suppressed. Second, the only area affected by fire suppression in 1932 was the shaded area.

Source: Reproduced from W. L. Baker (1993, p. 69, fig. 3) with permission of Blackwell Publishing Ltd.

BOX 10.1

Continued

control does not reach across the whole landscape until at least a fire rotation. Fire control is unique among land uses with spatial lags in requiring more time than a rotation, since only part of each fire is reduced.

Land uses thus often have complex, lagged, spatially heterogeneous effects on fires. For example, if fires are suppressed beginning at a particular time, landscapes become "a mosaic of patches that are not affected by suppression and that have been affected by suppression for differing periods of time" (W. L. Baker 1993, 70). The expected fraction of a landscape affected by a land use at a particular moment is the fire rotation divided by the duration of the land use. In a landscape with a fire rotation of 200 years, we expect 50 years of a land use to result in about one-quarter of the land area being affected by the land use. In the case of fire control, the fraction is smaller, as only part of each fire is controlled. Since much of the rotation occurs from a few large fires, the timing of large fires controls the timing of the arrival of lagged land-use effects. Land uses themselves also vary spatially, which adds to varying effects on fire regimes. The common assumption, that land uses have uniform and rapid effects on fire, is not correct.

1995; Hann et al. 1997; Kaufmann, Regan, and Brown 2000). Logged areas have higher tree density, more shrubs and shade-tolerant trees that are ladder fuels, and are thus more prone to crown fire.

Removing canopy trees leads to a hotter, drier, windier *microclimate*, the climate in a small area. Clear-cutting can increase temperatures in surface fuels by 50–100 percent (Hungerford 1980), reduce fuel moisture by half or more, and increase wind speed 10–20 times and rate of spread more than 5 times (Hornby 1935; Arnold and Buck 1954; Countryman 1956). Partial cutting or thinning has these effects to a lesser degree (Hungerford 1980). Removing half the volume of a western white pine stand in northern Idaho lowered fuel moisture about one-third (Hornby 1935); increased windspeed 6 to 10 times; and increased the number of critical fire days, with duff moisture less than 10 percent, by about 4 times (Jemison 1934). Modeling potential fire behavior in Colorado showed that creating open-canopy forests, by thinning closed-canopy forests, would lead to higher fireline intensity, even under moderate

Table 10.4. Effects of Logging Canopy Trees on Ignition, Spread, and Intensity of Fires

Ignition	Spread	Intensity	Effect
			ONLY INITIAL (SHORT-TERM) EFFECTS (i.e., less than a decade)
+	+	+	Increased flammable fine, dead fuels (activity fuels or slash) on the forest floor
			INITIAL AND LONG-LASTING EFFECTS (i.e., decades to more than a century)
+	+	+	Increased hot, dry, windy microclimate, extending as an edge effect into adjoining vegetation
	+	+	Increased small woody fuel for decades; longer-term decreased snags and down wood
			Fewer large-diameter boles, which have the thickest bark most resistant to cambium damage, and fewer tallest boles, which are most resistant to crown scorch and crown fire initiation
	+	+	More numerous short, small-diameter trees, which are more vulnerable to cambium damage, crown scorch, and crown fire initiation; also extending as an edge effect into adjoining forests
+	+	+	More young, shade-tolerant trees that are more flammable
+	+	+	Increased mortality of unlogged trees, also extending as an edge effect into adjoining vegetation, increasing snags and large deadwood
+	+	+	Increased understory shrub, forb, and graminoid growth, increasing fine fuels and ladder fuels in harvested areas, also extending as an edge effect into adjoining vegetation
+	+	+	Increased fuels and hot, dry, windy microclimate, as well as human-caused fires along temporary logging roads, also extending as an edge effect into adjoining vegetation
			PERSISTENT EFFECTS (i.e., more than a century)
+	+	+	Increased fuels and hot, dry, windy microclimate, as well as human-caused fires along permanent logging roads, also extending as an edge effect into adjoining vegetation

Sources: See text for citations and explanation of effects. + = increase.

weather, because of lower fuel moisture and higher wind speed (Platt, Veblen, and Sherriff 2006).

Logging produces *slash*, or *activity fuels*, including stumps, defective boles, branch wood, twigs, and needles–unlike a wildfire, in which fine fuels and some larger fuels are consumed. Since fuels are broken up in harvesting, logging typically brings loadings of small wood and fine fuels to exceptional levels and substantially reduces moisture content relative to live fuels in uncut forests (Anderson et al. 1966; Keyes and Varner 2006). Slash also has high fuelbed depths (Keyes and Varner 2006) and a packing ratio favorable for flaming combustion. Slash loadings vary with harvesting intensity and method; loadings were 60–100 metric tons per hectare at several sites in the northern Rockies (R. E. Benson and Schlieter 1980), about 120 metric tons per hectare at another (Albini and Brown 1978), and averaged about 290 metric tons per hectare in a clearcut Montana spruce-fir forest (J. K. Brown 1970). Slash varied from 14 percent of preharvest volume with close utilization, to 38 percent with conventional logging (harvesting only trees more than 23 cm in diameter) at several sites in the northern Rockies (R. E. Benson and Schlieter 1980). Slash loadings after precommercial thinning in moist forests in northern Idaho could add 20–40 percent, or 10–100 metric tons per hectare (Koski and Fischer 1979). These slash loadings are large relative to typical loadings in Rocky Mountain forests (table 2.3).

Slash creates extreme fire hazard for several years and elevated hazard for up to several decades. It loses some flammability by settling and decomposing. After five years, depth decreased by about half in a precommercial thin and about 30 percent in clear-cuts in the northern Rockies, and flammable foliage decreased by half or more (Albini and Brown 1978). Rate of spread declined more than 25 percent, after a year, in slash from many trees, particularly if needles dropped over the winter (Fahnestock and Dieterich 1962). Grand fir and larch slash declined rapidly, but slash of other trees still had a high rate of spread after five years (Fahnestock and Dieterich 1962). Conventional harvesting without fuel treatment can produce excessive hazard for up to 20 years in the northern Rockies (J. K. Brown 1980). Precommercial thinning has an extreme rate of spread and high resistance to control the first year, which may decline to a high rate of spread and low resistance to control in five years (Fahnestock 1968). Prescribed burning is one of the best methods of treating slash (Roe et al. 1971; J. K. Brown 1980), but slash fires commonly escape (Agee 1993). Escaped fires and natural ignitions in slash tend to become larger than other fires due to high flammability and locations remote from control forces (Pew and Larsen 2001). Slash fires that escape, and wildfires that burn through slash, extend the effects of logging into surrounding land areas; this extension is promoted by a microclimatic edge effect from the logged area.

Logging-increased flammability, from altered microclimate and fuels, extends as an edge effect into unharvested forests adjoining harvest units. Hotter, drier microclimate and increased flammability extend 20–70 meters into adjoining unharvested subalpine forests in Wyoming (Vaillancourt 1995) and up to 240 meters in Douglas-fir forests in the Northwest (Chen, Franklin, and

Spies 1995). Higher tree mortality, thus more snags and deadwood, extends up to 30 meters into unharvested forests in Wyoming (Vaillancourt 1995) and 125 meters in the Northwest (Chen, Franklin, and Spies 1995). Increased tree regeneration and thus higher tree density extend 30–50 meters into adjacent unharvested forests in Wyoming (Vaillancourt 1995) and up to 137 meters into Douglas-fir forests in the Northwest (Chen, Franklin, and Spies 1995). The net area affected by logging includes harvested areas, roads, and edge-affected areas, which together greatly exceed the area directly harvested (box 10.2).

Roads

Roads can act as fire barriers for low-intensity fire, but they are also among the most significant locations of human-caused fires in North America, because of smoking, escapes from activities along roads, and vehicular accidents (Show et al. 1941; C. C. Wilson 1979; Pew and Larsen 2001; Yang et al. 2007). For example, the 8,300-hectare Butte City fire in Idaho in 1994 started when a flat tire caught fire along a highway (B. W. Butler and Reynolds 1997). In national forests in Southern California in the 1930s, three-quarters of fires and 69

BOX 10.2

Landscape Effect of Logging and Roads

An active timber harvesting area in southeastern Wyoming illustrates the potential magnitude of landscape effects on flammability from the combined direct and indirect impacts of logging and roads (R. A. Reed, Johnson-Barnard, and Baker 1996). This 30,213-hectare subalpine landscape has areas that cannot be logged because of wilderness and other constraints (fig. B10.2.1). Clear-cuts from 1950 to 1993 occupy 3,268 hectares, or 16 percent of the 20,201-hectare loggable area. Assuming that a clear-cut alters microclimate and fuels 100 meters (between the estimates of Vaillancourt 1995 and Chen, Franklin, and Spies 1995) into the edge of the uncut forest, the edge-affected area adjacent to clear-cuts is about 15 percent of loggable area. The distance that microclimate is altered adjacent to roads was not measured but is likely shorter than that adjacent to clear-cuts, and the distance with increased tree regeneration is also less adjacent to roads than to clear-cuts (W. L. Baker and Dillon 2000). Assuming that microclimate and fuels are altered within about 50 meters of a road, the edge-affected area adjacent to roads is about 10 percent of

BOX 10.2
Continued

the loggable area. Thus, the combined area with increased flammability because of clear-cutting and roads includes about 16 percent of the loggable area from clear-cuts, about 15 percent from edge that adjoins clear-cuts, and about 10 percent from edge that adjoins roads, a total of about 41 percent of the loggable area, or about 2.6 times the area of the clear-cuts themselves.

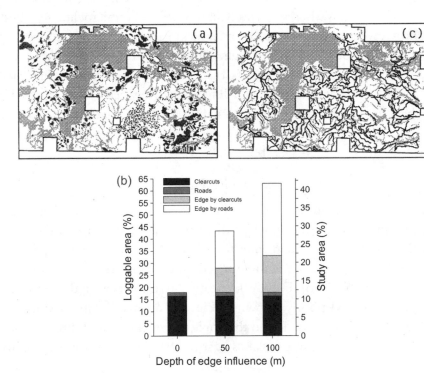

Figure B10.2.1. Landscape effect of logging and roads. Shown in the Tie Camp study area, which is 22.9 x 15.4 km, or 30,213 ha, and is located in the Medicine Bow Mountains in southeastern Wyoming, are (a) clear-cuts 1950–93 in black; (b) the percentage of the area affected by clear-cuts, roads, and edges, based on depth-of-edge influences 0–100 m; and (c) the road system (in black) on the study area. Shaded portions of (a) and (c) indicate a wilderness area and areas designated unsuitable for timber harvesting.

Source: Reproduced from R. A. Reed, Johnson-Barnard, and Baker (1996, p. 1101, fig. 1, and p. 1102, fig. 2) with permission of Wiley-Blackwell Publishing.

percent of burned area originated within 80 meters of a road (Show et al. 1941). In the Missoula Valley, Montana, human-caused fires are associated with roads and decline rapidly away from roads (fig. 10.2). The road system is extensive in the Rockies (e.g., fig. 10.4). Roads in national forests in Colorado, Wyoming, and South Dakota total 44,885 kilometers, enough to circle the earth (W. L. Baker and Knight 2000).

Roads increase flammability in adjacent forests by increasing fuels; creating a hotter, drier, windier microclimate; and increasing tree regeneration (W. L. Baker and Dillon 2000; W. L. Baker and Knight 2000). Slash from road construction and maintenance is a significant hazard (fig. 10.5), increasing fires in adjoining forests (Leiberg 1899a). Roadside firewood removal also leaves flammable activity fuels (W. L. Baker and Knight 2000). The microclimate is hotter, drier, and windier next to a road than inside a forest, just as it is adjacent to a clear-cut (Vaillancourt 1995). The higher-light environment adjacent to a road can increase tree regeneration (W. L. Baker and Dillon 2000), adding ladder fuels. Roads also are conduits for spreading nonnative weeds, some of which are flammable (see the next section, "Nonnative Plants"). Logging and roads together lead to more flammable landscapes, because of edge effects extending well beyond the harvested area (box 10.2).

Nonnative Plants

Invasive nonnative plants, a serious threat to native biological diversity, are abundant, but less so in the Rockies than in some other parts of the United States. Each Rocky Mountain state has 50–85 "noxious" weed species (e.g., P. M. Rice 2005). Only some of these are common after fires, based on an initial categorization (table 10.5). Fire-cycle creators, transformers, and most facultative transformers and poorly known species are on many state noxious weed lists. Zouhar et al. (2008) reviewed this topic in general. A useful review of fire and invasive plants in plains grasslands near the Rockies is also available (Grace et al. 2001).

If the invasion of a nonnative plant increases flammability and, ultimately, burned area, and the plant regenerates rapidly after fire, enhancing the possibility of future fire, a positive feedback loop may be created that can reduce native plants and lead to near monoculture of the invasive species. This positive feedback loop is a *grass-fire cycle* (D'Antonio and Vitousek 1992), since grasses often are the invaders, or, more broadly, an *invasive plant–fire regime cycle*, since other plants can also create a cycle (Brooks et al. 2004). In the Rockies, only cheatgrass (box 9.2) and medusahead, on uplands, and tamarisk, in wetlands and riparian areas, are known to have created invasive plant–fire regime cycles.

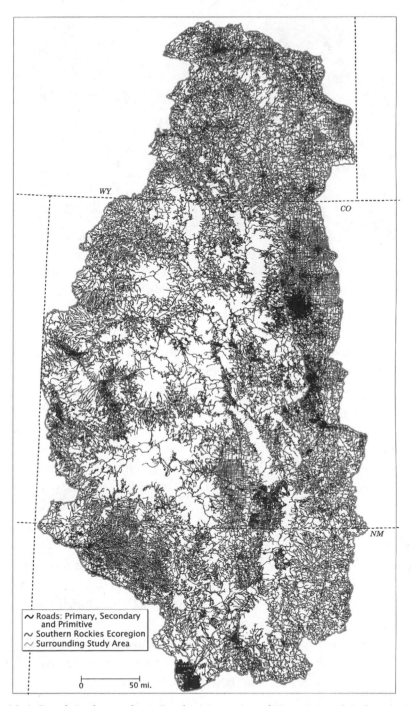

Figure 10.4. Roads in the southern Rocky Mountains of Wyoming and Colorado. Shown is the system of primary, secondary, and some primitive roads.
Source: Reproduced from Shinneman, McClellan, and Smith (2000, p. 55, map 4.10) with permission of the Southern Rockies Ecosystem Project.

Figure 10.5. Road construction slash along a forest road.
Source: Reproduced from C. C. Wilson and Dell (1971, p. 472, fig. 1), copyright 1971, with permission of the Society of American Foresters.

Six other invasive transformer species (table 10.5) can nearly completely dominate burned areas, significantly altering postfire succession. They have deep roots, allowing survival of even high-severity fire, or have heat-stimulated long-dormant seeds, or their growth is stimulated by fire. Similarly, five perennial and two annual nonnative grasses are seeded transformers. They are often seeded after fires to control erosion or establish cover (Keeley 2004; Parks et al. 2005), but they can significantly limit the recovery of native plants and escape into adjoining vegetation. The seeded annuals also increase flammability (Keeley 2004). Seeded transformers are not on noxious weed lists, where they belong because of their known adverse effects on recovery of native plants after fire. Eleven nonnatives are facultative transformers that can be reduced by high-severity fire or fire in a particular season but more often are serious transformers after fire. The remainder includes (1) small invaders that are common but seldom occupy much space, (2) temporary increasers that dominate postfire environments for a few years and then decline, and (3) potentially serious transformers whose responses are poorly known.

In general, invasive nonnative plants are most abundant in the Rocky Mountains in lower-elevation salt desert shrub, sagebrush, grassland, piñon-juniper, and ponderosa pine forests, as well as riparian areas and wetlands, although forests, grasslands, and disturbed areas at higher elevations also may have nonnatives (Parks et al. 2005). Nonnative plant species richness is often highest where native richness is highest, regardless of management (Fornwalt et al. 2003).

Table 10.5. Common Invasive Nonnative Plants That Occur in Burned Areas, and Their Response to Fire

Latin Name	Common Name	Generalized Response to Fire
FIRE-CYCLE CREATORS		
Bromus tectorum[a,b,c,d]	Cheatgrass	Can increase dramatically and increase fires.
Taeniatherum caput-medusae[d]	Medusahead	Can increase dramatically and increase fires.
Tamarix sp.	Tamarisk; saltcedar	Resprouts readily, and increases fire in riparian areas.
INVASIVE TRANSFORMERS		
Centaurea diffusa[d]	Diffuse knapweed	Germination and growth stimulated by fire.[e]
Centaurea solstitialis[d]	Yellow star-thistle	Fire kills live plants but stimulates seed germination; can take over burned areas.
Centaurea stoebe ssp. *micranthos* (*C. maculosa*)[a,b,c]	Spotted knapweed	Seed can survive, plants resprout; can take over.
Chondrilla juncea	Rush skeletonweed	Prolific seed production, deep roots allow resprouting and expansion.[f]
Euphorbia esula[a,d]	Leafy spurge	Readily survives; resprouts and growth may be stimulated, so can increase; has flammable oils.
Hypericum perforatum[d]	Common St. Johnswort	Survives, resprouts readily; seed is viable for decades and is stimulated by fire.
SEEDED TRANSFORMERS		
Agropyron cristatum	Crested wheatgrass	Burns lightly, resprouts; used in greenstrips.
Bromus inermis[a,d]	Smooth brome	Resprouts; variable, depending on season of burn.
Dactylis glomerata	Orchardgrass	Resprouts; remains stable or increases.
Lolium perenne ssp. *multiflorum*	Italian ryegrass	Seeded; flammable and prevents natural recovery.[g]
Phleum pratense	Timothy	Resprouts; remains stable if moderate severity.
Poa pratensis[a,d]	Kentucky bluegrass	Resprouts; variable, depending on season of burn.
Triticum aestivum	Common wheat	Seeded; flammable and prevents natural recovery.[h]
FACULTATIVE TRANSFORMERS		
Bromus arvensis (*B. japonicus*)[a,d]	Field brome	Reduced for 2 years after fire but recovers.
Carduus nutans[b,c,d]	Nodding plumeless thistle, Musk thistle	Can survive and resprout or disperse by wind; response depends on fire severity, competitors.
Cirsium arvense[b,c]	Canada thistle	Variable but can survive and resprout or disperse by seed, even if not present before fire.

Table 10.5. Continued

Latin Name	Common Name	Generalized Response to Fire
Cirsium vulgare[b,d]	Bull thistle	Variable; seed viable in soil and disperses well.
Cynoglossum officinale[a,b]	Gypsyflower, hounds-tongue	Variable; dispersed long distance by animals.
Halogeton glomeratus	Saltlover	Plants killed, but some seed can survive; readily disperses into burns and can become abundant.
Linaria dalmatica	Dalmatian toadflax	Survives, resprouts, but responses vary, sometimes increasing substantially.
Linaria vulgaris	Butter and eggs	Same as *L. dalmatica*.
Melilotus officinalis	Yellow sweetclover	Live plants are killed, but seed germination is stimulated; often disperses and increases in burns.
Potentilla recta[a,b,d]	Sulphur cinquefoil	Resprouts after light fires and can increase.
Sisymbrium altissimum[a]	Tall tumblemustard	Plants are killed; seed can survive and may increase.
Verbascum thapsus[a,b]	Common mullein	Seeds remain dormant for decades; can increase substantially after fire and alter succession.[i]
SMALL INVADERS		
Taraxacum officinale	Common dandelion	Prolific seed and wind dispersal; can increase.
Tragopogon dubius	Yellow salsify	No information in FEIS.
TEMPORARY INCREASERS		
Descurainia sophia	Herb sophia	Can increase from seed after fire, may decline?
Erodium cicutarium	Redstem stork's bill	Seed/seedlings survive light fires; can increase substantially in first few years, then may decline.
Lactuca serriola	Prickly lettuce	Increases after fire for a few years, then declines.
Salsola kali and *S. tragus*	Russian thistle	Flammable; can increase dramatically, then decline.
POORLY KNOWN		
Acroptilon repens	Hardheads, Russian knapweed	May resprout from root; effect on seeds unknown.
Bassia scoparia (formerly *Kochia*)	Burningbush	Little known; may colonize burns via dispersal.
Eleagnus angustifolia	Russian olive	Little known; resprouts observed in some cases.
Lythrum salicaria	Purple loosestrife	Wet sites difficult to burn; not likely to be affected
Thinopyrum intermedium	Intermediate wheatgrass	Little known; resprouts.

Table 10.5. Continued

Notes: The table applies to the Rocky Mountains only. Latin and common names are from the PLANTS online database (http://plants.usda.gov). Response to fire is from the Fire Effects Information System (FEIS) (http://www.fs.fed.us/database/feis), except where noted, but is only a generalized summary; please refer to FEIS or cited sources for more detailed information.

[a]Classified as a "strong invader" in western Montana ponderosa pine forests (Ortega and Pearson 2005).

[b]Classified as a "transformer" in western Montana ponderosa pine forests (Dodson and Fiedler 2006).

[c]Invasive species of most concern after fires in southwestern Colorado piñon-juniper and mixed mountain shrublands (M. L. Floyd et al. 2006).

[d]Identified as "particularly problematic" or covering large areas in the southern, middle or northern Rocky Mountains (C. G. Parks et al. 2005).

[e]Based on Wolfson et al.'s (2005) study of diffuse knapweed in northern Arizona.

[f]Based on Kinter et al. (2007).

[g]Based on Zedler, Gautier, and McMaster (1983).

[h]Based on Keeley (2004), plus observation by the author that this seeded annual has escaped in one case in Colorado, along the Poudre River.

[i]From University of California Extension Service summary: http://ucce.ucdavis.edu/datastore/detailreport.cfm?usernumber=87&surveynumber=182.

Landscape locations that harbor nonnative species may be especially vulnerable to postfire invasions (box 10.3). These include areas in and near roads, railroads, fuel breaks (Merriam, Keeley, and Beyers 2006), oil and gas pads, vegetation generally degraded by livestock (Shinneman and Baker, in press), and places recently disturbed by fires, fuel treatments, or land uses. High-severity fires or severe fires (e.g., pile burning) that are part of fuel treatments in vulnerable settings may strongly favor nonnatives (Dodson and Fiedler 2006; Hunter et al. 2006). Smaller fires with lower intensity may allow nonnative seed banks to survive (Getz and Baker 2008). I discuss the effects of fire control on nonnatives in the next section and review postfire activities that affect nonnatives in chapter 11.

Fire Control

Until the early 1970s, fire policy in the United States was generally intended to control (or suppress) wildland fires (table 10.6; Pyne 1982). Successful fire control depends on rapid arrival after ignition, which has become more possible with advances in mode of access (Hornby 1936; Hirsch, Corey, and Martell 1998; Martell 2001). I suggest three periods, each of about 50 years, of potentially improved fire control effectiveness: (1) from EuroAmerican settlement to 1910, fire control on foot or horse; (2) 1911–60, fire control using roads; and (3) 1961 to the present, aerial attack.

BOX 10.3

Factors Influencing Vulnerability to Postfire Invasive Plants

Vulnerability Factors before Fire
- High native species richness, especially in riparian areas
- High light, moisture, nutrients
- More vulnerable vegetation
  —Ponderosa pine forests
  —Piñon-juniper woodlands
  —Sagebrush
  —Salt-desert shrublands
- High cover or frequency of invasives
- Near human infrastructure
- Vegetation degraded by livestock
  —Bare ground common
  —Low cover of biological soil crust
  —High annual forb cover
- Vegetation degraded by logging, including fuel treatments
- Dispersal corridors into fire area (e.g., roads, trails, powerlines)
- Dispersal agents into fire area (e.g., roads with high traffic volume, ORVs, trail use)

Vulnerability Factors from Fire
- High fire intensity
- Smaller fires
- Soil-disturbing fire control methods (e.g., bulldozer firelines, fire staging areas)

Vulnerability Factors after Fire
- Postfire seeding
  —Intentional seeding of invasives
  —Seed contamination with invasives
  —Disturbance from seeding
  —Reduction of invasives by successful seeding
- Postfire erosion control (e.g., straw mulch)
- Postfire logging (e.g., logged areas, skid trails, log landings)

Note: Sources are cited in the text.

Table 10.6. Historical Changes in Fire Control

Years	Policy, Law, and Administration Changes, Important Fires	Equipment and Technique Developments	Sources
1886	Army assigned to protect national parks from fire.		Pyne (1982); J. F. Wilson (1989)
1897	Organic Administration Act passed—suppression policy.		T. C. Nelson (1979)
1905	Transfer Act created national forests.		Pyne (1982)
1910	Major fires.		Pyne (1982)
1919		First use of reconnaissance aircraft	A. A. Brown (1956); J. F. Wilson (1989); M. Frey (2007)
1922		First trial of reconnaissance helicopters	Johnston (1978)
1927–28		First air cargo delivery	J. F. Wilson (1989); M. Frey (2007)
1929		First use of transport aircraft	M. Frey (2007)
1935	10 AM policy enacted, to control fires by the day after they were reported.		Pyne (1982)
1939–40		First efforts at smoke jumping	A. A. Brown (1956); J. F. Wilson (1989); M. Frey (2007)
1945	Smokey Bear campaign began.		Pyne (1982)
1946–47		First use of helicopters in fire control	Johnston (1978); Dudley & Greenhoe (1998); M. Frey (2007)
1954–56		First retardant drops from large air tankers	Ely et al. (1957); H. K. Harris (1958); M. Frey (2007)
1958		First large-scale smoke jumping	L. M. Stewart (1958)
1958		First use of large helicopters	Johnston (1978)
1963	Leopold Report on national parks completed.		Pyne (1982)
1964	Wilderness Act passed.		Pyne (1982)
1965	Boise Interagency Fire Center created.		Pyne (1982); J. F. Wilson (1989)
1968		Use of larger transport aircraft	J. F. Wilson (1989)

Table 10.6. Continued

Years	Policy, Law, and Administration Changes, Important Fires	Equipment and Technique Developments	Sources
Early 1970s	"Fire management" policy is developed.		Brackebusch (1973)
Early 1970s	Incident command system (ICS) created.		J. F. Wilson (1989)
1972		National Fire Danger Rating System	Deeming, Burgan, & Cohen (1977)
1972		First rappelling from helicopters	M. Frey (2007)
1974	National Wildfire Coordinating Group (NWCG) created.		J. F. Wilson (1989)
1977	Formal change to "fire management" occurred.		T. C. Nelson (1979)
Early 1980s	National Interagency Incident Management System created.		J. F. Wilson (1989)
1988	Yellowstone fires occurred.		R. W. Gorte (2006)
1988	Formal end to "fire management" occurred.		R. W. Gorte (2006)
1994	Major western fires; Forest Health Conference occurred.		R. W. Gorte (2006)
1995	Federal Wildland Fire Management Policy published.		
1999	GAO reports released.		Hill (1999a,b)
2000	Major western fires (e.g., Cerro Grande, NM) occurred.		
2000	National Fire Plan released.		http://www.forestsandrangelands.gov/ NFP/index .shtml
2002	Major western fires (e.g., Hayman, Biscuit) occurred.		
2002	Healthy Forests Initiative created.		http://www.forestsandrangelands.gov/healthy_ forests/index.shtml
2002	National Fire Plan—10-year strategy— released.		http://www.forestsandrangelands.gov/plan/ documents/10-YearStrategyFinal_Dec2006.pdf
2003	Healthy Forests Restoration Act passed.		http://www.forestsandrangelands.gov/Healthy_ Forests/index.shtml

Until about 1910, fire control was limited at much distance from settlements and infrastructure, because access was by foot or horse and firefighting was by small crews with hand tools. The great fires of 1910 led to increasing efforts at fire control in the Rockies, facilitated by automobiles and roads, between 1911 and the 1950s. A basic fire lookout system was in place by 1915–19 (Flint 1928). In the northern Rockies, lookouts and roads for detection and initial attack covered about 15 percent of land area in 1910 but expanded to 70–75 percent by 1934 (Hornby 1936). Hornby explains that lookouts did detection and initial attack on small fires, but fast-spreading intense fires required reinforcement crews, from nearby towns, which traveled to fires by roads and on foot and could control fires that had spread to no more than about 20 hectares. However, 50 percent of land area in 1935 remained beyond the 8.5-hour travel time allowed for heavy reinforcement crews to arrive from major cities and achieve control by 10 AM the day after detection. By 1935, more than 90 percent of human-caused fires in the northern Rockies, common near roads and settlements, could be detected and attacked within two hours, but only about half of more distant lightning fires could be similarly reached (Hornby 1936). Fire control during this period was thus most effective on human-caused fires concentrated near roads and settlements.

Large-scale mechanized fire control, using aircraft to transport firefighters and drop smoke jumpers and retardant, began about 1957 (table 10.6). Accordingly, initial attack and mechanized control of fires have been available at much distance from roads and settlements for only about 50 years. In the 1960s, mechanization continued, along with increased interagency cooperation and centralized management of crews and equipment (table 10.6). The Leopold Report on national parks recognized fire as part of the ecosystems that parks were charged to protect and maintain, and the 1964 Wilderness Act increased interest in natural fires in wild landscapes (Stephens and Ruth 2005). Thus, fire began to be considered part of nature in these areas.

The 1970s ended fire control as an exclusive goal and replaced it with *fire management*, which recognized fire's multiple values, diverse management methods, and economic criteria. In the Rockies, a task force concluded that wildland fire has value for regenerating trees and other vegetation, maintaining wildlife habitat, and reducing fuels and is an appropriate part of wilderness (Roe et al. 1971). More attention was focused on fuels (C. C. Wilson and Dell 1971; Arno and Brown 1989), partly because they can be managed, but also because fire control had not prevented all large fires (Connaughton 1970). Thus, landscape mosaics with fuel breaks and treated areas were envisioned to prevent rapidly spreading, high-intensity fire (Connaughton 1970; Brackebusch 1973).

Concern about rising costs and limited effectiveness increased economic analyses of management options (e.g., J. K. Gorte and Gorte 1979). One new technique was situation analysis, in which a multidisciplinary team evaluated how to manage a particular fire based on behavior, positive and negative effects, management costs, and other factors, aiming to minimize costs (T. C. Nelson 1979). Agencies also designated fire management areas, in which particular management actions were planned before fires occurred (T. C. Nelson 1979; Arno and Brown 1989). Fallout from the 1988 Yellowstone fires essentially suspended fire management (R. W. Gorte 2006), but nearly all the management practices that emerged in the 1970s remain in play in some form today. Chapter 11 examines recent fire policy and management.

Control of a single wildland fire typically includes a series of steps: ignition, detection, reporting, dispatch, arrival, initial attack, and eventual control (fig. 10.6). The goal is to reduce or stop fire growth, which means a race between expansion of the fire's perimeter and construction of a fireline (Martell 2001). The sooner fire is detected and the faster a crew is dispatched and arrives, the more likely fire can be contained at a small size. The larger and more experienced and effective the initial attack crew is, the smaller the fire (Hornby 1936).

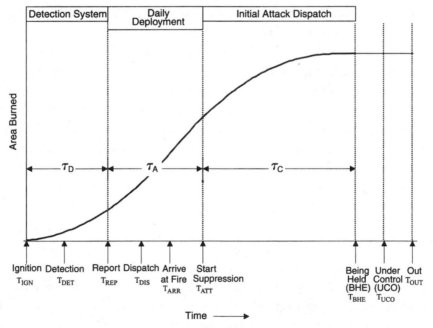

Figure 10.6. Steps to control a forest fire. Symbols with subscripts indicate the times for each step in fire control (e.g., T_{ATT} = time of initial attack).

Source: Reproduced from Martell (2001, p. 548, fig. 8), © Elsevier 2001, with permission of Elsevier Limited.

High daily ignitions can delay arrival and overwhelm resources, allowing more escapes (Cumming 2005). For example, 137 lightning fires started in two days in a 283,000-hectare national forest in the northern Rockies (Hornby 1936), overwhelming fire control crews.

Fire size (area or length of perimeter) and fireline intensity at crew arrival also influence how effective initial attack can be (Hirsch, Corey, and Martell 1998). Both size and intensity are affected by fuels and weather. Fuels can be rated by the probability of reaching uncontrollable size before initial attack, by using curves showing mean area burned by time elapsed since ignition (fig. 10.7) or percentage of fires by initial rate of spread (Hornby 1936). These two indicators have similar trends, but Hornby's curves include some additions and slight differences from those in fig. 10.7: (1) "cut over" is similar to grass and brush; (2) lodgepole is closer to Douglas-fir; and (3) western white pine, western red cedar, and western hemlock are similar to spruce. Expected fireline intensity can be estimated using models, given a particular fuel type and kind of weather (fig. 10.7). These curves vary substantially among Rocky Mountain vegetation types, suggesting that fires in grass, shrubs, and logged areas are most likely to escape control, because they commonly have a rapid initial spread, even though fireline intensity may be modest (fig. 10.7). Severe weather can boost daily ignitions and the fire size and intensity reached by the time crews arrive, increasing the probability of escape (Hornby 1936). If a fire becomes large or intense, crews can primarily only steer it with lateral firelines or back burns and must await lulls in weather to seek full control.

The idea of a temporally lagged effect from fire control leading to increased and more severe fires was presented by Weaver (1943), who thought fire control beginning about 1910 allowed the unnatural persistence of extensive tree regeneration after logging and tree mortality in the Northwest. Weaver thought that regeneration of trees after logging or natural events was normal, but their survival was not. He was misled by composite fire intervals into thinking fires were frequent and would have killed most regeneration, an unsupported inference (chap. 5).

The idea of fuel buildup in interludes without fire appears logical, but it is complex and not generally supported by the evidence. The Forest Service's fuel specialist, James Brown, suggested in the mid-1980s that buildup might occur in low-elevation forests but is much less likely in higher-elevation forests because of multiple processes that produce and consume fuels (J. K. Brown 1985). However, fuel buildup is not clearly evident, even in low-elevation forests (W. L. Baker, Veblen, and Sherriff 2007). See chapter 2 for more discussion of fuel changes, including the idea of buildup.

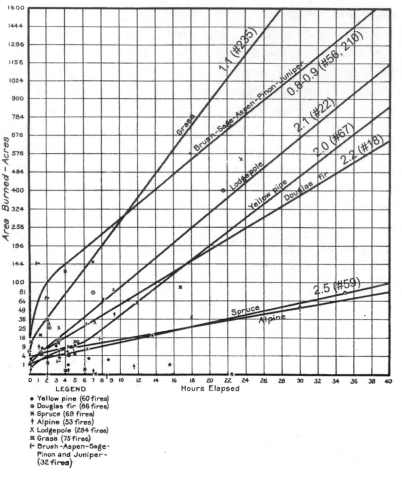

Figure 10.7. Area burned vs. time elapsed between discovery and initial attack. Data, shown by vegetation type, are for northern Colorado and Utah, except that Douglas-fir, lodgepole pine, and alpine data are also from the northern Rockies. Above or below each line, I added "flame length potential" (as a number from 0 to 10), which is proportional to fireline intensity, obtained by running the Fuel Characteristic Classification System (Ottmar et al. 2007) for the fuelbed number given in parentheses.
Source: Reproduced from Sparhawk (1925, p. 710, fig. 11).

If fire control did cause fuel buildup, that effect would be spatially lagged and confined to the part of the landscape where fire control had an actual effect (box 10.1). For example, if the fire rotation is 250 years and fire control has been effective for 50 years, on average only 50/250 or one-fifth (20 percent) of the landscape is expected to have burned in the 50 years of fire control. The maximum area on which a fire control effect or a fuel-buildup effect could have occurred is the actual area that would have burned in the period of fire control, but only part of that area would actually have been affected, because only part of the burned area would have been prevented.

DETECTING CHANGES IN ROCKY MOUNTAIN FIRE REGIMES

This section reviews plot-based fire histories and other data sources for detecting changes in fire regimes resulting from EuroAmerican settlement in the Rocky Mountains. Detecting change is difficult because data are needed for large land areas and a long period of time. The best available data suggest that fire did not decline overall or in most vegetation types in the Rockies after EuroAmerican settlement.

Plot-Based Fire History Studies

I tabulated fire history data used for testing changes in fire across the Rocky Mountains and attributing changes to land uses, and assessed the quality of data for detecting change (table 10.2). Data quality varies because statistically detecting a change requires adequate sample size, unbiased sampling, and valid statistical inference. Some composite fire interval (CFI) and fire rotation (FR) studies are not adequate for detecting change for several reasons (chap. 5). If a modern sample of scarred trees is smaller than a pre-EuroAmerican sample, mean CFI may be longer just because sample size is smaller, which precludes detecting change (W. L. Baker and Ehle 2001). Using deadwood can lead to a smaller sample of scarred trees since settlement, as shown by the extreme case in figure 10.8. Modern calibration shows that targeted fire history sites aimed at

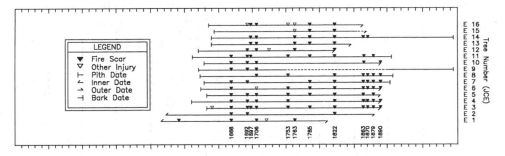

Figure 10.8. Fire history study that included many samples from dead trees and wood. This study included few records beyond 1900. JCE is the name of the sampled stand, and individual samples are numbered from E1 to E16.

Source: Reproduced from P. M. Brown and Sieg (1996, p. 100, fig. 2c), copyright *International Journal of Wildland Fire* 2001, with permission of CSIRO Publishing, Melbourne, Australia.

old trees and long fire records (avoiding recent burns) lead to underestimates of fire since EuroAmerican settlement (Heyerdahl, Morgan, and Riser 2008).

CFI or FR studies that are adequate but use small sample sites scattered over a study area are rated "scatter," because their ability to detect change in fire regimes is lower than FR studies from fire-year maps (chap. 5). FR studies using fire-year maps are rated "short" if they are adequate except for the modern period of data being shorter than the fire rotation under the HRV. To adequately estimate mean fire interval or fire rotation requires a study area several times the largest fire and a record lasting at least one fire rotation (W. L. Baker 1995). Time since EuroAmerican settlement in most of the Rockies is less than 150 years, but fire rotations under the HRV in many forests were longer than that, and sufficient time has not yet elapsed to enable us to detect whether change has occurred. "Best" studies are similar to "short" studies but without that limitation, as fire rotation under the HRV was shorter, or nearly so, than that of the modern period.

About 20 percent of 20 studies with adequate data, all in subalpine forests, documented increased fire during early EuroAmerican settlement. In contrast, 85 percent of these 20 studies, across many ecosystems, documented a decline in fire since EuroAmerican settlement (table 10.2). The table includes the identified causes of fire decline.

Other Data Sources

Other data sources provide information on fires and area burned since EuroAmerican settlement but do not alone provide evidence about whether modern fire regimes have changed relative to the HRV. Data have been collected by the government on burned area since about 1915 (see fig. 10.9) and on individual fires in fire atlases for specific areas (Rollins, Swetnam, and Morgan 2001). Digital data have been available since the 1960s but were more systematically collected after about 1980 (USDI Bureau of Land Management 2004). Data on burned area by state are poorer, particularly prior to about 1930, but nevertheless represent the longest available data set covering the entire Rockies and are used below (see "Causes of Change in Fire Regimes during EuroAmerican Settlement") in qualitatively evaluating land-use effects. Records are based on different land areas after 1990, so I could not use data after 1990 in this analysis.

Fire Rotation under the HRV vs. Twentieth-Century Estimates

Fire decline since EuroAmerican settlement is documented by 85 percent of small-plot studies with adequate data (see previous section, "Plot-Based Fire

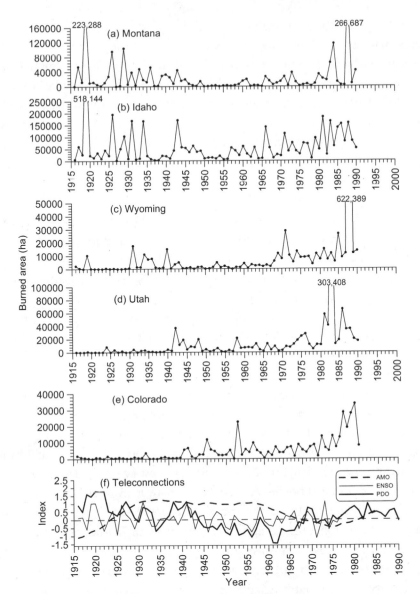

Figure 10.9. Annual area burned (1916–90) on federal, state, and private lands. Shown are data for (a) Montana; (b) Idaho; (c) Wyoming; (d) Utah; and (e) Colorado and (f) reconstructions of the Atlantic Multidecadal Oscillation (AMO), El Niño–Southern Oscillation (ENSO), and Pacific Decadal Oscillation (PDO). Data prior to 1931 are low estimates, as they do not include all land ownerships (Arno and Allison-Bunnell 2002).

Sources: Data for (a–e) are from the National Interagency Fire Center, Boise, Idaho. Data for (f) are from the NOAA Climate Reconstructions database (http://www.ncdc.noaa.gov/paleo/recons .html). The AMO reconstruction is from Gray et al. (2004), the PDO reconstruction is from Mac-Donald and Case (2005), and the ENSO reconstruction is the unpublished Niño 3 index, by Edward Cook, Lamont-Doherty Earth Observatory, Palisades, New York.

History Studies"), but the magnitude of change documented by these studies appears inaccurate. Many small-plot studies show little or no fire since Euro-American settlement (table 10.2), which is clearly inaccurate, as fire has continued (see chaps. 2 and 5, and later in this section). Small-plot studies might correctly detect decline but be inaccurate in estimating magnitude, because they sample small parts of large land areas, and fire occurs in a patchy manner; this is also a limitation of charcoal and pollen studies. In addition, small-plot studies often targeted old forests that lack recent fires, thus underestimating fire since EuroAmerican settlement, based on comparison with digital fire atlas data (Heyerdahl, Morgan, and Riser 2008).

To overcome the limitation of small-plot studies in estimating the magnitude of change, I more directly estimated the fire rotation for natural vegetation in the Rockies under the HRV to compare with twentieth-century estimates. I calculated fire rotation under the HRV (table 10.7) using the estimate for each major vegetation type (table 5.7), weighted by its fractional area in the Rockies (table 1.1). To include uncertainty, I used the low and high value from the range of rotation estimates (table 5.7). Resulting low and high estimates of fire rotation under the HRV for the Rockies are 140 years and 328 years, respectively (table 10.7).

These HRV estimates can be compared to two estimates of fire rotation for the twentieth century. First, an estimate of fire rotation was derived for most of the Rockies using a GIS data set of fire locations from 1980 to 2003 (see chap. 5). The 1980–2003 estimate is from too short a period to accurately estimate fire rotation (chaps. 5, 10), but it represents the best available data at the present time for most of the Rockies. A second estimate is from fire atlas data, which were digitized and analyzed for the period 1900–2003 across 12.1 million hectares in 12 national forests and a national park in Idaho and western Montana, including a cross-section of the vegetation of the northern Rockies (P. Morgan, Heyerdahl, and Gibson 2008). I used the fraction of fire atlas area of each vegetation type that burned in the 104-year period to estimate fire rotation by vegetation type (table 10.8), using equation 5.1 (chap. 5). I also estimated fire rotation for the whole fire atlas area, using the fraction of the atlas area covered by each vegetation type as weights. This is likely a low (short) estimate of fire rotation, as it could not be corrected for unburned area within fire perimeters (P. Morgan, Heyerdahl, and Gibson 2008). It is likely also inaccurate, as it is from only 104 years, but it is surely a better estimate than the one for 1980–2003 and is the best available estimate for this area.

These estimates suggest that the best current hypothesis is that fire has not declined in the Rockies as a whole or in the northern Rockies since

Table 10.7. Weighted Estimate of Fire Rotation across the Rockies under the HRV

Vegetation Type	Weight by Fraction of Rockies Area (table 1.1)	Low Fire Rotation Est. (yrs)	High Fire Rotation Est. (yrs)	Low Subtotal	High Subtotal
MONTANE WOODLANDS & FORESTS					
Piñon–juniper woodlands	0.039	400	600	15.6	23.4
Ponderosa pine–Douglas-fir forests	0.071	50	150	3.6	10.7
Douglas-fir forests	0.047	125	175	5.9	8.2
Moist northwestern forests	0.045	100	600	4.5	27.0
Mixed-conifer forests	0.100	100	250	10.0	25.0
Quaking aspen forests	0.027	150	150	4.1	4.1
Total montane woodlands & forests	0.329			43.7	98.4
Weighted estimate of fire rotation (years)				*133*	*299*
SUBALPINE FORESTS					
Lodgepole pine forests	0.081	135	280	10.9	22.7
Spruce-fir forests	0.098	175	700	17.2	68.6
Five-needle pine forests	0.001	250	350	0.3	0.4
Total subalpine forests	0.180			28.4	91.7
Weighted estimate of fire rotation (years)				*158*	*319*
SHRUBLANDS					
Salt desert shrubs	0.016	500	700	8.0	11.2
Sagebrush	0.121	150	350	18.2	42.4
Misc. shrublands	0.004	Unknown	—	—	—
Mixed mountain shrubs	0.053	100	100	5.3	5.3
Total shrublands	0.190			31.5	58.9
Weighted estimate of fire rotation (years)				*166*	*310*
GRASSLANDS					
Great Plains grasslands	0.080	140	220	11.2	17.6
Plateau grasslands	0.006	Unknown	—	—	—
Montane grasslands	0.024	50	115	1.2	2.8
Subalpine grasslands	0.036	50	170	1.8	6.1
Total grasslands	0.140			14.2	26.5
Weighted estimate of fire rotation (years)				*101*	*189*
RIPARIAN	0.016	Unknown	—	—	—
TOTAL VEGETATION	0.839			117.8	275.5
Weighted estimate of fire rotation (years)				**140**	**328**

Note: The analysis uses low and high estimates of fire rotation, based on table 5.7. Agriculture, urban, and unvegetated areas are left out of the analysis, as they do not include natural vegetation. Subtotal columns represent the fire rotation column multiplied by the weight column. The weighted estimate of fire rotation is the subtotal divided by the weight. Types with unknown weights are not included in the estimate. The estimate for the Rockies as a whole is at the bottom of the table.

Table 10.8. Estimated Fire Rotations in the Northern Rocky Mountains by Vegetation Type (1900–2003)

Vegetation Type (Morgan et al. 2008)	My Vegetation Types (tables 1.1, 10.7)	Fraction of Fire Atlas Area (Morgan et al. 2008)	Fraction Burned 1900–2003 (Morgan et al. 2008)	Estimated Fire Rotation 1900–2003 using equation 5.1 (yrs)
Cold forest	Lodgepole pine forests; spruce-fir forests	0.36	0.39	267
Mesic forest	Moist northwestern forests	0.09	0.59	176
Dry forest	Ponderosa pine–Douglas-fir forests; Douglas-fir forests	0.36	0.41	254
Woodland	Juniper woodlands; quaking aspen forests	0.04	0.53	196
Mesic shrub	Riparian shrublands	0.06	0.44	236
Dry shrub	Sagebrush	0.05	0.43	242
Grassland	Montane and subalpine grasslands	0.01	0.61	170
Other vegetation	None (rock, barren, water)	0.03	0.27	385

Source: P. Morgan, Heyerdahl, and Gibson (2008).

EuroAmerican settlement. The 1980–2003 estimate of fire rotation for the Rockies from the GIS data set is about 275 years (chap. 5). The 1900–2003 weighted estimate of fire rotation for the fire atlas analysis area in the northern Rockies is 247 years, which is a low estimate. Both estimates lie near the middle between the low (140 years) and high (328 years) estimates of fire rotation for the Rockies under the HRV (table 10.7). However, imprecise estimates of fire rotation under the HRV, combined with inadequate data for the Rockies as a whole before about 1980 and an inherently insufficient observation period since 1980, make it impossible to be sure how fire has changed relative to the HRV for the Rockies as a whole. The estimate of fire rotation for the northern Rockies for 1900–2003 is likely much better, but comparison with the HRV is still inconclusive, because of imprecision in the estimates under the HRV and because the HRV estimate is not just from the northern Rockies. Nonetheless, these comparisons suggest that substantial declines in fire have not occurred since EuroAmerican settlement, and the most logical hypothesis is that fire has not declined overall in the Rockies or in the northern Rockies since Euro-American settlement.

However, particular vegetation types may have had more or less fire since EuroAmerican settlement, as suggested by landscape-scale fire history studies (table 10.2). The 1900–2003 estimates for the northern Rockies by vegetation type (table 10.8) can be compared with the low and high estimates under the HRV by vegetation type (table 10.7). These comparisons show that twentieth-century fire rotations by vegetation type are between low and high estimates of fire rotation under the HRV, except in dry forest. My estimate of fire rotation in dry forest under the HRV is quite tentative (chap. 7). These comparisons suggest that the best current hypothesis is that substantial declines in fire have not occurred since EuroAmerican settlement in any of the major vegetation types of the northern Rockies, with the possible exception of dry forests, where better data are needed on fire rotation under the HRV.

CAUSES OF CHANGE IN FIRE REGIMES DURING EUROAMERICAN SETTLEMENT

Studies of land-use effects on fire regimes have been necessarily observational, founded on estimated changes in mean fire interval or fire rotation. Controlled experiments on land uses are nearly impossible. A few natural experiments allowed analysis of a particular land use (e.g., livestock grazing—Madany and West 1983), but replicates are rare. Understanding comes largely from uncontrolled observational studies with interpretation of potential land-use effects, based on logic and temporal coincidence. Potential causes of change since EuroAmerican settlement are complex (table 10.1), and most fire history studies (table 10.2) do not include detailed analysis of all the causes. A particular limitation, explained below (see "Past Analysis of EuroAmerican Effects on Fire"), is the inadequate consideration of confounded effects, alternative explanations, and temporal and spatial heterogeneity in burned area (fig. 10.9). Fire history studies have been focused on broad changes, not on the particulars of each land-use effect, but past explanations appear to fail to fit, or only partly fit, patterns in burned area.

Beyond Fire Exclusion

The common idea of fire exclusion does not match the known effects of land uses, reviewed previously in this chapter, as no land use or set of land uses consistently reduces fire (table 10.1). Harold Weaver (1943), perhaps the first to use the term *fire exclusion*, thought fire control was the primary source of fire decline. Fire exclusion later became a syndrome of livestock grazing, fire control, landscape fragmentation, or other human-caused factors that restrict fire spread or put out fires (Keane et al. 2002). Livestock grazing may reduce fire in the grazing season, with persistent effects if productivity is permanently reduced; but it also increases tree regeneration and fuel loads, and herders accidentally and intentionally set fires. Fire control logically reduced fire, but a lagged fuel-buildup effect could increase fire. Roads, power lines, agricultural fields, live-stock trails, settlements, and other barriers reduced fire spread, but flammability and ignitions are actually increased in these areas. *Fire exclusion* is the wrong term for land uses with immediate and long-term effects that can increase or decrease fire; thus land uses are considered individually.

Changes in Human-Caused Ignitions since EuroAmerican Settlement

None of the 20 fire history studies with adequate data suggested that a decline in ignitions by Indians contributed to a decline in fire after EuroAmerican set-tlement (table 10.2). However, focused analysis found burning by Indians in low-elevation camping, hunting, and travel zones (see the earlier section "Indi-ans and Fire: The Myth of the Humanized Landscape"), and removal of Indians could thus have contributed to fire decline in these areas.

The fraction of human-caused fires and burned area has changed little since about 1920. Lightning caused about two-thirds of fires in both an earlier and a later period (fig. 10.2b,c). Between 1980 and 2003, the rotation for all fires in the Rockies was about 275 years (chap. 5). In this period, an average of about 150,000 hectares per year burned, about 92,000 hectares (61 percent of burned area) from lightning and about 58,000 hectares (39 percent of burned area) from human causes. Human-caused fires thus increased burned area by about 63 percent, although some of this might have burned anyway in lightning fires. If these rates continue, human-caused fire could reduce the rotation from about 450 years to about 275 years. Although human-caused fire is well documented, and likely is spatially associated with human infrastructure, it is not detectable in state-level records of burned area (fig. 10.9).

Livestock Grazing

Stocking rates of sheep were high in northern New Mexico by the early 1800s but increased to higher levels in the 1870s and 1880s (H. A. Wallace 1936; Denevan 1967). By then, most ranges in the Rockies were fully stocked. Sheep peaked by about 1910, and total animals peaked during World War I, reflecting the war effort, but degradation continued during the 1930s drought (H. A. Wal-lace 1936).

Aldo Leopold (1924) hypothesized that livestock reduced grass, decreasing fire spread and lowering competition that held tree regeneration in check, based on qualitative observations of fire scars, remnant vegetation, and tree ages in piñon-juniper woodlands in central Arizona. Analysis in other areas confirmed some of his hypothesis. In the early 1900s, livestock grazing was reported to have reduced fires in seven western national forests by 50–75 percent (Hatton 1920), but the expected affected land area is much smaller because of lags (box 10.1). Fires became rare in the Chuska Mountains, in northwestern New Mexico, in the 1830s after sheep reached high levels (Savage and Swetnam 1990; Savage 1991). Fire decline in the Jemez Mountains of northern New Mexico occurred after grazing onset at two sites with different grazing histories, substantiating a grazing effect (Touchan, Swetnam, and Grissino-Mayer 1995).

Fire declines after EuroAmerican settlement were found in 17 of 20 fire histories in all major ecosystems in the Rockies (table 10.2). Five of the 17, primarily in ponderosa pine, Douglas-fir, and mixed-conifer forests, suggested that livestock played a major role in the decline, based on coincidence of the decline with onset of grazing; 7 found that livestock contributed; and 1 included livestock in a general syndrome. Lags, as suggested above, would delay affected land area. In 4 cases in subalpine forests with little livestock use, decline was not from livestock.

Direct effects of livestock grazing on fire likely continue, but stocking rates of livestock have declined since the 1930s. Given high burned area in the early 1900s in Montana and Idaho, a direct livestock-induced fire decline is not evident; but in Wyoming, Utah, and Colorado, low burned area until about 1930 is consistent with a livestock effect (fig. 10.9), which could have begun affecting sufficient land area after spatial lags. Indirect effects probably have increased since the 1930s, along with vegetation recovery. Increased burned area in Utah and Colorado in the 1940s could possibly be a lagged effect of earlier livestock grazing. Arno and Allison-Bunnell (2002) thought that low burned area through the late 1970s in the West might be partly the result of ongoing grazing reductions in fine fuels. Persistent reductions in productivity were evident by the 1930s in the Rockies in lower-elevation ecosystems (H. A. Wallace 1936) and could well have persisted into the 1970s, but a lagged effect cannot explain the fairly consistent rise in burned area in several states (fig. 10.9).

Logging and Roads

Logging and associated access roads led to more and larger fires in the 1900s and still create a high fire hazard in harvested areas, due to slash and increased flammability beyond the area of harvest units and roads (box 10.2). Slash left untreated over large areas in the Rockies in the early 1900s increased ignitions and large wildfires (Lyman 1947), as explained earlier. Large fires in the West in the 1900s burned at times exclusively in logging slash (Fahnestock 1968). Fire records from the 1930s for the northern Rockies showed that (1) 10 times as many fires ignited per unit area in harvested as in unharvested forests, (2) 2.4 times as many fires burned more than 4 hectares each in harvested as in unharvested forests, and (3) fires burned 10 times as much area per million hectares of land in harvested as in unharvested forests (Barrows 1951b). Even thinned stands with slash led to high rates of spread in wildfires (Dell and Franks 1971).

Timber harvesting was evident and even locally extensive in the Rockies in the late 1800s. Harvesting of old-growth western white pine in northern Idaho and locally abundant ponderosa pine was extensive into the early 1900s (Haig, Davis, and Weidman 1941). Early harvest volume and extent of the road system are poorly known, but a government report (USDA Forest Service 1973, 71) said of the Rockies: "A large portion of the timber inventory was not economically available until after World War II, when improved transportation in the region and growing markets led to increased values for diverse species and smaller sized trees." In 1953, about 4.8 million hectares of old-growth forests and 11.2 million hectares of "sawtimber" (dominant trees >30 cm in diameter) remained in 21.5 million hectares of commercial forest area in the Rockies; compared to other U.S. regions, these were large areas of remaining mature and old-growth forests (USDA Forest Service 1958).

Annual timber harvest on all ownerships in the Rockies rose rapidly from 1952 to 1970, and continued upward until about 1990 (fig. 10.10). A similar trend occurred with the U.S. road system, whose expansion was aided by government subsidies for logging and for moving people out of cities (Kennedy 2006). In 1953, about three-quarters of timber volume harvested in the Rockies was from Idaho and Montana, which had about two-thirds of total commer-

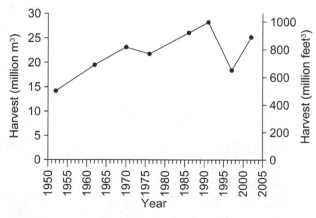

Figure 10.10. Annual harvest of timber in the Rocky Mountains (1952–2002). *Sources:* Data are from Haynes (2003, table 16) and Haynes et al. (2007, table 12).

cial forest area (USDA Forest Service 1958). Concern with untreated slash peaked in the early 1970s, after increased harvesting, and the Forest Service began research on how to address the problem (Barger 1980). Percentage of fires that were escaped fires in debris, including slash, on state and private lands in the United States rose gradually from the 1920s to 1944, then rapidly from 1945 to a peak in 1959 (fig. 10.11b), partly mirroring the rise in slash from logging.

This history can be compared with annual burned area by state (fig. 10.9). Given the larger timber industry in the northern Rockies, any effect is likely to be larger there. Although logging effects likely incur some spatial lags (box 10.1), extensive, untreated slash (Lyman 1947) could have contributed to large burned areas in Montana and Idaho until the 1930s. Lack of large burned area in Utah, Wyoming, and Colorado at that time is consistent with their smaller industry (fig. 10.9). Doubling of annual timber harvest and expansion of roads between 1950 and 1990 match general patterns of burned area in all states. In Idaho, burned area gradually rose from about 1950 to the 1980s (fig. 10.9b). In Montana, a similar, but subtler rise began in the late 1950s, extending to about 1990 (fig. 10.9a). Wyoming lacked a rise in burned area until the late 1960s, but then it, too, experienced a rise until about 1990 (fig. 10.9c). In Utah and Colorado, a modest rise in burned area occurred too early—in the 1940s (fig. 10.9d,e)—to match the logging trend, but an upward trend after 1950 is evident. Thus, annual burned area generally increased with the expansion of logging and roads, even though other factors likely also influenced burned area.

Invasive-Plant Fire Cycles

The most significant fire cycle is from cheatgrass (box 9.2), which shortened fire rotations primarily in sagebrush and salt desert shrublands on the margins of the Rockies (W. L. Baker, in press). The cheatgrass fire cycle was already evident in the Great Basin in the 1920s (Pickford 1932). This fire cycle was especially evident in the last two decades, perhaps because of further cheatgrass expansion and increased cheatgrass productivity from rising carbon dioxide (Ziska, Reeves, and Blank 2005), along with a longer fire season because of global warming (Westerling et al. 2006). Some of the rise in burned area in several states since the late 1970s (fig. 10.9) is likely from this cheatgrass fire cycle (W. L. Baker, in press).

Fire Control

Nearly every periodic assessment of the progress of fire control has questioned its effectiveness and noted rising costs. In the mid-1930s, L. G. Hornby wrote of the northern Rockies, "No pronounced tendency in area burned annually by

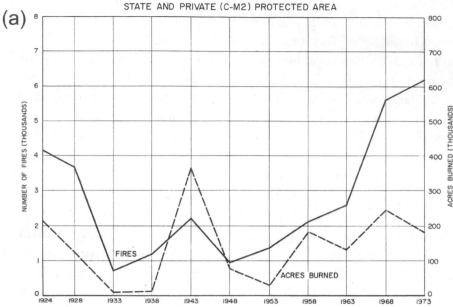

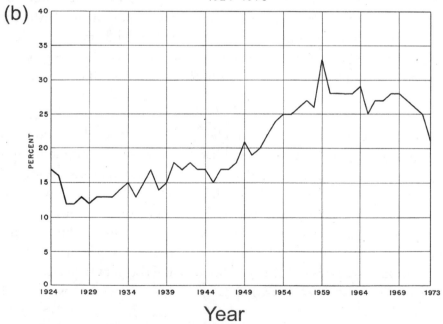

Figure 10.11. Effectiveness of fire control. Shown are (a) the trend in the number of fires and area burned (acres) on state and private land under government fire protection in the Rocky Mountains from 1924 to 1973; and (b) the percentage of fires on state and private land under government protection that are attributed to burning of slash and other debris. *Sources:* Part (a) is reproduced from USDA Forest Service (1973, p. 62); part (b) is reproduced from USDA Forest Service (1973, p. 70).

lightning fires can be stated. This means that the fire control organization has not gotten the upper hand in the fight" (1936, 32). Around the same time, E. Koch said, "The country would be shocked at the lack of results for the millions expended" (1935, 100). In 1958, according to the USDA Forest Service, "The trend of area burned on protected land is unsatisfactory. For the country as a whole, there has been no significant improvement since about 1940" (1958, 202). Five decades later, reviewing national trends, Stephens and Ruth said, "The average annual area burned is increasing. Further, this increase is occurring despite a parallel rise in resources and funds utilized to manage fuels and suppress fire" (2005, 539). Some fires are certainly put out before they spread far, but the question remains whether fire control has had much effect.

One hypothesis is that fire control has been very effective, substantially reducing burned area nationwide by the 1930s or perhaps by 1950 (e.g., Arno Allison-Bunnell 2002; Pyne 2004). A related hypothesis is that control indirectly increased burned area after a lag, as fuel built up on land that would have burned without fire control (H. A. Weaver 1943; Arno and Allison-Bunnell 2002). Some, however, suggest that fire control has not been very effective. A skeptical hypothesis, referring to California Forest Service lands, is that fire control reduced larger fires somewhat by 1950 but had little further effect on large fires, and only reduced medium-sized fires to small ones, with little net impact (Noste and Davis 1975). In Canada, fire control has been thought by some to have had little or no effect, because of the overwhelming importance of climate for fires (Bridge, Miyanishi, and Johnson 2005).

Unfortunately, no quantitative analyses are available for the Rockies, and hypotheses rest on qualitative analysis. Using data for the 11 western states, Arno and Allison-Bunnell (2002) suggested that low burned area between 1946 and 1978 was from continued heavy grazing and effective fire control since the 1935 10 AM policy, which had not yet been in effect long enough to produce a lagged increase in fire. They further suggested that rising burned area after 1978 was a lagged effect of fuel buildup from this past fire control, combined with regional drought and increasing wildland fire use in wilderness areas. Using similar data, Pyne (2004) agreed that the 1970s rise was from fuel buildup and drought, but he suggested that fire control led to rapid declines in burned area on federal land nationally by the early 1930s and continued declines on state and private lands until the 1950s.

Burned-area time series for the Rockies do not support these hypotheses in general, as patterns are inconsistent across states. Neither Arno and Allison-Bunnell (2002) nor Pyne (2004) realized that fire control effects would be lagged over decades (box 10.1). Nonetheless, patterns of burned area do not even match their expectations. Burned area did not decrease on state and private lands under government fire protection in the Rockies through 1950, and no decline is evident through 1973 (fig. 10.11a). Rapid decline in burned area into the 1930s is not evident in Montana or Idaho (fig. 10.9a,b). Little burned area occurred in Wyoming, Utah, and Colorado, consistent with Pyne's hypothesis, through about 1930 (fig. 10.9c–e), although data are poorer prior to 1931. Fire control was generally ineffective in the early years, yet burned area was low then (fig. 10.9). In Montana, annual burned area was high but declined slightly (fig. 10.9a) through the mid-1940s, weakly supporting Arno and Allison-Bunnell's (2002) hypothesis. However, in Idaho and Wyoming, burned area was high, but fluctuated from 1930 to 1950 (fig. 10.9b,c), and in Utah and Colorado burned area rose in the 1940s (fig. 10.9d,e). The data thus do not support the idea that fire control was generally effective in the Rockies prior to the 1930s or 1950. The quotes by Hornby and Koch, given previously, suggest that these managers, who were in charge of fire control planning, believed they had not achieved a significant reduction in burned area by the 1930s, and the doubt was repeated in 1958 by the Forest Service.

A lagged effect of fuel buildup from early fire control (Arno and Allison-Bunnell 2002) thus is unlikely, because early fire control was not very effective, but also because the amount of burned area rose at different times. It remained low in Montana (fig. 10.9a) and moderate in Utah and Colorado (fig. 10.9d,e) until the late 1970s; but in Idaho burned area was moderately high after 1965 (fig. 10.9b), and in Wyoming it remained high after 1968 (fig. 10.9c). Weaver (1943) thought that extensive fuel buildup occurred within 30–40 years after fire control began—that is, by the mid-1940s. Why would the lag in the Rockies be three to four decades longer, until the late 1970s (Arno and Allison-Bunnell 2002)?

Perhaps the late-1970s rise was a lag from more effective fire control after 1957? That seems improbable. The lag would have been from only two decades of fire control in Montana, Utah, and Colorado and about a decade in Idaho and Wyoming. Not enough land area would have been affected by fire control to explain the large rise in burned area in the late 1970s (fig. 10.9). If fire rotation in these states until the late 1970s was similar to that in the Rockies from 1980 to 2003, about 275 years (chap. 5), only 4–8 percent of the area (10/275 or 20/275) is expected, on average, to burn in 10–20 years. That is the maximum area available for fire control to affect, but not all of that burned area could have been prevented. Buildup of fuels on less than 4–8 percent of the land area could not explain the large late-1970s rise in burned area. Not even a longer period of fire control could explain it. Fire control did not prevent all fires and remained ineffective until at least 1930 (Hornby 1936). If fire control prevented 50 percent of burned area from 1930 to 1980, the expected affected area, on average, would be only about 9 percent of the Rockies. If even all the expected burned area was prevented, which it was not (fig. 10.9), 18 percent of the Rockies would have been affected, not enough to explain the large rise in burned area in the late 1970s.

Tree ring–based fire histories in the Rockies that reported fire decline and attributed it to fire control did not analyze it using dates that correspond with fire control history and included limited or no testing of alternative causes. Six of 17 medium- and high-quality fire histories that found fire decline suggested that fire control was a major cause; 8 of 17 suggested that control contributed to decline; and 1 included control in a syndrome of causes (table 10.2). However, these studies used the years from about 1900 to 1945 as the onset of fire control (fig. 10.12), dates that seldom correspond with changes in technology or policy (table 10.6). Many studies also were in remote areas, where fire control was likely weak until after 1957 (table 10.6), yet none used a date that late (fig. 10.12).

Some studies identified fire control as the cause without evaluating the three major alternatives: livestock grazing, natural recovery after nineteenth-century fires, and climate change. Of 14 studies that attributed fire decline in part to fire control, only 2 evaluated all three alternatives, 6 evaluated only two, and 6 evaluated one (table 10.2). A few studies may have valid evidence that fire control led to local fire decline (table 10.2), but evidence for the idea that fire control led to a general fire decline is weak. Fire control history does not match the timing of fire declines or spatial patterns in burned area, and alternative explanations were not evaluated.

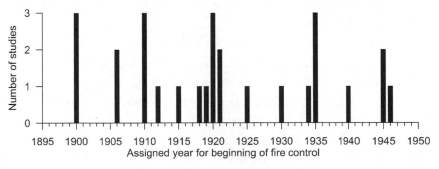

Figure 10.12. Year fire control is considered to have begun in the Rocky Mountains. The graph is based on all medium- to high-quality fire history studies (table 10.2) for which an exact year was specified. Studies specifying only a general period (e.g., the 1920s) were omitted.

Climatic Fluctuations, Late-Nineteenth-Century Fires, and Natural Recovery

Climatic fluctuations linked to fire in the Rockies (chap. 2) are a potentially important influence on burned area since EuroAmerican settlement. Global teleconnections between the atmosphere and oceans that have promoted drought since EuroAmerican settlement vary in timing and effect along the Rockies, reflecting the climate dipole between the northern and southern Rockies (see chap. 2). In the northern Rockies, burned area was greater during a positive Pacific Decadal Oscillation (PDO) and a negative Atlantic Multi-decadal Oscillation (AMO), from about 1900 to 1924 (fig. 10.9f), when major fire years (e.g., 1910, 1919) burned large areas. A negative PDO, unfavorable to fire in the northern Rockies, prevailed from about 1944 to 1977 (fig. 10.9f), a period of low fire (in Montana and Wyoming—fig. 10.9a,c) that Arno and Allison-Bunnell (2002) attributed to fire control and grazing. A positive PDO and negative AMO correspond well with the rise in fire after the late 1970s (fig. 10.9), which these authors attributed to fuel buildup from fire control.

In the southern Rockies, drought and fire are promoted during La Niñas, especially during negative PDO and positive AMO (chap. 2). From 1900 to 1924 a negative AMO was unfavorable for fire in Wyoming, Utah, and Colorado, and little burned area occurred (fig. 10.9c–e). From 1925 to 1944, positive PDO was unfavorable for fire in Utah and Colorado (fig. 10.9d,e). In contrast, from 1944 to the late 1970s, negative PDO with many La Niñas (low ENSO index) favored fire in Utah and Colorado (fig. 10.9d,e). Thus, teleconnections and drought appear better related to burned area since EuroAmerican settlement than do many land uses. Still, land uses likely had an effect, and studies of effects of climate on burned area either did not consider land uses (Collins, Omi, and Chapman 2006) or assumed they were minor (Westerling et al. 2006), which is not likely true.

EuroAmerican settlement in the last half of the nineteenth century produced many fires started by people (see the earlier section "Changes in Human-Caused Ignitions since EuroAmerican Settlement"), but that period also had high climatic variability and many years of drought favorable for fires (Veblen, Kitzberger, and Donnegan 2000). Teleconnections with the AMO, PDO, and ENSO were aligned in a way that favored fires in the southern Rockies (table 2.4; fig. 10.13a). Years with significant burned area (e.g., 1872, 1879, 1880) were often regional or national in extent (fig. 10.14; Plummer 1912), suggesting that climate played a significant role (Kipfmueller and Baker 2000), but both people and climate contributed, a combination that is difficult to disentangle.

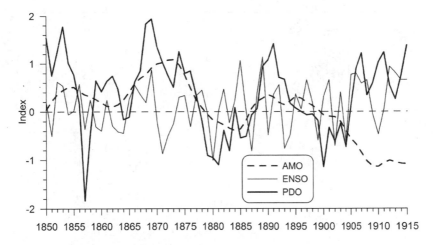

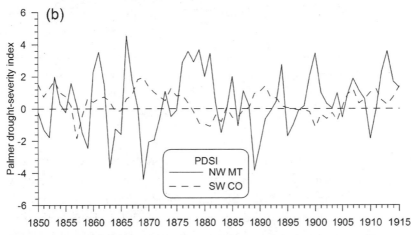

Figure 10.13. AMO, ENSO, PDO, and PDSI during the EuroAmerican settlement period (1850–1915). Shown are (a) reconstructions of the Atlantic Multidecadal Oscillation, El Niño–Southern Oscillation, and Pacific Decadal Oscillation, and (b) reconstruction of the Palmer drought-severity index for grid point 68 in northwestern Montana and grid point 118 in southwestern Colorado.

Sources: Data for (a) and (b) are from the NOAA Climate Reconstructions database (http://www.ncdc.noaa.gov/paleo/recons.html). The AMO reconstruction is from Gray et al. (2004), the PDO is from MacDonald and Case (2005), and the ENSO is the unpublished Niño 3 index by Edward Cook. The PDSI reconstructions were originally by E. R. Cook et al. (1999) but were later revised by Cook (shown here).

Extensive nineteenth- and early-twentieth-century fires (e.g., in 1910) may have reduced susceptibility to fire in the twentieth century. In the northern Rockies, these fires initially created grass- and shrub-dominated fuels that favored extensive reburns into the 1930s (J. A. Larsen 1925b; Gabriel 1976). However, as trees recovered, young forests with lower fuel loads led to lower

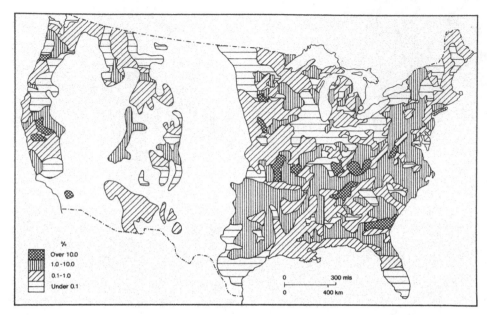

Figure 10.14. Proportion of land areas burned in the United States in 1880.
Source: Reproduced from M. Williams (1989, p. 451, fig 13.4), based on a map in Sargent 1884, with permission of Cambridge University Press.

ignition, spread, and severity relative to older forests, partly explaining mid- to late-twentieth-century fire decline (Veblen and Lorenz 1991; M. P. Murray, Bunting, and Morgan 1998; Sibold and Veblen 2006). Older, middle-aged forests can burn readily when fire occurs, as in the 2002 Hayman fire in central Colorado; much of this area had burned in 1851 or 1880 and had also been logged in the late 1800s (Jack 1900). Thus, reduction in susceptibility to fire during recovery is temporary and affects only areas that had fires or logging in the late 1800s and early 1900s, but that area is extensive.

Large areas were recovering from nineteenth-century disturbances during the twentieth century, but changes during natural recovery were instead attributed to fire exclusion. Recovery from nineteenth-century disturbances, including mining, human-caused and natural fires, logging, and insect outbreaks, was a dominant process that is evident in repeat photography of the Colorado Front Range (Veblen and Lorenz 1991) and the San Juan Mountains (Zier and Baker 2006) and in repeat mapping of western Colorado's aspen and mixed-conifer forests (Kulakowski, Veblen, and Drinkwater 2004; Kulakowski, Veblen, and Kurzel 2006). During recovery, trees increase in density, crowns expand, and canopies close, and conifers can increase relative to early-successional aspen or shrubs. Shrubs, such as sagebrush, may reappear in grasslands. Recovery after fire has the same processes and appearance as those expected from fire exclusion (chap. 2).

It is normal for a dominant process in Rocky Mountain vegetation to be recovery from past disturbances. Large disturbances tend to be at least partly synchronized, by regional fire years or other events (e.g., insect outbreaks— W. L. Baker and Veblen 1990), and recovery can thus have extensive regional synchrony. Nearly all Rocky Mountain forests have slow recovery relative to their fire rotations (table 5.7), which means that much of the forested area at any one time is in the process of recovering from a past disturbance. A substantial area in the Rockies is recovering, in rough synchrony, from nineteenth- and early-twentieth-century disturbances, and similar spatially extensive recovery likely also occurred under the HRV.

PAST ANALYSIS OF EUROAMERICAN EFFECTS ON FIRE

Past analysis of changes in fire regimes since EuroAmerican settlement, based on fire history data, includes a number of conclusions that are not supported; see table 10.9.

A significant problem was the assumption that mean composite fire intervals (CFIs) were equal to the fire rotation, which led to invalid comparisons of mean CFI estimates under the HRV to modern burned area. For example, Barrett, Arno, and Menakis (1997) used mostly mean CFI values to estimate that the fire rotation in a study area of 80 million hectares was about 33 years; thus, they thought that about 2.4 million hectares (1/33 × 80 million) burned annually on average under the HRV. However, in the largest fire years since 1900, only 0.8–1.2 million hectares of area burned, and they inferred from this that fire had been reduced substantially relative to the HRV. However, mean CFI is nearly always short relative to the fire rotation in calibrations and simulation analyses, typically by a factor of 3.6–16.0 (chap. 5). This means that the CFI of 33 years used by Barrett et al. is a fire rotation of about 120–530 years, which indicates that about 0.2–0.7 million hectares burned annually, on average, in 80 million hectares This would be likely to reach 0.8–1.2 million hectares in large fire years. Thus, the modern fire regime is probably not much different from that under the HRV. Barrett et al.'s unsupported conclusion was repeated in an overview of fire exclusion in the Rockies (Keane et al. 2002). In the national LANDFIRE/FRCC program, modeling commonly also used mean CFI values as though they were equal to fire rotations (W. L. Baker, personal observation from LANDFIRE/FRCC workshops), leading to similarly large errors.

Table 10.9. Unsupported Conclusions about Change in Fire Regimes since EuroAmerican Settlement

Unsupported Conclusion, Chapter Where Discussed, and Sources of Conclusion	Why Unsupported
Fire is significantly less now than under the HRV (chap. 5):	
J. K. Brown et al. (1994); Barrett, Arno, & Menakis (1997); LANDFIRE	Composite fire intervals are equated with fire rotation.
Several authors—see table 10.2	Composite fire interval sample is unbalanced.
Many authors—see table 10.2	Recent fire record is too short to estimate fire rotation.
Fire exclusion is the cause of fire decline:	
Many authors—see table 10.2	Natural recovery from 19th-c. fires is not considered.
Many authors—see table 10.2	Climatic fluctuations are not considered.
Many authors—see table 10.2	Land uses do not consistently cause fire decline.
Fire exclusion caused aspen decline (chap. 6):	
DeByle, Bevins, & Fischer (1987)	Recent fire records are too short to estimate fire rotation.
Fire exclusion caused whitebark pine decline (chap. 8):	
Keane, Morgan, & Menakis (1994); P. Morgan et al. (1994); M. P. Murray, Bunting, & Morgan (2000)	Episodic, extensive 19th-c. fires are misinterpreted to be the norm.
Fire exclusion caused more severe fires:	
N. T. Korb (2005)	Missed historical evidence of severe fires nearby.
Many authors—ponderosa pine forests (chap. 7)	Missed evidence of high-severity fires under the HRV.
Fire exclusion caused tree density to increase:	
Kaufmann et al. (2001); Gallant et al. (2003); Romme et al. (2003); Heyerdahl, Miller, & Parsons (2006); S. L. Powell & Hansen (2007)	Natural recovery from 19th-c. fires is not considered.
Many authors—ponderosa pine forests (chap. 7)	Historical records show tree density is not outside HRV.
Many authors—piñon-juniper woodlands (chap. 6)	Tree density fluctuates with climate.
Fire exclusion caused unnatural fuel buildup:	
Grissino-Mayer et al. (2004)	Fuel does not consistently build up during interludes.
Many authors—ponderosa pine forests (chap. 7)	Fuel buildup is assumed, but no actual evidence is presented.
Fire exclusion increased shade-tolerant trees:	
Many authors—ponderosa pine forests (chap. 7)	Logging, grazing, and fire increase these trees.

Sources: Based on studies conducted in the Rocky Mountains. Chapter references indicate where explanations of unsupported conclusions can be found.

Other problems in comparing the modern fire regime to that under the HRV include using an unbalanced CFI sample (see above, "Plot-Based Fire History Studies") and a short record of recent fires. At least a complete fire rotation, typically one to five centuries in Rocky Mountain vegetation, is needed to accurately estimate a recent fire rotation. Authors commonly used only a decade or two of data (e.g., DeByle, Bevins, and Fischer 1987). Short periods of data do indicate short-term rates of burning, but it is normal for interludes and episodes to lead to very different short-term estimates of fire rotation (box 3.1).

Fire exclusion is commonly cited as the explanation for a decline in fire or a particular vegetation change (e.g., aspen decline) without considering the two potential natural causes: natural recovery and climatic change (table 10.2). Many studies did not consider natural recovery as an explanation of fire decline or vegetation change. Natural recovery was omitted in 10 of 14 studies that found a fire decline and attributed it to fire control. Similarly, of all the human and natural factors that might explain patterns of burned area in the Rockies since 1916 (fig. 10.9), climatic fluctuations offer the best single match (see above, "Climatic Fluctuations, Late-Nineteenth-Century Fires, and Natural Recovery"). However, 8 of 14 studies that found fire decline and attributed it to fire control omitted any consideration of the role of climatic fluctuations in the decline. Many changes (tree density increase, tree invasion of meadows, conifers overtopping aspen, fuel buildup, severe fires) are commonly attributed to fire exclusion without considering that they might better represent recovery from nineteenth- and twentieth-century disturbances or climatic fluctuations. Some studies that attribute changes in fire to climatic fluctuations did not consider land-use effects (see above). Findings will have a sound basis only if all factors are analyzed.

REASSESSING EUROAMERICAN EFFECTS ON FIRE

As explained earlier, the most likely hypothesis is that fire rotation did not decline in the Rockies after EuroAmerican settlement relative to the HRV. However, burned area fluctuated in response to individual land uses and climatic variation after EuroAmerican settlement. Some land uses and climatic episodes increased fire; others decreased fire, at different times and in different places. There is no consistent effect of land use (e.g., fire exclusion) on fire.

Between the onset of EuroAmerican settlement and 1910, teleconnections (fig. 10.13a) and drought years (fig. 10.13b) in both the northern (e.g., 1889, 1910) and southern (1879–80, 1900) Rockies favored large fires. This period was thus climatically favorable to fire, and people added ignitions, especially in logging, mining, settlement, and sheep-grazing areas. Livestock grazing likely directly reduced fuels, especially in lower-elevation ecosystems, but effects were lagged (box 10.1). Fire control and logging effects on fire were likely minor in this period, given limited access, and were also lagged.

From 1910 to the mid-1940s, a positive PDO favored fire in the northern Rockies and discouraged fire in the southern Rockies (fig. 10.9). Direct reduction in fine fuels and persistent declines in productivity by livestock could also partly explain low burned area in the southern Rockies, which were more vulnerable to livestock because of a less extensive history of bison. Fire control also superficially matched low burned area in the southern Rockies, but burned area was low even in the early years, when fire control was probably less effective, and rose in the 1940s (fig. 10.9). In the northern Rockies, a favorable PDO, human-caused fires, and fire-susceptible landscapes may best explain burned area. Extensive areas that burned in 1889 and 1910 were vulnerable to reburns. Human-caused fires rose, as did landscape flammability from expanded logging and roads and lack of slash treatment. Fire control and livestock effects likely occurred locally but are not obvious at the state level in this period.

From about the mid-1940s to 1977, a negative PDO favored reduced fire in the northern Rockies and increased fire in the southern Rockies (fig. 10.9). Logging, roads, and fuel buildup from slash dominated this period in the northern Rockies, where logging expanded rapidly, but the effect may have been lagged until after 1977 and balanced somewhat by lower susceptibility to fire on large areas of forests recovering from 1889–1919 fires. In the southern Rockies, the negative PDO likely increased fire, but the increase is also consistent with expanded logging and a lagged indirect effect of livestock grazing. Fire control expanded with aerial attack after 1957, but immediate effect on total burned area was not expected or evident (fig. 10.9). A cheatgrass fire cycle (box 9.2) likely added to burned area in Utah and Colorado after 1940.

After 1977, a positive PDO and negative AMO, as in 1900–24, again favored large fires in the northern Rockies, but after 1982 the AMO was positive, favoring large fires in the southern Rockies. Teleconnection patterns do not match burned area as well as in earlier periods. However, in this period, a general increase in fire caused by global warming became apparent, especially in middle to upper elevations of the northern Rockies. This increase in fire is linked to earlier springs (Westerling et al. 2006) and to delays in the arrival of late-season cyclones with cool, wet weather that can put out large fires (Knapp and Soulé 2007). Logging, roads, and associated landscape flammability continued to rise until at least 1990, and forests that had burned in the late 1800s and early 1900s reached maturity, providing more fuels. Increased population and more recreational access increased human-caused ignitions. Fire control could not explain the rise in burned area, except through a lagged effect on fuel buildup, but lack of evidence of fire control effectiveness earlier makes this unlikely. In the southern Rockies, logging and road effects on flammability may also explain fuel buildup and increased burned area in this period, particularly at higher elevations. Increased fuel from earlier livestock grazing could have added to burned area, especially in lower elevations, but a multidecade lagged effect of grazing is difficult to detect. A cheatgrass fire cycle became common and accounted for substantial burned area, particularly at lower elevations in Idaho (W. L. Baker, in press). This period thus appears to reflect a combination of fuel buildup from logging and roads, cheatgrass and past livestock grazing, and increased ignitions by people, combined with warming climate and more drought; but these effects likely vary spatially.

SUMMARY

Indians lived in the Rockies for thousands of years but likely did not burn and significantly modify Rocky Mountain vegetation. Indians did burn at times for logical reasons, most commonly in low-elevation valleys, along travel routes, and in hunting areas; but Indians were few, and the Rockies were not gardens maintained by frequent burning by Indians.

The history of burned area since EuroAmerican settlement in the Rockies appears to represent a temporally and spatially varying set of confounded and interacting land uses and climatic fluctuations. Since settlement, burned area shows strong fluctuations and multidecadal episodes that appear linked to climatic fluctuations and varying landscape susceptibility to fire, which was largely a legacy of land uses but also of past fires and other disturbances. Fluctuation in drought linked to the Atlantic Multidecadal Oscillation, the Pacific Decadal Oscillation, and the El Niño–Southern Oscillation better match spatial and temporal fluctuations in burned area since EuroAmerican settlement than do most land-use changes. A general warming-related increase in fire is evident in the past two decades, particularly in the northern Rockies.

A general syndrome of fire exclusion does not match the history of land uses, which did not consistently reduce or increase fire. Fire control must have had some effect, but that effect must not have been large or extensive, as it is not evident at the state level. Logging, roads, and livestock grazing, along with increased human-caused fires, stand out as the land uses that most altered Rocky

Mountain fire regimes and vegetation over the past century. The best available data suggest that fire has not declined since EuroAmerican settlement, relative to the HRV, either overall in the Rockies or in most individual vegetation types. Instead, land uses have left a general legacy of increased flammability, at a time when the climate is warming and more of us are visiting or moving into fire-prone places, where we are lighting more fires.

CHAPTER 11

Emerging Threats and Tools for Living with Fire in Landscapes

Wildland fire management in the United States today has been shown to be expensive, ineffective, and logically misdirected, but it can be changed. More than $1 billion has been spent annually in recent years to control fire (Donovan and Brown 2007). Laws and policy (e.g., the 2001 Federal Wildland Fire Policy, the 2003 Healthy Forests Restoration Act) have supported fire control and fuel reduction as central approaches to fire, with federal funding of $921 million and $489 million, respectively, in FY 2006 (Steelman and Burke 2007). Ineffectiveness is documented by rising amounts of burned area in spite of increasing fire control and fuel reduction (chap. 10). Misdirected policy is evident in the lack of attention to emerging threats, including expanding development and global warming, that are exacerbating fire problems. Expanding development in fire-prone landscapes may even be facilitated by federal fire control and fuel-reduction programs. Government subsidies for expanding roads and infrastructure into rural areas a half century ago set the stage for this situation (Kennedy 2006). A USDA inspector general stated that federal agencies charged with managing a host of land resources are becoming fire protection agents for private developments in fire-prone settings (U.S. Department of Agriculture 2006). Since fire control and fuel reduction have been ineffective, and emerging threats are not being addressed, a new approach is needed.

Insufficient attention is being given to the creation of sustainable landscapes that could help maintain a natural role for fire in ecosystems while protecting people and communities. A sustainable landscape approach emphasizes *living with fire* using a variety of tools, rather than simply reducing fuel or controlling fire. Central to sustainability is redesigning landscapes to allow some fire in wildland portions, while increasing protection for landscape areas with people and their infrastructure, aiming for solutions that require low maintenance. Landscapes can be designed to simultaneously improve human communities and minimize adverse impacts on wild nature. Landscape design principles for creating a better relationship between people and fire are not yet fully developed, but sufficient knowledge is available to take steps now.

This chapter reviews current land-use legacies and emerging threats that must be addressed to create more sustainable landscapes; it critically evaluates existing tools for addressing legacies and threats and for living with fire and, finally, considers the use of these tools in living with fire along a gradient of land uses from interfaces to wild landscapes (fig. 11.1). Chapter 12 suggests immediate actions to move toward sustainable landscapes.

IDENTIFYING LEGACIES AND UNDERSTANDING EMERGING THREATS

Legacies from past land uses (chap. 10) and emerging threats, such as increased fire from global warming, are constraints to sustainability that must be understood and addressed in redesigning landscapes. Each land use provides a differ-

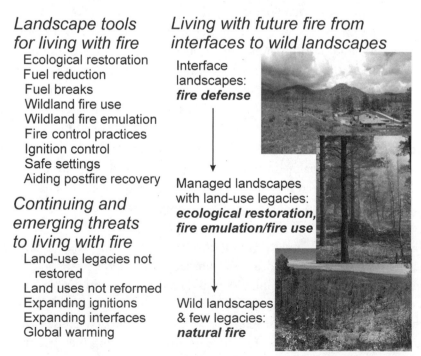

Landscape tools for living with fire
Ecological restoration
Fuel reduction
Fuel breaks
Wildland fire use
Wildland fire emulation
Fire control practices
Ignition control
Safe settings
Aiding postfire recovery

Continuing and emerging threats to living with fire
Land-use legacies not restored
Land uses not reformed
Expanding ignitions
Expanding interfaces
Global warming

Living with future fire from interfaces to wild landscapes
Interface landscapes: *fire defense*

Managed landscapes with land-use legacies: *ecological restoration, fire emulation/fire use*

Wild landscapes & few legacies: *natural fire*

Figure 11.1. Themes and topics of this chapter.

ent constraint, so it is important to know specific land-use histories. Local land-use histories are often distinct and require historical analysis to untangle competing effects on fire, but some generalities can be made. Few Rocky Mountain ecosystems are generally deficient in fire. Logging, grazing, and roads are the primary land uses, along with ignitions by people, that have left a legacy of increased flammability and fire, particularly in the more humanized portions of the Rockies over the past few decades (chap. 10). These are the land uses that should often be considered in evaluating ecological restoration needs, addressed later in this chapter.

Added to more flammable landscapes is an expanding *wildland-urban interface* (WUI), defined as "where structures and vegetation merge in a wildfire-prone environment" (International Code Council 2008, 11). Housing is expanding rapidly in fire-prone settings, aided by governments and people eager for growth (Kennedy 2006). Rocky Mountain states have low percentages of land area (0.7–3.0 percent) but 44–80 percent of the housing in WUI areas (table 11.1). One-third to one-half of the housing in WUI areas in the Rockies is *intermix* (fig. 11.2a), in which wildland vegetation is dominant. The remainder in WUI is *interface* (fig. 11.2b), which has less wildland vegetation (Radeloff et al. 2005; table 11.1). Intermix can have more fire than interface, because enough people are present to increase ignitions but development is too sparse to break up fuels and limit fire spread (Syphard et al. 2007). The highest WUI in the Rockies is in western Montana, the Colorado Front Range, and northern New Mexico (S. I. Stewart et al. 2007). WUI in the Rockies is projected to expand by about 20 percent by 2030 (table 11.1). WUI near forests is exposed to infrequent high-intensity fire in much of the Rockies (chaps. 6–9). Later in the chapter, I will discuss how to address the expanding WUI.

Global warming is also adding to fire risk. In western North America, including the Rockies, the fourth assessment of the Intergovernmental Panel on Climate Change (IPCC) projects the following by the years 2080–99: (1) an increase in mean annual temperature of 2.1–5.7° centigrade (median 3.4° centigrade), especially in summer; (2) a change from −3 to +18 percent (median +5 percent) in mean precipitation in all seasons but summer, when about a 1 percent decline is projected; (3) decreased snowpack and earlier snowmelt; and (4) northward displacement of midlatitude flow in autumn and winter (Christensen et al. 2007). Projections also suggest increased fire in the Rockies, but the geography is uncertain. Key aspects of climate that affect fire in the Rockies, including the North American monsoon, regional differences in climate, precipitation changes, and climate extremes, are not yet well simulated (Bachelet, Lenihan, and Neilson 2007).

(a)

(b)

Figure 11.2. Housing in the wildland–urban interface. Shown are (a) intermix in the upper Bitterroot Valley, Montana; and (b) interface in Dillon, Colorado.
Source: Photos by the author.

Global warming may increase fire in the Rockies, through its effects on vegetation, fuels, lightning, and fire spread. Within the next century, a decline in grasslands and shrublands and replacement by woodlands and savannas may occur (Lenihan et al. 2008). Sagebrush may decline to low levels in the Great Basin but persist on the margins of the Rockies in central Wyoming, southern Idaho, and south of the Uintas (Neilson et al. 2005). Warmer, wetter winters may allow expansion of ponderosa pine, western red cedar, western larch, and Gambel oak and lead to contraction of Engelmann spruce and whitebark pine in the Rockies (Bartlein, Whitlock, and Shafer 1997). These trends agree with models that suggest a reduction in subalpine forests and expansion of montane forests (Lenihan et al. 2008). Nonnative plants may expand (Ryan 2000), as some are favored by fire, and their productivity may also increase (Ziska, Reeves, and Blank 2005).

Table 11.1. Land and Houses in the Wildland–Urban Interface (WUI), by State

State	% Total Land Area in WUI[a]	Intermix WUI Housing Units[a]	Interface WUI Housing Units[a]	Total WUI Housing Units[a]	% Housing Units in WUI[a]	Intermix WUI Area (ha)[a]	Interface WUI Area (ha)[a]	Total WUI Area (ha)[a]	Total WUI Area (ha)[b]	% WUI Developed[c]	Projected Increase in WUI by 2030[b]
CO	3.0	298,543	538,418	836,961	46.3	603,200	197,500	800,700	805,200	20.6	21%
ID	1.6	85,872	147,693	233,565	44.3	209,100	134,900	344,000	302,300	9.8	17%
MT	1.0	90,666	163,771	254,437	61.7	245,900	135,300	381,200	331,600	9.0	19%
NM	2.1	292,444	320,640	613,084	78.5	541,000	126,100	667,100	404,800	17.2	—
UT	1.5	130,260	304,946	435,206	56.6	190,900	129,100	320,000	261,200	4.8	18%
WY	0.7	40,630	139,060	179,690	80.3	112,200	68,700	180,800	62,900	4.1	—

Sources:

[a]Data are from Radeloff et al. (2005, appendix A), which is available in Ecological Archives A015-020-A1 online at http://www.esapubs.org/archive.

[b]Data are from Theobald and Romme (2007, table 5). The definition of a WUI includes a 3,200-meter buffer, which represents a community protection zone in which fuel reduction to protect housing is envisioned.

[c]Data are from Headwaters Economics, http://www.headwaterseconomics.org/wildfire.

Warmer temperatures and less snowpack will decrease fuel moisture. Mortality from drought, insects, and disease may rise (Ryan 2000; McKenzie et al. 2004), changing fuels and fire (box 4.1). More severe fire could favor disperser trees over resisters (Keane, Ryan, and Finney 1998; table 3.2). A 5–6 percent increase in lightning may occur with each 1° centigrade of warming (Price and Rind 1994). A longer fire season will allow more and larger fires (Westerling et al. 2006). Northward displacement of the cyclonic storm track could decrease late-season wet storms that have at times stopped large fires in the northern Rockies (Knapp and Soulé 2007).

Five studies analyzed the potential warming effects on fire in the Rockies, assuming a doubling of CO_2. The first predicted an increase in lightning fires of about 30 percent per month (Price and Rind 1994). The second (Flannigan et al. 1998) used the fire weather index (FWI), which integrates the effects of humidity, temperature, precipitation, and wind on fire spread. FWI may increase one to two times on the western slope of the northern Rockies, two to five times on the eastern slope and south through Colorado, and more than five times in eastern Wyoming and Montana (fig. 11.3a,b). The third study (T. J. Brown, Hall, and Westerling 2004) used the energy release component (ERC) of the National Fire Danger Rating System (Bradshaw et al. 1983) to index large-fire potential. Projections suggest (1) fewer low-humidity days, except from southern Idaho to southwestern Colorado (fig. 11.3c); (2) about 7 fewer days of ERC 40–59 (common with large fires) in southern Idaho, but a few more in western Montana, southern Colorado, and northern New Mexico (fig. 11.3d); and (3) 7–14 more days with ERC ≥60 (extreme fire weather) along the western slope of the Rockies from northern Idaho to southwestern Colorado, with little change along the eastern slope (fig. 11.3e).

The fourth study (McKenzie et al. 2004) used the relationship between climate and area burned in the twentieth century to project that future area burned will increase by about 1.4 times in Idaho, 1.8 in Colorado, 2.6 in Montana, 3.5 in Wyoming, 4.0 in Utah, and 5.0 in New Mexico (fig. 11.4). The fifth study (Batchelet et al. 2007) used two climate models and emissions assumptions, in the most severe case predicting more than 5 times as much biomass would be burned by 2070–99 as in 1961–90, particularly over large parts of the central and northern Rockies. In the more modest case, a substantial area, with more than 5 times as much biomass burned, still occurs; but a larger area with 1.0–2.5 times as much biomass burned also occurs.

Increased fire is emerging in some of these areas, but also in other areas. Midelevations in the central and northern Rockies and parts of western Colorado have been most vulnerable to warming and increased large fires in the past two decades (Westerling et al. 2006). Increased fire since about 1980 has also been evident in sagebrush and salt desert shrublands in southern Idaho, possibly linked to global warming but also linked to cheatgrass (chap. 10). However, rising burned area has been evident over the past few decades, starting at different times, in all states in the Rockies (fig. 10.9).

These studies agree that fire will increase substantially in the Rockies, but they differ on where the greatest change is likely to occur. Regardless of which projection is most accurate, however, the potential impacts are large and include increased invasive species, declines in old-growth forests and sagebrush, adverse effects on sensitive species (McKenzie et al. 2004), and more fire risk to people and infrastructure. Later in the chapter I discuss how to address increased fire from global warming.

EVALUATING TOOLS FOR LIVING WITH FIRE IN LANDSCAPES

This section reviews the major tools available to help in living with fire, including ecological restoration, fuel reduction, fuel breaks, wildland fire use, wildland fire emulation, fire control practices, ignition control, safe settings for housing and infrastructure, and aiding postfire recovery (fig. 11.1). Ecological restoration aims to restore the natural role of fire, fuel reduction and fuel breaks seek to reduce fire intensity and slow or stop fire, wildland fire use allows natural fire to burn under observation, and wildland fire emulation is intentional burning to mimic natural fire. Other approaches include using best practices during fire control actions, changing the probability of human-set fires by controlling ignitions, and placing housing and infrastructure in safe settings.

Here I critically evaluate the ability of these tools to contribute to creation or maintenance of sustainable landscapes for people and fire. Questions include whether the tools work, where and how to use them in landscapes, whether they have beneficial or deleterious effects on nature and human communities, and to what extent they require ongoing maintenance.

Ecological Restoration

Ecological restoration and restoration ecology are complementary but have different meanings. *Ecological restoration* is a practice, defined as "an intentional activity that initiates or accelerates the recovery of an ecosystem with respect to its health, integrity and sustainability" (Society for Ecological Restoration 2004, 1). Some examples of restoration are controlling invasive species, slowing soil erosion, and reintroducing natural processes, such as fire. *Restoration*

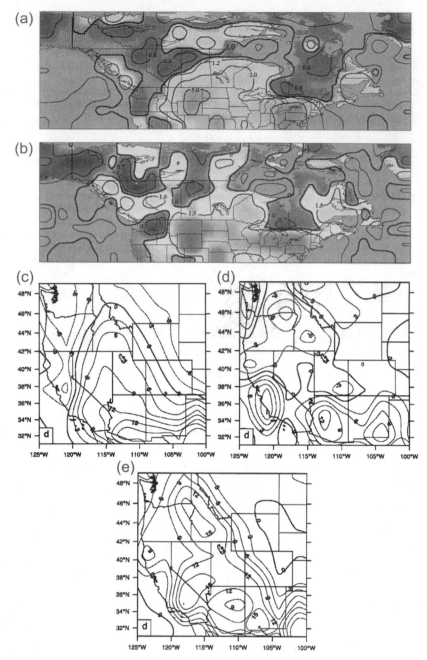

Figure 11.3. Projected changes in fire in the Rockies under a doubled CO_2 climate: (a) ratio of mean fire weather index (FWI) for $2 \times CO_2$ versus $1 \times CO_2$; (b) similar ratio for maximum FWI; (c) difference in the mean annual number of days with minimum relative humidity ≤30 percent for 2070–89 versus 1975–96; (d) difference in the mean annual number of days with energy release component (ERC) 40–59 for 2070–89 versus 1975–96; and (e) difference in the mean annual number of days with ERC ≥60 for 2070–89 versus 1975–96. *Sources:* Parts (a–b) are reproduced from Flannigan, Stocks, and Weber (2003, p. 109, fig. 4.5), copyright 2003, with kind permission of Springer Science and Business Media; (c–e) are reproduced from T. J. Brown, Hall, and Westerling (2004, figs. 3, 5, 7), copyright 2004, with kind permission of Springer Science and Business Media.

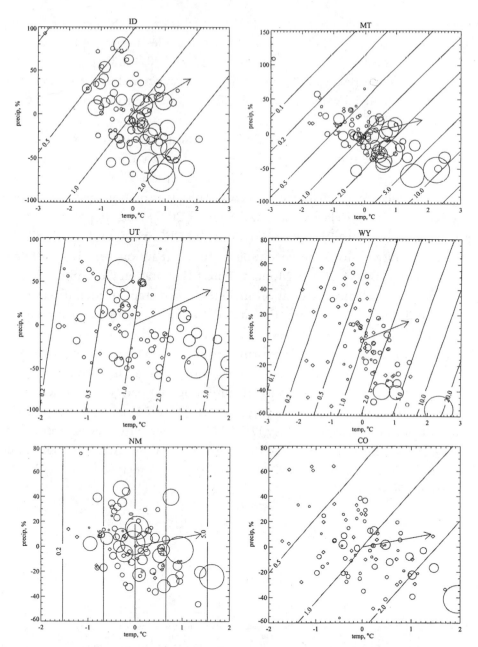

Figure 11.4. Projected changes in fire in the Rockies under a doubled CO_2 climate (continued). The figure shows the area burned by year (1916–2002) relative to mean precipitation and temperature in Rocky Mountain states and projected increase in mean area burned under doubled CO_2 (tip of arrow). Each circle or diamond represents one year. Circle size is proportional to area burned. Contours represent mean area burned, estimated by regression, relative to area burned under mean temperature and precipitation.
Source: Reproduced from McKenzie et al. (2004, pp. 895–97, fig. 1) with permission of Blackwell Publishing Ltd.

ecology is "the science upon which the practice is based" (Society for Ecological Restoration 2004, 11), and that science includes monographs (e.g., Falk, Palmer, and Zedler 2006; Clewell and Aronson 2007) and regional analyses (e.g., Paulson and Baker 2006). Ecological restoration and restoration ecology are both furthered by the Society for Ecological Restoration International (www.ser.org).

Ecological restoration is expanding in landscapes that have significant ecological legacies from past land uses. The theory is that if the structure and composition of vegetation are restored and fire is reintroduced, fire can begin playing a natural role in maintaining ecosystems. Restoration needs are being assessed (e.g., W. L. Baker, Veblen, and Sherriff 2007), and supportive national policy (Dombeck, Williams, and Wood 2004; D. Bosworth and Brown 2007a) and collaborative efforts are becoming common (Paulson and Baker 2006; D. Bosworth and Brown 2007b; Montana Forest Restoration Committee 2007).

Effective ecological restoration includes linked steps of identification, reversal, and reform: (1) identify the altered conditions and particular land uses that have led to the need for restoration, (2) restore the ecological conditions altered by the land uses, and (3) reform the land uses so they do not produce the altered conditions again. Restoration varies on a gradient from restoring structure (e.g., tree density), which often requires ongoing periodic treatments (e.g., thinning), to restoring processes (e.g., natural fire) that maintain that structure without periodic treatments (fig. 11.5). The most effective restoration is long lasting and has little ongoing need for maintenance, thus toward the right in fig. 11.5. Later in the chapter, I review restoration needs relative to fire in Rocky Mountain ecosystems.

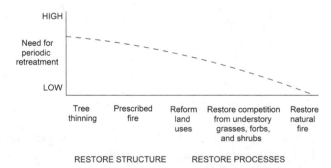

Figure 11.5. Ecological restoration variation along a gradient. The gradient runs from restoring structure, with necessarily regular retreatment, to restoring processes, which requires little or no retreatment.

Source: Reproduced from Noss et al. (2006, p. 485, fig. 4) with permission of the Ecological Society of America.

Fuel Reduction

Fuel reduction is commonly proposed for living with fire and has been a focus of federal fire policy. Here I review goals and methods of fuel reduction, consider its effectiveness, evaluate the ecological need for fuel reduction in Rocky Mountain forests, and review ecological impacts. Available evidence suggests that the ability of fuel reduction to lower fire risk has been overestimated; it has adverse ecological impacts, and it is a tool best used in limited locations, as it is not necessary for ecological restoration in most Rocky Mountain ecosystems.

GOALS AND METHODS OF FUEL REDUCTION

Fuel reduction in forests generally seeks to reduce surface fuels, increase canopy base height, decrease canopy bulk density and continuity (see chap. 2 for terms), and maintain the tallest, largest, most fire-resistant trees, with the goal of reducing fireline intensity and crown fire initiation and spread (Agee and Skinner 2005). These goals are best met by prescribed burning or low thinning (thinning from below), not silviculture that maintains multiple forest layers or high canopy bulk density, such as crown or selection thinning, individual tree selection, or seed-tree or shelterwood harvesting (Graham et al. 1999; D. L. Peterson et al. 2005).

How much to raise canopy base height or lower canopy bulk density can be analyzed for a particular wind speed, based on the minimum (critical) wind speed likely to lead to crown fire initiation and spread, given understory fuels (Cruz, Alexander, and Wakimoto 2004; Cruz, Butler, and Alexander 2006; Cruz et al. 2006; J. H. Scott 2006). Crown fires may not spread actively if canopy bulk density is low or fuels are discontinuous. Canopy bulk density below an empirical threshold (equation 2.8), or below about 0.1 kilogram per cubic meter, may prevent active crown fire spread (Agee 1996; Cruz, Alexander, and Wakimoto 2005).

Agreement is widespread that if slash from mechanical thinning is not burned, fire intensity and probability of crown fire are increased, not decreased (Alexander and Yancik 1977; Kalabokidis and Omi 1998; Graham et al. 1999, 2004; Raymond and Peterson 2005; Omi, Martinson, and Chong 2006). Broadcast burning is the least damaging, most ecologically compatible and effective on-site method of reducing understory fuels from treatments such as thinning (Stephens 1998). Other common methods, including mastication and pile-and-burn, are less effective and more damaging. *Mastication* (grinding, shredding, chunking, or chopping fuels into small pieces) significantly increased fuel depth and continuity, as well as loadings of fine and coarse woody fuels, and increased flame lengths and probability of crown fire relative to areas without mastication (Stephens and Moghaddas 2005b; T. Bradley, Gibson, and Bunn 2006). Burning masticated fuels also can sterilize soils and damage their physical properties (Busse et al. 2005). *Piling and burning* slash also can sterilize soil and damage its physical properties (Massman and Frank 2004; Massman, Frank, and Reisch 2008), increase erosion (Rhodes 2007), and favor invasive plants (see "Ecological Impacts of Fuel Reduction" below). Soils under burned slash piles in Colorado may have increased daily and seasonal cycles of soil heating and cooling to a meter or more depth, persisting for months to years (Massman, Frank, and Reisch 2008).

Is Fuel Reduction Effective?

Fuel reduction should theoretically reduce fire intensity and the probability that a surface fire will become a crown fire that spreads. Evidence to test fuel reduction theory includes simulation using computer models and observations of wildfires that burn into fuel reduction areas.

Unfortunately, common computer models, including BehavePlus (Andrews, Bevins, and Seli 2005), FARSITE (Finney 1998), FlamMap (Finney 2006), and NEXUS (J. H. Scott 1999), underestimate the probability of crown fire initiation and rate of spread at a particular wind speed (chap. 2). The only validated model (Cruz, Alexander, and Wakimoto 2004) suggests that crown fire would initiate in a sample Rocky Mountain ponderosa pine forest if wind speed 10 meters above the surrounding vegetation (U_{10}) exceeds 10 kilometers per hour (kph), and crown fire would initiate at even lower wind speeds in Douglas-fir and lodgepole pine forests (fig. 11.6). Active spread is projected above the 10-kph wind speed. To raise the critical wind speed for initiation to 20 kilometers per hour requires removing all understory trees and 75 percent of *basal area* (the total cross-sectional area of tree stems), a very heavy thinning. Active spread would then require 60-kilometers-per-hour winds (J. H. Scott 2006). Twenty-kilometers-per-hour winds were common and 60-kilometers-per-hour winds have occurred during large fires over the past century in these forests in the Rockies (table 2.5). This is important—even very heavy thinning may be ineffective with the large fires that account for most of the burned area in the Rockies.

Most modeling does not take into account that thinning that lowers canopy cover can actually increase, rather than decrease, fire risk. This can occur because reduction in canopy cover lowers fuel moisture and increases wind speed, potentially increasing, not decreasing, fireline intensity (Platt, Veblen, and Sherriff 2006). (See chapter 10 on the microclimatic effects of logging.)

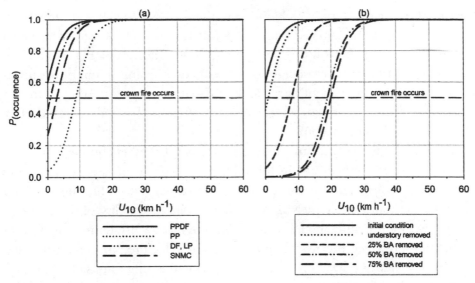

Figure 11.6. Probability that crown fire will occur. The probability is based on the Cruz, Alexander, and Wakimoto (2004) model, given particular wind speeds 10 m above the surrounding vegetation (U_{10}), in (a) five conifer stands (PPDF = ponderosa pine–Douglas-fir; PP = ponderosa pine; DF, LP = Douglas-fir, lodgepole pine; SNMC = Sierra Nevada mixed-conifer); and (b) a ponderosa pine–Douglas-fir forest, in which basal area (BA) is removed by low thinning.
Source: Reproduced from J. H. Scott (2006, p. 12, fig. 3).

Wildfires that burn into treated areas are often mentioned as demonstrating fuel reduction effectiveness, but they really provide little empirical evidence about effectiveness. Observation of a single fire (e.g., Cron 1969; McLean 1993) provides little evidence because of small treatment areas, limited control for topographic and weather effects, and high *spatial autocorrelation* (the tendency for things close together to be similar and, thus, not statistically independent). One study (Martinson and Omi 2003) controlled somewhat for topography and weather by focusing on periods of consistent weather and uniform topography, but other studies (Wilmes et al. 2002; Schoennagel, Veblen, and Romme 2004; Finney, McHugh, and Grenfell 2005) did not. If fire severity is autocorrelated to about three kilometers, as in Arizona (Finney, McHugh, and Grenfell 2005), studies have few observations (treatments in Wilmes et al. 2002; Martinson and Omi 2003; and data used by Schoennagel, Veblen, and Romme 2004).

Better evidence that fuel reduction can reduce fire severity comes from field studies that compare fire severity in large wildfires using adequately spaced replicates (low autocorrelation), each with adjacent treatments, which minimizes confounding by topography and weather. Six studies with replicated treatments sampled 16 fires, including 3 in the Rockies (Pollet and Omi 2002; Martinson and Omi 2003; Raymond and Peterson 2005; Cram, Baker, and Boren 2006; Omi, Martinson, and Chong 2006; Strom and Fulé 2007). All showed reduction in scorch height or crown scorch from thinning if slash was removed. Treatments had all been done less than 12 years (mostly less than 5 years) before fires. Treatment effects may not often last beyond 10 years (Omi, Martinson, and Chong 2006).

Individual fuel reductions might be effective if they encounter a fire, but this is rare in Rocky Mountain forests, and net effectiveness is consequently limited. Net effectiveness is small because, to have an effect, fire must reach a fuel reduction area while fuels are reduced; yet the probability of this happening is low. Fire rotations are moderate to long in most ecosystems in the Rockies (table 5.7), which means the annual probability of fire burning a particular fuel reduction area is low. For example, in ponderosa pine forests, the fire rotation for high- to moderate-severity fire on Forest Service lands is about 278 years in Region 1 (northern Rockies) and 244 years in Region 2 (central and southern Rockies), based on 1980–2003 data (Rhodes and Baker 2008). The effectiveness of fuel reduction has limited duration, since fuel reaccumulates. If a fuel reduction is assumed to be effective for 20 years, these fire rotations mean that the probability that fire will encounter a fuel reduction while it is effective is only 0.0693 in Region 1 and 0.0786 in Region 2. Thus, only about 7–8 percent of fuel reduction area can be effective, since 92–93 percent is not expected to encounter fire in the 20 years of effectiveness. Effectiveness is likely closer to 10–11 years (see preceding paragraph); at 11 years, only 3–4 percent of fuel reduction area can be effective (Rhodes and Baker 2008). Fuel reduction is ineffective and expensive, because fire is unlikely. Treatments must be repeated many times before fire occurs while fuels are reduced. The high cost of repeated treatments suggests that fuel reduction is best used in limited high-value areas, such as near homes and infrastructure.

Nonetheless, some envision a network of landscape treatments (fig. 11.7) to limit the large fires that cause most of the damage to infrastructure (Finney 2001; Finney et al. 2007). The network theoretically would slow the fire's rate of spread and reduce its size by targeting prevailing spread directions and places with high rate of spread. Finney et al. suggest that fuel reduction is needed on a massive scale, perhaps on 10–20 percent of the land per decade in perpetuity (Finney et al. 2007), equivalent to a 50- to 100-year rotation across entire landscapes. Impacts of this include a large road network and damage by fuel reduction equipment and escaped fires, as well as many ecological impacts (see "Ecological Impacts of Fuel Reduction" below). Large fires, which are targeted, do

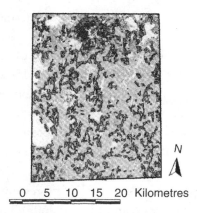

Figure 11.7. Landscape fuel reduction to slow large fires and decrease their size. This is a hypothetical example showing an aerial view of an area near Flagstaff, Arizona, in which about 20 percent of the landscape has fuel reduction treatments (dark areas) and the remainder is not treated (gray areas).
Source: Reproduced from Finney (2007, p. 709, fig. 7), © International Association of Wildland Fire, with permission from CSIRO Publishing, Melbourne, Australia.

most of the work in fire regimes, so the system would exacerbate fire problems. The system also would likely be ineffective in the Rockies. It is designed for directional fires, which do occur, but large fires last over days to weeks and incur widely varying wind directions (chap. 2). Spot fires were not considered in the design of Finney et al. (2007) but would likely overwhelm it, since spotting distances of up to several kilometers are common in large fires. Also, meadows and untreatable areas significantly diminish effectiveness, and both are common in the montane zone in the Rockies, where treatments would be most likely. Thus, the system would be costly and ineffective in the Rockies and would have significant deleterious ecological impacts across large land areas.

FUEL REDUCTION AND ECOLOGICAL RESTORATION

Fuel reduction can be used to lower fire risk near people and infrastructure, but it is seldom needed for ecological restoration in forests of the Rockies, based on their HRV (Schoennagel, Veblen, and Romme 2004; W. L. Baker, Veblen, and Sherriff 2007; chaps. 6–9). Fuel reduction, to reduce fire intensity and crown fires, is not inherently ecological restoration, which aims to restore vegetation structure, composition, and the natural role of fire. Ecological restoration uses the HRV as a frame of reference (box 1.1). In the Rockies, thinning to reduce fire severity generally is not ecological restoration, since forests usually had high-severity fire under the HRV (chaps. 6–9). Tree density also is probably not outside the HRV in most forests in the Rockies (Dillon, Knight, and Meyer

2005; C. B. Meyer, Knight, and Dillon 2005; Veblen and Donnegan 2005; Kulakowski and Veblen 2006), including ponderosa pine forests (W. L. Baker, Veblen, and Sherriff 2007). In Boulder County, Colorado, fuel reduction and ecological restoration were compatible on only 15–27 percent of the area, mostly at low elevations on private land (Platt, Veblen, and Sherriff 2006).

Removing trees for wood products or biomass (e.g., USDA Forest Service 2005) is not the goal of ecological restoration. Foresters have long used silviculture for timber management goals, such as obtaining tree regeneration, maximizing tree growth, and harvesting marketable products. Timber management criteria and methods appear in some ecological restoration proposals (Fiedler, Arno, and Harrington 1998; Arno and Fiedler 2005). These proposals may refer to historical information or ecological function but use timber management criteria—including stand density index or basal area—and also focus on wood products and removing trees, usually including some large, old trees (Fiedler, Arno, and Harrington 1998; Arno and Fiedler 2005). These approaches may thus address few of the numerous nontimber ecological restoration needs (see "Managed Landscapes with Land-Use Legacies: Ecological Restoration" below). Since the development of a government focus on fuel reduction, many proposals have focused on wood products and revenues (Hollenstein, Graham, and Shepperd 2001; Vissage and Miles 2003; Fiedler et al. 2004; USDA Forest Service 2005), rather than ecological restoration needs.

Similarly, optimal vegetation growth, often a goal of timber management or other product-oriented management, is not likely to represent ecological restoration. Some studies imply that forest restoration in and near the Rockies will be successful if tree growth, understory richness and cover, nutrient cycling, or tree seed production and regeneration increase (Fiedler, Arno, and Harrington 1998; Wienk, Sieg, and McPherson 2004; Sala et al. 2005; Metlen and Fiedler 2006; Youngblood, Metlen, and Coe 2006; Peters and Sala 2008). However, historical data (W. L. Baker, Veblen, and Sherriff 2007; chap. 7) show, for example, that under the HRV, ponderosa pine–Douglas-fir forests in the Rockies were not all open, with healthy trees growing at high rates or with diverse understories and high plant cover. In lodgepole pine forests, which were often historically dense (chap. 8), tree growth and understory richness were commonly suppressed for decades to centuries, which also refutes the notion that nature optimizes growth or diversity. Trees in historical forests were often dense, undergoing self-thinning, and thus were stressed by competition. Tree health and other optimal traits were not general characteristics of historical forests in the Rockies. It is important to use the HRV as a frame of reference for determining ecological restoration needs. A product focus can be avoided by using the HRV and by determining ecological restoration goals without, or prior to, analysis of products or revenue (e.g., Montana Forest Restoration Committee 2007).

ECOLOGICAL IMPACTS OF FUEL REDUCTION

Direct animal mortality from treatments may be low, except slow-moving and nesting species can suffer high mortality (Pilliod et al. 2006). Most adversely affected are species dependent on snags; down wood; shrubby or dense understory vegetation; and dense, closed-canopy forests (e.g., invertebrates, cavity-nesting or shrub-nesting birds, some small mammals), while species dependent on open forests (e.g., some raptors) may benefit—although predator-prey interactions may alter responses (Converse, Block, and White 2006; Converse et al. 2006; Pilliod et al. 2006). Thinning effects differ from the effects of prescribed fires (Tiedemann, Klemmedson, and Bull 2000). Effects of the same treatment vary among sites (Converse et al. 2006). To minimize the adverse impacts of fuel reduction on animals, it is logical to retain some of the following: (1) forest structures reduced by treatment; (2) forest structures slow to redevelop (e.g., large snags); and (3) forest structures that facilitate survival and recolonization, including large snags, dense groups of trees, and large down wood (Tiedemann, Klemmedson, and Bull 2000; Pilliod et al. 2006; Saab et al. 2006). Fuel reduction effects have been reviewed for mammals, reptiles, amphibians, invertebrates, birds (Pilliod et al. 2006), butterflies (Huntzinger 2003), and small mammals (Converse, Block, and White 2006; Converse et al. 2006). Prescribed fire effects have also been reviewed for amphibians and reptiles (Russell, Van Lear, and Guynn 1999; Pilliod et al. 2003; see also chap. 4).

Fuel reduction impacts soils, watershed functions, and downstream aquatic ecosystems because of increased runoff, erosion, and sedimentation (Rhodes 2007). Impacts particularly are from mechanical fuel treatments—including associated roads, landings, and soil compaction from equipment—and long-term reductions in soil organic matter and nutrients from removal of vegetation and wood. Thinning can raise within-stand wind speeds, increasing soil erosion (Whicker, Pinder, and Breshears 2008). Prescribed fires also have some of these effects. Impacts from fuel reduction are added to already damaged watersheds and aquatic ecosystems (Rhodes 2007).

Fuel reduction elicits diverse understory plant responses, reflecting species-specific responses, spatial heterogeneity in vegetation, impacts of confounding events (e.g., drought), and other factors. Rather minor changes have occurred in many cases, suggesting that plants are tolerant of treatments or that treatments were not intense enough to elicit much response (Griffis et al. 2001; Metlen, Fiedler, and Youngblood 2004; Kerns, Thies, and Niwa 2006; Youngblood et al.

2006). Annuals and other short-lived species may be favored after prescribed fire (Dodson, Metlen, and Fiedler 2007) as after wildfires (chaps. 6–9). Some functional groups may respond, but others may not, and drought can reverse initial productivity increases (M. M. Moore et al. 2006). Spatially diverse pretreatment vegetation can lead to diverse responses that cloud treatment effects (Fulé, Laughlin, and Covington 2005; Kerns, Thies, and Niwa 2006). Diverse responses to fuel reduction mirror the diverse responses to wildfire (chaps. 6–9).

An increase in invasive species after fuel reduction is ecologically damaging. Thinning and prescribed burning treatments in Montana ponderosa pine forests led to 5 times the cover of invasive transformers (see table 10.5), as in thin-only or burn-only treatments, and 17 times as much as in controls, likely because of higher fire intensity (Dodson and Fiedler 2006; Metlen and Fiedler 2006). Similarly, higher cover or richness of invasives with higher treatment intensity (i.e., thin + burn, higher-severity fire) was found in Oregon (Kerns, Thies, and Niwa 2006; Youngblood, Metlen, and Coe 2006) and Arizona (Griffis et al. 2001; Fulé, Laughlin, and Covington 2005; M. M. Moore et al. 2006). Soil under burned slash piles in Arizona was chemically altered, lacked viable native seeds or arbuscular mycorrhizae, and was vulnerable to invasive plants (Korb, Johnson, and Covington 2004). Five years after burning slash piles in piñon-juniper woodlands, mycorrhizae were present, but invasive species had 7 times as much cover as controls (Haskins and Gehring 2004). Optimism that an invasive increase is short term (Metlen and Fiedler 2006) is premature, as only the first few postfire years have been studied. Invasive response to treatments can be reduced using lower-intensity treatments, fewer treatments, or treatments in winter (Dodson and Fiedler 2006; Kerns, Thies, and Niwa 2006). Given the potentially irreversible nature of invasives, minimum-impact methods are clearly advisable.

Conclusions about Fuel Reduction

Fuel reduction could reduce fire intensity and the probability of crown fire in some cases. However, its potential has been overestimated because of errors in models and because the probability of fire reaching the fuel reduction during the period it is potentially effective is low, necessitating repeated treatments. Fuel reduction must be very heavy to affect the severe fires that have historically accounted for most of the burned area. Fuel reduction is not necessary for ecological restoration in most ecosystems in the Rockies and has adverse effects on native plants and animals, soils, and watersheds, including potentially irreversible increases in invasive species. From an ecological standpoint, fuel reduction is best used only in limited areas adjacent to high-value features such as homes.

Fuel Breaks and Green Forests

The desirability of not altering fire across landscapes and the ecological impacts of fuel treatments together suggest minimum-impact, low-maintenance methods targeted at high-value locations. If fuel reduction is limited to strategic locations and much of the vegetation is removed, the fuel reduction effectively becomes a *fuel break*, a wide block or strip with lighter, usually grass-dominated, fuels that aids fire control (fig. 11.8; Murphy, Green, and Bentley 1967; Green 1977). Fuel breaks are created by clearing vegetation over 65–100 meters, or even up to 400 meters (Green 1977; Agee et al. 2000). Leaving trees creates a *shaded fuel break*, which keeps fuel moisture higher (Agee et al. 2000). Fuel breaks may not stop high-intensity, rapidly spreading fires (Keeley, Fotheringham, and Moritz 2004), but they can lower fire intensity and provide access for fire control crews to backfire to slow approaching fire.

Fuel breaks are best placed in defensive positions adjacent to urban areas, infrastructure, or other valuable resources and are widely used in California (Green 1977; Merriam, Keeley, and Beyers 2006). Constructed breaks have the desirable attribute of, to some extent, mimicking or expanding upon natural fire breaks (fig. 2.8). However, constructed break systems require added roads, increasing traffic and logging operations, thus adversely affecting already damaged watersheds (Rhodes 2007). Fuel breaks also may favor invasive plants because of repeated disturbance during maintenance, providing corridors for invasive spread, and placing invasives into a position to expand after fire (Merriam, Keeley, and Beyers 2006). These concerns suggest that fuel-break systems

Figure 11.8. Example of a fuel break.
Source: Reproduced from Green (1977, p. 2, fig. 1a).

warrant minimization and careful design, placement, and maintenance (Merriam, Keeley, and Beyers 2006; Rhodes 2007). Construction that uses low-impact equipment, retention of trees, access and grazing closure, and weed control may reduce invasive species (Merriam, Keeley, and Beyers 2006). A fuel-break system does not require expansive landscape transformation. A 100-meter-wide break enclosing a square area of 2,500 hectares (5 km by 5 km) would require 2 percent of the area to be transformed versus 100 percent (up to 20 percent per decade) in perpetuity using the Finney et al. (2007) system (see "Is Fuel Reduction Effective?" above). Breaks need not be newly constructed, as they can link into and use existing fields, canals, and low-stature open space, as well as natural fire breaks.

Another low-impact defensive approach is to restore dense, old-growth forests, which would be ecological restoration and would also lower fire risk relative to middle-aged forests. Old-growth mixed-conifer forests without fuel reduction were modeled to be as resistant to crown fire as forests thinned from below, because high tree canopies resist crowning and shade keeps fuel moisture high (Stephens and Moghaddas 2005a). Long-unburned closed forests had the lowest proportion of high severity during major fires in California (Odion et al. 2004). Dense, green forests were observed to slow the rate of fire spread in Idaho, because these forests have less flashy fuels that also are kept cooler, moister, and better protected from the wind (Gisborne 1935).

Wildland Fire Use

The practice of allowing lightning fires to burn naturally under management guidance for resource benefit, called *wildland fire use* (WFU) in current parlance, is increasing. WFU has fluctuated since its general inception in 1968, undergoing name changes, temporary suspension, and refinement of methods (Agee 2000; D. J. Parsons, Landres, and Miller 2003; Aplet 2006; Zimmerman and Lasko 2006). It has been part of formal federal policy since publication of the 1995 Federal Wildland Fire Management Policy, which says that wildland fire should be allowed to function in its natural role as much as possible (Zimmerman and Bunnell 2000). Under current policy, wildland fires are suppressed if they are caused by humans or if a WFU fire escapes, but natural ignitions can be managed as WFU if they are allowed by approved land and resource management plans and are in an area designated for WFU in an approved fire management plan. Each natural ignition gets screened for risk to life, property, safety, and resources and for manageability. If approved for WFU, the fire then gets periodic management actions and assessments (e.g., overflights, mapping, fire modeling) and is either continued or suppressed (USDA and USDI 2006).

Wildland fire use does not represent a let-it-burn policy or fully natural use of fire, as there are limits based on weather, proximity to resources, and manage-ability; but it is more ecologically beneficial than other approaches to fire man-agement and may cost a fraction of what fire control costs (Dale, Aplet, and Wilmer 2005). Social and institutional barriers to WFU remain, including the need for planning documents, a strong fire control culture, uncertainty in man-aging WFU events, concern for political consequences and liability, insufficient personnel, and air-quality concerns; but some remedies exist (Aplet 2006; Doane et al. 2006; Black, Williamson, and Doane 2008), and agency support is growing (Zimmerman and Lasko 2006).

WFU requires landscapes that are large, remote, or well designed, as a cen-tral concern is fire burning into valuable resources or onto private lands. Mil-lions of hectares of federal land in the Rockies appear suitable, based on their remoteness from human infrastructure (Dale, Aplet, and Wilmer 2005). In 2005, about 25 percent of Forest Service lands in the United States were available for WFU (Sexton 2006), and WFU has expanded to include some nonwilder-ness areas (Zimmerman and Lasko 2006). Large wilderness areas, such as the Selway-Bitterroot in Idaho and Montana, more commonly have WFU (C. Miller 2006). In 2005, a 1.6-million-hectare area was used for managing some 25 WFU fires across parts of six national forests in Montana and Idaho (Parks 2006).

Wildland Fire Emulation and Prescribed Burning

In some settings, wildland fire use may not be possible, but prescribed burning may be. Prescribed burning is inherently part of the "nature is a garden" view more than the "nature is a wild landscape" view (chap. 1), but the two views can be brought closer together. In the western United States, prescribed burning is often conceived to achieve general goals, such as fuel reduction, with little at-tention to the resulting landscape mosaic of burn patches and successional stages that significantly influence biological diversity (chaps. 2–9). A focus on fuel reduction and simply getting fire back into landscapes has overshadowed the details of restoring fire using the HRV as a frame of reference (D. Cohen 2006), but a prescribed fire regime outside the HRV will likely adversely affect biological diversity. Fire that is too frequent can reduce species diversity (Laughlin and Grace 2006) and favor invasive species (M. A. Moritz and Odion 2004). Fire that is entirely low severity in ecosystems that historically experi-enced some high-severity fire may not favor germination of fire-dependent species (M. A. Moritz and Odion 2004) or provide habitat for key animals (Smucker, Hutto, and Steele 2005). Program success has been measured in cooperation, area burned, and fuels reduced (e.g., Sexton 2006)—worthy goals, but what is missing is how well prescribed burning emulates fire under the HRV, an essential concern for biological diversity.

Attributes of fire regimes under the HRV that warrant emulation include the rotation, size distribution, interval distribution (chap. 5), and variation in fire intensity or severity, as well as fire season and spatial patchiness (W. L. Baker 1992a). Fire rotations in the Rockies under the HRV have been estimated for most ecosystems (table 5.7). Fire rotations equal the average mean fire interval across a landscape and are appropriate intervals at which individual points or the whole landscape is burned. Composite fire intervals underestimate mean fire interval and fire rotation (chap. 5) and should not be used as prescribed burning intervals, as this would lead to too much fire and would be likely to adversely affect biological diversity (Laughlin and Grace 2006). Fire-size distributions under the HRV are unknown for many ecosystems in the Rockies, but they commonly have an inverse-J shape with numerous small and few large fires (chap. 5). Since most of the total area that burned under the HRV was from the few percent of fires that were large (chap. 5), emulating the HRV means landscapes generally burning in large fires at intervals near the fire rotation, with many small fires that total little land area burning in interludes between the large fires. Substantial variability was the norm.

Variation in fire severity, where characteristic of the HRV, cannot be achieved by burning exclusively in shoulder seasons (spring, fall) under modest weather conditions. A tendency toward low-severity fire is a limitation of prescribed burning. This tendency is not surprising because managers face liability issues and are understandably risk averse (Yoder, Engle, and Fuhlendorf 2004). A mosaic burn is often considered desirable, but patchiness or lack of it under the HRV is the best frame of reference (Parr and Andersen 2006). Overemphasizing patchy burns is inappropriate, since most of the area burned under the HRV was from large fires that likely were not very patchy (W. L. Baker 2006b).

It is inherently challenging to use prescribed fire to restore a landscape mosaic that has been altered by land uses that were not uniform (W. L. Baker 1993). Using prescribed fire is commonly assumed to restore vegetation structure and fuels altered by land uses. However, this result is unlikely because much more than fire needs attention (box 11.1), and the effects of land uses, including fire control, are localized in the landscapes where those uses occurred (box 10.1). Thus, a typical landscape contains a mosaic of patches with and without land-use effects on fire. To restore a mosaic in which fire control has occurred, prescribed fire must be applied in the specific locations where control took place. If fire is applied where no control occurred, the result is another alteration of

BOX 11.1

Examples of Land-Use Effects That Warrant Attention in Restoring Fire to Ecosystems, Based on Current Conditions Relative to the HRV

Logging
 Deficiency of snags
 Deficiency of coarse woody debris
 Deficiency of large, fire-resistant trees
 Deficiency of old-growth forest
 Surplus of small trees forming ladder fuels
 Surplus of young, dense post-logging forests
 Increased invasive species
 Increased landscape flammability from edge effects
 Increased soil compaction and erosion
 Reduced postfire recovery

Livestock Grazing
 Permanently reduced productivity
 Increased loadings of fine fuels
 Increased invasive species
 Increased soil compaction and erosion
 Reduced cover of shrubs, forbs, and grasses favors increased tree regeneration, forming ladder fuels

Roads
 Increased landscape flammability from edge effects
 Increased landscape flammability from activity fuels
 Increased human-caused ignitions
 Increased invasive species

 Reduced fire spread because road acts as fuel break

Fire Control
 Reduced fire
 Reduced tree regeneration
 Reduced coarse woody debris
 Reduced snags
 Increased invasive species on fire lines
 Increased mortality in aquatic ecosystems from toxic fire-control chemicals

Invasive Species
 Increased fire where fire cycle formed
 Increased competition with native plants

Development (e.g., WUI)
 Increased human-caused ignitions
 Reduced fire spread because of fuel breaks
 Reduced possibility of wild-land fire use
 Increased invasive species
 Increased flammability from edge effects

Global Warming
 Increased fire spread
 Increased fireline intensity

Human-Caused Fires
 Increased fire near people and infrastructure
 Reduced old forest, unburned shrublands

the mosaic (too much fire), not restoration. Similarly, if a logged area is deficient in large wood, putting large wood into an unlogged part of the landscape would produce a surplus of wood in that area and would not rectify the deficiency in the logged area. Doing the right things in the wrong place does not achieve restoration.

Prescribed burning is also used to create, modify, and restore habitat for particular species; an entire landscape mosaic may even be managed for a specific biodiversity goal. In tropical areas, plants and animals may be relatively insensitive to the details of the prescribed burning regime (Parr and Andersen 2006). In contrast, where fire rotations are long and postfire successional stages from fires a century ago are still evident, as in the Rockies, how well the prescribed burning program emulates the HRV is important for biological diversity. Emulating variability in key attributes of fire regimes under the HRV is likely to generally benefit biodiversity, while ignoring variability is likely to be detrimental.

Fire Control Practices

Fires must simply be put out in certain cases, but methods for controlling fire could be more ecologically sensitive. Fire control impacts soil, air, water, and vegetation from construction of firelines, roads, helicopter pads, and camps and from use of fire retardants (Backer, Jensen, and McPherson 2004). Erosion can be greater from bulldozer-constructed lines than from the fire itself; lines created with explosives may have the lowest adverse effects. Since firelines are dug to mineral soil, invasive weeds are a threat. Spotted knapweed, an invasive transformer (table 10.5), was denser on firelines in Montana (Backer, Jensen, and McPherson 2004). Bulldozed firelines in Glacier National Park had higher cover and frequency of 23 nonnatives, including 4 facultative transformers and 2 seeded transformers, than occurred in undisturbed or burned forests (N. C. Benson and Kurth 1995). Invasive weeds are also a threat in fire camps and staging areas disturbed during fire control.

Aerially applied fire control includes fire-suppressant foams, usually detergent based; water-enhancer gels; and fire-retardant chemicals, based on nitrogen and phosphorus (Backer, Jensen, and McPherson 2004; Giménez et al. 2004). These can add to eutrophication of streams and lakes and be directly toxic to aquatic organisms, including fish (R. Adams and Simmons 1999; Backer, Jensen, and McPherson 2004). Long-term fire retardants can create chemical fire breaks but significantly impact vegetation (Giménez et al. 2004), reducing or preventing seed germination (Cruz, Alexander, and Wakimoto 2005), killing some plants, and favoring weeds (Bell, Tolhurst, and Wouters 2005). The National Marine Fisheries Service has said that continued use of eight long-term retardants is likely to jeopardize the existence of 24 endangered and threatened fish in the West (Lecky 2007).

Minimum-impact suppression tactics (MIST), part of general light-hand-on-the-land tactics (Mohr 1989, 1992–93; T. K. Brown and Bright 1997; Mohr and Curtiss 1998), have been adopted by federal agencies; these include minimizing the fireline by using natural barriers and employing wet-line and cold-trail methods that minimize soil disturbance. If a line is constructed, good methods include using explosives, minimizing the width and depth necessary to check the fire, and contouring along slopes rather than up or down. Burned trees can be left uncut unless they are likely to allow fire spread or are a safety concern (Mohr 1989), and snags can be preserved without compromising fire hazard by removing low branches or fine fuels beneath them (T. K. Brown and Bright 1997). Crews can use no-trace camping, minimize and restore helspots, rehabilitate ecologically damaged areas, and remove visible evidence of control (Mohr 1992–93; Mohr and Curtiss 1998). Considerable field-tested wisdom about MIST can be found on a diversity of agency Web sites.

Ignition Control

Reducing the ignition and spread of human-caused fires makes sense in general. Almost 40 percent of burned area on federal and Indian lands in the Rockies between 1980 and 2003 was from human-set fires, concentrated near towns and cities, campgrounds, roads, and other places used by people (chap. 10). Global warming and expanding human land uses increase the chance that accidents and arson can lead to large fires (e.g., Cerro Grande, Hayman).

The number of human-set fires can be reduced by limiting the probability of ignition and spread in time and space though a combination of education, design, and policing. During periods of deep drought and fire danger, entire forests have sensibly been closed to access. Closure to public access needs to become an acceptable management tool, perhaps through education, just as various levels of fire bans have become accepted. Punishment for arson and accidents could be increased and posted at entry points. Many fires still arise from escaped campfires (fig. 10.1c), suggesting that more education is needed, along with fire bans as danger increases; but a design solution may be more effective. Why not create many free places to camp and picnic in the most naturally fire-safe settings or in other settings using constructed fuel breaks, rather than continue costly emergency firefighting?

Roads and railroads also can be managed to reduce the ignition and spread of human-caused fires (R. F. Johnson 1963; C. C. Wilson 1979). Permanently closing roads is appropriate for many resource benefits (W. L. Baker and Knight

2000), and it is one of the best methods for reducing human-caused fires and landscape flammability across the Rockies (chap. 10). Road closures alone can reduce ignitions, but flammability is not reduced without road removal. Along open roads, areas of flammable fuel can be treated or designed so there is no stopping place. Reduction of fuels within 10 meters of a road sharply reduced fires in Southern California (R. F. Johnson 1963). A fuel break about 6 meters wide on each side of railroad tracks also reduced ignitions (C. R. Crandall 1980). Other ideas include planting less flammable vegetation on the margin of the road or railroad track and using signs to warn people of fire danger (R. F. Johnson 1963; C. C. Wilson 1979).

Safe Settings for Housing and Infrastructure

Houses and infrastructure are too often sited without considering fire risk; but locations that are inherently fire prone could be avoided, and places that are safer could be favored. The most dangerous settings are where fire intensity is likely highest and fire rotation shortest. In the Rockies this includes tall shrublands and ponderosa pine and mixed-conifer forests (table 5.7). Settings subject to intense fire include all forests, but especially where wind is funneled or downsloping wind is favored. The safest settings are natural openings with shrubs or grass, where fire intensity is likely to be lower, although fire rotation may be short. Unfortunately, shrublands and grasslands are already degraded by land uses, and it would be ecologically detrimental for them to be further damaged. Adding houses inside existing development is ecologically most desirable, followed by clustering adjacent to existing development or in already altered or degraded vegetation, while minimizing fragmentation of the surrounding landscape (W. L. Baker 2000). It is also a good idea to surround development with constructed fuel breaks or to use natural fire breaks (fig. 2.8).

Aiding Postfire Recovery

The aftermath of fire has commonly been treated as an emergency that requires human assistance for recovery (e.g., seeding, erosion control), and postfire forests are often logged to offset loss of wood products, but postfire seeding and logging slow natural ecological recovery after fire. For postfire management to be more sustainable, approaches that aid natural recovery could be emphasized.

POSTFIRE SEEDING

Virtually all recent scientific assessments conclude that postfire seeding is ecologically damaging and relatively ineffective for erosion control (Robichaud, Beyers, and Neary 2000; Beschta et al. 2004; Beyers 2004; Keeley et al. 2006;

Noss et al. 2006). Postfire seeding is expensive; the Bureau of Land Management approved $11.4 million in 2007 toward seeding more than 195,000 hectares of sagebrush burned in northern Nevada and southern Idaho (Miller 2007). Nevertheless, postfire landscapes are often seeded, sometimes to combat invasive weeds but usually to reduce soil erosion.

Erosion can be high the first year after fire, but often returns to prefire levels within three to four years (Robichaud and Brown 2006). Some postfire erosion is natural and maintains upland landscapes and riparian areas, as it likely did under the HRV. Reducing postfire erosion may make sense if it is likely to be outside the HRV because the prefire ecosystem was damaged by land uses (Beschta et al. 2004), or if particular human resources (e.g., water sources) or natural values (e.g., preserving a special fish species) have an unacceptable risk of damage.

Seeding has had limited success in reducing erosion because seeded species are seldom abundant enough to substantially reduce erosion by the year after seeding, when erosion is highest (Robichaud, Beyers, and Neary 2000, 2006; Keeley et al. 2006). Success of seeding even a few years after fire has varied, with failure common in arid settings (L. E. Hughes 2006; Robichaud, Lillybridge, and Wagenbrenner 2006; Jessop and Anderson 2007).

Seeding has also introduced invasive species, which reduce the recovery of native plants. Nonnative grasses, which are cheap and widely available, are often seeded (e.g., Frischknecht and Plummer 1955; Clary 1988). Many are seeded transformers (table 10.5), which suppress the recovery of native plants, including native trees (table 11.2), and can escape and invade unburned vegetation (Keeley 2006). Mixes may also be contaminated with invasives (Keeley et al. 2006). Over one billion cheatgrass seeds contaminated 360,000 kilograms of Italian ryegrass seeded after the Cerro Grande fire (Barclay, Betancourt, and Allen 2004). Contamination may explain the higher cover of invasive species in seeded than unseeded areas in piñon-juniper fire sites in western Colorado (Getz and Baker 2008), and the positive association of nonnative cover with seeded grass cover in the Cerro Grande and Hayman fires (Hunter et al. 2006). Seeding invasive grasses had limited success in reducing postfire cheatgrass and nonnative forbs (e.g., Clary 1988; Goodrich and Rooks 1999; Ott, McArthur, and Roundy 2003; Jessop and Anderson 2007). Seeded nonnative annuals (e.g., Italian ryegrass, wheat) may only delay postfire weeds (Keeley 2004).

Seeding can also damage soils and increase fire hazard (table 11.2). Drill seeding turns over substantial fractions of soil surface, and tires or tracks damage or destroy soil crusts and soil structure (Evangelista et al. 2004; Hilty et al. 2004; Jessop and Anderson 2007). Seeded grasses are generally flammable, but the most flammable are annual grasses, which produce dense swards that dry up

Table 11.2. Adverse Impacts of Postfire Seeding, by Vegetation Type

Vegetation and Sources	Location	Seed Mix Contained Invasives	Seed Mix Contaminated with Invasives	Suppressed Native Understory Recovery	Suppressed Native Tree Regeneration	Damaged Biological Soil Crust	Increased Fire Hazard
Salt desert shrublands							
Beavers (2001)	W CO	Y	—	—	—	—	—
Jessop & Anderson (2007)	W UT	Y	—	—	—	Y	—
Sagebrush							
Clary (1988)	C UT	Y	—	N[a]	—	—	—
Hilty et al. (2004)	SW ID	?	—	—	—	Y	—
Ratzlaff & Anderson (1995)	S ID	Y	—	Y	—	—	—
Shinneman (2006)	W CO	Y	—	—	—	—	—
T. W. Thompson et al. (2006)	C UT	NY	—	—	—	—	—
Piñon-juniper woodlands							
Goodrich & Rooks (1999)	NE UT	Y	—	N[a]	—	—	—
Clary (1988)	C UT	Y	—	—	—	—	—
T. W. Thompson et al. (2006)	C UT	NY	Y[b]	—	—	—	—
Getz & Baker (2008)	W CO	N	—	—	—	—	—
Shinneman (2006)	W CO	Y	—	—	—	—	—
M. L. Floyd, Hanna, & Romme (2004)	SW CO	N	—	—	—	—	—
Ponderosa pine forests							
Geier-Hayes (1997)	C ID	Y	—	Y	—	—	—
Chong et al. (2003)	C CO	Y	—	—	—	—	—
Hunter et al. (2006)	C CO, N NM	Y	Y	—	—	—	Y
Barclay, Betancourt, & Allen (2004)	N NM	Y	Y	Y	Y	—	Y
Hunter & Omi (2006)	N NM	Y	—	Y[c]	Y[c]	—	—

Douglas-fir forests						
Geier-Hayes (1997)	C ID	Y	—	Y	—	—
N. Rockies mixed-conifer forests						
Crane, Habeck, & Fischer (1983)	W MT	Y	—	Y	—	Y
Toth (1992)	W MT	Y	—	Y	—	—
Lodgepole pine forests						
Lyon (1976, 1984)	W MT	Y	—	Y	—	—
Spruce-fir forests						
Geier-Hayes (1997)	C ID	Y	—	Y	—	—
Mixed mountain shrublands						
Frischknecht & Plummer (1955)	C UT	Y	—	—	—	—
Poreda (1992)	NE UT	Y	—	—	—	—
M. L. Floyd, Hanna, & Romme (2004)	SW CO	N	—	—	—	—

Notes: Based on studies conducted in and near the Rocky Mountains. Within a vegetation type, studies are arranged from north to south. A dash (—) means not studied.

[a]No controls, thus reported finding is inconclusive.

[b]One of two possible explanations.

[c]Likely, based on seed banks.

early, providing flammable material for a large part of the fire season (Keeley 2004; Barclay, Betancourt, and Allen 2004).

POSTFIRE LOGGING

Postfire logging further damages the physical environment and slows natural recovery. Logging varies, but typical ground-based logging increases soil compaction, erosion, and nutrient loss, which slow recovery and reduce productivity (McIver and Starr 2001; Karr et al. 2004; G. H. Reeves et al. 2006). Increased runoff and erosion lead to earlier peaks and higher flows in streams and increased sediment, particularly from logging access roads, that together damage aquatic systems (McIver and Starr 2001; Karr et al. 2004; G. H. Reeves et al. 2006).

Damage to biological legacies and recovering plants and animals slows natural recovery and favors invasive species. Unburned trees and shrubs, standing dead trees, down wood, and unburned patches are aboveground biological legacies that aid recovery by harboring surviving organisms and stabilizing the postfire environment (J. F. Franklin et al. 2000). Logging removes or damages legacies. Ground-based postfire logging damaged or destroyed tree regeneration (Roy 1956; Sexton 1998; Donato et al. 2006). Winter postfire logging reduced growth of tree seedlings by 17 percent, aboveground biomass by 73 percent, species richness by 30 percent, and shrub density by 41 percent two years after fire, relative to unlogged ponderosa pine in Oregon (Sexton 1998). These effects likely occurred because of a hotter, drier microclimate and soil compaction or physical damage. Many animals are adapted to, or dependent upon, standing dead trees after fire (chap. 4). With few exceptions, logging eliminates or reduces the birds most restricted to burned forests (Kotliar et al. 2002; Hutto 2006; Saab, Russell, and Dudley 2007). Invasive species may also be increased. Helicopter logging after fire in an Idaho ponderosa pine forest led to more cover of nonnative plants, including invasives (table 10.5), five years later (Acton 2002). Nonnatives were five times more frequent the year after winter logging than in unlogged ponderosa pine in Oregon, but this difference disappeared two years after fire (Sexton 1998).

Postfire logging also increases fire risk. Logging large dead trees leads to a hotter, drier, windier microclimate than exists in unlogged areas (chap. 10; Sexton 1998; Karr et al. 2004). Drier, windier microclimate can extend 15–30 meters farther into unburned forests if logging occurs (Hanson and Stuart 2005). High-severity fire in Oregon montane forests initially reduced finer woody fuels (less than 7.6 cm in diameter) that contribute most to fire spread, but logging brought loadings of these finer fuels near or above loadings in the prefire forest (Donato et al. 2006; McIver and Ottmar 2007). Elevated finer fuels may decline to levels of the unlogged forest within about 25 years (McIver and Ottmar 2007). Down coarse woody fuels (more than 7.6 cm in diameter) initially increased above both pre- and postfire levels, because of leftover branchwood (Donato et al. 2006; McIver and Ottmar 2007). Down coarse woody fuels were modeled to be below prefire levels by about 25 years after fire and logging (McIver and Ottmar 2007). Thus, logging may elevate the risk of a spreading fire for 10–25 years, because of more fine woody fuels, but reduce flame length, smoldering combustion, and soil heating thereafter, because of reduced coarse woody fuels (Reinhardt and Ryan 1998; McIver and Ottmar 2007).

PROMOTING POSTFIRE NATURAL RECOVERY

The year after fire often sees few recovering plants (fig. 6.3a), but this is normal and not usually an ecological emergency. Many native plants are damaged by fire but recover from surviving underground roots, stems, or seeds within a few years (chaps. 3, 6–9). Early postfire habitat supports considerable biodiversity, but natural postfire habitat has become rare because of widespread postfire treatments (Hutto 2006; Lindenmayer and Noss 2006; Noss et al. 2006), and it would be worthwhile to explicitly designate these areas for natural recovery.

Seeding is not generally ecologically advisable after fire. If native plants were healthy and common before a fire and weed populations were not high, facilitating natural recovery is likely the best postfire management (Clary 1988; Ratzlaff and Anderson 1995; M. L. Floyd et al. 2006; Keeley et al. 2006; Freeman et al. 2007). It is substantially less costly and optimizes recovery of native plants adapted to the site. Natural recovery of postfire ecosystems can be enhanced by reducing effects of land uses, such as livestock grazing and access by roads (Karr et al. 2004).

Postfire seeding of native plants may be appropriate in two situations. First, seeding may supplement direct control of invasive plants in vulnerable postfire settings. Vulnerable locations are generally known (box 10.3), and a weed-risk model can assess vulnerability to particular invasives (M. L. Floyd et al. 2006). Weed populations and native plant diversity in unburned areas near a burn may also suggest whether postfire weeds are likely to be problematic, and damage and likelihood of recovery of perennial plants can be examined inside the burn (fig. 11.9). Field methods for evaluating burn severity are presented in Ryan and Noste (1985) and Goodwin and Sheley (2001). Invasives often initially appear in small patches, which expand and coalesce within a few years after fire, unless the invasives are controlled (Shinneman 2006). Seeding alone is unlikely

Figure 11.9. Fire border with unburned native bunchgrasses. The unburned bunchgrasses are on the right, and charred but unconsumed bases of bunchgrasses are on the left, indicating a relatively low-intensity burn and the likelihood of grass survival. This is the site of the Murphy Complex fires of 2007 in northern Nevada and southern Idaho.
Source: Photo by the author.

to fully control invasives—direct attack is needed (M. L. Floyd et al. 2006; Freeman et al. 2007)—but seeding might help. Seeded native grasses reduced postfire nonnative forbs, without limiting recovery of native plants, after fires in Mesa Verde National Park, Colorado (M. L. Floyd et al. 2006).

The second situation in which postfire seeding of native plants may be appropriate is to allow missing or reduced species to be restored in degraded vegetation. Competition is low after fire, and burned seedbeds may favor seed germination, providing an enhanced opportunity for successful seeding. Specific missing species for seeding a particular site can be identified based on the flora in *reference areas*, areas less affected by land uses (e.g., Shinneman, Baker, and Lyon 2008).

Where seeding is considered, agencies and scientists are increasingly favoring locally adapted native plants (Richards, Chambers, and Ross 1998) and low-impact seeding methods. "Native" species grown from nonlocal seeds can contaminate local gene complexes; thus, broad postfire seeding of nonlocal natives is inadvisable (Hunter et al. 2006; Keeley et al. 2006). Although growing native seed from a variety of settings can partly offset this concern, it is better to collect and use local seed from wild native plants, which could be facilitated by identifying and protecting wildland seed-growing areas. Further research is needed to improve seed mixes, by identifying and testing native species for competitive ability against particular invasives (Getz and Baker 2008). Aerial seeding may be less effective, but it is less damaging to soils (Evangelista et al. 2004). Seeding requires longer-term removal of livestock grazing, recreation, and other land uses that can damage emerging populations (Evangelista et al. 2004; Eiswerth and Shonkwiler 2006).

Where postfire erosion control is warranted because of high local human or natural resources, the best technique is to fell a fraction of dead trees, lay them along contours, and add needles from scorched trees (Robichaud and Brown 2006). Straw mulch is effective (Robichaud and Brown 2006) but reduces native plant recovery, including tree regeneration, and can be a source of invasives (Chong et al. 2003; Kruse, Bend, and Bierzychudek 2004). Water-repellent soils may occur naturally after fire in some cases but are short-lived; water repellency cannot be alleviated by postfire mechanical treatments without leaving much longer-lasting damage from soil compaction and increased erosion (Beschta et al. 2004). In semiarid ecosystems, such as piñon–juniper and sagebrush, the biological soil crust needs protection and restoration, given its ability to limit soil erosion and discourage invasives (Evangelista et al. 2004). Several authors further discuss how to enhance postfire recovery (see Beschta et al. 2004; Karr et al. 2004; Lindenmayer and Noss 2006; Noss et al. 2006; G. H. Reeves et al. 2006).

LIVING WITH FUTURE FIRE FROM INTERFACES TO WILD LANDSCAPES

This section reviews methods for living with fire along a gradient of land uses (fig. 11.1), focusing on fire defense in wildland-urban interfaces, ecological restoration in managed landscapes, and wildland fire use in wild landscapes.

Interface Landscapes: Fire Defense

Interfaces occur between wildlands and many land uses (e.g., agricultural, industrial, roads, power lines), but the wildland-urban interface (WUI) is very significant for fire, as it contains high-value flammable property, is a source of ignitions, and is expanding (table 11.1). WUIs have expanded 52 percent since 1970 (Theobald and Romme 2007). Government policy after World War II favored moving people out of cities, and this led to expanded WUIs and few restrictions on the placement of houses in fire-prone settings (Kennedy 2006).

Law, policy, and science failed to define where development is inappropriate relative to fire and, until recently, failed to require or encourage fire-safe design. Thus, much of the WUI is poorly located and badly designed relative to fire. This poorly placed WUI also limits wildland fire use on adjoining public lands and is a source of ignitions that burn public lands.

There are three approaches to fire and the WUI. The first is to reduce further expansion by means of regulations and incentives, which are necessary since about 65 percent of total WUI area in the West is privately owned (Theobald and Romme 2007). Limiting further development of WUIs is important, because only small percentages (4–21 percent) of potential WUI in the Rockies are developed (table 11.1). Risk maps can help identify key parts of potential WUI that have the highest fire risk (e.g., Haight et al. 2004; Theobald and Romme 2007), where policy and law might best prevent or reshape development. Some insurance companies no longer insure homes in high-risk settings, and others charge higher rates and require fire-safe construction (Flaccus 2007). Local land-use planning can limit expansion of WUIs by zoning, designated growth boundaries, transfer of development rights, development fees and incentives, and other means (Kennedy 2006). Conservation NGOs can acquire easements that limit development. The high costs of fire control in the WUI, including the costs of fuel reduction and fuel breaks, have been externalized to taxpayers but should be internalized as part of development or home cost (Theobald and Romme 2007).

The second approach is that new and existing development in the WUI can be designed or retrofitted to lower fire risk. New development can be located in ways that minimize fire risk and ecological damage (see "Safe Settings for Housing and Infrastructure" above). A buffer or community fire planning zone (CFPZ) of about 400–800 meters (one-fourth to one-half a mile) around existing and new development is a key zone for planning protection from fire. Within the CFPZ, fuel reduction can reduce the possibility of sustained crown fire, reduce flaming firebrands that can ignite homes, and give firefighters space to control approaching fires (Wilmer and Aplet 2005). The CFPZ can also intentionally contain greener or wetter vegetation (irrigated agriculture, golf courses, parks, greenbelts, constructed wetlands, canals), roads, fuel breaks, or other areas that can potentially limit fire spread into and out of the development (Plevel 1997; Keeley, Fotheringham, and Moritz 2004). The CFPZ need not be a dead zone; it can be beneficial to developments, providing open space, recreational opportunities, and nature protection (see chap. 12).

Finally, and in many ways most important, individual home owners are most responsible for lowering fire risk using firewise construction and design

(Long, Wade, and Beall 2005; www.firewise.org). Modeling and experiments show that the house itself and the 40 meters around it determine home ignitability, essentially independent of fire behavior in the surrounding wildlands (J. D. Cohen 2000). As this home ignition zone controls ignitability, lowering the risk of home ignition is the responsibility of individual home owners or groups of homes (J. D. Cohen 2000). Homes lacking firewise construction tend to ignite, increasing fire intensity and fire spread and leading to the ignition of other homes and vegetation across WUI landscapes (Spyratos, Bourgeron, and Ghil 2007).

The upshot is to minimize WUI and instead cluster development where it already exists. As Roger Kennedy (2006, 260) put it, "It is time to lock the doors to the treasury against further raids to finance dumb growth, the kind of growth that destroys precious places, including natural communities as well as human communities, and puts lives at risk."

Managed Landscapes with Land-Use Legacies: Ecological Restoration

In managed landscapes, where land uses (e.g., logging, grazing) are ongoing and land-use legacies exist, ecological restoration may be the most effective approach to increase the sustainability of the people-fire relationship. Legacies can be complex, but they must be addressed for restoration to be effective. For the Rockies, assessments of ecological restoration needs relative to fire and vegetation are available (W. L. Baker and Shinneman 2004; W. L. Baker 2006b; W. L. Baker, Veblen, and Sherriff 2007). Ecological conditions that have been altered or are outside the HRV are primarily from human-caused fires, logging, livestock grazing, roads, and development, with contributions from fire suppression, invasive species, and, increasingly, global warming (box 11.1; chap. 10). Ecological restoration requires rectifying alterations produced by these land uses and reforming these land uses so that restoration does not become part of an endless cycle of degradation and restoration (W. L. Baker, Veblen, and Sherriff 2007). Alterations are diverse and require many kinds of actions (e.g., controlling invasive species, reducing ignitions, closing roads), including both passive (e.g., reducing the stocking rate of livestock grazing) and active approaches (e.g., planting missing species). The current focus on fuel reduction is misdirected, as it is not a general need in the Rockies, as discussed previously, and, even where appropriate, is only one of many actions necessary to achieve restoration (box 11.1).

Given the major land-use effects (chap. 10), the most commonly needed restoration actions and land-use reforms include (1) in logging areas, broadcast burning of slash and avoiding logging that creates high-contrast edges,

fragments forests, removes large fire-resistant trees, or damages vegetation recovering after fire; (2) in grazing areas, reducing stocking rates of livestock and
refocusing grazing to maintain bunchgrasses and forbs needed for fire spread
and for competition that limits tree regeneration; (3) closing and rehabilitating
roads to reduce human-caused fires and landscape flammability from edge effects; (4) using minimum-impact methods for fire control; and (5) controlling
invasive species before fires occur, to reduce potential postfire expansion, and
using direct control after fire.

Increased fire from global warming could hamper effective ecological restoration, and more attention to this threat is warranted in areas where restoration is undertaken. Landscapes with enhanced flammability (e.g., slash, cheatgrass, edge-affected forests near logging and roads) or increased potential
ignitions (e.g., roads) are some of the places where increased fire is appearing
from global warming (chap. 10) and that most need attention near restoration
areas. Intentionally reducing flammability and the possibility of ignition is logical in such areas (see "Ignition Control" above). Land managers can also reduce
vulnerable fuels near housing, infrastructure, roads, and other locations where
human-set fires could spread into restoration areas, as well as restricting access
to restoration areas during droughts.

Wild Landscapes with Few Legacies: Natural Fire

Compared to roaded areas, wilderness and parks have few human-caused fires
(DellaSala et al. 1995; USDA Forest Service 2000; Yang et al. 2007), and their
remoteness, topographic complexity, and higher elevations have led to less effective fire control (DellaSala and Frost 2001). Their fire regimes and vegetation
remain relatively unaltered compared to those of logged and roaded areas (DellaSala and Frost 2001; Rollins, Swetnam, and Morgan 2001). The Interior Columbia Basin Assessment, in Idaho, western Montana, and western Wyoming,
concluded that "subbasins having the highest forest integrity index were largely
unroaded" (Hann et al. 1997, 833–34).

Rocky Mountain wilderness ecosystems may have experienced minor effects from fire control, but they are not generally outside their HRV because of
fire exclusion (chaps. 6–10). Fire exclusion in ponderosa pine forests in wilderness was thought to have led to unnaturally dense forests with too many shade-
tolerant trees and with abnormal high-severity fire (Barrett 1999; Keane et al.
2006). However, evidence shows that most ponderosa pine forests in the Rockies are not outside the HRV for tree density or fire severity and do not require
thinning (W. L. Baker, Veblen, and Sherriff 2007; chap. 7). Intentional fire con-

trol did occur in parks and wilderness in the Rockies but was not very effective (chap. 10) and did not cause clearly detectable departures from the HRV (Romme and Despain 1989a; Rollins, Swetnam, and Morgan 2001; chap. 10). Other land uses also were seldom extensive, and alterations from the HRV were generally minor. There is little conflict between wildness and naturalness in managing wilderness fire in the Rockies (D. J. Parsons, Landres, and Miller 2003), as little active management is needed. Minor active management, such as restoring areas directly damaged by people (e.g., from campsites) or controlling invasive species, is needed but can be minimized.

Calls for active management of resources, including resources in parks and wilderness, have been part of American culture since its inception and often have been promoted as ecologically based (Bonnicksen 1989; Arno and Fiedler 2005). Unfortunately, unsupported conclusions about fire history and land-use effects (table 10.9) and a utilitarian tendency to want to tinker with and fix things (C. E. Bock and Bock 2000) have fed calls for active management, which in some cases has damaged and could further damage wilderness, parks, and re-search natural areas. For example, researchers thought thinning was needed in piñon-juniper in wilderness in a national monument (e.g., Sydoriak, Allen, and Jacobs 2000) because of a misinterpretation of fire history (W. L. Baker and Shinneman 2004; W. L. Baker 2006a). J. K. Brown et al. (1994) compared com-posite fire intervals directly to fire rotations, which is not supported (chaps. 5, 10), wrongly concluding that the Selway-Bitterroot Wilderness had been af-fected by fire exclusion in a paper cited as evidence that wilderness areas are threatened by fire exclusion (D. J. Parsons, Landres, and Miller 2003). Uncritical acceptance of the idea of fire exclusion (chap. 10) has led to misdirected and potentially damaging active restoration proposals.

Although wilderness and park landscapes are less altered, these landscapes will most benefit if our influences are minimized and wildland fire use is ex-panded. Management-ignited fires, other human-set fires, and suppression of natural ignitions are all inappropriate, as these would not leave wilderness gen-erally untrammeled by human impacts. Indians likely caused some accidental or intentional fires over the thousands of years they lived in the Rockies, and oc-casional accidental fires in today's wilderness would likely be within the HRV (chap. 10). However, human-set fires were historically uncommon outside a few particular settings (W. L. Baker 2002), and prescribed burning programs are thus incompatible with the HRV. In parks and wilderness, wildland fire use is the best current policy, even though some, perhaps many, of the large fires that historically shaped Rocky Mountain landscapes were inherently unstable, beyond control, and truly wild. Wildland fire use is like having a grizzly bear on a loose leash. It is not natural fire, but it is the next best thing. To allow truly un-controllable, severe fire requires foresight and courage.

Good design would help. Most parks and wilderness areas were not de-signed with maintaining natural fire or other natural disturbances in mind. Some expansion and redesign would benefit nearly all of them and would also minimize threats to people and infrastructure. Some landscape design principles for large disturbances in reserves are outlined in W. L. Baker (1992a). First, it is logical that if reserves are larger, fewer fires will burn from them onto adjoining lands. In the Rockies, this means several tens of thousands of hectares are best, given that fires historically and over the last century reached 50,000–100,000 hectares (table 5.2). Simulation modeling can aid in estimating the needed re-serve size (Leroux et al. 2007). Second, it is best to have fire run-out zones, in which reserve areas extend to natural fire breaks (fig. 2.8), meadows, or other places likely to have low-severity fire or to constructed fuel breaks, perhaps in the most probable direction of spread under severe weather. Third, high fire-initiation or preferred spread locations for fires affecting the reserve are best contained inside the reserve, or the reserve's fire regime will be difficult to maintain. A reserve design that includes a buffer zone of less developed land is preferable to one that has a sharp boundary with intensive land uses.

SUMMARY

More sustainable landscapes for people and fire are needed now more than ever, given the high cost and relative ineffectiveness of the current emphasis on fuel reduction and fire control. We have a legacy of increased ignitions by people and elevated flammability from past land uses, at a time of expanding wildland-urban interfaces and increasing fire from global warming. I suggest modifying a variety of tools to help create landscapes that function better for both people and nature. Fuel reduction and fuel breaks are shown to have limited effective-ness and deleterious impacts on nature and are best used only in defensive po-sitions adjacent to housing and infrastructure. In landscapes away from homes and infrastructure, wildland fire use is cost effective and beneficial for nature. Prescribed burning, where wildland fire use is not possible, could contribute to sustainability if it is refocused on emulating natural fire regimes. Minimum-impact suppression tactics make a significant contribution to sustainability where fire must be controlled. Human-caused ignitions have added to fire in the Rockies but could be reduced by treating fuels near roads and infrastructure, by channeling camping into safe sites, and through education and policing. People could also be educated about the severe danger of placing a home in a forest where fire can be high intensity, and policies could channel development into safer settings. Wildland–urban interfaces can be limited by using planning tools (e.g., growth boundaries), and fire risk can be lowered by using fuel breaks, green forests, open space, and natural fire breaks, along with firewise construction by home owners.

After fire, seeding and logging reduce or prevent natural recovery, but recovery can be enhanced by controlling invasive species and reducing livestock grazing and other land uses. Ecological restoration, with attention to increased fire from global warming, is appropriate in managed forests. Most do not need fuel reduction but would benefit from a variety of other restoration actions. The Rockies have parks, wilderness, and other protected areas where fire regimes remain fundamentally natural, little active restoration is needed, and wildland fire use is possible and would be ecologically beneficial.

CHAPTER 12

Toward a Better Relationship between People and Fire

The Rocky Mountains are a favorable setting for developing a better relationship between people and fire. These mountains have a relatively sparse human population and have long contained more nature-dominated landscapes than most of the United States outside Alaska. Wildland fire in the Rockies is still strongly influenced by natural processes, at least away from human infrastructure (chap. 10). And, after a century of scientific research, we know more about natural fire in the Rockies than in many places. In chapter 11, I reviewed the tools available today for living with fire and their application across a gradient of land uses. Here, I highlight what we have learned about Rocky Mountain fire regimes and the response of plants and animals to fire from a century of scientific research. I suggest how to use some of the tools to begin immediately to create more sustainable landscapes that will allow us to have a better relationship with fire.

FIRE ECOLOGY IN ROCKY MOUNTAIN LANDSCAPES

Themes that have emerged from a century of scientific study of Rocky Mountain wildland fire include (1) the need for a multicentury landscape perspective to understand fire; (2) a large variability in fire at multiple spatial and temporal scales, including fluctuating landscapes from drought-linked episodes of extensive fire followed by long interludes of recovery; (3) the ability of native Rocky Mountain plants and animals to survive or recover well after fire; and (4) a general lack of evidence for a fire-exclusion effect, although climatic and land-use changes match fluctuations in area burned in the last century.

Historical evidence shows that we cannot understand fire in these mountains without observing a large land area over a long time period (chaps. 5–9). The landscape spatial scale and the multicentury temporal scale, well beyond the usual scales of human perception, are essential to understanding fire in the Rockies. A single person can experience only part of the spectrum of variation in fire, as fire rotations are commonly in the 100- to 500-year range in Rocky Mountain ecosystems (table 5.7). Most fires experienced in a human lifetime in a particular area are small, but fires in the Rockies have reached 100,000 hectares or more many times in the last century and a half (table 5.2). Relatively few of these large fires, patchily distributed across the region, make up most of the burned area. Only about 50 fires in excess of 15,000 hectares accounted for more than half the burned area in the Rockies over the past few decades (chap. 5). Observations in small areas or over short periods can consequently be quite misleading (table 10.9), as they can easily miss the infrequent fires that account for most of the burned area. To accurately reconstruct the historical range of variability (HRV) of fire requires a large and long view.

Scientific research and land management at these more extensive scales of thinking and understanding have expanded greatly but remain incomplete, in part because a lingering scientific tradition of small-plot studies and frequency-based measures of fire has held back understanding. Fast computers, remote sensing, geographic information systems, and the emergence of landscape ecology have facilitated scientific inquiry over large spatial extents in the last two decades, but fire history research at landscape scales is still limited (chaps. 5–9). Variation in fire across landscapes has been missed by past methods, which focused on counting and dating fires in small plots in intentionally chosen locations, often in old forests with long fire-scar records.

Frequency-based measures of fire, used in these small-plot studies, have also fostered extensive misunderstanding of the role of fire, including the idea of frequent fire (chap. 5). This idea derives from counting fires, which makes little sense, as it fails to take into account the enormous variation in fire size (fig. 5.3). Simply counting fires tells us that there are indeed many of them (e.g., 56,350 fires between 1980 and 2003 in the Rockies), but 55,228 of those 56,350 fires (98 percent) together accounted for only about 4 percent of the total area burned over the period. This is a fire rotation, and average mean fire interval across all points in the landscape, of about 6,900 years (equation 5.1 and data in chap. 5). Summing the areas of the 55,228 small fires shows that this apparently large number really means only 1 fire per 6,900 years on average at any particular point in the Rocky Mountains, not at all a frequent-fire regime. This example illustrates that counting fires, or using frequency-based measures (e.g., composite fire intervals), provides a misleading perspective on the importance of fire. Unfortunately, the idea that fires were frequent and of low severity across points in dry forests under the HRV is common, and much effort and funding have gone into efforts to restore and maintain frequent fires (chap. 7). However, fires were not actually frequent, on average, at points in any Rocky Mountain ecosystem under the HRV, based on population mean fire intervals and fire rotations (table 5.7). It is time for small-plot studies and frequency-based measures, along with the idea of frequent fire, to be replaced by area-based measures and methods (chap. 5).

A primary finding of the limited landscape-scale studies in the Rockies is the extensive variability of fire in space and time. Variation in fire size, fire severity, fire intervals, and nearly every other attribute of fire is manifest when large land areas and long time periods have been studied (chaps. 6–9). Variability occurs at several scales in the Rockies. The location of large fires varies regionally from year to year (fig. 5.4), shaped by patterns of drought, lightning, and vegetation. Within a landscape, patchy fire location and fire-size variation can lead to large short-term variation in time since fire and fire interval (fig. 5.10c). Individual fires may be of high severity over large areas, but they also can vary greatly in severity and leave surviving trees or tree groups, as well as a zone of scorched trees or scattered spot fires (figs. 3.15, 6.8), potentially creating thousands of discrete patches from a single fire (fig. 7.2). Fires occur every year, but most of the total burned area is from a small percentage of the fires that are large (box 3.1) and that tend to occur in drought years. Decadal episodes and individual years of severe drought (e.g., 2000, 2002), linked to oceanic and atmospheric conditions, historically and recently promoted large fires that are responsible for most of the total burned area in the Rockies (chaps. 2, 5).

Variation in fire sizes and intervals leads to fluctuating landscapes characterized by episodes of extensive fire followed by long interludes of small fires and general recovery (box 3.1). Rocky Mountain ecosystems that recover quickly relative to their fire rotation (e.g., prairies—table 5.7) spend long interludes in a mature condition, but slow-recovery ecosystems (e.g., many forests) may be recovering during large parts of the interludes. Episodes of extensive fire are often synchronized during regional droughts, leading to postfire forests recovering in unison across large land areas. Large areas of Rocky Mountain forests are still recovering from severe mid- to late-nineteenth-century fires. Recovery is natural, but the pattern of change expected from fire exclusion is similar, creating confusion. Fluctuating landscapes are normal in fire regimes (box 3.1) and mean that wildlife habitat, clean water, old-growth forests, and other ecosystem services desired by people quite normally change rapidly after long interludes of relative stability.

Native animals and plants survive or recover well after fire and appear generally adjusted to fluctuating landscapes. Some clearly are stimulated by fire, but fire dependence is rare; most plants and animals simply have functional traits that allow individuals or populations to persist or recolonize after fire. Animals, particularly those that are less mobile are sometimes killed by fast-moving fires, but most survive by moving away or taking refuge (chap. 4). Fires, no matter how severe, are not catastrophes for most animals, and postfire habitat after even severe fires may favor a flush of insects and birds. However, not just any fire or mosaic of burned patches is beneficial to animals, as they may have specific landscape requirements. Plants cannot escape approaching fires, of course. Most Rocky Mountain trees are killed and must recolonize by dispersing seed from nearby unburned areas. Resprouting and seed banks are rare among Rocky Mountain trees, but many common shrubs, graminoids, and forbs resprout readily after fire, and forbs and some shrubs may have seed banks (chap. 3). Rapid resprouting of surviving plants dominates after fires in most Rocky Mountain ecosystems; seed is a rarer source of postfire plants. Rocky Mountain plants generally have traits that allow survival and recovery within a few years after fires of any size or severity. However, many trees and some dominant shrubs that rely on postfire dispersal may recolonize more slowly, at times leading to multidecadal lags in regeneration that allow openings to persist for decades. These semipersistent, fire-created openings add diversity to Rocky Mountain landscapes.

People have altered fire regimes, but the best available data suggest that fire may not have declined in the Rockies relative to the pre-EuroAmerican era, either overall or in most individual vegetation types (chap. 10). Fire rotation has been estimated as about 275 years in the Rockies as a whole since 1980 and about 247 years in the northern Rockies over the last century, and both figures are near the middle between the low (140 years) and high (328 years) estimates of fire rotation for the Rockies under the HRV (chap. 10). These estimates suggest that since EuroAmerican settlement, fire control and other activities may have reduced fire somewhat in particular places, but a general syndrome of fire exclusion is lacking. Fire exclusion also does not accurately characterize the effects of land uses on fire or match the pattern of change in area burned at the state level over the last century (fig. 10.9). In contrast, fluctuation in drought linked to oceanic and atmospheric conditions appears to match many state-level patterns in burned area over the last century. Land uses that also match fluctuations include logging, livestock grazing, roads, and development, which have generally increased flammability and ignition at a time when the climate is warming and more fire is coming.

GOOD INTERFACES FOR PEOPLE AND NATURE

Faced with global warming and more flammable landscapes, at least three dreams and a nightmare are before us. Pyne's (2004) "tending fire" story is the dream of people preventing large, severe fire through continual management. The story had Indians managing, and thus taming, fire everywhere through ongoing use, which we are advised to emulate today with expansive prescribed burning. However, this story and dream fail in the Rockies, where Indians were generally few and impacts of their burning likely minor except in a few places (chap. 10). The second dream is the fire control dream, which promises comfortable living in fire-prone settings, backed by expansive, taxpayer-funded, fuel-reduction programs in the surrounding forests and a cadre of well-trained firefighters with tankers, foams, and other suppressants. This dream is challenged by the lack of any clearly detectable decline in burned area in the Rockies (chap. 10) and the enormous fires (table 5.2) that at times still burn into communities. The third is a John Muir dream of nature unleashed, of wildly spreading fires with enormous, beautiful, and powerful flames watched safely from beneath old-growth trees on a wilderness mountaintop. Unfortunately, the realities of expanding human-caused fires, cheatgrass, and diminished wild landscapes constrain this dream today. All three dreams have pleasing overtones but fail in some way. What is more realistically before us is a potential nightmare: large-scale insect outbreaks, drought-killed vegetation, expanding invasive plant populations, arson and accidental ignitions, and wildland–urban interface fire ramped up by global warming, with houses in flames, fleeing residents, exploding propane tanks, sirens, hoses, foam everywhere—all live on the nightly news.

Something better is needed soon. A mosaic of people and nature at landscape and regional scales could be created, a mosaic better designed for fire than current landscapes, which have been shaped willy-nilly by individual property sales and disparate land-use decisions. We have limited time, as an increase of burned area by two to five times could enable fire to sweep across much of the Rockies in as little as the next half century. Individual communities, of course, could burn sooner, which provides incentive for action. Although action can also be taken to improve conditions in managed and wild landscapes (chap. 11), the most important immediate place for action is in interfaces between people and nature. The adage "Good fences make good neighbors" could be changed to "Good interfaces make good neighbors of people and nature."

Designing better interfaces requires taking steps on both the nature and the community sides. On the nature side, some previous actions no longer make sense. Fuel reduction is not ecologically needed in most Rocky Mountain eco-systems (chap. 10), and it is also rather futile, given the magnitude of the pro-jected increase in fire and the inability of fuel reduction to prevent large fires under extreme weather (table 5.2; chap. 11). Fuel reduction outside of commu-nity fire-planning zones (chap. 11) is a waste of funding that would be better di-rected at interfaces. Also, since fire may not have declined relative to the HRV and may increase several times in coming decades, there is no need to conduct prescribed burns to offset a perceived fire deficit or to regenerate shrubs or trees. A focus on restoring tree populations (e.g., tree density) may also be mis-directed, as restored trees may burn anyway. Postfire logging and seeding are generally inadvisable; as global warming continues, recovering vegetation may face a drier, more stressful postfire environment; logging, seeding, and other treatments may further hamper natural recovery (table 11.2). In addition, post-fire recovery of graminoids, forbs, and shrubs in many Rocky Mountain eco-systems is generally from resprouting, with seed playing a minor role (chap. 3).

Since fire is increasing, and postfire seeding often slows natural recovery, ecological restoration is needed well before fire occurs, to increase the ability of ecosystems to recover after fire (chap. 11). An immediate need is to control in-vasive weeds that might expand with increased fire. Many other restoration ac-tions and land-use reforms also can increase the ability of vegetation to recover after fire, and the most effective actions, such as closing and rehabilitating roads or reducing livestock grazing, are underused (box 11.1). It may also be worth-while to use fuel breaks to directly protect particular high-value sites (e.g., old-growth remnants, vulnerable populations of sensitive species) that could be damaged by increased fire.

On the community side, reducing risk is most needed in and around pri-vate and community property right away. Most of the need is inside and on the periphery of communities and developments. Open spaces with low fire risk (future parks, recreational fields, golf courses, low-stature greenspace) could soon be created on peripheries and linked into fuel breaks or moist, green forests where possible (chap. 11). Dedicated open spaces, with low-risk features, could provide initial and enduring protection. A linked system of low-risk open space and natural or constructed fuel breaks or other fire-resistant features (e.g., wide cleared trails, clearings) also serves as a logical community growth boundary, which can prevent an expanding WUI. Reducing the expansion of its WUI is one of the most powerful steps a community can take to allow its members to coexist with wildland fire. Fuel reduction outside the community fire-planning zone is not needed. Fuel reduction on public land inside the zone is worthwhile, but adjoining property owners should have to adopt fire-wise construction methods (chap. 11). Communities in general would be wise to adopt fire-wise methods, particularly if they are near forests or where it is difficult to obtain or construct suggested defensive open spaces or breaks.

Intermix cannot be saved. It was foolish to build homes in the midst of natural vegetation in the Rockies, where nearly all the vegetation is subject to high-severity fire. Like smoking, it may be a right, but it is now recognized to have negative public consequences. Intermix is a large potential source of human-caused fires that may burn public or other private lands. When fires approach housing in intermix, firefighters will likely continue to practice triage, as unprotectable housing is common, access often poor, and danger for firefighters high. A total ban on intermix is warranted. Preventing or reducing intermix serves both people and nature.

It is sensible to reduce human-caused fires in the WUI and in wildlands, as such fires already account for almost 40 percent of burned area in the Rockies (chap. 11). Ideas for reducing fire risk along roads, in campgrounds, and near other infrastructure (chap. 11) could be implemented quickly, providing immediate and long-term benefits. Increased warning signs, phased reductions in fire use, smoking bans, and intentional closures of public lands will likely be needed. The public could be educated about the potential magnitude of increased fire that may be coming, to gain support for closures and other inconveniences needed to reduce risk.

The current fire control workforce could be redirected to help create and maintain sustainable landscapes. Firefighters will be needed during the decades of global warming, even if communities are well protected, to set backfires and extinguish spot fires in the WUI. Perhaps they can eventually become wildland-fire-use teams, who will always be needed to monitor fires and to control some of them, particularly until we have people-nature mosaics that allow more wildland fire use.

The historical evidence is compelling that wildland fire in the Rockies has nearly always been too powerful to be controlled during severe fire weather, when most of the burning occurs. Industrial-scale, brute-force control of fire is very costly, requires large-scale alteration of nature, and is unlikely to generally succeed in creating sustainable landscapes. The alternative I propose is immediate action and enduring changes that seek instead to keep people and their infrastructure some distance from nature and wildland fire, minimize alterations of nature, and create landscapes that require relatively little ongoing, costly maintenance. The needed actions that can help us live sustainably with fire are generally beneficial for both people and nature, as they can help us to contain growth, include open space in our communities, and protect nature. Communities can quickly and often relatively cheaply redesign and limit their interfaces with wildlands. If we begin now to create landscapes that keep us some distance away, protected by good fences, then nature, fire, and people can become good neighbors.

Appendix A. Common Rocky Mountain Trees and Their Functional Traits That May Increase Persistence

Common (& Latin) Names	I				P		C		D			
	Resprouting[a]	Mature Height (m)[b]	Foliage Flammability[c]	Bark Thickness[d]	Decay Resistance[f]	Serotiny	Shade Tolerance[f]	Typical/Max. Age (yrs)[e]	Seed Mass (mg)[b]	Seed Crop Period (yrs)[b]	Age of First Seed (yrs)[b]	Seedbed[g]
Black cottonwood (*Populus balsamifera* ssp. *trichocarpa*)	—	31	—	—	1		1	150/250	—	1[h]	10–20	—
Blue spruce (*Picea pungens*)		51	—	—	1		3	150/350	4.3	2–3[h]	20	—
Bristlecone pine (*Pinus aristata*)		15	—	2	—		1	500/2500[i]	25.1(A)	—	20	—
Corkbark fir (*Abies lasiocarpa* var. *arizonica*)		34	4	1	—		5	—	20.4	2–3	50	—
Douglas-fir (*Pseudotsuga menziesii* var. *glauca*)		40	4	5	2		2	750/1200	19.6[a]	3–11	20	2
Engelmann spruce (*Picea engelmannii* & hybrids)		30	5	2	1		2	450/550	3.3	2–3	15–25	2
Grand fir (*Abies grandis*)		62	3	4	1		4	200/400	20.0	3	20	1
Limber pine (*Pinus flexilis*)		25	—	3	—		2	200/400	92.6(A)	2–4	20	—
Lodgepole pine (*Pinus contorta* var. *latifolia*)	R	46	2	1	2	X	1	120/300	4.8	1	5–10	2
Mountain hemlock (*Tsuga mertensiana*)		46	3	3	1		5	400/800	4.0	1–5	20–30	—
Narrowleaf cottonwood (*Populus angustifolia*)	X	20	—	—	—		—	100/200[a]	—	—	—	—

463

Appendix A. Continued

Common (& Latin) Names	Resprouting[a]	I: Mature Height (m)[b]	P: Foliage Flammability[c]	Bark Thickness[d]	Decay Resistance[e]	Serotiny	Shade Tolerance[f]	C: Typical/Max. Age (yrs)[e]	D: Seed Mass (mg)[b]	Seed Crop Period (yrs)[b]	Age of First Seed (yrs)[b]	Seedbed[g]
Oneseed juniper (*Juniperus monosperma*)	–	8	–	–	–		–	–	25.0(A)	2–5	10–20	–
Pacific yew (*Taxus brevifolia*)	R	12	3	–	3		5	250/350	29.4	1	–	–
Paper birch (*Betula papyrifera*)	X	21	–	1	1		4	100/140	0.3	2	15	–
Plains cottonwood (*Populus deltoides* ssp. *monilifera*)	W	30[a]	–	5[h]	1		1	100/200[a]	0.9–1.8[h]	1[h]	10[h]	–
Ponderosa pine (*Pinus ponderosa* var. *ponderosa*)		71	1	5	2		1	600/726	37.9	2–5	15–20	2
Ponderosa pine (*Pinus ponderosa* var. *scopulorum*)		35	1	5	2		1	–	34.6	2–5	6–20	2
Quaking aspen (*Populus tremuloides*)	X	31	1	–	1		1	70/200	0.1	4–5	10–20	2
Rio Grande cottonwood (*Populus deltoides* ssp. *wislizeni*)	X	31	–	5[h]	–		–	100/200[a]	–	1[h]	5	–
Rocky Mountain juniper (*Juniperus scopulorum*)		15	–	2	3		1	250/300	16.7(A)	2–5	10–20	–
Southwestern white pine (*Pinus strobiformis*)		39	–	–	–		2	–	169.5(A)	3–4	15	–
Subalpine fir (*Abies lasiocarpa* var. *lasiocarpa*)		34	4	1	1		5	150/250	13.2	2–4	20	2

Species										
Subalpine larch (*Larix lyallii*)	12	—	2	—	1	450/1000[h]	3.2	1–10	30	—
Thinleaf alder (*Alnus incana* ssp. *tenuifolia*)	9	X	—	—	—	—	0.7	—	—	—
Twoneedle piñon (*Pinus edulis*)	12	2	2	—	—	350/540	238.1(A)	2–5	25–75	—
Utah juniper (*Juniperus osteosperma*)	12	—	—	3	—	650/800	90.9(A)	2	30[h]	—
Western hemlock (*Tsuga heterophylla*)	77	5	3	1	5	400/600	1.8	2–8	20–30	1
Western larch (*Larix occidentalis*)	55	3	5	2	1	700/915	3.2	1–10	25	2
Western red cedar (*Thuja plicata*)	62	4	2	—	5	1000/1200	0.8–2.2[a]	3–4	15–25	1
Western white pine (*Pinus monticola*)	62	4	3	—	2	400/615	16.8	3–7	7–20	2
White fir (*Abies concolor*)	55	3	3	—	3	150/500	28.6	2–4	40	—
Whitebark pine (*Pinus albicaulis*)	33	—	1	1	1	500/1000[h]	175.4(A)	3–5	20–30	—

Notes: Traits can be identified as increasing individual tree persistence during or after a fire (I); population persistence after a fire (P); persistence in the postfire community through competitive ability (C); or persistence across the landscape, based on capacity to disperse seed into burned areas and germinate on burned seedbeds (D). Mature height, placed in the I category, also applies at the community level. Shading indicates values that increase persistence during or after fire. A dash indicates no information, while a blank indicates that a trait is lacking. Common and Latin names are generally from the USDA PLANTS online database (http://plants.usda.gov).

[a] Data are from the Fire Effects Information System, http://www.fs.fed.us/database/feis. R = rare, W = weak, X = known to occur.

[b] Data are from J. A. Young and Young (1992). Animal-dispersed seeds are indicated by (A).

[c] Data represent the means of ranks, increasing from 1 to 5, for foliage flammability and epiphyte receptivity derived from data in Minore (1979). Foliage flammability rates the ability of foliage to catch fire, which is increased by higher epiphyte (e.g., lichen) load, so the two are combined here.

[d] Data are from Flint (1925), Burns and Honkala (1990), Keeley and Zedler (1998), and the Fire Effects Information System. Bark thickness of mature trees increases from 1 to 5. The lowest value of 1 corresponds to an average of about 1 cm, and the highest value of 5 corresponds to about 7.5 cm.

[e] Data are from Loehle (1988). Decay resistance increases from 1 to 3.

[f] Data are from Loehle (1988) and Minore (1979), whose estimate is accepted first where available. A few values not given in these two sources are estimated from data in Burns and Honkala (1990). Shade tolerance increases from 1 to 5.

[g] Data are from Minore (1979) and R. B. Boyce (1985) and represent germination of seeds and early survival of seedlings on an undisturbed organic seedbed or charred postfire seedbed: 1 = favored by organic seedbed relative to charred seedbed, 2 = favored by charred seedbed relative to organic seedbed.

[h] Data are from Burns and Honkala (1990).

[i] Data are from W. L. Baker (1992b).

Appendix B. Common Rocky Mountain Shrubs and Their Response to Fire

Common Name	Latin Name	Seed Source[b]	Seed Weight (mg)[c]	No Sprouting	Branch	Bud on Stem or Root	Lignotuber	Rhizome	Root Crown	Stolon	High-Severity Fire Reduces Sprouts	Postfire Flush?	Time to Recover (yrs)
Alderleaf mountain mahogany	Cercocarpus montanus	Rare	7.7						X		Yes	No	5+
Antelope bitterbrush	Purshia tridentata	Off	29.4				V		V		Yes	No	5, 15–30
Basin big sagebrush	Artemisia tridentata ssp. tridentata	Resid., off	0.3	X							None	No	—
Big sagebrush	Artemisia tridentata ssp. spiciformis	—	—	—							—	No	—
Big sagebrush	Artemisia tridentata ssp. xericensis	—	—	—							—	No	—
Bigelow sage	Artemisia bigelovii	—	—	X							None	No	—
Bigtooth maple	Acer grandidentatum	—	71.4						X		Yes	No	—
Black sagebrush	Artemisia nova	—	0.2	X							None	No	—
Broom snakeweed	Gutierrezia sarothrae	Bank	—			R			R		None	No	5–25
Chokecherry	Prunus virginiana	Resid., off	94.3		R			X	X		Yes	Yes	2+
Common juniper	Juniperus communis	Off-slow	12.5	X							None	No	Decades
Common snowberry	Symphoricarpos albus	—	5.9						X		Yes	Poss.	3–5
Creeping barberry	Mahonia repens	Resid.	7.9					X			yes	no	5–10
Curlleaf mountain mahogany	Cercocarpus ledifolius	—	8.7	X							None	No	60+

Seeding

Sprouting[a]

Common name	Scientific name	Occurrence	Cover (%)						Palatable		Age range
Dwarf bilberry	*Vaccinium caespitosum*	Off-rare	0.09					X	Yes	Poss.	—
Dwarf rose	*Rosa gymnocarpa*	Off-rare	16.7					X	Yes	No	—
Fendler's ceanothus	*Ceanothus fendleri*	—	—		—				—	No	—
Fourwing saltbush	*Atriplex canescens*	—	—			V		V	—	No	—
Gambel oak	*Quercus gambelii*	—	—			X		X	—	No	2–18
Gardner's saltbush	*Atriplex gardneri*	Off	—					X	—	No	2–3
Gooseberry currant	*Ribes montigenum*	Bank	3.2					V	—	No	—
Greasewood	*Sarcobatus vermiculatus*	—	—					X	Yes	No	2–3
Greene's rabbitbrush	*Chrysothamnus greenei*	—	—						—	No	—
Greenleaf manzanita	*Arctostaphylos patula*	Bank	—		—	X			—	No	—
Grouse whortleberry	*Vaccinium scoparium*	Rare	—					X	Yes	No	5+
Kinnikinnick	*Arctostaphylos uva-ursi*	Bank	7.8	X				X	Yes	No	Variable
Little sagebrush	*Artemisia arbuscula*	Off	—	X	R				None	No	Unknown
Longflower rabbitbrush	*Chrysothamnus depressus*	—	—	—					—	No	—
Mallow ninebark	*Physocarpus malvaceus*	—	0.6	—				X	Yes	No	4+
Mormon tea	*Ephedra viridis*	—	—					X	—	No	—
Mountain big sagebrush	*Artemisia tridentata* ssp. *vaseyana*	Resic., off	0.2	X					None	No	25–100[d]
Mountain snowberry	*Symphoricarpos oreophilus*	—	—					V	Yes	No	5–15
Oceanspray	*Holodiscus discolor*	Bank	0.1					X	No	Poss.	5–10
Oregon boxleaf	*Paxistima myrsinites*	Bank	—		X			X	Yes	No	<10
Parry's rabbitbrush	*Ericameria parryi*	—	—	—					—	No	—
Redosier dogwood	*Cornus sericea*	Bank	24.4			X		X	Yes	No	3+
Redstem ceanothus	*Ceanothus sanguineus*	Bank	3.4					X	No	Yes	5–15
Rocky Mountain maple	*Acer glabrum*	Off	33.8					X	Yes	Yes	3–5
Rubber rabbitbrush	*Ericameria nauseosa*	Off	0.7					X	Yes	No	3–5
Russet buffaloberry	*Shepherdia canadensis*	Off	50.0					X	Yes	No	3+
Saskatoon serviceberry	*Amelanchier alnifolia*	Rare	5.6			X		X	No	Yes	3–5
Scouler's willow	*Salix scouleriana*	Off	—					X	No	Yes	<5

Appendix B. Common Rocky Mountain Shrubs and Their Response to Fire

Common Name	Latin Name	Seed Source[b]	Seed Weight (mg)[c]	No Sprouting	Branch	Bud on Stem or Root	Lignotuber	Rhizome	Root Crown	Stolon	High-Severity Fire Reduces Sprouts	Postfire Flush?	Time to Recover (yrs)
Shadscale saltbush	Atriplex confertifolia	—	—	X							None	No	—
Shrubby cinquefoil	Dasiphora floribunda	—	—						X		Yes	No	—
Silver buffaloberry	Shepherdia argentea	—	50.0						X		Yes	No	—
Silver sagebrush	Artemisia cana ssp. cana	—	—					X	X		Yes	No	2–6
Silver sagebrush	Artemisia cana ssp. viscidula	—	—					X	X		Yes	No	2–6
Skunkbush sumac	Rhus trilobata	—	22.2						X		—	No	3+
Snowbush ceanothus	Ceanothus velutinus	Bank	5.0						X		No	Yes	2–5
Spineless horsebrush	Tetradymia canescens	Off	3.8						X		No	Yes	3–15
Spiny hopsage	Grayia spinosa	—	—						V		Yes	No	—
Stansbury cliffrose	Purshia stansburiana	Off	22.2						V		Yes	No	Decades
Thimbleberry	Rubus parviflorus	Bank, off	—					X			Yes	Yes	2+
Thinleaf huckleberry	Vaccinium membranaceum	—	—					X	X		Yes	No	3–20
Threetip sagebrush	Artemisia tripartita ssp. tripartita	—	—						V		Yes	No	25–40
Timberline sagebrush	Artemisia rothrockii	—	—	X							None	No	—
Twinberry honeysuckle	Lonicera involucrata	—	1.4	—							—	No	—
Utah serviceberry	Amelanchier utahensis	—	—						X		—	No	25
Valley saltbush	Atriplex cuneata	—	—	—							—	No	—

Common name	Latin name	Seed[b]							Fire[a]	Years
Water birch	*Betula occidentalis*	Off	—		X			—	No	—
Wax currant	*Ribes cereum*	Bank	1.8				W	—	No	—
White spirea	*Spiraea betulifolia*	—	—			X	X	No	Yes	—
Whortleberry	*Vaccinium myrtilus*	Rare	—			X	X	Yes	No	10+
Willow species	*Salix* spp.	—	0.1–0.9			X		Yes	No	—
Winterfat	*Krascheninnikovia lanata*	Off-slow	2.2–4.2		V	V	V	Yes	No	Decades
Woods' rose	*Rosa woodsii*	Rare	9.1			X	X	Yes	No	2+
Wyoming big sagebrush	*Artemisia tridentata* ssp. *wyomingensis*	Resid., off	0.3	X				None	No	Unknown
Wyoming threetip sagebrush	*Artemisia tripartita* ssp. *rupicola*	—	—		V		V	Yes	No	—
Yellow rabbitbrush	*Chrysothamnus viscidiflorus*	Off	0.6			X	X	Yes	No	3–5

Notes: Data are from the Fire Effects Information System (http://www.fs.fed.us/database/feis) except where noted. Common and Latin names are from the USDA PLANTS online database (http://plants.usda.gov). A dash indicates no information, and a blank indicates that a trait is lacking.

[a] X = has the trait; R = rarely; V = variable, occurring at some times or in some places; W = weak response.

[b] Bank = seed bank, off = off site, resid. = residual unburned seed on site.

[c] Data are from Young and Young (1992) or occasionally from the Fire Effects Information System, http://www.fs.fed.us/database/feis.

[d] Data are from W. L. Baker (2006b and in press).

Appendix C. Common Rocky Mountain Graminoids and Their Response to Fire

Common Name	Latin Name	Seed Bank	Off-Site Seed	Rhizome	Bunchgrass	Stolon	Stimulated	Unchanged	Depressed	Litter and Graminoid Recovery (years)
Alkali sacaton	Sporobolus airoides				X			X	X	2–4
Arizona fescue	Festuca arizonica				X					
Baltic rush	Juncus balticus			X			X	X		
Basin wildrye	Leymus cinereus			X	X		X	X		
Big bluestem	Andropogon gerardii			X	X		X	X		2–5
Blue grama	Bouteloua gracilis			X	X	X	X	X	X	2–3
Blue wildrye	Elymus glaucus	X			X			X	X	3–4
Bluebunch wheatgrass	Pseudoroegneria spicata				X		X			1–5
Bluejoint	Calamagrostis canadensis		X	X			X	X		1
Buffalograss	Buchloe dactyloides					X	X	X		1
Foxtail barley	Hordeum jubatum		X		X			X	X	
Geyer's sedge	Carex geyeri	—	—	—	—	—	—	—	—	—
Greenleaf fescue	Festuca viridula	—	—	—	—	—	—	—	—	—
Idaho fescue	Festuca idahoensis				X				X	3–30
Indian ricegrass	Achnatherum hymenoides		X		X			X	X	4
Inland saltgrass	Distichlis spicata	X		X			X	X		1
James' galleta	Pleuraphis jamesii			X			X	X	X	2
Little bluestem	Schizachyrium scoparium				X		X	X		1–5
Mountain muhly	Muhlenbergia montana				X			X	X	3+
Muttongrass	Poa fendleriana	—	—	—	—	—	—	—	—	—
Needle and thread	Hesperostipa comata		X		X		X	X	X	1+
New Mexico feathergrass	Hesperostipa neomexicana	—	—	—	—	—	—	—	—	—

Appendix C. Continued

Common Name	Latin Name	Source of Regen.					Response			Litter and Graminoid Recovery (years)
		Seed Bank	Off-Site Seed	Rhizone	Bunchgrass	Stolon	Stimulated	Unchanged	Depressed	
Parry's oatgrass	*Danthonia parryi*	—	—	—	—	—	X	—	—	—
Pinegrass	*Calamagrostis rubescens*	—	X	X	—	—	—	—	—	1
Plains rough fescue	*Festuca hallii*	—	—	—	—	—	—	—	—	—
Ross' sedge	*Carex rossii*	X	—	X	—	—	X	X		2–10
Rough fescue	*Festuca campestris*	—	—	—	X	—	—	—	—	—
Saline wildrye	*Leymus salinus*			X	X		X	X	X	
Sand dropseed	*Sporobolus cryptandrus*				X		X	X	X	
Sandberg bluegrass	*Poa secunda*				X		X	X	X	1+
Sideoats grama	*Bouteloua curtipendula*		X	X			X	X	X	2–3
Sixweeks fescue	*Vulpia octoflora*	X	X				X			
Slender wheatgrass	*Elymus trachycaulus*		X		X			X	X	2+
Slimstem muhly	*Muhlenbergia filiculmis*			—	—	—		—	—	—
Spike fescue	*Leucopoa kingii*			X	X		X	X		
Squirreltail	*Elymus elymoides*		X		X		X	X	X	variable
Switchgrass	*Panicum virgatum*			X				X	X	
Thurber's fescue	*Festuca thurberi*				X					
Timber oatgrass	*Danthonia intermedia*		X		X			X	X	5–10
Tufted hairgrass	*Deschampsia caespitosa*	X			X			X	X	3+
Water sedge	*Carex aquatilis*		X	X			X	X	X	1–13
Western wheatgrass	*Pascopyrum smithii*			X			X	X	X	1–5
White Mountain sedge	*Carex geophila*	—	—	—	—	—	—	—	—	—

Notes: A dash indicates no information. Data are from the Fire Effects Information System, http://www.fs.fed.us/database/feis. Common and Latin names are from the USDA PLANTS online database, http://plants.usda.gov.

Appendix D. Modeling Studies of Mortality in Rocky Mountain Trees

Sources by Tree	Place	n	Time since Fire (yrs)	Tree Diameter	Bark Thickness	Stem Char	% Crown Scorch	Scorch Height	Other
Douglas-fir									
Bevins (1980)	C MT	176	1	X				X	
Peterson & Arbaugh (1986)	E ID, W MT, NW WY	282	1				X		X
Ryan & Reinhardt (1988)	N ID, W MT	1,488	2		X		X		
Ryan, Peterson, & Reinhardt (1988)	W MT	166	1, 8		X		X		X
Wyant, Omi, & Laven (1986)	N CO	103	1, 2			X	X		
Engelmann spruce									
Ryan & Reinhardt (1988)	N ID, W MT	96	2		X		X		
Lodgepole pine									
Peterson & Arbaugh (1986)	E ID, W MT, NW WY	197	1			X	X		
Ryan & Reinhardt (1988)	N ID, W MT	144	2		X		X		
Ponderosa pine									
Bakken (1981)	N ID	443	1	X					X
J. H. Bock & Bock (1984)	W SD	—	1, 2	X					X
Gallup (1998)	N CO	477	1	X					X
Harrington (1987)	SW CO	526	1, 5	X		X	X	X	
Harrington (1993)	SW CO	526	10	X			X		
Keyser et al. (2006)	W SD	963	1–5	X		X	X		
Saveland & Neuenschwander (1989)	N ID	194	1	X				X	

Study	Location	N							
Saveland & Neuenschwander (1990)	N ID	194	1				X		
Sieg et al. (2006)	AZ, CO, MT, SD	5.083	1–3				X		X
Wyant, Omi, & Laven (1986)	N CO	95	1,2			X	X		
Quaking aspen									
J. K. Brown & DeByle (1987)	ID, WY	1.198	1–4	X		X		X	
Subalpine fir									
Ryan & Reinhardt (1988)	N ID, W MT	172	2		X		X		
Western hemlock									
Ryan & Reinhardt (1988)	N ID, W MT	100	2		X		X		
Western larch									
Ryan & Reinhardt (1988)	N ID, W MT	287	2		X		X		
Western red cedar									
Ryan & Reinhardt (1988)	N ID, W MT	69	2		X		X		

Note: An X indicates that mortality for a particular species was found to be related to that mortality predictor.

Appendix E. Animal Species Mentioned in the Text

Animal	Latin Name
AMPHIBIANS	
Boreal toad	*Anaxyrus boreas*
Columbia spotted frog	*Rana luteiventris*
Long-toed salamander	*Ambystoma macrodactylum*
Rocky Mountain tailed frog	*Ascaphus montanus*
BIRDS	
American robin	*Turdus migratorius*
American three-toed woodpecker	*Picoides dorsalis*
Bald eagle	*Halieetus leucocephalus*
Black-backed woodpecker	*Picoides arcticus*
Black-headed grosbeak	*Pheucticus melanocephalus*
Blue grouse	*Dendragapus obscurus*
Brewer's sparrow	*Spizella breweri*
Chipping sparrow	*Spizella passerina*
Clark's nutcracker	*Nucifraga columbiana*
Common raven	*Corus corax*
Cordilleran flycatcher	*Empidonax occidentalis*
Dark-eyed junco	*Junco hyemalis*
Downy woodpecker	*Picoides pubescens*
Dusky flycatcher	*Empidonax oberholseri*
Ferruginous hawk	*Buteo regalis*
Golden-crowned kinglet	*Regulus satrapa*
Gray jay	*Perisoreus canadensis*
Greater sage-grouse	*Centrocercus urophasianus*
Green-tailed towhee	*Pipilo chlorurus*
Gunnison sage-grouse	*Centrocercus minimus*
Hairy woodpecker	*Picoides villosus*
Hammond's flycatcher	*Empidonax hammondii*
Hermit thrush	*Catharus guttatus*
Horned lark	*Eremophila alpestris*
House wren	*Troglodytes aedon*
Lazuli bunting	*Passerina amoena*
Lewis's woodpecker	*Melanerpes lewis*
Mountain bluebird	*Sialia currucoides*
Mountain chickadee	*Poecile gambeli*
Northern flicker	*Colaptes auratus*
Northern goshawk	*Accipiter gentilis*
Olive-sided flycatcher	*Contopus cooperi*
Pine grosbeak	*Pinicola enucleator*
Pine siskin	*Carduelis pinus*
Plumbeous vireo	*Vireo plumbeus*
Red crossbill	*Loxia curvirostra*
Red-headed woodpecker	*Melanerpes erythrocephalus*

Appendix E. Continued

Animal	Latin Name
Ruby-crowned kinglet	*Regulus calendula*
Sage sparrow	*Amphispiza belli*
Sage thrasher	*Oreoscoptes montanus*
Spotted towhee	*Pipilo maculatus*
Steller's jay	*Cyanocitta stelleri*
Swainson's thrush	*Catharus ustulatus*
Townsend's solitaire	*Myadestes townsendi*
Tree swallow	*Tachycineta bicolor*
Vesper sparrow	*Pooecetes gramineus*
Virginia's warbler	*Vermivora virginiae*
Warbling vireo	*Vireo gilvus*
Western bluebird	*Sialia mexicana*
Western meadowlark	*Sturnella neglecta*
Western tanager	*Piranga ludoviciana*
Western wood-pewee	*Contopus sordidulus*
White-crowned sparrow	*Zonotrichia leucophrys*
MAMMALS	
American badger	*Taxidea taxus*
American bison	*Bison bison*
American marten	*Martes americana*
Bighorn sheep	*Ovis canadensis*
Canada lynx	*Lynx canadensis*
Cinereus shrew	*Sorex cinereus*
Coyote	*Canis latrans*
Deer mouse	*Peromyscus maniculatus*
Elk	*Cervus elaphus*
Fisher	*Martes pennanti*
Grizzly bear	*Ursus arctos* ssp. *horribilis*
Least chipmunk	*Tamias minimus*
Long-tailed vole	*Microtus longicaudus*
Meadow vole	*Microtus pennsylvanicus*
Montane vole	*Microtus montanus*
Moose	*Alces alces*
Mule deer	*Odocoileus hemionus*
Northern flying squirrel	*Glaucomys sabrinus*
Pinyon mouse	*Peromyscus truei*
Prairie vole	*Microtus ochrogaster*
Pronghorn	*Antilocapra americana*
Pygmy rabbit	*Brachylagus idahoensis*
Red squirrel	*Tamiasciurus hudsonicus*
Snowshoe hare	*Lepus americanus*
Southern red-backed vole	*Clethrionomys gapperi*
Thirteen-lined ground squirrel	*Spermophilus tridecemlineatus*

Appendix E. Continued

Animal	Latin Name
Townsend's ground squirrel	*Spermophilus townsendii*
Vagrant shrew	*Sorex vagrans*
Western heather vole	*Phenacomys intermedius*
White-tailed deer	*Odocoileus virginianus*
Yellow-pine chipmunk	*Tamias amoenus*

Notes: Amphibian names are from the American Museum of Natural History's Amphibian Species of the World, version 5.2 (http://research.amnh.org/herpetology/amphibia/index.php). Bird names are from the *American Ornithologists' Union Check-list of North American Birds*, 7th ed. (http://www.aou .org/checklist/index.php3). Mammal names are from the Smithsonian's *Mammal Species of the World* (http://vertebrates.si.edu/mammals/msw).

Glossary*

Active crown fire: A fire that combusts both surface and canopy fuels.

Activity fuels: Slash, including stumps, defective boles, branch wood, twigs, and needles left after an activity, usually logging.

Age structure: The number of trees by age within a particular area of forest.

All-tree fire interval: A new unbiased and accurate method for estimating mean fire interval in small plots using all the trees, not just scarred trees.

AMO: See *Atlantic Multidecadal Oscillation*.

Annual probability of fire: The average fraction of a landscape expected to burn each year, or the probability, on average, that a point will burn each year, calculated as the inverse of the fire rotation or population mean fire interval.

Atlantic Multidecadal Oscillation (AMO): A 65- to 80-year cycle in Atlantic sea surface temperatures north of the equator.

Backfire: A fire that burns into the wind.

Basal area: The total cross-sectional area of tree stems.

Biological soil crust: A three-dimensional community of cyanobacteria (or blue-green algae), green algae, mosses, and lichens formed on soil surfaces and in the upper layers of mineral soil.

Bunchgrass: A grass plant with leaves and stems that originate in a cluster or bunch from fibrous roots.

Chronosequence: Information from a set of locations, differing in time since fire, used to infer postfire successional trends under the assumption that space is a reasonable substitute for time.

Cohort: A group of individuals born at the same time.

---

*The online glossary of the National Wildfire Coordinating Group (http://www.nwcg.gov/pms/pubs/glossary/index.htm) may also be useful.

Composite fire interval: The interval between fires, documented by fire scars, in a list of all the fires occurring within a sample area.

Conduction: Transfer of heat by direct molecule-to-molecule contact.

Contingent succession: Postfire plant succession in which the pathway depends on the composition of the forest around a burn, the burn size, the environment, and time since the most recent fire.

Convection: Transfer of heat by movement of air parcels, such as rising air on a warm day.

Crown base height: The lowest canopy base above ground with enough canopy fuels to allow fire to carry vertically through the rest of the canopy

Crown fire: A fire that combusts the canopies of the top vegetation layer; usually refers to tree canopies in a forest.

Crown-fire initiation: The process by which a spreading surface fire becomes a crown fire.

Crown scorch: Crown foliage killed by combustion or by convective or radiative heat from a fire.

Duff: Partially or fully decomposed organic matter above mineral soil and below litter.

Ecotone: The area where two ecosystems meet.

Edge effect: Creation of an opening or other modification that alters environmental and biological conditions in the edge of an adjoining forest, often extending 50 meters or more into the forest.

El Niño–Southern Oscillation (ENSO): Interannual variation between an El Niño with a warm eastern equatorial Pacific and a La Niña that is cool in the same area, cycling over a period of a few years.

ENSO: See *El Niño–Southern Oscillation*.

Facilitation: The process by which early postfire species make the environment more favorable for later-arriving species.

Fine fuels: Fuels less than 1/4 inch in diameter, including grass, leaves, needles, small wood, and other materials that ignite readily and are consumed rapidly by fire when dry, with a time lag of one hour or less.

Fire atlas: Sets of maps of fires, often maintained in agency offices.

Fire break: A physical or biological feature that can reduce fire intensity or slow or stop fire spread.

Fire control: The process of intentionally putting out fires; see also *Fire suppression*.

Fire cycle: See *Fire rotation*.

Fire density: The number of fires per unit area.

Fire-episode frequency: The number of fire episodes, which may represent more than one fire, per unit period; used in charcoal fire studies.

Fire exclusion: A syndrome of livestock grazing, fire control, landscape fragmentation, or other human-caused factors that restrict fire spread or put out fires.

Fire frequency: The number of fires per unit period.

Fire hazard: The properties of fuels and other conditions that determine the probability of ignition and fire spread.

Fire intensity: A general term for the rate of energy output from a fire; see also *Fireline intensity*.

Fire-interval distribution: Amount of land area burned in relation to time elapsed since the most recent fire, usually shown in a bar graph.

Fireline: A line constructed on the perimeter of a fire to control it; also, the line of fire itself.

Fireline intensity: The rate of energy output from a 1-meter-wide strip parallel to the fireline, extending over the flame depth from the front to the back of the active (flaming) combustion zone.

Fire lookout: A tower and enclosed platform used to identify and locate fires from a high point of land.

Fire management: A type of policy enacted in the 1970s that recognized fire's multiple values, economic criteria, and diverse methods of management.

Fire regime: The overall variability of fire, characterized by attributes of all the patches of burned area within a particular landscape (usually thousands of hectares) over a particular time period (usually several hundred years or more).

Fire risk: The probability that a fire will ignite and spread.

Fire rotation: The expected time required to burn land area equal to a particular study area.

Fire scar: A usually triangular wound formed when heat kills part of the cambium of a woody plant.

Fire severity: A measure of the ecological effects of fire on ecosystems, often in three qualitative categories (low, moderate, high).

Fire-size distribution: The number of fires by size.

Fire suppression: The process or result of intentionally putting out fires; see also *Fire control*.

Fire-year map: A map, usually reconstructed using tree rings and other methods, that shows an estimate of the original extent of fire in a particular year within a study area.

Five-needle pine: A member of the white pines, which features five needles in a bunch.

Flame height: The vertical distance from the tip of flames to the ground.

Flame length: The sloping distance from the tip of flames to the point where the flames contact the ground.

Flaming combustion: Pyrolysis (thermal breakdown) and oxidation of fuels by flames.

Flammability: Tendency of a material to catch fire and burn.

Foliar moisture content: The moisture content of foliage, usually expressed as a percentage of the dry weight.

Forb: Nongraminoid herbaceous plant with relatively showy flowers.

Forest reserve reports: Reports written about 1900 by government scientists sent to evaluate the public lands that later became the national forests.

Fuel break: A low-fuel area that can slow or stop fire; see also *Fire break*.

Fuel buildup: The tendency for fuel to accumulate and loadings to increase.

Fuel model: A set of estimates of average fuel conditions (e.g., loadings) for a limited and standardized set of vegetation types and conditions.

Fuel reduction: The process of lowering fuel loadings, often also reducing the probability of crown-fire initiation.

Fuelbed: A homogeneous vegetation unit, and its fuel, on the landscape.

Functional trait: An adaptation that allows plants to resist or respond to a particular functional need, such as surviving fire.

Glowing combustion: Pyrolysis (thermal breakdown) and oxidation of fuels without flames.

Graminoid: A grass or grasslike plant, including sedges and rushes.

Ground fire: A fire that combusts organic soils by glowing combustion.

Head fire: A fire that burns in the direction of the wind.

High-severity fire: A fire that kills or topkills a high percentage of canopy plants.

Historical range of variability (HRV): "The ecological conditions, and the spatial and temporal variation in these conditions, that are relatively unaffected by people, within a period of time and geographical area" (Landres, Morgan, and Swanson 1999, 1180).

Holdover fire: A fire that ignites, then smolders for some time in down wood, duff, or other fuels before it reinitiates fire spread by flaming combustion.

HRV: See *Historical range of variability*.

Ignition ratio: The ratio of lightning strikes to fire starts.

Independent crown fire: A fire that burns only in shrub or tree canopies, not in surface fuels.

Inhibition: The process by which survivors or first-arriving plants may control postfire succession by preventing other species from dominating; also called *initial floristics*.

Initial attack: The time at which the first fire control crews arrive at a wildland fire.

Initial floristics: See *Inhibition*.

Interface: Wildland-urban interface (WUI) in which housing and associated infrastructure are dominant.

Intermix: Wildland-urban interface (WUI) in which wildland vegetation is dominant.

Invasive plant: A nonnative plant with the ability and tendency to invade and become abundant in natural vegetation.

Ladder fuels: Shrubs, small trees, or standing deadwood that allows a surface fire to climb into the canopy.

Landscape ecology: The science of structure, function, and change in the biosphere on the landscape scale, typically a scale of hundreds of meters to kilometers.

Landscape snapshot: The characteristics of a landscape at a particular moment in time.

Life history: An individual trait or the suite of functional traits that characterize an organism.

Lignotuber: A set of large underground stem or root swellings that store energy and have buds, which help a plant to survive fire.

Litter: The top layer of the forest floor, composed of branches, twigs, and recently fallen leaves or needles that have been little altered by decomposition.

Loading: The mass of a particular fuel per unit area.

Low-severity fire: A fire that burns beneath, but kills few or no, canopy trees (or few or no shrubs in a shrubland).

Mastication: The process of grinding, shredding, chunking or chopping fuels into small pieces.

Mean fire interval: The average interval between fires at a point (see *Point mean fire interval*); also the mean of a set of point mean fire intervals.

Mixed-severity fire: A fire that burns some area at low severity and another area at moderate or high severity.

Moderate-severity fire: A fire that kills a medium fraction of canopy trees (or canopy shrubs).

NAO: See *North Atlantic Oscillation*.

Nature as a garden: The view that, prior to EuroAmerican settlement, nature was effectively a garden managed by the Indians.

Nature as wild landscape: The view that, prior to EuroAmerican settlement, nature was shaped primarily by nonhuman forces.

North Atlantic Oscillation (NAO): Variation in the strength and position of the subtropical anticyclone over the North Atlantic.

Nurse plant: A plant that facilitates the regeneration or survival of another plant by shading or otherwise protecting it.

Pacific Decadal Oscillation (PDO): An oscillation over 20- to 30-year periods in Pacific sea surface temperatures poleward of 20 degrees N latitude.

Pacific–North America pattern (PNA): A deep winter Aleutian low linked to a blocking ridge over the northwestern United States and parts of the northern Rockies.

Paleoecological study: Research on ecological conditions over millennia.

Palmer drought-severity index (PDSI): A popular local index of dryness or precipitation deficits.

Passive crown fire: A fire that combusts surface fuels and torches individual canopy stems or groups of stems.

PDO: See *Pacific Decadal Oscillation*.

PDSI: See *Palmer drought-severity index*.

Plume-dominated fire: A fire dominated by a convection column rather than the wind, which can lead to turbulent indrafts, increased radiation, and unexpectedly high spread rates; see *Wind-driven fire*, in contrast.

PNA: See *Pacific–North America pattern*.

Point fire frequency: The number of fires at a point in the landscape per unit period, also equal to the inverse of the point mean fire interval.

Point mean fire interval: The mean interval between fires at a point in the landscape, also equal to the inverse of the point fire frequency.

Population mean fire interval: The average mean fire interval across all points in a landscape.

Postfire logging: Logging in the first few years after fire.

Postfire succession: Succession after fire.

QBO: See *Quasi-Biennial Oscillation*.

Quasi-Biennial Oscillation (QBO): A biennial period linked with winter alteration in tropical stratospheric winds.

Radiation: Transfer of heat through space by electromagnetic waves rather than direct molecule-to-molecule transfer (conduction) or movement of air parcels (convection).

Reburn: The occurrence of another fire a short time (e.g., <30 years) after a moderate- to high-severity fire.

Recorder tree: Fire-scarred tree previously thought to be more likely to be scarred again in subsequent fires.

Regeneration: A general term for young plants, such as seedlings or saplings of trees; also the process by which they are produced.

Replacement fire: A high-severity fire that leads to initiation of a new stand of vegetation.

Reproduction: See *Regeneration*.

Residence time: The time that flaming or glowing combustion is present at a point.

Rhizome: A horizontal belowground stem that can allow plants to survive and resprout after fire.

Scorch height: The height at which plant material is killed by combustion or by radiant or convective heat.

Seed bank: Viable seed stored in litter, duff, or mineral soil or in sealed cones in tree canopies.

Shifting mosaic: A landscape pattern of patches, produced by disturbances (e.g., fires), that shifts over time.

Slash: Waste materials, including stumps, defective boles, branch wood, twigs, and needles, left after an activity, usually logging; see also *Activity fuels*.

Smoldering combustion: Pyrolysis (thermal breakdown) and oxidation of fuels without flames.

Spatial autocorrelation: The tendency for things closer together to be more similar than things farther apart.

Spatially contingent succession: Postfire plant succession in which the pathway depends on the composition of the forest around a burn, the burn size, the environment, and time since the most recent fire.

Spot fire: A fire started by dislodged, burning fuel particles that rise in the fire's convection column, potentially landing and igniting a fire ahead of the fireline.

Stand: An area within a forest that is relatively uniform in vegetation composition and structure and in environment.

Stand development: The process by which stand structure (e.g., tree density, spatial pattern) reforms after a natural or human disturbance.

Stand-origin date: The estimated year that a particular stand of trees originated after a stand-replacing disturbance, such as a crown fire.

Stand-replacement fire: A high-severity surface or crown fire that kills most of the canopy trees or shrubs, leading to their eventual replacement by new trees or shrubs.

Stolon: An aboveground horizontal stem that allows asexual reproduction.

Succession: The complex process by which vegetation recovers from disturbance, such as fire, or colonizes newly exposed land.

Surface fire: A fire that combusts only surface fuels below a shrub or tree canopy that generally survives the fire.

Synoptic: Pertaining to weather conditions over a large area.

Targeted sampling: Fire history sampling in which the researcher purposely chooses locations of landscapes, plots, and evidence to sample, leading to a biased sample.

Teleconnection: An atmospheric connection, usually between distant regions, that leads to large-scale linkage of global weather and climate, particularly linking temperature and pressure conditions in the ocean-atmosphere system to weather and climate on the continents.

Thermal belt: An elevational zone potentially having the driest and warmest conditions because of nighttime temperature inversions and cold-air drainage.

Time-lag fuel moisture class: A class reflecting the time for dead fuels with a particular moisture content to adjust about 63 percent of the way to equilibrium with atmospheric conditions.

Tolerance: The concept of late-successional species tolerating the conditions (e.g., low resource levels) imposed by early-successional species but eventually outgrowing them to dominate.

Torching: Fire climbing from the surface into the canopy of individual trees or groups of trees, and then returning to the surface, as in a passive crown fire.

Wildland fire use: Allowing fires to burn naturally under management guidance for resource benefit.

Wildland-urban interface (WUI): A zone or interface between wildland vegetation and housing and its associated infrastructure.

Wind-driven fire: A fire whose spread is shaped primarily by wind; see *Plume-dominated fire* for a contrast.

WUI: See *Wildland-urban interface.*

References

Abella, S. R., and P. Z. Fulé. 2008. *Fire effects on Gambel oak in southwestern ponderosa pine–oak forests.* USDA Forest Service Research Note RMRS-RN-34. Fort Collins, Colorado: Rocky Mountain Research Station.

Abella, S. R., J. D. Springer, and W. W. Covington. 2007. Seed banks of an Arizona *Pinus ponderosa* landscape: Responses to environmental gradients and fire cues. *Canadian Journal of Forest Research* 37:552–67.

Acton, S. 2002. Fire severity and salvage logging effects on exotics in ponderosa pine dominated forests. M.S. thesis, Colorado State University, Fort Collins.

Adams, D. K., and A. C. Comrie. 1997. The North American monsoon. *Bulletin of the American Meteorological Society* 78:2197–213.

Adams, K. R. 1993. *Second annual report, Mesa Verde fire effects: An ecological and ethnobotanical study of vegetation recovery after the Long Mesa fire of July 1989.* Unpublished report to Mesa Verde National Park. Cortez, Colorado: Crow Canyon Archaeological Center.

Adams, R., and D. Simmons. 1999. Ecological effects of fire fighting foams and retardants. *Conference Proceedings: Australian Bushfire Conference, Albury, July 1999,* pp. 1–8.

Agee, J. K. 1993. *Fire ecology of Pacific Northwest forests.* Washington, D.C.: Island Press.

———. 1996. *The influence of forest structure on fire behavior.* Unpublished manuscript from a presentation at the 17th Annual Forest Vegetation Management Conference, Redding, California, January 16–18, 1996.

———. 2000. Wilderness fire science: A state-of-knowledge review. Pp. 5–22 in *Wilderness Science in a Time of Change Conference.* Vol. 5, *Wilderness ecosystems, threats, and management,* ed. D. N. Cole, S. F. McCool, W. T. Borrie, and J. O'Loughlin. USDA Forest Service Proceedings RMRS-P-15. Fort Collins, Colorado: Rocky Mountain Research Station.

Agee, J. K., B. Bahro, M. A. Finney, P. N. Omi, D. B. Sapsis, C. N. Skinner, J. W. van Wagtendonk, and C. P. Weatherspoon. 2000. The use of shaded fuelbreaks in landscape fire management. *Forest Ecology and Management* 127:55–66.

Agee, J. K., and C. N. Skinner. 2005. Basic principles of forest fuel reduction treatments. *Forest Ecology and Management* 211:83–96.

Ahlgren, I. F. 1974. The effect of fire on soil organisms. Pp. 47–72 in *Fire and ecosystems*, ed. T. T. Kozlowski and C. E. Ahlgren. New York: Academic Press.

Akinsoji, A. 1988. Postfire vegetation dynamics in a sagebrush steppe in southeastern Idaho, USA. *Vegetatio* 78:151–55.

Albini, F. A. 1976. *Estimating wildfire behavior and effects*. USDA Forest Service General Technical Report INT-30. Ogden, Utah: Intermountain Forest and Range Experiment Station.

———. 1983. *Potential spotting distance from wind-driven surface fires*. USDA Forest Service Research Paper INT-309. Ogden, Utah: Intermountain Forest and Range Experiment Station.

Albini, F. A., M. R. Amin, R. D. Hungerford, W. H. Frandsen, and K. C. Ryan. 1996. *Models for fire-driven heat and moisture transport in soils*. USDA Forest Service General Technical Report INT-GTR-335. Ogden, Utah: Intermountain Forest and Range Experiment Station.

Albini, F. A., and J. K. Brown. 1978. *Predicting slash depth for fire modeling*. USDA Forest Service Research Paper INT-206. Ogden, Utah: Intermountain Forest and Range Experiment Station.

Albini, F. A., J. K. Brown, E. D. Reinhardt, and R. D. Ottmar. 1995. Calibration of a large fuel burnout model. *International Journal of Wildland Fire* 5:173–92.

Albini, F. A., and E. D. Reinhardt. 1995. Modeling ignition and burning rate of large woody natural fuels. *International Journal of Wildland Fire* 5:81–91.

———. 1997. Improved calibration of a large fuel burnout model. *International Journal of Wildland Fire* 7:21–28.

Alder, G. M. 1970. Age-profiles of aspen forests in Utah and northern Arizona. M.S. thesis, Salt Lake City, University of Utah.

Alexander, M. E. 1978. *Photo series of natural residues and conditions in Colorado lodgepole pine fuel complexes*. Fort Collins, Colorado: Colorado State University, College of Forestry and Natural Resources, Department of Forestry and Wood Science, Fire Science and Technology Program.

———. 1979. Fuels description in lodgepole pine stands of the Colorado Front Range. M.S. thesis, Colorado State University, Fort Collins.

———. 1981. Four fire scar records on lodgepole pine (*Pinus contorta* Dougl.) in north-central Colorado. *Southwestern Naturalist* 25:432–34.

———. 1982. Calculating and interpreting forest fire intensities. *Canadian Journal of Botany* 60:349–57.

———. 2004. What triggered the Brewer fire blowup remains the mystery. *Fire Management Today* 64(2): 56–57.

Alexander, M. E., and M. G. Cruz. 2006. Evaluating a model for predicting active crown fire rate of spread using wildfire observations. *Canadian Journal of Forest Research* 36:3015–28.

Alexander, M. E., M. G. Cruz, and A. M. G. Lopes. 2006. CFIS: A software tool for simulating crown fire initiation and spread. *Forest Ecology and Management* 234S:S133.

Alexander, M. E., and F. G. Hawksworth. 1975. *Wildland fires and dwarf mistletoes: A literature review of ecology and prescribed burning.* USDA Forest Service General Technical Report RM-14. Fort Collins, Colorado: Rocky Mountain Forest and Range Experiment Station.

Alexander, M. E., B. J. Stocks, and B. D. Lawson. 1991. Fire behavior in black spruce-lichen woodland: The Porter Lake Project. Information Report NOR-X-310. Edmonton, Alberta: Forestry Canada, Northwest Region, Northern Forestry Centre.

Alexander, M. E., and R. F. Yancik. 1977. The effect of precommercial thinning on fire potential in a lodgepole pine stand. *Fire Management Notes* 38(3): 7–9, 20.

Alftine, K. J., G. P. Malanson, and D. B. Fagre. 2003. Feedback-driven response to multidecadal climatic variability at an alpine treeline. *Physical Geography* 24:520–35.

Alington, C. 1998. Fire history and landscape pattern in the Sangre de Cristo Mountains, Colorado. Ph.D. dissertation, Colorado State University, Fort Collins.

Allen, C. D. 1984. Montane grasslands in the landscape of the Jemez Mountains, New Mexico. M.S. thesis, University of Wisconsin, Madison.

———. 1989. Changes in the landscape of the Jemez Mountains, New Mexico. Ph.D. dissertation, University of California, Berkeley.

Allen, C. D., R. S. Anderson, R. B. Jass, J. L. Toney, and C. H. Baisan. 2008. Paired charcoal and tree-ring records of high-frequency Holocene fire from two New Mexico bog sites. *International Journal of Wildland Fire* 17:115–30.

Allen, C. D., M. Savage, D. A. Falk, K. F. Suckling, T. W. Swetnam, T. Schulke, P. B. Stacey, P. Morgan, M. Hoffman, and J. T. Klingel. 2002. Ecological restoration of southwestern ponderosa pine ecosystems: A broad perspective. *Ecological Applications* 12:1418–33.

Alvarado, E., D. V. Sandberg, and S. G. Pickford. 1998. Modeling large forest fires as extreme events. *Northwest Science* 72:66–75.

Ament, R. J. 1995. Pioneer plant communities five years after the 1988 Yellowstone fires. M.S. thesis, Montana State University, Bozeman.

Amman, G. D., and K. C. Ryan. 1991. *Insect infestation of fire-injured trees in the Greater Yellowstone area.* USDA Forest Service Research Note INT-398. Ogden, Utah: Intermountain Research Station.

Amman, G. D., and R. F. Schmitz. 1988. Mountain pine beetle–lodgepole pine interactions and strategies for reducing tree losses. *Ambio* 17:62–68.

Andersen, H.-E., R. J. McGaughey, and S. E. Reutebuch. 2005. Estimating forest canopy fuel parameters using LIDAR data. *Remote Sensing of the Environment* 94:441–49.

Andersen, M. D., and W. L. Baker. 2006. Reconstructing landscape-scale tree invasion using survey notes in the Medicine Bow Mountains, Wyoming, USA. *Landscape Ecology* 21:243–58.

Anderson, H. E. 1968. *Sundance fire: An analysis of fire phenomena.* USDA Forest Service Research Paper INT-56. Ogden, Utah: Intermountain Forest and Range Experiment Station.

———. 1974. *Appraising forest fuels: A concept.* USDA Forest Service Research Note INT-187. Ogden, Utah: Intermountain Forest and Range Experiment Station.

———. 1982. *Aids to determining fuel models for estimating fire behavior.* USDA Forest Service

General Technical Report INT-122. Ogden, Utah: Intermountain Forest and Range Experiment Station.

———. 1990. *Predicting equilibrium moisture content of some foliar forest litter in the northern Rocky Mountains.* USDA Forest Service Research Paper INT-429. Ogden, Utah: Intermountain Forest and Range Experiment Station.

Anderson, H. E., A. P. Brackebusch, R. W. Mutch, and R. C. Rothermel. 1966. *Mechanisms of fire spread research progress report No. 2.* USDA Forest Service Research Paper INT-28. Ogden, Utah: Intermountain Forest and Range Experiment Station.

Anderson, J. E., M. Ellis, C. D. von Dohlen, and W. H. Romme. 2004. Establishment, growth, and survival of lodgepole pine in the first decade. Pp. 55–101 in *After the fires: The ecology of change in Yellowstone National Park*, ed. L. L. Wallace. New Haven, Connecticut: Yale University Press.

Anderson, J. E., and W. H. Romme. 1991. Initial floristics in lodgepole pine (*Pinus contorta*) forests following the 1988 Yellowstone fires. *International Journal of Wildland Fire* 1:119–24.

Anderson, L., C. E. Carlson, and R. H. Wakimoto. 1987. Forest fire frequency and western spruce budworm outbreaks in western Montana. *Forest Ecology and Management* 22:251–60.

Anderson, P. G. 1994. Conifer forest dynamics in Grand Teton National Park, Wyoming. Ph.D. dissertation, University of Utah, Salt Lake City.

Anderson, R. S., C. D. Allen, J. L. Toney, R. B. Jass, and A. N. Bair. 2008. Holocene vegetation and fire regimes in subalpine and mixed conifer forests, southern Rocky Mountains, USA. *International Journal of Wildland Fire* 17:96–114.

Andrews, P. L. 1986. *BEHAVE: Fire behavior prediction and fuel modeling system—BURN subsystem, part 1.* USDA Forest Service General Technical Report INT-194. Ogden, Utah: Intermountain Research Station.

Andrews, P. L., C. D. Bevins, and R. C. Seli. 2005. *BehavePlus fire modeling system, version 3.0: User's guide.* USDA Forest Service General Technical Report RMRS-GTR-106WWW. Ogden, Utah: Rocky Mountain Research Station.

Antos, J. A., and J. R. Habeck. 1981. Successional development in *Abies grandis* (Dougl.) Forbes forests in the Swan Valley, western Montana. *Northwest Science* 55:26–39.

Antos, J. A., B. McCune, and C. Bara. 1983. The effect of fire on an ungrazed western Montana grassland. *American Midland Naturalist* 110:354–64.

Aplet, G. H. 2006. Evolution of wilderness fire policy. *International Journal of Wilderness* 12:9–13.

Aplet, G. H., R. D. Laven, and F. W. Smith. 1988. Patterns of community dynamics in Colorado Engelmann spruce–subalpine fir forests. *Ecology* 69:312–19.

Aplet, G. H., F. W. Smith, and R. D. Laven. 1989. Stemwood biomass and production during spruce-fir stand development. *Journal of Ecology* 77:70–77.

Armour, C. D. 1982. Fuel and vegetation succession in response to mountain pine beetle epidemics in northwestern Montana. M.S. thesis, University of Idaho, Moscow.

Armour, C. D., S. C. Bunting, and L. F. Neuenschwander. 1984. Fire intensity effects on the understory in ponderosa pine forests. *Journal of Range Management* 37:44–49.

Arno, S. F. 1976. *The historical role of fire on the Bitterroot National Forest.* USDA Forest Service Research Paper INT-187. Ogden, Utah: Intermountain Forest and Range Experiment Station.

———. 1979. *Forest regions of Montana.* USDA Forest Service Research Paper INT-218. Ogden, Utah: Intermountain Forest and Range Experiment Station.

———. 1986. Whitebark pine cone crops: A diminishing source of wildlife food? *Western Journal of Applied Forestry* 1:92–94.

———. 2001. Community types and natural disturbance processes. Pp. 74–88 in *Whitebark pine communities: Ecology and restoration,* ed. D. F. Tomback, S. F. Arno, and R. E. Keane. Washington, D.C.: Island Press.

Arno, S. F., and S. Allison-Bunnell. 2002. *Flames in our forest: Disaster or renewal?* Washington, D.C.: Island Press.

Arno, S. F., and J. K. Brown. 1989. Managing fire in our forests: Time for a new initiative. *Journal of Forestry* 87:44–46.

Arno, S. F., and D. H. Davis. 1980. Fire history of western redcedar/hemlock forests in northern Idaho. Pp. 21–26 in *Proceedings of the fire history workshop,* ed. M. A. Stokes and J. H. Dieterich. USDA Forest Service General Technical Report RM-81. Fort Collins, Colorado: Rocky Mountain Forest and Range Experiment Station.

Arno, S. F., and C. E. Fiedler. 2005. *Mimicking nature's fire: Restoring fire-prone forests in the West.* Washington, D.C.: Island Press.

Arno, S. F., and G. E. Gruell. 1983. Fire history at the forest-grassland ecotone in southwestern Montana. *Journal of Range Management* 36:332–36.

———. 1986. Douglas-fir encroachment into mountain grasslands in southwestern Montana. *Journal of Range Management* 39:272–76.

Arno, S. F., M. G. Harrington, C. E. Fiedler, and C. E. Carlson. 1995. Restoring fire-dependent ponderosa pine forests in western Montana. *Restoration & Management Notes* 13:32–36.

Arno, S. F., and R. J. Hoff. 1989. *Silvics of whitebark pine (*Pinus albicaulis*).* USDA Forest Service General Technical Report INT-253. Ogden, Utah: Intermountain Research Station.

Arno, S. F., D. J. Parsons, and R. E. Keane. 2000. Mixed-severity fire regimes in the northern Rocky Mountains: Consequences of fire exclusion and options for the future. Pp. 225–32 in *Wilderness Science in a Time of Change Conference.* Vol. 5, *Wilderness ecosystems, threats, and management,* ed. D. N. Cole, S. F. McCool, W. T. Borrie, and J. O'Loughlin. USDA Forest Service Proceedings RMRS-P-15-VOL-5. Fort Collins, Colorado: Rocky Mountain Research Station.

Arno, S. F., and T. D. Petersen. 1983. *Variation in estimates of fire intervals: A closer look at fire history on the Bitterroot National Forest.* USDA Forest Service Research Paper INT-301. Ogden, Utah: Intermountain Forest and Range Experiment Station.

Arno, S. F., E. D. Reinhardt, and J. H. Scott. 1993. *Forest structure and landscape patterns in the subalpine lodgepole pine type: A procedure for quantifying past and present conditions.* USDA Forest Service General Technical Report INT-294. Ogden, Utah: Intermountain Research Station.

Arno, S. F., J. H. Scott, and M. G. Hartwell. 1995. *Age-class structure of old growth ponderosa pine/Douglas-fir stands and its relationship to fire history.* USDA Forest Service Research Paper INT-RP-481. Ogden, Utah: Intermountain Research Station.

Arno, S. F., D. G. Simmermann, and R. E. Keane. 1985. *Forest succession on four habitat types in western Montana.* USDA Forest Service General Technical Report INT-GTR-177. Ogden, Utah: Intermountain Forest and Range Experiment Station.

Arno, S. F., H. Y. Smith, and M. A. Krebs. 1997. *Old growth ponderosa pine and western larch stand structures: Influences of pre-1900 fires and fire exclusion.* USDA Forest Service Research Paper INT-RP-495. Ogden, Utah: Intermountain Research Station.

Arno, S. F., and K. M. Sneck. 1977. *A method for determining fire history in coniferous forests of the mountain west.* USDA Forest Service General Technical Report INT-42. Ogden, Utah: Intermountain Forest and Range Experiment Station.

Arno, S. F., and A. E. Wilson. 1986. Dating past fires in curlleaf mountain-mahogany communities. *Journal of Range Management* 39:241–43.

Arnold, R. K., and C. C. Buck. 1954. Blow-up fires: Silviculture or weather problems? *Journal of Forestry* 52:408–11.

Ash, M., and R. J. Lasko. 1990. Postfire vegetative response in a whitebark pine community, Bob Marshall Wilderness, Montana. Pp. 360–61 in *Proceedings—Symposium on whitebark pine ecosystems: Ecology and management of a high-mountain resource,* ed. W. C. Schmidt and K. J. McDonald. USDA Forest Service General Technical Report INT-270. Ogden, Utah: Intermountain Research Station.

Ashton, R. E. 1930. Preliminary observations on revegetation of the Twinsisters Burn in Rocky Mountain National Park. *Journal of the Colorado-Wyoming Academy of Sciences* 1:13.

Asselin, H., and S. Payette. 2005. Detecting local-scale fire episodes on pollen slides. *Review of Palaeobotany and Palynology* 137:31–40.

Ayres, H. B. 1900a. The Flathead Forest Reserve. House of Representatives, 56th Cong., 1st sess., document no. 5, *Annual Reports of the Department of the Interior for the Fiscal Year Ended June 30, 1899, 20th Annual Report of the United States Geological Survey, Part V— Forest Reserves,* pp. 245–316.

———. 1900b. Lewis and Clark Forest Reserve, Montana. House of Representatives, 56th Cong., 2nd Sess., document no. 5, *Annual Reports of the Department of the Interior for the Fiscal Year Ended June 30, 1900, 21st Annual Report of the United States Geological Survey, Part V—Forest Reserves,* pp. 35–80.

Bachelet, D., J. M. Lenihan, and R. P. Neilson. 2007. *Wildfires & global climate change: The importance of climate change for future wildfire scenarios in the western United States.* Arlington, Virginia: Pew Center on Global Climate Change.

Backer, D. M., S. E. Jensen, and G. R. McPherson. 2004. Impacts of fire-suppression activities on natural communities. *Conservation Biology* 18:937–46.

Bailey, R. G. 2002. *Ecoregion-based design for sustainability.* New York: Springer.

Bair, A. N. 2004. A 15,000 year vegetation and fire history record, southern Sangre de Cristo Mountains, New Mexico. M.S. thesis, Northern Arizona University, Flagstaff.

Baisan, C. H., and T. W. Swetnam. 1990. Fire history on a desert mountain range: Rincon Mountain Wilderness, Arizona, U.S.A. *Canadian Journal of Forest Research* 20:1559–69.

Baker, F. S. 1918. Aspen as a temporary forest type. *Journal of Forestry* 16:294–303.

———. 1925. *Aspen in the central Rocky Mountain region.* USDA Department Bulletin 1291. Washington, D.C.: U.S. Government Printing Office.

Baker, F. S., and C. F. Korstian. 1931. *Suitability of brush lands in the intermountain region for the growth of natural or planted western yellow pine forests.* USDA Technical Bulletin No. 256. Washington, D.C.: U.S. Government Printing Office.

Baker, W. L. 1949. Soil changes associated with recovery of scrub oak, *Quercus gambelii,* after fire. M.S. thesis, University of Utah, Salt Lake City.

———. 1987. Recent change in the riparian vegetation of the montane and subalpine zones of western Colorado, U.S.A. Ph.D. dissertation, University of Wisconsin, Madison.

———. 1989a. Effect of scale and spatial heterogeneity on fire-interval distributions. *Canadian Journal of Forest Research* 19:700–706.

———. 1989b. Landscape ecology and nature reserve design in the Boundary Waters Canoe Area, Minnesota. *Ecology* 70:23–35.

———. 1991. Livestock grazing alters succession after fire in a Colorado subalpine forest. Pp. 84–90 in *Fire and the environment: Ecological and cultural perspectives. Proceedings of an international symposium, Knoxville, Tennessee, March 20–24, 1990,* ed. S. C. Nodvin and T. A. Waldrop. USDA Forest Service General Technical Report SE-69. Asheville, North Carolina: Southeastern Forest Experiment Station.

———. 1992a. The landscape ecology of large disturbances in the design and management of nature reserves. *Landscape Ecology* 7:181–94.

———. 1992b. Structure, disturbance, and change in the bristlecone pine forests of Colorado, U.S.A. *Arctic and Alpine Research* 24:17–26.

———. 1993. Spatially heterogenous multi-scale response of landscapes to fire suppression. *Oikos* 66:66–71.

———. 1995. Longterm response of disturbance landscapes to human intervention and global change. *Landscape Ecology* 10:143–59.

———. 1999. Progress and future directions in spatial modeling of forest landscapes. Pp. 333–49 in *Spatial modeling of forest landscape change: Approaches and applications,* ed. D. J. Mladenoff and W. L. Baker. Cambridge, England: Cambridge University Press.

———. 2000. Measuring and analyzing forest fragmentation in the Rocky Mountains and western United States. Pp. 55–94 in *Forest fragmentation in the southern Rocky Mountains,* ed. R. L. Knight, F. W. Smith, S. W. Buskirk, W. H. Romme, and W. L. Baker. Boulder: University Press of Colorado.

———. 2002. Indians and fire in the Rocky Mountains: The wilderness hypothesis renewed. Pp. 41–76 in *Fire, native peoples, and the natural landscape,* ed. T. R. Vale. Washington, D.C.: Island Press.

———. 2003. Fires and climate in forested landscapes of the U.S. Rocky Mountains. Pp. 120–57 in *Fire and climatic change in temperate ecosystems of the western Americas,* ed. T. T. Veblen, W. L. Baker, G. Montenegro, and T. W. Swetnam. New York: Springer.

————. 2006a. Fire history in ponderosa pine landscapes of Grand Canyon National Park: Is it reliable enough for management and restoration? *International Journal of Wildland Fire* 15:433–37.

————. 2006b. Fire and restoration of sagebrush ecosystems. *Wildlife Society Bulletin* 34:177–85.

————. In press. Pre-EuroAmerican and recent fire in sagebrush ecosystems. *Studies in Avian Biology.*

Baker, W. L., and Y. Cai. 1992. The r.le programs for multiscale analysis of landscape structure using the GRASS geographical information system. *Landscape Ecology* 7:291–302.

Baker, W. L., and G. K. Dillon. 2000. Plant and vegetation responses to edges in the southern Rocky Mountains. Pp. 221–45 in *Forest fragmentation in the southern Rocky Mountains,* ed. R. L. Knight, F. W. Smith, S. W. Buskirk, W. H. Romme, and W. L. Baker. Boulder: University Press of Colorado.

Baker, W. L., and D. Ehle. 2001. Uncertainty in surface-fire history: The case of ponderosa pine forests in the western United States. *Canadian Journal of Forest Research* 31:1205–26.

————. 2003. Uncertainty in fire history and restoration of ponderosa pine forests in the western United States. Pp. 319–33 in *Fire, fuel treatments, and ecological restoration: Conference proceedings, April 16–18, 2002, Fort Collins, Colorado,* ed. P. N. Omi and L. A. Joyce. USDA Forest Service Proceedings RMRS-P-29. Fort Collins, Colorado: Rocky Mountain Research Station.

Baker, W. L., P. H. Flaherty, J. D. Lindemann, T. T. Veblen, K. S. Eisenhart, and D. W. Kulakowski. 2002. Effect of vegetation on the impact of a severe blowdown in the southern Rocky Mountains, USA. *Forest Ecology and Management* 168:63–75.

Baker, W. L., J. J. Honaker, and P. J. Weisberg. 1995. Using aerial photography and GIS to map the forest-tundra ecotone in Rocky Mountain National Park, Colorado, for global change research. *Photogrammetric Engineering and Remote Sensing* 61:313–20.

Baker, W. L., and K. F. Kipfmueller. 2001. Spatial ecology of pre-Euro-American fires in a southern Rocky Mountain subalpine forest landscape. *Professional Geographer* 53:248–62.

Baker, W. L., and R. L. Knight. 2000. Roads and forest fragmentation in the southern Rocky Mountains. Pp. 221–45 in *Forest fragmentation in the southern Rocky Mountains,* ed. R. L. Knight, F. W. Smith, S. W. Buskirk, W. H. Romme, and W. L. Baker. Boulder: University Press of Colorado.

Baker, W. L., J. A. Munroe, and A. E. Hessl. 1997. The effects of elk on aspen in the winter range in Rocky Mountain National Park. *Ecography* 20:155–65.

Baker, W. L., and D. J. Shinneman. 2004. Fire and restoration of piñon-juniper woodlands in the western United States: A review. *Forest Ecology and Management* 189:1–21.

Baker, W. L., and T. T. Veblen. 1990. Spruce beetles and fires in the nineteenth-century subalpine forests of western Colorado, U.S.A. *Arctic and Alpine Research* 22:65–80.

Baker, W. L., T. T. Veblen, and R. L. Sherriff. 2007. Fire, fuels and restoration of ponderosa pine–Douglas fir forests in the Rocky Mountains, USA. *Journal of Biogeography* 34:251–69.

Baker, W. L., and P. J. Weisberg. 1995. Landscape analysis of the forest-tundra ecotone in Rocky Mountain National Park, Colorado. *Professional Geographer* 47:361–75.

Bakken, S. R. 1981. Predictions of fire behavior, fuel reduction, and tree damage from understory prescribed burning in Douglas-fir/ninebark habitat type of northern Idaho. M.S. thesis, University of Idaho, Moscow.

Balling, R. C. Jr., G. A. Meyer, and S. G. Wells. 1992. Climate change in Yellowstone National Park: Is the drought-related risk of wildfires increasing? *Climatic Change* 22:35–45.

Barclay, A. D., J. L. Betancourt, and C. D. Allen. 2004. Effects of seeding ryegrass (*Lolium multiflorum*) on vegetation recovery following fire in a ponderosa pine (*Pinus ponderosa*) forest. *International Journal of Wildland Fire* 13:183–94.

Barclay, H. J., C. Li, L. Benson, S. Taylor, and T. Shore. 2005. Effects of fire return rates on traversability of lodgepole pine forests for mountain pine beetle (Coleoptera: Scolytidae) and the use of patch metrics to estimate traversability. *Canadian Entomologist* 137:566–83.

Barger, R. L. 1980. The forest residues utilization program in brief. Pp. 7–25 in *Environmental consequences of timber harvesting in Rocky Mountain coniferous forests,* ed. USDA Forest Service. USDA Forest Service General Technical Report INT-90. Ogden, Utah: Intermountain Forest and Range Experiment Station.

Barlow, M., S. Nigam, and E. H. Berbery. 2001. ENSO, Pacific decadal variability, and U.S. summertime precipitation, drought, and stream flow. *Journal of Climate* 14:2105–28.

Barney, M. A., and N. C. Frischknecht. 1974. Vegetation changes following fire in the pinyon-juniper type of west-central Utah. *Journal of Range Management* 27:91–96.

Barrett, S. W. 1980a. Indian fires in the pre-settlement forests of western Montana. Pp. 35–41 in *Proceedings of the Fire History Workshop,* ed. M. A. Stokes and J. H. Dieterich. USDA Forest Service General Technical Report RM-81. Fort Collins, Colorado: Rocky Mountain Forest & Range Experiment Station.

———. 1980b. Indians & fire. *Western Wildlands* 6:17–21.

———. 1981. Relationship of Indian-caused fires to the ecology of western Montana forests. M.S. thesis, University of Montana, Missoula.

———. 1988. Fire suppression's effects on forest succession within a central Idaho wilderness. *Western Journal of Applied Forestry* 3:76–80.

———. 1994. Fire regimes on andesitic mountain terrain in northeastern Yellowstone National Park, Wyoming. *International Journal of Wildland Fire* 4:65–76.

———. 1999. Why burn wilderness? *Fire Management Notes* 59(4): 18–21.

———. 2000. Fire history along the ancient Lolo Trail. *Fire Management Today* 60(3): 21–28.

———. 2002. Moose fire: The historical perspective. *Fire Management Today* 62(4): 42–44.

———. 2004. Fire regimes in the northern Rockies. *Fire Management Today* 64(2): 32–38.

Barrett, S. W., and S. F. Arno. 1982. Indian fires as an ecological influence in the northern Rockies. *Journal of Forestry* 80:647–51.

———. 1988. *Increment-borer methods for determining fire history in coniferous forests.* USDA Forest Service General Technical Report INT-244. Ogden, Utah: Intermountain Research Station.

———. 1991. Classifying fire regimes and defining their topographic controls in the Selway-Bitterroot Wilderness. *Proceedings of the Conference on Fire and Forest Meteorology* 11:299–307.

———. 1999. Indian fires in the northern Rockies: Ethnohistory and ecology. Pp. 50–64 in *Indians, fire, and the land in the Pacific Northwest,* ed. R. Boyd. Corvallis: Oregon State University Press.

Barrett, S. W., S. F. Arno, and C. H. Key. 1991. Fire regimes of western larch–lodgepole pine forests in Glacier National Park, Montana. *Canadian Journal of Forest Research* 21:1711–20.

Barrett, S. W., S. F. Arno, and J. P. Menakis. 1997. *Fire episodes in the inland northwest (1540–1940) based on fire history data.* USDA Forest Service General Technical Report INT-GTR-370. Ogden, Utah: Intermountain Research Station.

Barrett, S. W., T. W. Swetnam, and W. L. Baker. 2005. Indian fire use: Deflating the legend. *Fire Management Today* 65(3): 31–34.

Barrows, J. S. 1951a. *Fire behavior in northern Rocky Mountain forests.* USDA Forest Service Station Paper 29. Missoula, Montana: Northern Rocky Mountain Forest & Range Experiment Station.

———. 1951b. *Forest fires in the northern Rocky Mountains.* USDA Forest Service Station Paper 28. Missoula, Montana: Northern Rocky Mountain Forest & Range Experiment Station.

Barrows, J. S., D. V. Sandberg, and J. D. Hart. 1976. *Lightning fires in northern Rocky Mountain forests.* Report to the USDA Forest Service, Intermountain Forest & Range Experiment Station, Northern Forest Fire Laboratory. Fort Collins: Colorado State University, Department of Forest and Wood Science.

Barth, R. C. 1970. Revegetation after a subalpine wildfire. M.S. thesis, Colorado State University, Fort Collins.

Bartlein, P. J., C. Whitlock, and S. L. Shafer. 1997. Future climate in the Yellowstone National Park region and its potential impact on vegetation. *Conservation Biology* 11:782–92.

Bartos, D. L. 2001. Landscape dynamics of aspen and conifer forests. Pp. 5–14 in *Sustaining aspen in western landscapes: Symposium proceedings,* ed. W. D. Shepperd, D. Binkley, D. L. Bartos, T. J. Stohlgren, and L. G. Eskew. USDA Forest Service Proceedings RMRS-P-18. Fort Collins, Colorado: Rocky Mountain Research Station.

Bartos, D. L., J. K. Brown, and G. D. Booth. 1994. Twelve years of biomass response in aspen communities following fire. *Journal of Range Management* 47:79–83.

Bartos, D. L., and R. B. Campbell Jr. 1998. Decline of quaking aspen in the interior West: Examples from Utah. *Rangelands* 20:17–24.

Bartos, D. L., W. F. Mueggler, and R. B. Campbell Jr. 1991. *Regeneration of aspen by suckering on burned sites in western Wyoming.* USDA Forest Service Research Paper INT-448. Ogden, Utah: Intermountain Research Station.

Bates, C. G. 1917. Forest succession in the central Rocky Mountains. *Journal of Forestry* 15:587–92.

Battaglia, M. A., J. M. Dodson, W. D. Shepperd, M. J. Platten, and O. M. Tallmadge. 2005. *Colorado Front Range fuel photo series.* USDA Forest Service General Technical Report

RMRS-GTR-155WWW. Fort Collins, Colorado: Rocky Mountain Research Station.

Baughman, R. G. 1981. *Why windspeeds increase on high mountain slopes at night.* USDA Forest Service General Technical Report INT-276. Ogden, Utah: Intermountain Forest and Range Experiment Station.

Baumeister, D., and R. M. Callaway. 2006. Facilitation by *Pinus flexilis* during succession: A hierarchy of mechanisms benefits other plant species. *Ecology* 87:1816–30.

Beaty, R. M., and A. H. Taylor. 2001. Spatial and temporal variation of fire regimes in a mixed conifer forest landscape, southern Cascades, California, USA. *Journal of Biogeography* 28:955–66.

Beavers, A. M. 2001. Vegetation recovery following fire in west-central Colorado. MS thesis, Colorado State University, Fort Collins.

Bebi, P., D. Kulakowski, and T. T. Veblen. 2003. Interactions between fire and spruce beetles in a subalpine Rocky Mountain forest landscape. *Ecology* 84:362–71.

Beck, J. L., J. W. Connelly, and K. P. Reese. In press. Recovery of greater sage-grouse habitat features in Wyoming big sagebrush following prescribed fire. *Restoration Ecology.*

Beetle, D. E. 1997. Recolonization of burned aspen groves by land snails. *Yellowstone Science* 5:6–8.

Beidleman, R. G. 1957. Fire burn succession studies in the northern Colorado Rockies. *Proceedings of the Colorado-Wyoming Academy of Science* 4:39.

———. 1967. Twenty-year successional studies in northern Colorado. *Proceedings of the Colorado-Wyoming Academy of Science* 5:63–64.

Beighley, M., and J. Bishop. 1990. Fire behavior in high-elevation timber. *Fire Management Notes* 51(2): 23–28.

Bell, T., K. Tolhurst, and M. Wouters. 2005. Effects of the fire retardant Phos-Chek on vegetation in eastern Australian heathlands. *International Journal of Wildland Fire* 14:199–211.

Belnap, J., and O. L. Lange, eds. 2001. *Biological soil crusts: Structure, function, and management.* New York: Springer-Verlag.

Belsky, A. J., and D. M. Blumenthal. 1997. Effects of livestock grazing on stand dynamics and soils in upland forests of the interior West. *Conservation Biology* 11:315–27.

Bendell, J. F. 1974. Effects of fire on birds and mammals. Pp. 73–138 in *Fire and ecosystems,* ed. T. T. Kozlowski and C. E. Ahlgren. New York: Academic Press.

Benedict, J. B. 2000. Game drives of the Devil's Thumb Pass area. Pp. 18–94 in *This land of shining mountains: Archeological studies in Colorado's Indian Peaks Wilderness Area,* ed. E. S. Cassells. Research Report No. 8. Ward, Colorado: Center for Mountain Archeology.

Benkman, C. W. 1995. Wind dispersal capacity of pine seeds and the evolution of different seed dispersal modes in pines. *Oikos* 73:221–24.

Benkman, C. W., and A. M. Siepielski. 2004. A keystone selective agent? Pine squirrels and the frequency of serotiny in lodgepole pine. *Ecology* 85:2082–87.

Bennetts, R. E., G. C. White, F. G. Hawksworth, and S. E. Severs. 1996. The influence of dwarf mistletoe on bird communiies in Colorado ponderosa pine forests. *Ecological Applications* 6:899–909.

Benoit, J. W., and D. J. Strauss. 1994. Preliminary development of a prediction model for lightning-caused forest fires. *Proceedings of the Conference on Fire and Forest Meteorology* 12:422–27.

Benson, L. A., C. E. Braun, and W. C. Leininger. 1991. Sage grouse response to burning in the big sagebrush type. Pp. 97–104 in *Issues and technology in the management of impacted wildlife: Proceedings of a national symposium, Snowmass Resort, Colorado, April 8–10, 1991,* ed. R. D. Comer, P. R. Davis, S. Q. Foster, C. V. Grant, S. Rush, O. Thorne II, and J. Todd. Boulder, Colorado: Thorne Ecological Institute.

Benson, N. C., and L. L. Kurth. 1995. Vegetation establishment on rehabilitated bulldozer lines after the 1988 Red Bench fire in Glacier National Park. Pp. 164–67 in *Proceedings: Symposium on fire in wilderness and park management (March 30–April 1, 1993, Missoula, MT),* ed. J. K. Brown, R. W. Mutch, C. W. Spoon, and R. H. Wakimoto. USDA Forest Service General Technical Report INT-GTR-320. Ogden, Utah: Intermountain Research Station.

Benson, R. E., and J. A. Schlieter. 1980. Woody material in northern Rocky Mountain forests: Volume, characteristics, and changes with harvesting. Pp. 27–36 in *Environmental consequences of timber harvesting in Rocky Mountain coniferous forests,* ed. USDA Forest Service. USDA Forest Service General Technical Report INT-90. Ogden, Utah: Intermountain Forest and Range Experiment Station.

Beschta, R. L., J. J. Rhodes, J. B. Kauffman, R. E. Gresswell, G. W. Minshall, J. R. Karr, D. A. Perry, F. R. Hauer, and C. A. Frissell. 2004. Postfire management on forested public lands of the western United States. *Conservation Biology* 18:957–67.

Bessie, W. C., and E. A. Johnson. 1995. The relative importance of fuels and weather on fire behavior in subalpine forests. *Ecology* 76:747–62.

Betters, D. R., and R. F. Woods. 1981. Uneven-aged stand structure and growth of Rocky Mountain aspen. *Journal of Forestry* 79:673–76.

Bevins, C. D. 1980. *Estimating survival and salvage potential of fire-scarred Douglas-fir.* USDA Forest Service Research Note INT-287. Ogden, Utah: Intermountain Forest and Range Experiment Station.

Beyers, J. L. 2004. Postfire seeding for erosion control: Effectiveness and impacts on native plant communities. *Conservation Biology* 18:947–56.

Bigler, C., D. Kulakowski, and T. T. Veblen. 2005. Multiple disturbance interactions and drought influence fire severity in Rocky Mountain subalpine forests. *Ecology* 86:3018–29.

Bigler, R. L. 1976. Age and size class distribution in the *Abies lasiocarpa/Luzula hitchcockii–Menziesa ferruginea* habitat type in northwestern Montana. M.S. thesis, University of Idaho, Moscow.

Billings, W. D. 1969. Vegetational pattern near alpine timberline as affected by fire-snowdrift interactions. *Vegetatio* 19:192–207.

———. 1990. *Bromus tectorum,* a biotic cause of ecosystem impoverishment in the Great Basin. Pp. 301–22 in *The earth in transition: Patterns and processes of biotic impoverishment,* ed. G. M. Woodwell. Cambridge, England: Cambridge University Press.

Black, A., M. Williamson, and D. Doane. 2008. Wildland fire use barriers and facilitators. *Fire Management Today* 68(1): 10–14.

Blackford, J. L. 1955. Woodpecker concentration in burned forest. *The Condor* 53:28–30.

Blaisdell, J. P. 1953. *Ecological effects of planned burning of sagebrush-grass range on the Upper Snake River Plains.* USDA Technical Bulletin 1075. Washington, D.C.: U.S. Government Printing Office.

Blaisdell, J. P., and W. F. Mueggler. 1956. Sprouting of bitterbrush (*Purshia tridentata*) following burning or top removal. *Ecology* 37:365–70.

Blanchard, B. M., and R. R. Knight. 1990. Reactions of grizzly bears, *Ursus arctos horribilis*, to wildfire in Yellowstone National Park, Wyoming. *Canadian Field Naturalist* 104:592–94.

———. 1996. Effects of wildfire on grizzly bear movements and food habits. Pp. 117–22 in *The ecological implications of fire in Greater Yellowstone. Proceedings second biennial conference on the Greater Yellowstone ecosystem, Yellowstone National Park, Sept. 19–21, 1993,* ed. J. M. Greenlee. Fairfield, Washington: International Association of Wildland Fire.

Blank, R. R., and J. A. Young. 1998. Heated substrate and smoke: Influence on seed emergence and plant growth. *Journal of Range Management* 51:577–83.

Blodgett, J. T., and K. F. Sullivan. 2004. First report of white pine blister rust on Rocky Mountain bristlecone pine. *Plant Disease* 88:311.

Bock, C. E., and J. H. Bock. 1983. Responses of birds and deer mice to prescribed burning in ponderosa pine. *Journal of Wildlife Management* 47:836–40.

———. 1987. Avian habitat occupancy following fire in a Montana shrubsteppe. *Prairie Naturalist* 19:153–58.

———. 2000. *The view from Bald Hill: Thirty years in an Arizona grassland.* Berkeley: University of California Press.

Bock, J. H., and C. E. Bock. 1984. Effect of fires on woody vegetation in the pine-grassland ecotone of the southern Black Hills. *American Midland Naturalist* 112:35–42.

Boe, K. N. 1947–48. Is Douglas-fir replacing ponderosa pine in the cut-over stands in western Montana? *Proceedings of the Montana Academy of Sciences* 7–8: 42–50.

———. 1954. Periodicity of cone crops for five Montana conifers. *Proceedings of the Montana Academy of Sciences* 14:5–9.

Bollinger, W. H. 1973. The vegetation patterns after fire at the alpine forest-tundra ecotone in the Colorado Front Range. Ph.D. dissertation, University of Colorado, Boulder.

Boltz, M. 1994. Factors influencing postfire sagebrush regeneration in south-central Idaho. Pp. 281–90 in *Proceedings: Ecology and management of annual rangelands,* ed. S. B. Monsen and S. G. Kitchen. USDA Forest Service General Technical Report INT-GTR-313. Ogden, Utah: Intermountain Research Station.

Bond, W. J., and J. J. Midgley. 2001. Ecology of sprouting in woody plants: The persistence niche. *Trends in Ecology and Evolution* 16:45–51.

Bonnet, V. H., A. W. Schoettle, and W. D. Shepperd. 2005. Postfire environmental conditions influence the spatial pattern of regeneration for *Pinus ponderosa. Canadian Journal of Forest Research* 35:37–47.

Bonnicksen, T. M. 1989. Nature vs. man(agement). *Journal of Forestry* 87:41–43.

Bormann, F. H., and G. E. Likens. 1979. *Pattern and process in a forested ecosystem.* New York: Springer-Verlag.

Bosworth, D., and H. Brown. 2007a. After the timber wars: Community-based stewardship. *Journal of Forestry* 105:271–73.

———. 2007b. Investing in the future: Ecological restoration and the USDA Forest Service. *Journal of Forestry* 105:208–11.

Bosworth, N. 1999. Dead and down fuel consumption resulting from prescribed burning in Black Hills ponderosa pine. M.S. thesis, Colorado State University, Fort Collins.

Bothwell, P. D. 2000. Predicting lightning associated with dry thunderstorms in the western United States. Unpublished report. Norman, Oklahoma: Storm Prediction Center. www.thelightningpeople.com/htm/about/events/ildc/ildc2000/docs/25*Bothwell*E File.pdf.

Botkin, D. B., J. F. Janak, and J. R. Wallis. 1972. Some ecological consequences of a computer model of forest growth. *Journal of Ecology* 60:849–72.

Bova, A. S., and M. B. Dickinson. 2005. Linking surface-fire behavior, stem heating, and tissue necrosis. *Canadian Journal of Forest Research* 35:814–22.

Bowker, M. A., J. Belnap, R. Rosentreter, and B. Graham. 2004. Wildfire-resistant biological soil crusts and fire-induced loss of soil stability in Palouse prairies, USA. *Applied Soil Ecology* 26:41–52.

Boyce, M. S., and E. H. Merrill. 1991. Effects of the 1988 fires on ungulates in Yellowstone National Park. *Proceedings of the Tall Timbers Fire Ecology Conference* 17:121–32.

Boyce, R. B. 1985. Conifer germination and seedling establishment on burned and unburned seedbeds. M.S. thesis, University of Idaho, Moscow.

Boyce, R. B., and L. F. Neuenschwander. 1989. Douglas-fir germination and seedling establishment on burned and unburned seedbeds. Pp. 69–74 in *Prescribed fire in the Intermountain region: Forest site preparation and range improvement, symposium proceedings,* ed. D. M. Baumgartner, D. W. Breuer, B. A. Zamora, L. F. Neuenschwander, and R. H. Wakimoto. Pullman, Washington: Washington State University, Cooperative Extension Service.

Boyden, S., D. Binkley, and W. Shepperd. 2005. Spatial and temporal patterns in structure, regeneration, and mortality of an old-growth ponderosa pine forest in the Colorado Front Range. *Forest Ecology and Management* 219:43–55.

Brackebusch, A. P. 1973. Fuel management: A prerequisite not an alternative to fire control. *Journal of Forestry* 71:637–39.

Bradley, A. F., W. C. Fischer, and N. V. Noste. 1992. *Fire ecology of the forest habitat types of eastern Idaho and western Wyoming.* USDA Forest Service General Technical Report INT-290. Ogden, Utah: Intermountain Research Station.

Bradley, A. F., N. V. Noste, and W. C. Fischer. 1992. *Fire ecology of forests and woodlands in Utah.* USDA Forest Service General Technical Report INT-287. Ogden, Utah: Intermountain Research Station.

Bradley, T., J. Gibson, and W. Bunn. 2006. Fire severity and intensity during spring burning in natural and masticated mixed shrub woodlands. Pp. 419–28 in *Fuels management— How to measure success: Conference proceedings, 28–30 March 2006, Portland, Oregon,* ed. P. L. Andrews and B. W. Butler. USDA Forest Service Proceedings RMRS-P-41. Fort Collins, Colorado: Rocky Mountain Research Station.

Bradshaw, L. S., J. E. Deeming, R. E. Burgan, and J. D. Cohen. 1983. *The 1978 national fire-danger rating system: Technical documentation.* USDA Forest Service General Technical Report INT-169. Ogden, Utah: Intermountain Research Station.

Branson, D. H. 2005. Effects of fire on grasshopper assemblages in a northern mixed-grass prairie. *Environmental Entomology* 34:1109–13.

Breshears, D. D., N. S. Cobb, P. M. Rich, K. P. Price, C. D. Allen, R. G. Balice, W. H. Romme, J. H. Kastens, M. L. Floyd, J. Belnap, J. J. Anderson, O. B. Myers, and C. W. Meyer. 2005. Regional vegetation die-off in response to global-change–type drought. *Proceedings of the National Academy of Sciences* 102:15144–48.

Bridge, S. R. J., K. Miyanishi, and E. A. Johnson. 2005. A critical evaluation of fire suppression effects in the boreal forest of Ontario. *Forest Science* 51:41–50.

Britton, C. M., and R. G. Clark. 1985. Effects of fire on sagebrush and bitterbrush. Pp. 22–26 in *Rangeland fire effects: A symposium,* ed. K. Sanders and J. Durham. Boise, Idaho: Bureau of Land Management, Idaho State Office.

Britton, C. M., B. L. Karr, and F. A. Sneva. 1977. A technique for measuring rate of fire spread. *Journal of Range Management* 30:395–97.

Brockway, D. G., R. G. Gatewood, and R. B. Paris. 2002. Restoring fire as an ecological process in shortgrass prairie ecosystems: Initial effects of prescribed burning during the dormant and growing seasons. *Journal of Environmental Management* 65:135–52.

Brooks, M. L., C. M. D'Antonio, D. M. Richardson, J. B. Grace, J. E. Keeley, J. M. DiTomaso, R. J. Hobbs, M. Pellant, and D. Pyke. 2004. Effects of invasive alien plants on fire regimes. *BioScience* 54:677–88.

Brotak, E. A. 1983. Weather conditions associated with major wildland fires in the western United States. *Proceedings of the Conference on Fire and Forest Meteorology* 7:7–8.

Brown, A. A. 1940. Lessons of the McVey fire, Black Hills National Forest. *USDA Forest Service, Fire Control Notes* 4(2): 63–67.

———. 1956. Prospects of extinguishing fires from the air. *Fire Control Notes* 17(1): 5–8.

———. 2003. The factors and circumstances that led to the Blackwater fire tragedy. *Fire Management Today* 63(3): 11–14.

Brown, D. E. 1994. *Biotic communities: Southwestern United States and northwestern Mexico.* Salt Lake City: University of Utah Press.

Brown, H. E. 1958. Gambel oak in west-central Colorado. *Ecology* 39:317–27.

Brown, J. K. 1970. *Vertical distribution of fuel in spruce-fir logging slash.* USDA Forest Service Research Paper INT-81. Ogden, Utah: Intermountain Forest and Range Experiment Station.

———. 1975. Fire cycles and community dynamics in lodgepole pine forests. Pp. 429–56 in *Management of lodgepole pine ecosystems: Symposium proceedings, Pullman, Washington, Oct. 9–11, 1973,* ed. D. M. Baumgartner. Pullman: Washington State University, Cooperative Extension Service.

———. 1980. Influence of harvesting and residues on fuels and fire management. Pp. 417–32 in *Environmental consequences of timber harvesting in Rocky Mountain coniferous forests,* ed. USDA Forest Service. USDA Forest Service General Technical Report INT-90. Ogden, Utah: Intermountain Forest and Range Experiment Station.

———. 1981. Bulk densities of nonuniform surface fuels and their application to fire modeling. *Forest Science* 27:667–83.

———. 1982. *Fuel and fire behavior prediction in big sagebrush.* USDA Forest Service Research Paper INT-290. Ogden, Utah: Intermountain Forest and Range Experiment Station.

———. 1985. The "unnatural fuel buildup" issue. Pp. 127–28 in *Proceedings: Symposium and Workshop on Wilderness Fire,* ed. J. E. Lotan, B. M. Kilgore, W. C. Fischer, and R. W. Mutch. USDA Forest Service General Technical Report INT-182. Ogden, Utah: Intermountain Forest & Range Experiment Station.

Brown, J. K., S. F. Arno, S. W. Barrett, and J. P. Menakis. 1994. Comparing the prescribed natural fire program with presettlement fires in the Selway-Bitterroot Wilderness. *International Journal of Wildland Fire* 4:157–68.

Brown, J. K., and N. V. DeByle. 1987. Fire damage, mortality and suckering in aspen. *Canadian Journal of Forest Research* 17:1100–09.

———. 1989. *Effects of prescribed fire on biomass and plant succession in western aspen.* USDA Forest Service Research Paper INT-412. Ogden, Utah: Intermountain Research Station.

Brown, J. K., M. A. Marsden, K. C. Ryan, and E. D. Reinhardt. 1985. *Predicting duff and woody fuel consumption by prescribed fire in the northern Rocky Mountains.* USDA Forest Service Research Paper INT-337. Ogden, Utah: Intermountain Forest and Range Experiment Station.

Brown, J. K., R. D. Oberheu, and C. M. Johnston. 1982. *Handbook for inventorying surface fuels and biomass in the interior West.* USDA Forest Service General Technical Report INT-129. Ogden, Utah: Intermountain Forest and Range Experiment Station.

Brown, J. K., E. D. Reinhardt, and K. A. Kramer. 2003. *Coarse woody debris: Managing benefits and fire hazard in the recovering forest.* USDA Forest Service General Technical Report RMRS-GTR-105. Fort Collins, Colorado: Rocky Mountain Research Station.

Brown, J. K., and T. E. See. 1981. *Downed dead woody fuel and biomass in the northern Rocky Mountains.* USDA Forest Service General Technical Report INT-117. Ogden, Utah: Intermountain Forest and Range Experiment Station.

Brown, J. K., and D. G. Simmerman. 1986. *Appraising fuels and flammability in western aspen: A prescribed fire guide.* USDA Forest Service General Technical Report INT-205. Ogden, Utah: Intermountain Forest and Range Experiment Station.

Brown, J. K., and J. K. Smith, eds. 2000. *Wildland fire in ecosystems: Effects of fire on flora.* USDA Forest Service General Technical Report RMRS-GTR-42, vol. 2. Fort Collins, Colorado: Rocky Mountain Research Station.

Brown, K. J., J. S. Clark, E. C. Grimm, J. J. Donovan, P. G. Mueller, B. C. S. Hansen, and I. Stefanova. 2005. Fire cycles in North American interior grasslands and their relation to prairie drought. *Proceedings of the National Academy of Sciences* 102:8865–70.

Brown, P. M. 2006. Climate effects on fire regimes and tree recruitment in Black Hills ponderosa pine forests. *Ecology* 87:2500–10.

Brown, P. M., and B. Cook. 2005. Early settlement forest structure in Black Hills ponderosa pine forests. *Forest Ecology and Management* 223:284–90.

Brown, P. M., D. R. D'Amico, A. T. Carpenter, and D. Andrews. 2001. Restoration of montane ponderosa pine forests in the Colorado Front Range. *Ecological Restoration* 19:19–26.

Brown, P. M., E. K. Heyerdahl, S. G. Kitchen, and M. H. Weber. 2008. Climate effects on historical fires (1630–1900) in Utah. *International Journal of Wildland Fire* 17:28–39.

Brown, P. M., M. R. Kauffman, and W. D. Sheppard. 1999. Long-term, landscape patterns of past fire events in a montane ponderosa pine forest of central Colorado. *Landscape Ecology* 14:513–32.

Brown, P. M., M. G. Ryan, and T. G. Andrews. 2000. Historical surface fire frequency in ponderosa pine stands in research natural areas, central Rocky Mountains and Black Hills, USA. *Natural Areas Journal* 20:133–39.

Brown, P. M., and A. W. Schoettle. 2008. Fire and stand history in two limber pine (*Pinus flexilis*) and Rocky Mountain bristlecone pine (*Pinus aristata*) stands in Colorado. *International Journal of Wildland Fire* 17:339–47.

Brown, P. M., and W. D. Shepperd. 2001. Fire history and fire climatology along a 5 degree gradient in latitude in Colorado and Wyoming, USA. *Paleobotanist* 50:133–40.

Brown, P. M., W. D. Shepperd, C. C. Brown, S. A. Mata, and D. L. McClain. 1995. *Oldest known Engelmann spruce.* USDA Forest Service Research Note RM-RN-534. Fort Collins, Colorado: Rocky Mountain Forest and Range Experiment Station.

Brown, P. M., W. D. Shepperd, S. A. Mata, and D. L. McClain. 1998. Longevity of windthrown logs in a subalpine forest of central Colorado. *Canadian Journal of Forest Research* 28:932–36.

Brown, P. M., and C. H. Sieg. 1996. Fire history in interior ponderosa pine communities of the Black Hills, South Dakota, USA. *International Journal of Wildland Fire* 6:97–105.

———. 1999. Historical variability in fire at the ponderosa pine–Northern Great Plains prairie ecotone, southeastern Black Hills, South Dakota. *Ecoscience* 6:539–47.

Brown, P. M., and R. Wu. 2005. Climate and disturbance forcing of episodic tree recruitment in a southwestern ponderosa pine landscape. *Ecology* 86:3030–38.

Brown, T. J., B. L. Hall, and A. L. Westerling. 2004. The impact of twenty-first century climate change on wildland fire danger in the western United States: An applications perspective. *Climatic Change* 62:365–88.

Brown, T. K., and L. Bright. 1997. Wildlife habitat preservation and enrichment during and after fires. Pp. 65–67 in *Proceedings: First conference on fire effects on rare and endangered species and habitats,* ed. J. M. Greenlee. Fairfield, Washington: International Association of Wildland Fire.

Brunelle, A., and C. Whitlock. 2003. Postglacial fire, vegetation, and climate history in the Clearwater Range, northern Idaho, USA. *Quaternary Research* 60:307–18.

Brunelle, A., C. Whitlock, P. Bartlein, and K. Kipfmueller. 2005. Holocene fire and vegetation along environmental gradients in the northern Rocky Mountains. *Quaternary Science Reviews* 24:2281–300.

Brunner Jass, R. M. 1999. Fire occurrence and paleoecology at Alamo Bog and Chihuahueños Bog, Jemez Mountains, New Mexico. M.S. thesis, Northern Arizona University, Flagstaff.

Buechling, A., and W. L. Baker. 2004. A fire history from tree rings in a high-elevation forest of Rocky Mountain National Park. *Canadian Journal of Forest Research* 34:1259–73.

Bumstead, A. P. 1943. Sunspots and lightning fire. *Journal of Forestry* 41:69–70.

Burgan, R. E., R. W. Klaver, and J. M. Klaver. 1998. Fuel models and fire potential from satellite and surface observations. *International Journal of Wildland Fire* 8:159–70.

Burgan, R. E., and R. C. Rothermel. 1984. *BEHAVE: Fire behavior prediction and fuel modeling system.* USDA Forest Service General Technical Report INT-167. Ogden, Utah: Intermountain Forest and Range Experiment Station.

Burkhardt, J. W., and E. W. Tisdale. 1976. Causes of juniper invasion in southwestern Idaho. *Ecology* 57:472–84.

Burns, R. M., and B. H. Honkala. 1990. *Silvics of North America.* Vol. 1, *Conifers.* USDA Forest Service Agriculture Handbook No. 654. Washington, D.C.: U.S. Government Printing Office.

Busch, D. E. 1995. Effects of fire on southwestern riparian plant community structure. *Southwestern Naturalist* 40:259–67.

Busse, M. D., K. R. Hubbert, G. O. Fiddler, C. J. Shestak, and R. F. Powers. 2005. Lethal soil temperatures during burning of masticated forest residues. *International Journal of Wildland Fire* 14:267–76.

Butler, B. W., R. A. Bartlette, L. S. Bradshaw, J. D. Cohen, P. L. Andrews, T. Putnam, and R. J. Mangan. 1998. *Fire behavior associated with the 1994 South Canyon fire on Storm King Mountain, Colorado.* USDA Forest Service Research Paper RMRS-RP-9. Fort Collins, Colorado: Rocky Mountain Research Station.

Butler, B. W., M. A. Finney, P. L. Andrews, and F. A. Albini. 2004. A radiation-driven model for crown fire spread. *Canadian Journal of Forest Research* 34:1588–99.

Butler, B. W., and T. D. Reynolds. 1997. *Wildfire case study: Butte City Fire, Southeastern Idaho, July 1, 1994.* USDA Forest Service General Technical Report INT-GTR-351. Ogden, Utah: Intermountain Research Station.

Butler, D. R. 1986. Conifer invasion of subalpine meadows, central Lemhi Mountains, Idaho. *Northwest Science* 60:166–73.

Butts, D. B. 1985. Case study: The Ouzel fire, Rocky Mountain National Park. Pp. 248–51 in *Proceedings: Symposium and workshop on wilderness fire,* ed. J. E. Lotan, B. M. Kilgore, W. C. Fischer, and R. W. Mutch. USDA Forest Service General Technical Report INT-182. Ogden, Utah: Intermountain Forest and Range Experiment Station.

Byram, G. M. 1959. Forest fire behavior. Pp. 90–123 in *Forest Fire Control and Use,* ed. K. P. Davis. New York: McGraw-Hill.

Byrne, M. W. 2002. Habitat use by female greater sage grouse in relation to fire at Hart Mountain National Antelope Refuge, Oregon. M.S. thesis, Oregon State University, Corvallis.

Cain, M. D. 1984. Height of stem-bark char underestimates flame length in prescribed burns. *Fire Management Notes* 45(1): 17–21.

Callaway, R. M. 1998. Competition and facilitation on elevation gradients in subalpine forests of the northern Rocky Mountains, USA. *Oikos* 82:561–73.

Campbell, G. S., J. D. Jungbauer Jr., K. L. Bristow, and R. D. Hungerford. 1995. Soil temperature and water content beneath a surface fire. *Soil Science* 159:363–74.

Castrale, J. S. 1982. Effects of two sagebrush control methods on nongame birds. *Journal of Wildlife Management* 46:945–52.

Caton, E. L. 1996. Effects of fire and salvage logging on the cavity-nesting bird community in northwestern Montana. Ph.D. dissertation, University of Montana, Missoula.

Cattelino, P. J., I. R. Noble, R. O. Slatyer, and S. R. Kessell. 1979. Predicting the multiple pathways of plant succession. *Environmental Management* 3:41–50.

Cayan, D. R. 1996. Interannual climate variability and snowpack in the western United States. *Journal of Climate* 9:928–48.

Chambers, J. C., S. B. Vander Wall, and E. W. Schupp. 1999. Seed and seedling ecology of piñon and juniper species in the pygmy woodlands of western North America. *Botanical Review* 65:1–38.

Changnon, D., T. B. McKee, and N. J. Doesken. 1993. Annual snowpack patterns across the Rockies: Long-term trends and associated 500-mb synoptic patterns. *Monthly Weather Review* 121:633–47.

Charron, I., and E. A. Johnson. 2006. The importance of fires and floods on tree ages along mountainous gravel-bed streams. *Ecological Applications* 16:1757–70.

Chen, J., J. F. Franklin, and T. A. Spies. 1995. Growing-season microclimatic gradients from clearcut edges into old-growth Douglas-fir forests. *Ecological Applications* 5:74–86.

Cheney, N. P. 1981. Fire behaviour. Pp. 151–75 in *Fire and the Australian biota,* ed. A. M. Gill, R. H. Groves, and I. R. Noble. Canberra: Australian Academy of Sciences.

Cheney, N. P., J. S. Gould, and W. R. Catchpole. 1998. Prediction of fire spread in grasslands. *International Journal of Wildland Fire* 8:1–13.

Chew, R. M., B. B. Butterworth, and R. Grechman. 1959. The effects of fire on the small mammal populations of chaparral. *Journal of Mammalogy* 40:253.

Cholewa, A. F. 1977. Successional relationships of vegetation composition to logging, burning, and grazing in the Douglas-fir/Physocarpus habitat type of northern Idaho. M.S. thesis, University of Idaho, Moscow.

Chong, G., T. Stohlgren, C. Crosier, S. Simonson, G. Newman, and E. Petterson. 2003. Ecological effects of the Hayman fire. Part 7. Key invasive nonnative plants. Pp. 244–49 in *Hayman fire case study,* ed. R. T. Graham. USDA Forest Service General Technical Report RMRS-GTR-114. Fort Collins, Colorado: Rocky Mountain Research Station.

Christensen, E. M. 1964. Succession in a mountain brush community in central Utah. *Proceedings of Utah Academy of Science, Arts, and Letters* 41:10–13.

Christensen, J. H., B. Hewitson, A. Busuioc, A. Chen, X. Gao, I. Held, R. Jones, R. K. Kolli, W.-T. Kwon, R. Laprise, V. Magaña Rueda, L. Mearns, C. G. Menéndez, J. Räisänen, A. Rinke, A. Sarr, and P. Whetton. 2007. Regional climate projections. Pp. 847–940 in *Climate change 2007: The physical science basis. Contribution of working group 1 to the fourth assessment report of the intergovernmental panel on climate change,* ed. S. Solomon, D. Qin, M. Manning, Z. Chen, M. Marquis, K. B. Averyt, M. Tignor and H. L. Miller. Cambridge, England: Cambridge University Press.

Christiansen, T. A., and R. J. Lavigne. 1996. Habitat requirements for the reestablishment of litter invertebrates following the 1988 Yellowstone National Park fires. Pp. 147–50 in *The ecological implications of fire in Greater Yellowstone: Proceedings second biennial conference* on the Greater Yellowstone ecosystem, Yellowstone National Park, Sept. 19–21, 1993, ed. J. M. Greenlee. Fairfield, Washington: International Association of Wildland Fire.

Clagg, H. B. 1975. Fire ecology in high-elevation forests in Colorado. Master's thesis, Colorado State University, Fort Collins.

Clark, D. L. 1991. The effect of fire on Yellowstone ecosystem seed banks. M.S thesis, Montana State University, Bozeman.

Clark, J. S. 1988. Particle motion and the theory of charcoal analysis: Source area, transport, deposition, and sampling. *Quaternary Research* 30:67–80.

Clary, W. P. 1988. *Plant density and cover response to several seeding techniques following wildfire.* USDA Forest Service Research Note INT-384. Ogden, Utah: Intermountain Research Station.

Clements, C. B., S. Zhong, S. Goodrick, J. Li, B. E. Potter, X. Bian, W. E. Heilman, J. J. Charney, R. Perna, M. Jang, D. Lee, S. Street, and G. Aumann. 2007. Observing the dynamics of wildland grass fires: Fireflux—A field validation experiment. *Bulletin of the American Meteorological Society* 88:1369–82.

Clements, F. E. 1910. *The life history of lodgepole burn forests.* USDA Bulletin No. 79. Washington, D.C.: U.S. Government Printing Office.

———. 1916. *Plant succession: An analysis of the development of vegetation.* Publication No. 242. Washington, D.C.: Carnegie Institute of Washington.

Clewell, A. F., and J. Aronson. 2007. *Ecological restoration: Principles, values, and structure of an emerging profession.* Washington, D.C.: Island Press.

Clifton, N. A. 1981. Response to prescribed fire in a Wyoming big sagebrush/bluebunch wheatgrass habitat type. M.S. thesis, University of Idaho, Moscow.

Close, K. R., and R. H. Wakimoto. 1998. Geographic information systems (GIS) for hazard and risk analysis, and defensible space assessment, in the wildland/urban interface. Pp. 237–46 in *Fire management under fire (adapting to change): Proceedings of the 1994 Interior West Fire Council Meeting and Program, November 1–4, 1994, Coeur d'Alene, Idaho,* ed. K. Close and R. A. Bartlette. Fairfield, Washington: International Association of Wildland Fire.

Cochran, T. R. 1937. The Sundance fire. *USDA Forest Service, Rocky Mountain Region Bulletin* 20:35–36.

Coen, J., S. Mahalingam, and J. Daily. 2004. Infrared imagery of crown-fire dynamics during FROSTFIRE. *Journal of Applied Meteorology* 43:1241–59.

Coffin, D. P., and W. K. Lauenroth. 1989. Spatial and temporal variation in the seed bank of a semiarid grassland. *American Journal of Botany* 76:53–58.

Cohen, D. 2006. Wildland fire use as a prescribed fire primer. *Fire Management Today* 66(4): 47–49.

Cohen, J. D. 2000. Preventing disaster: Home ignitability in the wildland–urban interface. *Journal of Forestry* 98:15–21.

Cohen, S., and D. Miller. 1978. *The big burn: The Northwest's forest fire of 1910.* Missoula, Montana: Pictorial Histories Publishing.

Colket, E. C. 2003. Long-term vegetation dynamics and post-fire establishment patterns of sagebrush steppe. M.S. thesis, University of Idaho, Moscow.

Collins, B. M., P. N. Omi, and P. L. Chapman. 2006. Regional relationships between climate and wildfire-burned area in the Interior West, USA. *Canadian Journal of Forest Research* 36:699–709.

Connaughton, C. A. 1936. Fire damage in the ponderosa pine type in Idaho. *Journal of Forestry* 34:46–51.

———. 1970. Fire related research and development needs. Pp. 42–64 in *The role of fire in the Intermountain West: Proceedings of a symposium, Missoula, Montana, Oct. 27–29, 1970*, ed. R. Taber. Missoula, Montana: Intermountain Fire Research Council and University of Montana.

Connelly, J. W., W. J. Arthur, and O. D. Markham. 1981. Sage grouse leks on recently disturbed sites. *Journal of Range Management* 34:153–54.

Connelly, J. W., K. P. Reese, R. A. Fischer, and W. L. Wakkinen. 2000. Response of a sage grouse breeding population to fire in southeastern Idaho. *Wildlife Society Bulletin* 28:90–96.

Converse, S. J., W. M. Block, and G. C. White. 2006. Small mammal population and habitat responses to forest thinning and prescribed fire. *Forest Ecology and Management* 228:263–73.

Converse, S. J., G. C. White, K. L. Farris, and S. Zack. 2006. Small mammals and forest fuel reduction: National-scale responses to fire and fire surrogates. *Ecological Applications* 16:1717–29.

Cook, E. R., D. M. Meko, D. W. Stahle, and M. K. Cleaveland. 1999. Drought reconstructions for the continental United States. *Journal of Climate* 12:1145–62.

Cook, E. R., D. M. Meko, and C. W. Stockton. 1997. A new assessment of possible solar and lunar forcing of the bidecadal drought rhythm in the western United States. *Journal of Climate* 10:1343–56.

Cook, E. R., C. A. Woodhouse, C. M. Eakin, D. M. Meko, and D. W. Stahle. 2004. Long-term aridity changes in the western United States. *Science* 306:1015–18.

Cook, J. G., T. J. Hershey, and L. L. Irwin. 1994. Vegetative response to burning on Wyoming mountain-shrub big game ranges. *Journal of Range Management* 47:296–302.

Coop, J. D., and T. J. Givnish. 2007. Spatial and temporal patterns of recent forest encroachment in montane grasslands of the Valles Caldera, New Mexico, USA. *Journal of Biogeography* 34:914–27.

Cooper, S. V., P. Lesica, and G. M. Kudray. 2007. Post-fire recovery of Wyoming big sagebrush shrub-steppe in central and southeast Montana. Unpublished report to USDI Bureau of Land Management, Montana State Office. Helena: Montana Natural Heritage Program.

Cooper, W. S. 1913. The climax forest of Isle Royale, Lake Superior, and its development, part 1. *Botanical Gazette* 55:1–44.

Coppock, D. L., and J. K. Detling. 1986. Alteration of bison and black-tailed prairie dog grazing interaction by prescribed burning. *Journal of Wildlife Management* 50:452–55.

Countryman, C. M. 1956. Old-growth conversion also converts fire climate. *Fire Control Notes* 17(4): 15–19.

Covington, W. W., and M. M. Moore. 1994. Southwestern ponderosa pine forest structure: Changes since Euro-American settlement. *Journal of Forestry* 92(1): 39–47.

Cram, D. S., T. T. Baker, and J. C. Boren. 2006. *Wildland fire effects in silviculturally treated vs. untreated stands of New Mexico and Arizona*. USDA Forest Service Research Paper RMRS-RP-55. Fort Collins, Colorado: Rocky Mountain Research Station.

Crandall, C. R. 1980. Firebreaks for railroad right-of-way. *Fire Management Notes* 41(4): 9–10.

Crandall, C. S. 1897a. Natural reforestation on the mountains of northern Colorado. I. *Garden and Forest* 10:437.

———. 1897b. Natural reforestation on the mountains of northern Colorado. II. *Garden and Forest* 10:446–47.

———. 1901. *Natural reforestation and tree growth on the mountains of northern Colorado*. Washington, D.C.: U.S. Government Printing Office (copy at Colorado State University, Fort Collins).

Crane, M. F., and W. C. Fischer. 1986. *Fire ecology of the forest habitat types of central Idaho.* USDA Forest Service General Technical Report INT-218. Ogden, Utah: Intermountain Research Station.

Crane, M. F., J. R. Habeck, and W. C. Fischer. 1983. *Early postfire revegetation in a western Montana Douglas-fir forest.* USDA Forest Service Research Paper INT-319. Ogden, Utah: Intermountain Forest and Range Experiment Station.

Crawford, J. A., R. A. Olson, N. E. West, J. C. Mosley, M. A. Schroeder, T. D. Whitson, R. F. Miller, M. A. Gregg, and C. S. Boyd. 2004. Ecology and management of sage-grouse and sage-grouse habitat. *Journal of Range Management* 57:2–19.

Crimmins, M. A. 2006. Synoptic climatology of extreme fire-weather conditions across the southwest United States. *International Journal of Climatology* 26:1001–16.

Critchfield, W. B. 1957. *Geographic variation in* Pinus contorta. Maria Moors Cabot Foundation Publication No. 3. Cambridge, Massachusetts: Harvard University.

Cron, R. H. 1969. Thinning as an aid to fire control. *Fire Control Notes* 30(1): 1.

Cruz, A., M. Serrano, E. Navarro, B. Luna, and J. M. Moreno. 2005. Effect of a long-term fire retardant (Fire Trol 934) on the germination of nine Mediterranean-type shrub species. *Environmental Toxicology* 20:543–48.

Cruz, M. G., M. E. Alexander, and R. H. Wakimoto. 2004. Modeling the likelihood of crown fire occurrence in conifer forest stands. *Forest Science* 50:640–58.

———. 2005. Development and testing of models for predicting crown fire rate of spread in conifer forest stands. *Canadian Journal of Forest Research* 3:1626–39.

Cruz, M. G., B. W. Butler, and M. E. Alexander. 2006. Predicting the ignition of crown fuels above a spreading surface fire. Part II: Model evaluation. *International Journal of Wildland Fire* 15:61–72.

Cruz, M. G., B. W. Butler, M. E. Alexander, J. M. Forthofer, and R. H. Wakimoto. 2006. Predicting the ignition of crown fuels above a spreading surface fire. Part I: Model idealization. *International Journal of Wildland Fire* 15:47–60.

Cui, W., and A. H. Perera. 2008. What do we know about forest fire size distribution, and why is this knowledge useful for forest management? *International Journal of Wildland Fire* 17:234–44.

Cummer, M. R., and C. W. Painter. 2007. Three case studies of the effect of wildfire on the Jemez Mountains salamander (*Plethodon neomexicanus*): Microhabitat temperatures, size distributions, and a historical locality perspective. *Southwestern Naturalist* 52:26–37.

Cumming, S. G. 2001. A parametric model of the fire-size distribution. *Canadian Journal of Forest Research* 31:1297–1303.

———. 2005. Effective fire suppression in boreal forests. *Canadian Journal of Forest Research* 35:772–86.

Cunningham, C. A., M. J. Jenkins, and D. W. Roberts. 2005. Attack and brood production by the Douglas-fir beetle (Coleoptera: Scolytidae) in Douglas-fir, *Pseudotsuga menziesii* var. *glauca* (Pinaceae), following a wildfire. *Western North American Naturalist* 65:70–79.

Currie, P. O., C. B. Edminster, and F. W. Knott. 1978. *Effects of cattle grazing on ponderosa pine regeneration in central Colorado.* USDA Forest Service Research Paper RM-201. Fort Collins, Colorado: Rocky Mountain Forest and Range Experiment Station.

Currie, R. G. 1984. Evidence for 18.6-year lunar nodal drought in western North America during the past millennium. *Journal of Geophysical Research* 89:1295–1308.

Dale, L., G. Aplet, and B. Wilmer. 2005. Wildland fire use and cost containment: A Colorado case study. *Journal of Forestry* 103:314–18.

Dalzell, C. R. 2004. Post-fire establishment of vegetation communities following reseeding on southern Idaho's Snake River plain. M.S. thesis, Boise State University, Boise, Idaho.

Daniels, L. D., S. Butler, P. Clinton, B. Ganesh-Babu, K. Hrinkevich, A. Kanoti, E. Keeling, S. Powell, and J. H. Speer. 2005. Whitebark pine stand dynamics at Morrell Mountain, Montana. Pp. 15–29 in *Field-based educational investigations: Examples from the 14th annual North American Dendroecological Fieldweek,* ed. J. H. Speer. Professional Paper Series No. 23. Terre Haute: Indiana State University, Department of Geography, Geology, and Anthropology.

D'Antonio, C. M., and P. M. Vitousek. 1992. Biological invasions by exotic grasses, the grass/fire cycle, and global change. *Annual Review of Ecology and Systematics* 23:63–87.

Daubenmire, R. 1943. Vegetation zonation in the Rocky Mountains. *Botanical Review* 9:325–93.

Daubenmire, R. F. 1981. Subalpine parks associated with snow transfer in the mountains of northern Idaho and eastern Washington. *Northwest Science* 55:124–35.

Daubenmire, R. F., and J. B. Daubenmire. 1968. *Forest vegetation of eastern Washington and northern Idaho.* Washington Agricultural Experiment Station Technical Bulletin No. 60. Pullman: Washington State University.

Davis, J. N. 1976. Ecological investigations in *Cercocarpus ledifolius* Nutt. communities of Utah. M.S. thesis, Brigham Young University, Provo, Utah.

Davis, K. M. 1980. Fire history of a western larch/Douglas-fir forest type in northwestern Montana. Pp. 69–74 in *Proceedings of the Fire History Workshop,* ed. M. A. Stokes and J. H. Dieterich. USDA Forest Service General Technical Report RM-81. Fort Collins, Colorado: Rocky Mountain Forest and Range Experiment Station.

Davis, P. R. 1976. Response of vertebrate fauna to forest fire and clearcutting in south-central Wyoming. Ph.D. dissertation, University of Wyoming, Laramie.

———. 1977. Cervid response to forest fire and clearcutting in southeastern Wyoming. *Journal of Wildlife Management* 41:785–88.

DeBano, L. F., D. G. Neary, and P. F. Ffolliott. 1998. *Fire's effects on ecosystems.* New York: John Wiley.

DeByle, N. V., C. D. Bevins, and W. C. Fischer. 1987. Wildfire occurrence in aspen in the interior western United States. *Western Journal of Applied Forestry* 2:73–76.

DeByle, N.V., P. J. Urness, and D. L. Blank. 1989. *Forage quality in burned and unburned aspen communities.* USDA Forest Service Research Paper INT-404. Ogden, Utah: Intermountain Research Station.

DeCesare, N. J., and D. H. Pletscher. 2006. Movements, connectivity, and resource selection of Rocky Mountain bighorn sheep. *Journal of Mammalogy* 87:531–38.

Deeming, J. E., R. E. Burgan, and J. D. Cohen. 1977. *The national fire-danger rating system–1978.* USDA Forest Service General Technical Report INT-39. Ogden, Utah: Intermountain Forest and Range Experiment Station.

Defossé, G. E., and R. Robberecht. 1996. Effects of competition on the postfire recovery of 2 bunchgrass species. *Journal of Range Management* 49:137–42.

Dell, J. D., and D. E. Franks. 1971. Thinning slash contributes to eastside Cascade wildfires. *Fire Control Notes* 32(1): 4–6.

DellaSala, D. A., and E. Frost. 2001. An ecologically based strategy for fire and fuels management in national forest roadless areas. *Fire Management Today* 61(2): 12–23.

DellaSala, D. A., D. M. Olson, S. E. Barth, S. L. Crane, and S. A. Primm. 1995. Forest health: Moving beyond the rhetoric to restore healthy landscapes in the inland Northwest. *Wildlife Society Bulletin* 23:346–56.

Denevan, W. M. 1967. Livestock numbers in nineteenth-century New Mexico, and the problem of gullying in the Southwest. *Annals of the Association of American Geographers* 57:691–703.

———. 1992. The pristine myth: The landscape of the Americas in 1492. *Annals of the Association of American Geographers* 82:369–85.

Despain, D. G. 1973. Vegetation of the Big Horn Mountains, Wyoming, in relation to substrate and climate. *Ecological Monographs* 43:329–55.

———. 1983. Nonpyrogenous climax lodgepole pine communities in Yellowstone National Park. *Ecology* 64:231–34.

Despain, D. G., D. L. Clark, and J. J. Reardon. 1996. Simulation of crown fire effects on canopy seed bank in lodgepole pine. *International Journal of Wildland Fire* 6:45–49.

Despain, D. G., and R. E. Sellers. 1977. Natural fire in Yellowstone National Park. *Western Wildlands* 4:20–24.

Dettinger, M. D., D. R. Cayan, H. F. Diaz, and D. M. Meko. 1998. North-south precipitation patterns in western North America on interannual-to-decadal timescales. *Journal of Climate* 11:3095–111.

Dickinson, M. B., and E. A. Johnson. 2001. Fire effects on trees. Pp. 477–525 in *Forest fires: Behavior and ecological effects,* ed. E. A. Johnson and K. Miyanishi. New York: Academic Press.

Dick-Peddie, W. A. 1993. *New Mexico vegetation: Past, present, and future.* Albuquerque: University of New Mexico Press.

Dieterich, J. H. 1979. *Recovery potential of fire-damaged southwestern ponderosa pine.* USDA

Forest Service Research Note RM-379. Fort Collins, Colorado: Rocky Mountain Forest and Range Experiment Station.

———. 1980. The composite fire interval: A tool for more accurate interpretation of fire history. Pp. 8–14 in *Proceedings of the Fire History Workshop,* ed. M. A. Stokes and J. H. Dieterich. USDA Forest Service General Technical Report RM-81. Fort Collins, Colorado: Rocky Mountain Forest and Range Experiment Station.

Dieterich, J. H., and T. W. Swetnam. 1984. Dendrochronology of a fire-scarred ponderosa pine. *Forest Science* 30:238–47.

Dillon, G. K., D. H. Knight, and C. B. Meyer. 2005. *Historic range of variability for upland vegetation in the Medicine Bow National Forest, Wyoming.* USDA Forest Service General Technical Report RMRS-GTR-139. Fort Collins, Colorado: Rocky Mountain Research Station.

Doane, D., J. O'Laughlin, P. Morgan, and C. Miller. 2006. Barriers to wildland fire use: A preliminary problem analysis. *International Journal of Wilderness* 12:36–38.

Dodson, E. K. 2004. Monitoring change in exotic plant abundance after fuel reduction/restoration treatments in ponderosa pine forests of western Montana. M.S. thesis, University of Montana, Missoula.

Dodson, E. K., and C. E. Fiedler. 2006. Impacts of restoration treatments on alien plant invasion in *Pinus ponderosa* forests, Montana, USA. *Journal of Applied Ecology* 43:887–97.

Dodson, E. K., K. L. Metlen, and C. E. Fiedler. 2007. Common and uncommon understory species differentially respond to restoration treatments in ponderosa pine/Douglas-fir forests, Montana. *Restoration Ecology* 15:696–708.

Doering, W. R., and R. G. Reider. 1992. Soils of Cinnabar Park, Medicine Bow Mountains, Wyoming, U.S.A.: Indicators of park origin and persistence. *Arctic and Alpine Research* 24:27–39.

Dombeck, M. P., J. E. Williams, and C. A. Wood. 2004. Wildfire policy and public lands: Integrating scientific understanding with social concerns across landscapes. *Conservation Biology* 18:883–89.

Donato, D. C., J. B. Fontaine, J. L. Campbell, W. D. Robinson, J. B. Kauffman, and B. E. Law. 2006. Post-wildfire logging hinders regeneration and increases fire risk. *Science* 311:352.

Donnegan, J. A., and A. J. Rebertus. 1999. Rates and mechanisms of subalpine forest succession along an environmental gradient. *Ecology* 80:1370–84.

Donnegan, J. A., T. T. Veblen, and J. S. Sibold. 2001. Climatic and human influences on fire history in Pike National Forest, central Colorado. *Canadian Journal of Forest Research* 31:1526–39.

Donovan, G. H., and T. C. Brown. 2007. Be careful what you wish for: The legacy of Smokey Bear. *Frontiers in Ecology and the Environment* 5:73–79.

Dordel, J., M. C. Feller, and S. W. Simard. 2008. Effects of mountain pine beetle (*Dendroctonus ponderosae* Hopkins) infestations on forest stand structure in the southern Canadian Rocky Mountains. *Forest Ecology and Management* 255:3563–70.

Doyle, K. M. 1994. Succession following the 1974 Waterfalls Canyon fire, Grand Teton National Park, Wyoming. M.S. thesis, University of Wyoming, Laramie.

————. 1997. Fire, environment and early forest succession in a heterogeneous Rocky Mountain landscape, northwestern Wyoming. Ph.D. dissertation, University of Wyoming, Laramie.

————. 2004. Early postfire forest succession in the heterogeneous Teton landscape. Pp. 235–78 in *After the fires: The ecology of change in Yellowstone National Park,* ed. L. L. Wallace. New Haven, Connecticut: Yale University Press.

Doyle, K. M., D. H. Knight, D. L. Taylor, W. J. Barmore Jr., and J. M. Benedict. 1998. Seventeen years of forest succession following the Waterfalls Canyon fire in Grand Teton National Park, Wyoming. *International Journal of Wildland Fire* 8:45–55.

Druckenbrod, D. L. 2005. Dendroecological reconstructions of forest disturbance history using time-series analysis with intervention detection. *Canadian Journal of Forest Research* 35:868–76.

Drury, W. H., and I. C. T. Nisbet. 1973. Succession. *Journal of the Arnold Arboretum* 54:331–68.

DuBois, C. 1903. Report on the proposed San Juan Forest Reserve, Colorado. Unpublished report. Durango, Colorado: San Juan National Forest.

Dudley, M., and G. S. Greenhoe. 1998. Fifty years of helicopter firefighting. *Fire Management Notes* 58(4): 6–7.

Dunwiddie, P. W. 1977. Recent tree invasion of subalpine meadows in the Wind River Mountains, Wyoming. *Arctic and Alpine Research* 9:393–99.

Dwire, K. A., and J. B. Kauffman. 2003. Fire and riparian ecosystems in landscapes of the western USA. *Forest Ecology and Management* 178:61–74.

Eckert, S. A. 2004. "Brewer fire mystery" not so mysterious. *Fire Management Today* 64(2): 56–57.

Egler, F. E. 1954. Vegetation science concepts I. Initial floristic composition: A factor in old-field vegetation development. *Vegetatio* 4:412–17.

Ehle, D. S., and W. L. Baker. 2003. Disturbance and stand dynamics in ponderosa pine forests in Rocky Mountain National Park, USA. *Ecological Monographs* 73:543–66.

Eichhorn, L. C., and C. R. Watts. 1984. Plant succession of burns in the River Breaks of central Montana. *Proceedings of the Montana Academy of Sciences* 43:21–34.

Eisenhart, K. S. 2004. Historic range of variability and stand development in piñon-juniper woodlands of western Colorado. Ph.D. dissertation, University of Colorado, Boulder.

Eiswerth, M. E., and J. S. Shonkwiler. 2006. Examining post-wildfire reseeding on arid rangeland: A multivariate Tobit modelling approach. *Ecological Modelling* 192:286–98.

Elliott, J. G., and R. S. Parker. 2001. Developing a post-fire flood chronology and recurrence probability from alluvial stratigraphy in the Buffalo Creek watershed, Colorado, USA. *Hydrological Processes* 15:3039–51.

Ellis, L. 2001. Short-term response of woody plants to fire in a Rio Grande riparian forest, central New Mexico, USA. *Biological Conservation* 97:159–70.

Ellis, M., C. D. von Dohlen, J. E. Anderson, and W. H. Romme. 1994. Some important factors affecting density of lodgepole pine seedlings following the 1988 Yellowstone fires. Pp. 139–50 in *Plants and their environments: Proceedings of the first biennial scientific conference on the Greater Yellowstone Ecosystem, Sept. 16–17, 1991, Yellowstone National Park, Wyoming,* ed. D. G. Despain. Denver, Colorado: USDI National Park Service, Natural Resources Publication Office.

Ellison, L. 1960. Influence of grazing on plant succession of rangelands. *Botanical Review* 26:1–78.

Ely, J. B., A. W. Jensen, L. R. Chatten, and H. W. Jori. 1957. Air tankers: A new tool for forest fire fighting. *Fire Control Notes* 18(3): 103–9.

Enfield, D. B. 1989. El Niño, past and present. *Reviews of Geophysics* 27:159–87.

Enfield, D. B., A. M. Mestas-Nuñez, and P. J. Trimble. 2001. The Atlantic multidecadal oscillation and its relation to rainfall and river flows in the continental U.S. *Geophysical Research Letters* 28:2077–80.

Ensign, E. T. 1888. Report on the forest conditions of the Rocky Mountains, especially in the state of Colorado, the territories of Idaho, Montana, Wyoming, and New Mexico. Pp. 41–152 in *Report on the forest conditions of the Rocky Mountains and other papers; with a map showing the location of forest areas on the Rocky Mountain range,* ed. B. E. Fernow. Washington, D.C.: U.S. Government Printing Office.

Erdman, J. A. 1969. Pinyon-juniper succession after fires on residual soils of the Mesa Verde, Colorado. Ph.D. dissertation, University of Colorado, Boulder.

———. 1970. Pinyon-juniper succession after natural fires on residual soils of Mesa Verde, Colorado. *Brigham Young University Science Bulletin Biological Series* 11:1–26.

Evangelista, P., T. J. Stohlgren, D. Guenther, and S. Stewart. 2004. Vegetation response to fire and postburn seeding treatments in juniper woodlands of the Grand Staircase–Escalante National Monument, Utah. *Western North American Naturalist* 64:293–305.

Evanko, A. B., and R. A. Peterson. 1955. Comparisons of protected and grazed mountain rangelands in southwestern Montana. *Ecology* 36:71–82.

Fahnestock, G. R. 1968. *Fire hazard from precommercial thinning of ponderosa pine.* USDA Forest Service Research Paper PNW-57. Portland, Oregon: Pacific Northwest Forest and Range Experiment Station.

Fahnestock, G. R., and J. H. Dieterich. 1962. *Logging slash flammability after 5 years.* USDA Forest Service Research Paper No. 70. Ogden, Utah: Intermountain Forest and Range Experiment Station.

Fahnestock, G. R., and R. C. Hare. 1964. Heating of tree trunks in surface fires. *Journal of Forestry* 62:799–805.

Falk, D. A. 2004. Scaling rules for fire regimes. Ph.D. dissertation, University of Arizona, Tucson.

Falk, D. A., C. Miller, D. McKenzie, and A. E. Black. 2007. Cross-scale analysis of fire regimes. *Ecosystems* 10:809–23.

Falk, D. A., M. A. Palmer, and J. B. Zedler, eds. 2006. *Foundations of restoration ecology.* Washington, D.C.: Island Press.

Falkowski, M. J., P. E. Gessler, P. Morgan, A. T. Hudak, and A. M. S. Smith. 2005. Characterizing and mapping forest fire fuels using ASTER imagery and gradient modeling. *Forest Ecology and Management* 217:129–46.

Fall, J. G. 1998. Reconstructing the historical frequency of fire: A modeling approach to developing and testing methods. M.S. thesis, Simon Fraser University, Burneby, B.C., Canada.

Fall, P. L. 1997. Fire history and composition of the subalpine forest of western Colorado during the Holocene. *Journal of Biogeography* 24:309–25.

Farris, C. A., L. F. Neuenschwander, and S. L. Boudreau. 1998. Monitoring initial plant succession following fire in a subalpine spruce-fir forest. *Proceedings of the Tall Timbers Fire Ecology Conference* 20:298–305.

Fayt, P., M. M. Machmer, and C. Steeger. 2005. Regulation of spruce bark beetles by woodpeckers: A literature review. *Forest Ecology and Management* 206:1–14.

Fechner, G. H., and J. S. Barrows. 1976. *Aspen stands as wildfire fuel breaks.* Eisenhower Consortium Bulletin 4. Fort Collins: Colorado State University, College of Forestry and Natural Resources, Department of Forest and Wood Science.

Fenneman, N. M. 1931. *Physiography of western United States.* New York: McGraw-Hill.

Fiedler, C. E., S. F. Arno, and M. G. Harrington. 1998. Reintroducing fire in ponderosa pine–fir forests after a century of fire exclusion. *Proceedings of the Tall Timbers Fire Ecology Conference* 20:245–49.

Fiedler, C. E., C. E. Keegan III, C. W. Woodall, and T. A. Morgan. 2004. *A strategic assessment of crown fire hazard in Montana: Potential effectiveness and costs of hazard reduction treatments.* USDA Forest Service General Technical Report PNW-GTR-622. Portland, Oregon: Pacific Northwest Research Station.

Finegan, B. 1984. Forest succession. *Nature* 312:109–14.

Finklin, A. I. 1973. *Meteorological factors in the Sundance fire run.* USDA Forest Service General Technical Report INT-6. Ogden, Utah: Intermountain Forest and Range Experiment Station.

Finney, M. A. 1998. *FARSITE: Fire area simulator—Model development and evaluation.* USDA Forest Service Research Paper RMRS-RP-4. Fort Collins, Colorado: Rocky Mountain Research Station.

———. 2001. Design of regular landscape fuel treatment patterns for modifying fire growth and behavior. *Forest Science* 47:219–28.

———. 2005. The challenge of quantitative risk analysis for wildland fire. *Forest Ecology and Management* 211:97–108.

———. 2006. An overview of FlamMap fire modeling capabilities. Pp. 213–20 in *Fuels management—How to measure success: Conference proceedings, 28–30 March 2006, Portland, Oregon,* ed. P. L. Andrews and B. W. Butler. USDA Forest Service Proceedings RMRS-P-41. Fort Collins, Colorado: Rocky Mountain Research Station.

———. 2007. A computational method for optimising fuel treatment locations. *International Journal of Wildland Fire* 16:702–11.

Finney, M. A., C. W. McHugh, R. Bartlette, K. Close, and P. Langowski. 2003. Fire behavior, fuel treatments, and fire suppression on the Hayman Fire. Part 2. Description and interpretations of fire behavior. Pp. 59–95 in *Hayman fire case study,* ed. R. T. Graham. USDA Forest Service General Technical Report RMRS-GTR-114. Fort Collins, Colorado: Rocky Mountain Research Station.

Finney, M. A., C. W. McHugh, and I. C. Grenfell. 2005. Stand- and landscape-level effects of prescribed burning on two Arizona wildfires. *Canadian Journal of Forest Research* 35:1714–22.

Finney, M. A., R. C. Seli, C. W. McHugh, A. A. Ager, B. Bahro, and J. K. Agee. 2007. Simulation of long-term landscape-level fuel treatment effects on large wildfires. *International Journal of Wildland Fire* 16:712–27.

Fischer, R. A., K. P. Reese, and J. W. Connelly. 1996. An investigation on fire effects within xeric sage grouse brood habitat. *Journal of Range Management* 49:194–98.

Fischer, R. A., W. L. Wakkinen, K. P. Reese, and J. W. Connelly. 1997. Effects of prescribed fire on movements of female sage grouse from breeding to summer ranges. *Wilson Bulletin* 109:82–91.

Fischer, W. C. 1980. Prescribed fire and bark beetle attack in ponderosa pine forests. *Fire Management Notes* 41(2): 10–12.

———. 1981a. *Photo guide for appraising downed woody fuels in Montana forests: Grand fir–larch–Douglas-fir, western hemlock, western hemlock–western redcedar, and western redcedar cover types.* USDA Forest Service General Technical Report INT-96. Ogden, Utah: Intermountain Forest and Range Experiment Station.

———. 1981b. *Photo guide for appraising downed woody fuels in Montana forests: Interior ponderosa pine, ponderosa pine–larch–Douglas-fir, larch–Douglas-fir, and interior Douglas-fir cover types.* USDA Forest Service General Technical Report INT-97. Ogden, Utah: Intermountain Forest and Range Experiment Station.

———. 1981c. *Photo guide for appraising downed woody fuels in Montana forests: Lodgepole pine and Engelmann spruce–subalpine fir cover types.* USDA Forest Service General Technical Report INT-98. Ogden, Utah: Intermountain Forest and Range Experiment Station.

Fischer, W. C., and A. F. Bradley. 1987. *Fire ecology of western Montana forest habitat types.* USDA Forest Service General Technical Report INT-223. Ogden, Utah: Intermountain Research Station.

Fischer, W. C., and B. D. Clayton. 1983. *Fire ecology of Montana forest habitat types east of the continental divide.* USDA Forest Service General Technical Report INT-141. Ogden, Utah: Intermountain Forest and Range Experiment Station.

Fischer, W. C., M. Miller, C. M. Johnston, J. K. Smith, D. G. Simmerman, and J. K. Brown. 1996. *Fire effects information system: User's guide.* USDA Forest Service General Technical Report INT-GTR-327. Ogden, Utah: Intermountain Research Station.

Fisher, J. T., and L. Wilkinson. 2005. The response of mammals to forest fire and timber harvest in the North American boreal forest. *Mammal Review* 35:51–81.

Fisher, R. F., M. J. Jenkins, and W. F. Fisher. 1987. Fire and the prairie-forest mosaic of Devils Tower National Monument. *American Midland Naturalist* 117:250–57.

Fisk, H., K. Megown, and L. M. Decker. 2004. *Riparian area burn analysis: Process and applications.* USDA Forest Service—Engineering RSAC-0057-TIP 1. Salt Lake City, Utah: Remote Sensing Applications Center.

Flaccus, G. 2007. Insurers try to prevent losses in wildfire hotspots. *Insurance Journal* (May 21). http://www.insurancejournal.com.

Flannigan, M. D., Y. Bergeron, O. Engelmark, and B. M. Wotton. 1998. Future wildfires in circumboreal forests in relation to global warming. *Journal of Vegetation Science* 9:469–76.

Flannigan, M., B. Stocks, and M. Weber. 2003. Fire regimes and climatic change in Canadian forests. Pp. 97–119 in *Fire and climatic change in temperate ecosystems of the western Americas,* ed. T. T. Veblen, W. L. Baker, G. Montenegro, and T. W. Swetnam. New York: Springer.

Flinn, M. A., and J. K. Pringle. 1983. Heat tolerance of rhizomes of several understory species. *Canadian Journal of Botany* 61:452–57.

Flint, H. R. 1925. Fire resistance of northern Rocky Mountain conifers. *Idaho Forester* 7:7–10, 41–43.

———. 1928. Adequate fire control. *Journal of Forestry* 26:624–38.

Floyd, M. E. 1982. The interaction of pinyon pine and Gambel oak in plant succession near Dolores, Colorado. *Southwestern Naturalist* 27:143–47.

Floyd, M. L., D. D. Hanna, and W. H. Romme. 2004. Historical and recent fire regimes in piñon-juniper woodlands on Mesa Verde, Colorado, USA. *Forest Ecology and Management* 198:269–89.

Floyd, M. L., D. Hanna, W. H. Romme, and T. E. Crews. 2006. Predicting and mitigating weed invasions to restore natural post-fire succession in Mesa Verde National Park, Colorado, USA. *International Journal of Wildland Fire* 15:247–59.

Floyd, M. L., W. H. Romme, and D. D. Hanna. 2000. Fire history and vegetation pattern in Mesa Verde National Park, Colorado, USA. *Ecological Applications* 10:1666–80.

Foiles, M. W., and J. D. Curtis. 1973. *Regeneration of ponderosa pine in the northern Rocky Mountain–intermountain region.* USDA Forest Service Research Paper INT-145. Ogden, Utah: Intermountain Forest and Range Experiment Station.

Forcella, F., and T. Weaver. 1977. Biomass and productivity of the subalpine *Pinus albicaulis–Vaccinium scoparium* assocation in Montana, USA. *Vegetatio* 35:95–105.

Ford, P. L. 1999. Response of buffalograss (*Buchloë dactyloides*) and blue grama (*Bouteloua gracilis*) to fire. *Great Plains Research* 9:261–76.

Ford, P. L., and G. V. Johnson. 2006. Effects of dormant- vs. growing-season fire in shortgrass steppe: Biological soil crust and perennial grass responses. *Journal of Arid Environments* 67:1–14.

Forde, J. D. 1983. The effect of fire on bird and small mammal communities in the grasslands of Wind Cave National Park. M.S. thesis, Michigan Technological University, Houghton.

Forester, J. D., D. P. Anderson, and M. G. Turner. 2007. Do high-density patches of coarse wood and regenerating saplings create browsing refugia for aspen (*Populus tremuloides* Michx.) in Yellowstone National Park (USA)? *Forest Ecology and Management* 253:211–19.

Fornwalt, P. J., M. R. Kaufmann, L. S. Huckaby, J. M. Stoker, and T. J. Stohlgren. 2003. Non-native plant invasions in managed and protected ponderosa pine/Douglas-fir forests of the Colorado Front Range. *Forest Ecology and Management* 177:515–27.

Fowler, J. F., and C. H. Sieg. 2004. *Postfire mortality of ponderosa pine and Douglas-fir: A review of methods to predict tree death.* USDA Forest Service General Technical Report RMRS-GTR-132. Fort Collins, Colorado: Rocky Mountain Research Station.

Fowler, P. M., and D. O. Asleson. 1984. The location of lightning-caused wildland fires, northern Idaho. *Physical Geography* 5:240–52.

Fox, J. F. 1978. Forest fires and the snowshoe hare–Canada lynx cycle. *Oecologia* 31:349–74.

Foxx, T. S. 1984. La Mesa fire symposium. Pp. 1–10 in *La Mesa fire symposium, Los Alamos, New Mexico, Oct. 6–7, 1981,* ed. T. S. Foxx. Publication LA-9236-NERP. Los Alamos, New Mexico: Los Alamos National Laboratory.

———. 1996. Vegetation succession after the La Mesa fire at Bandelier National Monument. Pp. 47–69 in *Fire effects in southwestern forests: Proceedings of the second La Mesa fire symposium,* ed. C. D. Allen. USDA Forest Service General Technical Report RM-GTR-286. Fort Collins, Colorado: Rocky Mountain Forest and Range Experiment Station.

Foxx, T. S., and L. D. Potter. 1984. Fire ecology at Bandelier National Monument. Pp. 11–37 in *La Mesa fire symnposium, Los Alamos, New Mexico, Oct. 6–7, 1981,* ed. T. S. Foxx. Publication LA-9236-NERP. Los Alamos, New Mexico: Los Alamos National Laboratory.

Fraas, W. W., C. A. L. Wambolt, and M. R. Frisina. 1992. Prescribed fire effects on a bitterbrush–mountain big sagebrush–bluebunch wheatgrass community. Pp. 212–16 in *Symposium on ecology and management of riparian shrub communities,* ed. W. Clary, E. D. McArthur, D. Bedunah, and C. L. Wambolt. USDA Forest Service General Technical Report INT-289. Ogden, Utah: Intermountain Forest and Range Experiment Station.

Frandsen, W. H. 1987. The influence of moisture and mineral soil on the combustion limits of smoldering forest duff. *Canadian Journal of Forest Research* 17:1540–44.

Frandsen, W. H., and K. C. Ryan. 1986. Soil moisture reduces belowground heat flux and soil temperatures under a burning fuel pile. *Canadian Journal of Forest Research* 16:244–48.

Franklin, J. F., D. Lindenmayer, J. A. MacMahon, A. McKee, J. Magnuson, D. A. Perry, R. Waide, and D. Foster. 2000. Threads of continuity. *Conservation Biology in Practice* 1:8–16.

Franklin, T. L., and R. D. Laven. 1991. Fire influences on central Rocky Mountain lodgepole pine stand structure and composition. *Proceedings of the Tall Timbers Fire Ecology Conference* 17:183–96.

Freedman, J. D., and J. R. Habeck. 1985. Fire, logging, and white-tailed deer interrelationships in the Swan Valley, northwestern Montana. Pp. 23–35 in *Fire's effects on wildlife habitat: Symposium proceedings,* ed. J. E. Lotan and J. K. Brown. USDA Forest Service General Technical Report INT 186. Ogden, Utah: Intermountain Research Station.

Freeman, J. F., T. J. Stohlgren, M. E. Hunter, P. N. Omi, E. J. Martinson, G. W. Chong, and C. S. Brown. 2007. Rapid assessment of postfire plant invasions in coniferous forests of the western United States. *Ecological Applications* 17:1656–65.

French, M. G., and S. P. French. 1996. Large mammal mortality in the 1988 Yellowstone fires. Pp. 113–15 in *The ecological implications of fire in Greater Yellowstone: Proceedings second biennial conference on the Greater Yellowstone ecosystem, Yellowstone National Park, Sept. 19–21, 1993,* ed. J. M. Greenlee. Fairfield, Washington: International Association of Wildland Fire.

Frey, B. R., V. J. Lieffers, S. M. Landhäusser, P. G. Comeau, and K. J. Greenway. 2003. An analysis of sucker regeneration of trembling aspen. *Canadian Journal of Forest Research* 33:1169–79.

Frey, M. 2007. The humble beginnings of aircraft in the Forest Service. *Fire Management Today* 67(2): 6–9.

Frischknecht, N. C., and A. P. Plummer. 1955. A comparison of seeded grasses under grazing and protection on a mountain brush burn. *Journal of Range Management* 8:170–75.

Fulé, P. Z., W. W. Covington, M. M. Moore, T. A. Heinlein, and A. E. M. Waltz. 2002. Natural variability in forests of the Grand Canyon, USA. *Journal of Biogeography* 29:31–47.

Fulé, P. Z., T. A. Heinlein, W. W. Covington, and M. M. Moore. 2003. Assessing fire regimes on Grand Canyon landscapes with fire-scar and fire-record data. *International Journal of Wildland Fire* 12:129–45.

Fulé, P. Z., D. C. Laughlin, and W. W. Covington. 2005. Pine-oak forest dynamics five years after ecological restoration treatments, Arizona, USA. *Forest Ecology and Management* 218:129–45.

Fuller, B. 2007. Does prescribed burning on high-elevation, cool-season meadows impact forage quality, diet selection, and performance of grazers? *Rangelands* 29:6–9.

Fuquay, D. M. 1962. Mountain thunderstorms and forest fires. *Weatherwise* 15:148–52.

Fuquay, D. M., R. G. Baughman, and D. J. Latham. 1979. *A model for predicting lightning fire ignition in wildland fuels.* USDA Forest Service Research Paper INT-217. Ogden, Utah: Intermountain Forest and Range Experiment Station.

Fuquay, D. M., R. G. Baughman, A. R. Taylor, and R. G. Hawe. 1967. Characteristics of seven lightning discharges that caused forest fires. *Journal of Geophysical Research* 72:6371–73.

Fuquay, D. M., A. R. Taylor, R. G. Hawe, and C. W. Schmid Jr. 1972. Lightning discharges that caused forest fires. *Journal of Geophysical Research* 77:2156–58.

Furniss, M. M. 1965. Susceptibility of fire-injured Douglas-fir to bark beetle attack in southern Idaho. *Journal of Forestry* 63:8–11.

Fye, F. K., D. W. Stahle, E. R. Cook, and M. K. Cleaveland. 2006. NAO influence on subdecadal moisture variability over central North America. *Geophysical Research Letters* 33:1–5.

Gabriel, H. W. III. 1976. Wilderness ecology: The Danaher Creek drainage, Bob Marshall Wilderness, Montana. Ph.D. dissertation, University of Montana, Missoula.

Gaffney, W. S. 1941. The effects of winter elk browsing, South Fork of the Flathead River, Montana. *Journal of Wildlife Management* 5:427–53.

Gallant, A. L., A. J. Hansen, J. S. Councilman, D. K. Monte, and D. W. Betz. 2003. Vegetation dynamics under fire exclusion and logging in a Rocky Mountain watershed, 1856–1996. *Ecological Applications* 13:385–403.

Gallup, S. M. 1998. Ponderosa pine mortality following two prescribed fires in Boulder, Colorado. M.S. thesis, Colorado State University, Fort Collins.

Gara, R. I., J. K. Agee, W. R. Littke, and D. R. Geiszler. 1986. Fire wounds and beetle scars: Distinguishing between the two can help reconstruct past disturbances. *Journal of Forestry* 84:47–50.

Gardner, W. J. 1905. Results of a Rocky Mountain forest fire studied fifty years after its occurrence. *Proceedings of the Society of American Foresters* 1:102–9.

Gartner, F. R., and W. W. Thompson. 1972. Fire in the Black Hills forest-grass ecotone. *Proceedings of the Tall Timbers Fire Ecology Conference* 12:37–68.

Gary, H. L., and P. O. Currie. 1977. *The Front Range pine type: A 40-year photographic record of plant recovery on an abused watershed.* USDA Forest Service General Technical Report RM-46. Fort Collins, Colorado: Rocky Mountain Forest and Range Experiment Station.

Gates, R. J. 1983. Sage grouse, lagomorph, and pronghorn use of a sagebrush grassland burn site on the Idaho National Engineering Laboratory. M.S. thesis, Montana State University, Bozeman.

Gedalof, Z., D. L. Peterson, and N. J. Mantua. 2005. Atmospheric, climatic, and ecological controls on extreme wildfire years in the northwestern United States. *Ecological Applications* 15:154–74.

Geier-Hayes, K. 1997. The impact of post-fire seeded grasses on native vegetative communities in central Idaho. Pp. 19–26 in *Proceedings: First conference on fire effects on rare and endangered species and habitats,* ed. J. M. Greenlee. Fairfield, Washington: International Association of Wildland Fire.

Geils, B. W., J. C. Tovar, and B. Moody. 2002. *Mistletoes of North American conifers.* USDA Forest Service General Technical Report RMRS-GTR-98. Fort Collins, Colorado: Rocky Mountain Research Station.

Geluso, K. N., G. D. Schroder, and T. B. Bragg. 1986. Fire-avoidance behavior of meadow voles (*Microtus pennsylvanicus*). *American Midland Naturalist* 116:202–5.

Gentry, D. J., and K. T. Vierling. 2007. Old burns as source habitats for Lewis's woodpeckers breeding in the Black Hills of South Dakota. *The Condor* 109:122–31.

Getz, H. L., and W. L. Baker. 2008. Initial invasion of cheatgrass (*Bromus tectorum*) into burned piñon-juniper woodlands in western Colorado. *American Midland Naturalist* 159:489–97.

Gibson, C. E. 2006. A northern Rocky Mountain polygon fire history: Accuracy, limitations, strengths, applications, and protocol of fire perimeter data. M.S. thesis, University of Idaho, Moscow.

Gibson, K., E. Lieser, and B. Ping. 1999. *Bark beetle outbreaks following the Little Wolf fire.* Forest Health Protection Report 99–7. Missoula, Montana: USDA Forest Service Northern Region.

Giménez, A., E. Pastor, L. Zárate, E. Planas, and J. Arnaldos. 2004. Long-term forest fire retardants: A review of quality, effectiveness, application and environmental considerations. *International Journal of Wildland Fire* 13:1–15.

Gisborne, H. T. 1927. Meteorological factors in the Quartz Creek forest fire. *Monthly Weather Review* 55:56–60.

———. 1931. A five-year record of lightning storms and forest fires. *Monthly Weather Review* 59:139–50.

———. 1935. Shaded fire breaks. *Journal of Forestry* 33:86–87.

Gleason, H. A. 1917. The structure and development of the plant association. *Bulletin of the Torrey Botanical Club* 44:463–81.

Goens, D. W. 1990. Meteorological factors contributing to the Canyon Creek fire blowup September 6 and 7, 1988. Pp. 180–86 in *Proceedings of the 5th Conference on Mountain Meteorology, June 25–29, 1990, Boulder, Colorado.* Boston, Massachusetts: American Meteorological Society.

Goldblum, D., and T. T. Veblen. 1992. Fire history of a ponderosa pine–Douglas-fir forest, Colorado Front Range. *Physical Geography* 13:133–48.

Goodrich, S., and B. Barber. 1999. Return interval for pinyon-juniper following fire in the Green River Corridor, near Dutch John, Utah. Pp. 391–93 in *Proceedings: Ecology and management of pinyon-juniper communities within the Interior West, Sept. 15–18, 1997, Provo, Utah,* ed. S. B. Monsen and R. Stevens. USDA Forest Service Proceedings RMRS-P-9. Fort Collins, Colorado: Rocky Mountain Research Station.

Goodrich, S., and D. Rooks. 1999. Control of weeds at a pinyon-juniper site by seeding grasses. Pp. 403–7 in *Proceedings: Ecology and management of pinyon-juniper communities within the interior West: Sustaining and restoring a diverse ecosystem, Sept. 15–18, 1997, Provo, Utah,* ed. S. B. Monsen and R. Stevens. USDA Forest Service Proceedings RMRS-P-9. Fort Collins, Colorado: Rocky Mountain Research Station.

Goodwin, K. M., and R. L. Sheley. 2001. What to do when fires fuel weeds: A step-by-step guide for managing invasive plants after a wildfire. *Rangelands* 23:15–21.

Gordon, F. A. 1974. Spring burning in an aspen-conifer stand for maintenance of moose habitat, West Boulder River, Montana. *Proceedings of the Tall Timbers Fire Ecology Conference* 14:501–38.

Gorte, J. K., and R. W. Gorte. 1979. *Application of economic techniques to fire management: A status review and evaluation.* USDA Forest Service General Technical Report INT-56. Ogden, Utah: Intermountain Forest and Range Experiment Station.

Gorte, R. W. 2006. *Forest fire/wildfire protection.* Report for Congress, Order Code RL30755. Washington, D.C.: Congressional Research Service.

Grace, J. B., M. D. Smith, S. L. Grace, S. L. Collins, and T. J. Stohlgren. 2001. Interactions between fire and invasive plants in temperate grasslands of North America. Pp. 40–65 in *Proceedings of the invasive species workshop: The role of fire in the control and spread of invasive species,* ed. K. E. M. Galley and T. P. Wilson. Miscellaneous Publication No. 11. Tallahassee, Florida: Tall Timbers Research Station.

Graham, R. T., A. E. Harvey, T. B. Jain, and J. R. Tonn. 1999. *The effects of thinning and similar stand treatments on fire behavior in western forests.* USDA Forest Service General Technical Report PNW-GTR-463. Portland, Oregon: Pacific Northwest Research Station.

Graham, R. T., A. E. Harvey, M. F. Jurgensen, T. B. Jain, J. R. Tonn, and D. S. Page-Dumroese. 1994. *Managing coarse woody debris in forests of the Rocky Mountains.* USDA Forest Service Research Paper INT-RP-477. Ogden, Utah: Intermountain Research Station.

Graham, R. T., T. B. Jain, R. T. Reynolds, and D. A. Boyce. 1997. The role of fire in sustaining northern goshawk habitat in Rocky Mountain forests. Pp. 69–76 in *Proceedings: First conference on fire effects on rare and endangered species and habitats,* ed. J. M. Greenlee. Fairfield, Washington: International Association of Wildland Fire.

Graham, R. T., S. McCaffrey, and T. B. Jain. 2004. *Science basis for changing forest structure to modify wildfire behavior and severity.* USDA Forest Service General Technical Report RMRS-GTR-120. Fort Collins, Colorado: Rocky Mountain Research Station.

Graves, H. S. 1899. Black Hills forest reserve. Pp. 67–164 in *Nineteenth annual report of the United States Geological Survey. Part V: Forest reserves,* ed. H. Gannett. Washington, D.C.: U.S. Government Printing Office.

Gray, S. T., L. J. Graumlich, J. L. Betancourt, and G. D. Pederson. 2004. A tree-ring based reconstruction of the Atlantic Multidecadal Oscillation since 1567 A.D. *Geophysical Research Letters* 31:L12205.

Green, L. R. 1977. *Fuelbreaks and other fuel modification for wildland fire control.* USDA Agricultural Handbook No. 499. Washington, D.C.: U.S. Government Printing Office.

Greene, D. F., and E. A. Johnson. 1996. Wind dispersal of seeds from a forest into a clearing. *Ecology* 77:595–609.

Greene, D. F., J. C. Zasada, L. Sirois, D. Kneeshaw, H. Morin, I. Charron, and M.-J. Simard. 1999. A review of the regeneration dynamics of North American boreal forest tree species. *Canadian Journal of Forest Research* 29:824–39.

Griffis, K. L., J. A. Crawford, M. R. Wagner, and W. H. Moir. 2001. Understory response to management treatments in northern Arizona ponderosa pine forests. *Forest Ecology and Management* 146:239–45.

Grissino-Mayer, H. D. 1999. Modeling fire interval data from the American Southwest with the Weibull distribution. *International Journal of Wildland Fire* 9:37–50.

Grissino-Mayer, H. D., W. H. Romme, M. L. Floyd, and D. D. Hanna. 2004. Climatic and human influences on fire regimes of the southern San Juan Mountains, Colorado, USA. *Ecology* 85:1708–24.

Groves, C. R., and K. Steenhof. 1988. Responses of small mammals and vegetation to wildfire in shadscale communities of southwestern Idaho. *Northwest Science* 62:205–10.

Gruell, G. E. 1980a. *Fire's influence on wildlife habitat on the Bridger-Teton National Forest, Wyoming.* Vol. 1, *Photographic record and analysis.* USDA Forest Service General Technical Report INT-235. Ogden, Utah: Intermountain Forest and Range Experiment Station.

———. 1980b. *Fire's influence on wildlife habitat on the Bridger-Teton National Forest, Wyoming.* Vol. 2, *Changes and causes, management implications.* USDA Forest Service General Technical Report INT-252. Ogden, Utah: Intermountain Forest and Range Experiment Station.

———. 1983. *Fire and vegetative trends in the northern Rockies: Interpretations from 1871–1982 photographs.* USDA Forest Service General Technical Report INT-158. Ogden, Utah: Intermountain Forest and Range Experiment Station.

Gruell, G. E., S. Bunting, and L. Neuenschwander. 1985. Influence of fire on curlleaf mountain-mahogany in the intermountain West. Pp. 58–72 in *Fire's effects on wildlife habitat: Symposium proceedings,* ed. J. E. Lotan and J. K. Brown. USDA Forest Service General Technical Report INT-186. Ogden, Utah: Intermountain Research Station.

Gruell, G. E., and L. L. Loope. 1974. Relationships among aspen, fire, and ungulate browsing in Jackson Hole, Wyoming. Unpublished report. Ogden, Utah: USDA Forest Service, Intermountain Region; and Denver, Colorado: USDI National Park Service, Rocky Mountain Region.

Gruell, G. E., W. C. Schmidt, S. F. Arno, and W. J. Reich. 1982. *Seventy years of vegetative changes in a managed ponderosa pine forest in western Montana: Implications for resource management.* USDA Forest Service General Technical Report INT-130. Ogden, Utah: Intermountain Forest and Range Experiment Station.

Gundale, M. J., S. Sutherland, and T. H. DeLuca. 2008. Fire, native species, and soil resource interactions influence the spatio-temporal invasion pattern of *Bromus tectorum. Ecography* 31:201–10.

Guthrie, D. A. 1984. Effects of fire on small mammals within Bandelier National Monument. Pp. 115–34 in *La Mesa fire symnposium, Los Alamos, New Mexico, Oct. 6–7, 1981*, ed. T. S. Foxx. Publication LA-9236-NERP. Los Alamos, New Mexico: Los Alamos National Laboratory.

Gutsell, S. L., and E. A. Johnson. 1996. How fire scars are formed: Coupling a disturbance process to its ecological effect. *Canadian Journal of Forest Research* 26:166–74.

Guyette, R. P., M. C. Stambaugh, R.-M. Muzika, and E. R. McMurry. 2006. Fire history at the southwestern Great Plains margin, Capulin Volcano National Monument. *Great Plains Research* 16:161–72.

Habeck, J. R. 1968. Forest succession in the Glacier Park cedar-hemlock forests. *Ecology* 49:872–80.

———. 1969. A gradient analysis of a timberline zone at Logan Pass, Glacier Park, Montana. *Northwest Science* 32:65–73.

———. 1970. Fire ecology investigations in Glacier National Park. Unpublished report. Missoula: University of Montana.

———. 1976. Forests, fuels and fire in the Selway-Bitterroot Wilderness, Idaho. *Proceedings of the Tall Timbers Fire Ecology Conference* 14:305–53.

———. 1988. Present-day vegetation in the northern Rocky Mountains. *Annals of the Missouri Botanical Garden* 74:804–40.

———. 1990. Old-growth ponderosa pine–western larch forests in western Montana: Ecology and management. *Northwest Environmental Journal* 6:271–92.

Habeck, J. R., and R. W. Mutch. 1973. Fire-dependent forests in the northern Rocky Mountains. *Quaternary Research* 3:408–24.

Hadley, K. S., and T. T. Veblen. 1993. Stand response to western spruce budworm and Douglas-fir bark beetle outbreaks, Colorado Front Range. *Canadian Journal of Forest Research* 23:479–91.

Haig, I. T., K. P. Davis, and R. H. Weidman. 1941. *Natural regeneration in the western white pine type*. USDA Technical Bulletin No. 767. Washington, D.C.: U.S. Government Printing Office.

Haight, R. G., D. T. Cleland, R. B. Hammer, V. C. Radeloff, and T. S. Rupp. 2004. Assessing fire risk in the wildland-urban interface. *Journal of Forestry* 102:41–48.

Haines, D. A. 1988. Downbursts and wildland fires: A dangerous combination. *Fire Management Notes* 49(3): 8–10.

Haining, R. 1990. *Spatial data analysis in the social and environmental sciences*. Cambridge, England: Cambridge University Press.

Haley, J. E. 1929. Grass fires of the southern Great Plains. *West Texas Historical Year Book* 5:23–42.

Halford, D. K. 1981. Repopulation and food habits of *Peromyscus maniculatus* on a burned sagebrush desert in southeastern Idaho. *Northwest Science* 55:44–49.

Hall, B. L. 2007. Precipitation associated with lightning-ignited wildfires in Arizona and New Mexico. *International Journal of Wildland Fire* 16:242–54.

Hammer, K. J. 2000. Ponderosa poster child: U.S. Forest Service misrepresenting the historic condition of western forests and the effects of fire suppression and logging.

Unpublished report. Kalispell, Montana: Friends of the Wild Swan and Swan View Coalition.

Handley, C. O. Jr. 1969. Fire and mammals. *Proceedings of the Tall Timbers Fire Ecology Conference* 9:151–59.

Hann, W. J., J. L. Jones, M. G. Karl, P. F. Hessburg, R. E. Keane, D. G. Long, J. P. Menakis, C. H. McNicoll, S. G. Leonard, R. A. Gravenmier, and B. G. Smith. 1997. Landscape dynamics of the basin. Pp. 337–1057 in *An assessment of ecosystem components in the interior Columbia Basin and portions of the Klamath and Great Basins,* vol. 2, ed. T. M. Quigley and S. J. Arbelbide. USDA Forest Service General Technical Report PNW-GTR-405. Portland, Oregon: Pacific Northwest Research Station.

Hansen, K., W. Wyckoff, and J. Banfield. 1995. Shifting forests: Historical grazing and forest invasion in southwestern Montana. *Forest & Conservation History* 39:66–76.

Hanson, J. J., and J. D. Stuart. 2005. Vegetation responses to natural and salvage logged fire edges in Douglas-fir/hardwood forests. *Forest Ecology and Management* 214:266–78.

Harmon, M. E. 2002. Moving toward a new paradigm for woody detritus management. Pp. 929–44 in *Proceedings of the symposium on the ecology and management of dead wood in western forests,* ed. W. F. Laudenslayer, P. J. Shea, B. E. Valentine, C. P. Weatherspoon, and T. E. Lisle. USDA Forest Service General Technical Report PSW-GTR-181. Berkeley, California: Pacific Southwest Research Station.

Harmon, M. E., J. F. Franklin, F. J. Swanson, P. Sollins, S. V. Gregory, J. D. Lattin, N. M. Anderson, S. P. Cline, N. G. Aumen, J. R. Sedell, G. W. Lienkaemper, K. Cromack Jr., and K. W. Cummins. 1986. Ecology of coarse woody debris in temperate ecosystems. *Advances in Ecological Research* 15:133–302.

Harniss, R. O., and K. T. Harper. 1982. *Tree dynamics in seral and stable aspen stands of central Utah.* USDA Forest Service Research Paper INT-297. Ogden, Utah: Intermountain Forest and Range Experiment Station.

Harniss, R. O., and R. B. Murray. 1973. 30 years of vegetal change following burning of sagebrush-grass range. *Journal of Range Management* 26:322–25.

Harper, K. T., F. J. Wagstaff, and L. M. Kunzler. 1985. *Biology and management of the Gambel oak vegetative type: A literature review.* USDA Forest Service General Technical Report INT-179. Ogden, Utah: Intermountain Forest and Range Experiment Station.

Harrington, M. G. 1985. The effects of spring, summer, and fall burning on Gambel oak in a southwestern ponderosa pine stand. *Forest Science* 31:156–63.

———. 1987. Ponderosa pine mortality from spring, summer, and fall crown scorching. *Western Journal of Applied Forestry* 2:14–16.

———. 1993. Predicting *Pinus ponderosa* mortality from dormant season and growing season fire injury. *International Journal of Wildland Fire* 3:65–72.

———. 1996. *Fall rates of prescribed fire-killed ponderosa pine.* USDA Forest Service Research Paper INT-489. Ogden, Utah: Intermountain Research Station.

Harrington, M. G., and F. G. Hawksworth. 1990. Interactions of fire and dwarf mistletoe on mortality of southwestern ponderosa pine. Pp. 234–40 in *Effects of fire management of southwestern natural resources: Proceedings of the symposium, Nov. 15–17, 1988, Tucson, AZ,* ed. J. S. Krammes. USDA Forest Service General Technical Report RM-191. Fort Collins, Colorado: Rocky Mountain Forest and Range Experiment Station.

Harrington, M. G., and R. G. Kelsey. 1979. *Influence of some environmental factors on initial establishment and growth of ponderosa pine seedlings.* USDA Forest Service Research Paper INT-230. Ogden, Utah: Intermountain Forest and Range Experiment Station.

Harris, H. K. 1958. Fighting fires with airplanes and sodium calcium borate in western Montana and northern Idaho—1957. *Fire Control Notes* 19(2): 66–67.

Harris, M. A. 1982. Habitat use among woodpeckers in forest burns. M.S. thesis, University of Montana, Missoula.

Hart, R. H., and J. A. Hart. 1997. Rangelands of the Great Plains before European settlement. *Rangelands* 19:4–11.

Harvey, A. E., J. W. Byler, G. I. McDonald, L. F. Neuenschwander, and J. R. Tonn. 2008. *Death of an ecosystem: Perspectives on western white pine ecosystems of North America at the end of the twentieth century.* USDA Forest Service General Technical Report RMRS-GTR-208. Fort Collins, Colorado: Rocky Mountain Research Station.

Haskins, K. E., and C. A. Gehring. 2004. Long-term effects of burning slash on plant communities and arbuscular mycorrhizae in a semi-arid woodland. *Journal of Applied Ecology* 41:379–88.

Hassan, M. A., and N. E. West. 1986. Dynamics of soil seed pools in burned and unburned sagebrush semi-deserts. *Ecology* 76:269–72.

Hatton, J. H. 1920. *Live-stock grazing as a factor in fire protection on the national forests.* USDA Department Circular No. 134 Washington, D.C.: U.S. Government Printing Office.

Hayes, G. L. 1941. *Influence of altitude and aspect on daily variations in factors of forest-fire danger.* USDA Circular No. 591. Washington, D.C.: U.S. Government Printing Office.

Haynes, R. W. 2003. *An analysis of the timber situation in the United States: 1952 to 2050.* USDA Forest Service General Technical Report PNW-GTR-560. Portland, Oregon: Pacific Northwest Research Station.

Haynes, R. W., D. M. Adams, R. J. Alig, P. J. Ince, J. R. Mills, and X. Zhou. 2007. *The 2005 RPA timber assessment update.* USDA Forest Service General Technical Report PNW-GTR-699. Portland, Oregon: Pacific Northwest Research Station.

Heerwagen, A. 1954. The effect of grazing use upon ponderosa pine reproduction in the Rocky Mountain area. *Proceedings of the Society of American Foresters* 1954:206–7.

Hemphill, M. L. 1983. Fire, vegetation and people—Charcoal and pollen analyses of Sheep Mountain bog, Montana: The last 2800 years. M.A. thesis. Washington State University, Pullman.

Henderson, R. C. 1967. Thinning ponderosa pine in western Montana with prescribed fire. M.S. thesis, University of Montana, Missoula.

Henderson, S. J. 1997. Effects of fire on avian distributions and patterns of abundance over two vegetation types in southwest Montana: Implications for managing fire for biodiversity. M.S. thesis, Montana State University, Bozeman.

Hessl, A. E., and L. J. Graumlich. 2002. Interactive effects of human activities, herbivory, and fire on quaking aspen (*Populus tremuloides*) age structures in western Wyoming. *Journal of Biogeography* 29:889–902.

Hessl, A. E., D. McKenzie, and R. Schellhaas. 2004. Drought and Pacific decadal oscillation linked to fire occurrence in the inland Pacific Northwest. *Ecological Applications* 14:425–42.

Hessl, A., J. Miller, J. Kernan, D. Keenum, and D. McKenzie. 2007. Mapping paleo-fire boundaries from binary point data: Comparing interpolation methods. *Professional Geographer* 59:87–104.

Hester, D. A. 1952. The piñon-juniper fuel type can really burn. *USDA Forest Service Fire Control Notes* 13(1): 26–29.

Heyerdahl, E. K., L. B. Brubaker, and J. K. Agee. 2002. Annual and decadal climate forcing of historical fire regimes in the interior Pacific Northwest, USA. *The Holocene* 12:597–604.

Heyerdahl, E. K., and S. J. McKay. 2001. Condition of live fire-scarred ponderosa pine trees six years after removing partial cross sections. *Tree-Ring Research* 57:131–39.

Heyerdahl, E. K., D. McKenzie, L. D. Daniels, A. E. Hessl, J. Littell, and N. J. Mantua. 2008. Climate drivers of regionally synchronous fires in the inland Northwest (1651–1900). *International Journal of Wildland Fire* 17:40–49.

Heyerdahl, E. K., R. F. Miller, and R. A. Parsons. 2006. History of fire and Douglas-fir establishment in a savanna and sagebrush-grassland mosaic, southwestern Montana, USA. *Forest Ecology and Management* 230:107–18.

Heyerdahl, E. K., P. Morgan, and J. P. Riser II. 2008. Multi-season climate synchronized historical fires in dry forests (1650–1900), northern Rockies, USA. *Ecology* 89:705–16.

Higgins, K. F. 1984. Lightning fires in North Dakota grasslands and in pine-savanna lands of South Dakota and Montana. *Journal of Range Management* 37:100–103.

———. 1986. *Interpretation and compendium of historical fire accounts in the Northern Great Plains.* Resource Publication 161. Washington, D.C.: USDI Fish and Wildlife Service.

Higgins, K. F., A. D. Kruse, and J. L. Piehl. 1989. *Effects of fire in the northern Great Plains.* U.S. Fish and Wildlife Service and Cooperative Extension Service Circular 761. Brookings: South Dakota State University.

Higgins, R. W., K. C. Mo, and Y. Yao. 1998. Interannual variability of the U.S. summer precipitation regime with emphasis on the southwestern monsoon. *Journal of Climate* 11:2582–606.

Higuera, P. E., M. E. Peters, L. B. Brubaker, and D. G. Gavin. 2007. Understanding the origin and analysis of sediment-charcoal records with a simulation model. *Quaternary Science Reviews* 26:1790–809.

Higuera, P. E., D. G. Sprugel, and L. B. Brubaker. 2005. Reconstructing fire regimes with charcoal from small-hollow sediments: A calibration with tree-ring records of fire. *The Holocene* 15:238–51.

Hill, B. T. 1999a. *Western national forests: Nearby communities are increasingly threatened by catastrophic wildfires.* U.S. General Accounting Office, Testimony before the Subcommittee on Forests and Forest Health, Committee on Resources, House of Representatives. GAO/T-RCED-99–79. Washington, D.C.: U.S. Government Printing Office.

———. 1999b. *Western national forests: Status of Forest Service's efforts to reduce catastrophic wildfire threats.* U.S. General Accounting Office, Testimony before the Subcommittee on

Forests and Forest Health, Committee on Resources, House of Representatives. GAO/T-RCED-99–241. Washington, D.C.: U.S. Government Printing Office.

Hilty, J. H., D. J. Eldridge, R. Rosentreter, M. C. Wicklow-Howard, and M. Pellant. 2004. Recovery of biological soil crusts following wildfire in Idaho. *Journal of Range Management* 57:89–96.

Hirsch, K. G., P. N. Corey, and D. L. Martell. 1998. Using expert judgment to model initial attack fire crew effectiveness. *Forest Science* 44:539–49.

Hobbs, N. T. 1996. Modification of ecosystems by ungulates. *Journal of Wildlife Management* 60:695–713.

Hobbs, N. T., D. S. Schimel, C. E. Owensby, and D. S. Ojima. 1991. Fire and grazing in the tallgrass prairie: Contingent effects on nitrogen budgets. *Ecology* 72:1374–82.

Hobbs, N. T., and R. A. Spowart. 1984. Effects of prescribed fire on nutrition and mountain sheep and mule deer during winter and spring. *Journal of Wildlife Management* 48:551–60.

Hodson, E. R., and J. H. Foster. 1910. *Engelmann spruce in the Rocky Mountains, with special reference to growth, volume, and reproduction.* USDA Forest Service Circular No. 170. Washington, D.C.: U.S. Government Printing Office.

Hoff, R. J., D. E. Ferguson, G. I. McDonald, and R. E. Keane. 2001. Strategies for managing whitebark pine in the presence of white pine blister rust. Pp. 346–66 in *Whitebark pine communities,* ed. D. F. Tomback, S. F. Arno, and R. E. Keane. Washington, D.C.: Island Press.

Hoffman, T. L. 1996. An ecological investigation of mountain big sagebrush in the Gardiner Basin. M.S. thesis, Montana State University, Bozeman.

Holland, D. G. 1986. The role of forest insects and diseases in the Yellowstone ecosystem. *Western Wildlands* 12:19–23.

Hollenstein, K., R. L. Graham, and W. D. Shepperd. 2001. Biomass flow in western forests: Simulating the effects of fuel reduction and presettlement restoration treatments. *Journal of Forestry* 99:12–19.

Holmes, T. P., J. P. Prestemon, D. T. Butry, D. E. Mercer, and K. L. Abt. 2004. Using size-frequency distributions to analyze fire regimes in Florida. *Proceedings of the Tall Timbers Fire Ecology Conference* 22:88–94.

Hopkins, A. D. 1909. *Practical information on the Scolytid beetles of North American forests. I. Bark-beetles of the genus Dendrotonus.* USDA Bureau of Entomology Bulletin No. 83, part I. Washington, D.C.: U.S. Government Printing Office.

Hornby, L. G. 1935, Fuel type mapping in Region One. *Journal of Forestry* 33:67–71.

———. 1936. *Fire control planning in the northern Rocky Mountain region.* Progress Report No. 1. Missoula, Montana: Northern Rocky Mountain Forest and Range Experiment Station.

Hossack, B. R., and P. S. Corn. 2006. *Species-specific responses of amphibians to wildfire in Glacier National Park.* Paper presented at the 2006 Annual Meeting, Idaho Chapter of the Wildlife Society, Boise, Idaho.

Hough, F. B. 1882. *Report on forestry, submitted to Congress by the Commissioner of Agriculture.* Washington, D.C.: U.S. Government Printing Office.

Houston, D. B. 1973. Wildfires in northern Yellowstone National Park. *Ecology* 54:1111–17.

———. 1982. *The northern Yellowstone elk: Ecology and management.* New York: Macmillan.

Howard, W. E., R. L. Fenner, and H. E. Childs Jr. 1959. Wildlife survival in brush burns. *Journal of Range Management* 12:230–34.

Howe, E., and W. L. Baker. 2003. Landscape heterogeneity and disturbance interactions in a subalpine watershed in northern Colorado, USA. *Annals of the Association of American Geographers* 93:797–813.

Howe, G. E. 1976. The evolutionary role of wildfire in the northern Rockies and implications for resource managers. *Proceedings of the Tall Timbers Fire Ecology Conference* 14:257–65.

Howe, H. F., and J. Smallwood. 1982. Ecology of seed dispersal. *Annual Review of Ecology and Systematics* 13:201–28.

Huberman, M. A. 1935. The role of western white pine in forest succession in northern Idaho. *Ecology* 16:137–51.

Huckaby, L. S. 1991. Forest regeneration following fire in the forest-alpine ecotone in the Colorado Front Range. M.S. thesis, Colorado State University, Fort Collins.

Huckaby, L. S., M. R. Kaufmann, J. M. Stoker, and P. J. Fornwalt. 2001. Landscape patterns of montane forest age structure relative to fire history at Cheesman Lake in the Colorado Front Range. Pp. 19–27 in *Ponderosa pine ecosystems restoration and conservation: Steps toward stewardship, conference proceedings, Flagstaff, AZ, April 25–27, 2000,* ed. R. K. Vance, C. B. Edminster, W. W. Covington, and J. A. Blake. USDA Forest Service Proceedings RMRS-P-22. Fort Collins, Colorado: Rocky Mountain Research Station.

Huckaby, L. S., and W. H. Moir. 1995. Fire history of subalpine forests at Fraser Experimental Forest, Colorado. Pp. 153–58 in *Proceedings: Symposium on fire in wilderness and park management,* ed. J. K. Brown, R. W. Mutch, C. W. Spoon, and R. H. Wakimoto. USDA Forest Service General Technical Report INT-GTR-320. Ogden, Utah: Intermountain Research Station.

Huffines, G. R., and R. E. Orville. 1999. Lightning ground flash density and thunderstorm duration in the continental United States: 1989–96. *Journal of Applied Meteorology* 38:1013–19.

Huffman, D. W., P. Z. Fulé, K. M. Pearson, and J. E. Crouse. 2008. Fire history of pinyon-juniper woodlands at upper ecotones with ponderosa pine forests in Arizona and New Mexico. *Canadian Journal of Forest Research* 38:2097–108.

Huffman, D. W., P. Z. Fulé, K. M. Pearson, J. E. Crouse, and W. W. Covington. 2006. *Pinyon-juniper fire regime: Natural range of variability.* Final report to Rocky Mountain Research Station from the Ecological Restoration Institute, Northern Arizona University, Flagstaff, Arizona.

Hughes, J. 2002. Modeling the effect of landscape pattern on mountain pine beetles. M.S. thesis, Simon Fraser University, Burnaby, B.C., Canada.

Hughes, L. E. 2006. Wildfire's place in land management: A case study. *Rangelands* 28:17–21.

Humes, H. R. 1960. The ecological effects of fire on natural grassland in western Montana. Master's thesis, University of Montana, Missoula.

Humphrey, H. B., and J. E. Weaver. 1915. Natural reforestation in the mountains of northern Idaho. *Plant World* 18:31–47.

Humphrey, L. D. 1984. Patterns and mechanisms of plant succession after fire on Artemisia-grass sites in southeastern Idaho. *Vegetatio* 57:91–101.

Humphrey, L. D., and E. W. Schupp. 2001. Seed banks of *Bromus tectorum*–dominated communities in the Great Basin. *Western North American Naturalist* 61:85–92.

Hungerford, R. D. 1980. Microenvironmental response to harvesting and residue management. Pp. 37–73 in *Environmental consequences of timber harvesting in Rocky Mountain coniferous forests,* ed. USDA Forest Service. USDA Forest Service General Technical Report INT-90. Ogden, Utah: Intermountain Forest and Range Experiment Station.

Hungerford, R. D., W. H. Frandsen, and K. C. Ryan. 1995. Ignition and burning characteristics of organic soils. *Proceedings of the Tall Timbers Fire Ecology Conference* 19:78–91.

Hungerford, R. D., M. G. Harrington, W. H. Frandsen, K. C. Ryan, and G. J. Niehoff. 1991. Influence of fire on factors that affect site productivity. Pp. 32–50 in *Proceedings: Management and productivity of western-montane forest soils, April 10–12, 1990, Boise, Idaho,* ed. A. E. Harvey and L. F. Neuenschwander. USDA Forest Service General Technical Report INT-280. Ogden, Utah: Intermountain Research Station.

Hunt, C. B. 1974. *Natural regions of the United States and Canada.* San Francisco: W. H. Freeman.

Hunter, M. E., and P. N. Omi. 2006. Seed supply of native and cultivated grasses in the southwestern United States and the potential for vegetation recovery following wildfire. *Plant Ecology* 183:1–8.

Hunter, M. E., P. N. Omi, E. J. Martinson, and G. W. Chong. 2006. Establishment of nonnative plant species after wildfires: Effects of fuel treatments, abiotic and biotic factors, and post-fire grass seeding treatments. *International Journal of Wildland Fire* 15:271–81.

Huntzinger, M. 2003. Effects of fire management practices on butterfly diversity in the forested western United States. *Biological Conservation* 113:1–12.

Hutto, R. L. 1995. The composition of bird communities following stand-replacement fires in northern Rocky Mountain conifer forests. *Conservation Biology* 9:1041–58.

———. 2006. Toward meaningful snag-management guidelines for postfire salvage logging in North American conifer forests. *Conservation Biology* 20:984–93.

Ingram, D. C. 1931. Vegetative changes and grazing use on Douglas fir cut-over land. *Journal of Agricultural Research* 43:387–417.

International Code Council. 2008. *The blue ribbon panel report on wildland urban interface fire.* Washington, D.C.: Author.

Irwin, L. L., and J. M. Peek. 1983. Elk habitat use relative to forest succession in Idaho. *Journal of Wildlife Management* 47:664–72.

Ives, R. L. 1941. Forest replacement rates in the Colorado headwaters area. *Bulletin of the Torrey Botanical Club* 68:407–8.

Jack, J. G. 1900. Pikes Peak, Plum Creek, and South Platte Reserves. House of Representatives, 56th Cong., 1st sess., document no. 5, *Annual Reports of the Department of the Interior for the Fiscal Year Ended June 30, 1899, 20th Annual Report of the United States Geological Survey, Part V—Forest Reserves,* pp. 39–115.

Jackson, A. S. 1965. Wildfires in the Great Plains grasslands. *Proceedings of the Tall Timbers Fire Ecology Conference* 4:241–59.

Jakubos, B., and W. H. Romme. 1993. Invasion of subalpine meadows by lodgepole pine in Yellowstone National Park, Wyoming, U.S.A. *Arctic and Alpine Research* 25:382–90.

Jehle, G., J. A. Savidge, and N. B. Kotliar. 2006. Green-tailed towhee response to prescribed fire in montane shrubland. *The Condor* 108:634–46.

Jemison, G. M. 1932. Meteorological conditions affecting the Freeman Lake (Idaho) fire. *Monthly Weather Review* 60:1–2.

———. 1934. The significance of the effect of stand density upon the weather beneath the canopy. *Journal of Forestry* 32:446–51.

Jenkins, M. J., C. A. Dicus, and E. G. Hebertson. 1998. Postfire succession and disturbance interactions on an intermountain subalpine spruce-fir forest. *Proceedings of the Tall Timbers Fire Ecology Conference* 20:219–29.

Jenkins, M. J., E. Hebertson, W. Page, and C. A. Jorgensen. 2008. Bark beetles, fuels, fires and implications for forest management in the Intermountain West. *Forest Ecology and Management* 254:16–34.

Jeske, B. W., and C. D. Bevins. 1979. Spatial and temporal distribution of natural fuels in Glacier Park. Pp. 1219–24 in *Proceedings of the first conference on scientific research in the national parks,* vol. 2, ed. R. M. Linn. Transactions and Proceedings Series No. 5. Washington, D.C.: USDI National Park Service.

Jessop, B. D., and V. J. Anderson. 2007. Cheatgrass invasion in salt desert shrublands: Benefits of postfire reclamation. *Rangeland Ecology and Management* 60:235–43.

Jia, G. J., I. C. Burke, M. R. Kaufmann, A. F. H. Goetz, B. C. Kindel, and Y. Pu. 2006. Estimates of forest canopy fuel attributes using hyperspectral data. *Forest Ecology and Management* 229:27–38.

Jirik, S. J., and S. C. Bunting. 1994. Postfire defoliation response of *Agropyron spicatum* and *Sitanion hystrix*. *International Journal of Wildland Fire* 4:77–82.

Johansen, J. R. 2001. Impacts of fire on biological soil crusts. Pp. 385–400 in *Biological soil crusts: Structure, management and function,* ed. J. Belnap and O. Lange. Berlin: Springer-Verlag.

Johansen, J. R., J. Ashley, and W. R. Rayburn. 1993. Effects of rangefire on soil algal crusts in semiarid shrub-steppe of the lower Columbia Basin and their subsequent recovery. *Great Basin Naturalist* 53:73–88.

Johansen, J. R., A. Javakul, and S. R. Rushforth. 1982. Effects of burning on the algal communities of a high desert soil near Wallsburg, Utah. *Journal of Range Management* 35:598–600.

Johnson, C. A. 1989. Early spring prescribed burning of big game winter range in the Snake River canyon of westcentral Idaho. Pp. 151–55 in *Prescribed fire in the Intermountain region: Forest site preparation and range improvement—Symposium proceedings,* ed. D. M. Baumgartner, D. W. Breuer, B. A. Zamora, L. F. Neuenschwander, and R. H. Wakimoto. Pullman: Washington State University, Cooperative Extension Service.

Johnson, C. G. Jr. 1998. *Vegetation response after wildfires in national forests of northeastern*

Oregon. USDA Forest Service R6-NR-ECOL-TP-06-98. Portland, Oregon: Pacific Northwest Region.

Johnson, E. A., and S. L. Gutsell. 1993. Heat budget and fire behaviour associated with the opening of serotinous cones in two *Pinus* species. *Journal of Vegetation Science* 4:745–50.

———. 1994. Fire frequency models, methods and interpretations. *Advances in Ecological Research* 25:239–87.

Johnson, E. A., and K. Miyanishi. 2008. Testing the assumptions of chronosequences in succession. *Ecology Letters* 11:419–31.

Johnson, M. 1994. Changes in southwestern forests: Stewardship implications. *Journal of Forestry* 92:16–19.

Johnson, P. C. 1966. *Some causes of natural tree mortality in old growth ponderosa pine stands in western Montana.* USDA Forest Service Research Note INT-51. Ogden, Utah: Intermountain Forest and Range Experiment Station.

Johnson, R. F. 1963. The roadside fire problem. *Fire Control Notes* 24(1): 5–7.

Johnson, V. J. 1982. The dilemma of flame length and intensity. *Fire Management Notes* 43(4): 3–7.

Johnson, W. M. 1956. The effect of grazing intensity on plant composition, vigor, and growth of pine-bunchgrass ranges in central Colorado. *Ecology* 37:790–98.

Johnston, R. G. 1978. Helicopter use in forest fire suppression—3 decades. *Fire Management Notes* 39(4): 14–18.

Jones, A. 2000. Effects of cattle grazing on North American arid ecosystems: A quantitative review. *Western North American Naturalist* 60:155–64.

Jones, M. H. 1995. Do shade and shrubs enhance natural regeneration of Douglas-fir in south-central Idaho? *Western Journal of Applied Forestry* 10:24–28.

Jordan, G. J., M.-J. Fortin, and K. P. Lertzman. 2005. Assessing spatial uncertainty associated with forest fire boundary delineation. *Landscape Ecology* 20:719–31.

Jourdonnais, C. S., and D. J. Bedunah. 1990. Prescribed fire and cattle grazing on an elk winter range in Montana. *Wildlife Society Bulletin* 18:232–40.

Kalabokidis, K., and P. N. Omi. 1998. Reduction of fire hazard through thinning/residue disposal in the urban interface. *International Journal of Wildland Fire* 8:29–35.

Karr, J. R., J. J. Rhodes, G. W. Minshall, F. R. Hauer, R. L. Beschta, C. A. Frissell, and D. A. Perry. 2004. The effects of postfire salvage logging on aquatic ecosystems in the American West. *BioScience* 54:1029–33.

Karsian, A. E. 1995. A 6800-year vegetation and fire history in the Bitterroot Mountain Range, Montana. Master's thesis, University of Montana, Missoula.

Kashian, D. M., W. H. Romme, and C. M. Regan. 2007. Reconciling divergent interpretations of quaking aspen decline on the northern Colorado Front Range. *Ecological Applications* 17:1296–1311.

Kashian, D. M., D. B. Tinker, M. G. Turner, and F. L. Scarpace. 2004. Spatial heterogeneity of lodgepole pine sapling densities following the 1988 fires in Yellowstone National Park, Wyoming, USA. *Canadian Journal of Forest Research* 34:2263–76.

Kashian, D. M., M. G. Turner, and W. H. Romme. 2005. Variability in leaf area and stemwood increment along a 300-year lodgepole pine chronosequence. *Ecosystems* 8:48–61.

Kashian, D. M., M. G. Turner, W. H. Romme, and C. G. Lorimer. 2005. Variability and convergence in stand structural development on a fire-dominated subalpine landscape. *Ecology* 86:643–54.

Kaufman, D. W., S. K. Gurtz, and G. A. Kaufman. 1988. Movements of the deer mouse in response to prairie fire. *Prairie Naturalist* 20:225–29.

Kaufmann, M. R., P. J. Fornwalt, L. S. Huckaby, and J. M. Stoker. 2001. Cheesman Lake: A historical ponderosa pine landscape guiding restoration in the South Platte watershed of the Colorado Front Range. Pp. 9–18 in *Ponderosa pine ecosystems restoration and conservation: Steps toward stewardship—Conference proceedings, Flagstaff, AZ, April 25–27, 2000,* ed. R. K. Vance, C. B. Edminster, W. W. Covington, and J. A. Blake. USDA Forest Service Proceedings RMRS-P-22. Fort Collins, Colorado: Rocky Mountain Research Station.

Kaufmann, M. R., P. Z. Fulé, W. H. Romme, and K. C. Ryan. 2004. Restoration of ponderosa pine forests in the interior western U.S. after logging, grazing, and fire suppression. Pp. 481–500 in *Restoration of boreal and temperate forests,* ed. J. A. Stanturf and P. Madsen. Boca Raton, Florida: CRC Press.

Kaufmann, M. R., L. S. Huckaby, P. J. Fornwalt, J. M. Stoker, and W. H. Romme. 2003. Using tree recruitment patterns and fire history to guide restoration of an unlogged ponderosa pine/Douglas-fir landscape in the southern Rocky Mountains after a century of fire suppression. *Forestry* 76:231–41.

Kaufmann, M. R., C. Regan, and P. M. Brown. 2000. Heterogeneity in ponderosa pine/Douglas-fir forests: Age and size structure in unlogged and logged landscapes of central Colorado. *Canadian Journal of Forest Research* 30:698–711.

Kay, C. E. 1993. Aspen seedlings in recently burned areas of Grant Teton and Yellowstone National Park. *Northwest Science* 67:94–104.

———. 1997. Is aspen doomed? *Journal of Forestry* 95:4–11.

———. 2001. Evaluation of burned aspen communities in Jackson Hole, Wyoming. Pp. 215–23 in *Sustaining aspen in western landscapes: Symposium proceedings, June 13–15, 2000, Grand Junction, Colorado,* ed. W. D. Shepperd, D. Binkley, D. L. Bartos, T. J. Stohlgren, and L. G. Eskew. USDA Forest Service Proceedings RMRS-P-18. Fort Collins, Colorado: Rocky Mountain Research Station.

———. 2007. Are lightning fires unnatural? A comparison of aboriginal and lightning ignition rates in the United States. *Proceedings of the Tall Timbers Fire Ecology Conference* 23:16–28.

Kay, C. E., and F. H. Wagner. 1996. Response of shrub-aspen to Yellowstone's 1988 wildfires: Implications for "natural regulation" management. Pp. 107–11 in *The ecological implications of fire in Greater Yellowstone: Proceedings second biennial conference on the Greater Yellowstone ecosystem, Yellowstone National Park, Sept. 19–21, 1993,* ed. J. M. Greenlee. Fairfield, Washington: International Association of Wildland Fire.

Kayll, A. J. 1968. Heat tolerance of tree seedlings. *Proceedings of the Tall Timbers Fire Ecology Conference* 8:89–105.

Keane, R. E. 2008. *Surface fuel litterfall and decomposition in the northern Rocky Mountains, U.S.A.* USDA Forest Service Research Paper RMRS-RP-70. Fort Collins, Colorado: Rocky Mountain Research Station.

Keane, R. E., and S. F. Arno. 2001. Restoration concepts and techniques. Pp. 367–400 in *Whitebark pine communities: Ecology and restoration,* ed. D. F. Tomback, S. F. Arno, and R. E. Keane. Washington, D.C.: Island Press.

Keane, R. E., S. F. Arno, and J. K. Brown. 1989. *FIRESUM: An ecological process model for fire succession in western conifer forests.* USDA Forest Service General Technical Report INT-266. Ogden, Utah: Intermountain Forest and Range Experiment Station.

———. 1990. Simulating cumulative fire effects in ponderosa pine/Douglas-fir forests. *Ecology* 71:189–203.

Keane, R. E., S. Arno, and L. J. Dickinson. 2006. The complexity of managing fire-dependent ecosystems in wilderness: Relict ponderosa pine in the Bob Marshall Wilderness. *Ecological Restoration* 24:71–78.

Keane, R. E., R. Burgan, and J. van Wagtendonk. 2001. Mapping wildland fuels for fire management across multiple scales: Integrating remote sensing, GIS, and biophysical modeling. *International Journal of Wildland Fire* 10:301–19.

Keane, R. E., G. J. Cary, and R. Parsons. 2003. Using simulation to map fire regimes: An evaluation of approaches, strategies, and limitations. *International Journal of Wildland Fire* 12:309–22.

Keane, R. E., and L. J. Dickinson. 2007. *The photoload sampling technique: Estimating surface fuel loadings from downward-looking photographs of synthetic fuelbeds.* USDA Forest Service General Technical Report RMRS-GTR-190. Fort Collins, Colorado: Rocky Mountain Research Station.

Keane, R. E., and P. Morgan. 1994. Decline of whitebark pine in the Bob Marshall Wilderness Complex of Montana, U.S.A. Pp. 193–98 in *Proceedings—International workshop on subalpine stone pines and their environment: The status of our knowledge, Sept. 5–11, 1992, St. Moritz, Switzerland,* ed. W. C. Schmidt and F.-K. Holtmeier. USDA Forest Service General Technical Report INT-GR-309. Ogden, Utah: Intermountain Research Station.

Keane, R. E., P. Morgan, and J. P. Menakis. 1994. Landscape assessment of the decline of whitebark pine (*Pinus albicaulis*) in the Bob Marshall Wilderness Complex, Montana, USA. *Northwest Science* 68:213–29.

Keane, R. E., R. A. Parsons, and P. F. Hessburg. 2002. Estimating historical range and variation of landscape patch dynamics: Limitations of the simulation approach. *Ecological Modelling* 151:29–49.

Keane, R. E., E. D. Reinhardt, and J. K. Brown. 1994. FOFEM: A first order fire effects model for predicting the immediate consequences of wildland fire in the United States. *Proceedings of the Conference on Fire and Forest Meteorology* 12:628–31.

Keane, R. E., K. C. Ryan, and M. A. Finney. 1998. Simulating the consequences of fire and climate regimes on a complex landscape in Glacier National Park, Montana. *Proceedings of the Tall Timbers Fire Ecology Conference* 20:310–24.

Keane, R. E., K. C. Ryan, and S. W. Running. 1996. Simulating effects of fire on northern Rocky Mountain landscapes with the ecological process model FIRE-BGC. *Tree Physiology* 16:319–31.

Keane, R. E., K. C. Ryan, T. T. Veblen, C. D. Allen, J. A. Logan, and B. Hawkes. 2002. The cascading effects of fire exclusion in Rocky Mountain ecosystems. Pp. 133–52 in *Rocky Mountain futures: An ecological perspective,* ed. J. S. Baron. Washington, D.C.: Island Press.

Kearns, H. S. J., and W. R. Jacobi. 2007. The distribution and incidence of white pine blister rust in central and southeastern Wyoming and northern Colorado. *Canadian Journal of Forest Research* 37:462–72.

Kearns, H. S. J., W. R. Jacobi, and D. W. Johnson. 2005. Persistence of pinyon pine snags and logs in southwestern Colorado. *Western Journal of Applied Forestry* 20:247–52.

Keay, J. A., and J. M. Peek. 1980. Relationships between fires and winter habitat of deer in Idaho. *Journal of Wildlife Management* 44:372–80.

Keeley, J. E. 2004. Ecological impacts of wheat seeding after a Sierra Nevada wildfire. *International Journal of Wildland Fire* 13:73–78.

———. 2006. Fire management impacts on invasive plants in the western United States. *Conservation Biology* 20:375–84.

Keeley, J. E., C. D. Allen, J. Betancourt, G. W. Chong, C. J. Fotheringham, and H. D. Safford. 2006. A 21st century perspective on postfire seeding. *Journal of Forestry* 104:103–4.

Keeley, J. E., and C. J. Fotheringham. 2003. Impact of past, present, and future fire regimes on North American Mediterranean shrublands. Pp. 218–62 in *Fire and climatic change in temperate ecosystems of the western Americas,* ed. T. T. Veblen, W. L. Baker, G. Montenegro, and T. W. Swetnam. New York: Springer.

Keeley, J. E., C. J. Fotheringham, and M. A. Moritz. 2004. Lessons from the October 2003 wildfires in Southern California. *Journal of Forestry* 102:26–31.

Keeley, J. E., and T. W. McGinnis. 2007. Impact of prescribed fire and other factors on cheatgrass persistence in a Sierra Nevada ponderosa pine forest. *International Journal of Wildland Fire* 16:96–106.

Keeley, J. E., and P. H. Zedler. 1998. Evolution of life histories in Pinus. Pp. 219–49 in *Ecology and biogeography of Pinus,* ed. D. M. Richardson. Cambridge, England: Cambridge University Press.

Keeling, E. G., A. Sala, and T. H. DeLuca. 2006. Effects of fire exclusion on forest structure and composition in unlogged ponderosa pine/Douglas-fir forests. *Forest Ecology and Management* 237:418–28.

Kendall, D. M. 2003. Effects of fire on insect communities in piñon-juniper woodlands in Mesa Verde country. Pp. 279–85 in *Ancient piñon-juniper woodlands: A natural history of Mesa Verde country,* ed. M. L. Floyd. Boulder: University Press of Colorado.

Kendall, K. C., and S. F. Arno. 1990. Whitebark pine: An important but endangered wildlife resource. Pp. 264–73 in *Proceedings: Symposium on whitebark pine ecosystems—Ecology and management of a high-mountain resource,* ed. W. C. Schmidt and K. J. McDonald. USDA Forest Service General Technical Report INT-270. Ogden, Utah: Intermountain Research Station.

Kennedy, R. G. 2006. *Wildfire and Americans: How to save lives, property, and your tax dollars.* New York: Hill and Wang.

Kerley, L. L. 1994. Bird responses to habitat fragmentation caused by sagebrush management in a Wyoming sagebrush steppe ecosystem. Ph.D. dissertation, University of Wyoming, Laramie.

Kerley, L. L., and S. H. Anderson. 1995. Songbird responses to sagebrush removal in a high elevation sagebrush steppe ecosystem. *Prairie Naturalist* 27:129–46.

Kerns, B. K., W. G. Thies, and C. G. Niwa. 2006. Season and severity of prescribed burn in ponderosa pine forests: Implications for understory native and exotic plants. *Ecoscience* 13:44–55.

Kessell, S. R. 1979. *Gradient modelling: Resource and fire management.* New York: Springer-Verlag.

Kessell, S. R., and M. W. Potter. 1980. A quantitative succession model for nine Montana forest communities. *Environmental Management* 4:227–40.

Key, C. H., and N. C. Benson. 2006. *Landscape assessment (LA): Sampling and analysis methods.* USDA Forest Service General Technical Report RMRS-GTR-164-CD. Fort Collins, Colorado: Rocky Mountain Research Station.

Keyes, C. R., and J. M. Varner. 2006. Pitfalls in the silvicultural treatment of canopy fuels. *Fire Management Today* 66(3): 46–50.

Keyser, T. L., L. B. Lentile, F. W. Smith, and W. D. Shepperd. 2008. Changes in forest structure after a large, mixed-severity wildfire in ponderosa pine forests of the Black Hills, South Dakota, USA. *Forest Science* 54:328–38.

Keyser, T. L., F. W. Smith, L. B. Lentile, and W. D. Shepperd. 2006. Modeling postfire mortality of ponderosa pine following a mixed-severity wildfire in the Black Hills: The role of tree morphology and direct fire effects. *Forest Science* 52:530–39.

Keyser, T. L., F. W. Smith, and W. D. Shepperd. 2005. Trembling aspen response to a mixed-severity wildfire in the Black Hills, South Dakota, USA. *Canadian Journal of Forest Research* 35:2679–84.

Kinter, C. L., B. A. Mealor, N. L. Shaw, and A. L. Hild. 2007. Postfire invasion potential of rush skeletonweed (*Chondrilla juncea*). *Rangeland Ecology and Management* 60:386–94.

Kipfmueller, K. F. 2003. Fire-climate-vegetation interactions in subalpine forests of the Selway-Bitterroot Wilderness Area, Idaho and Montana, USA. Ph.D. dissertation, University of Arizona, Tucson.

Kipfmueller, K. F., and W. L. Baker. 1998a. A comparison of three approaches to date stand-replacing fires in lodgepole pine forests. *Forest Ecology and Management* 104:171–77.

———. 1998b. Fires and dwarf mistletoe in a Rocky Mountain lodgepole pine ecosystem. *Forest Ecology and Management* 108:77–84.

———. 2000. A fire history of a subalpine forest in south-eastern Wyoming, USA. *Journal of Biogeography* 27:71–85.

Kipfmueller, K. F., and J. A. Kupfer. 2005. Complexity of successional pathways in subalpine forests of the Selway-Bitterroot Wilderness Area. *Annals of the Association of American Geographers* 95:495–510.

Kipfmueller, K. F., and T. W. Swetnam. 2000. Fire-climate interactions in the Selway-Bitterroot Wilderness Area. Pp. 270–75 in *Wilderness science in a time of change conference,* vol. 5, ed. D. N. Cole, S. F. McCool, W. T. Borrie, and J. O'Loughlin. USDA Forest Service Proceedings RMRS-P-15-VOL-5. Fort Collins, Colorado: Rocky Mountain Research Station.

Kitzberger, T., P. M. Brown, E. K. Heyerdahl, T. W. Swetnam, and T. T. Veblen. 2007. Contingent Pacific-Atlantic ocean influence on multicentury wildfire synchrony over western North America. *Proceedings of the National Academy of Sciences* 104:543–48.

Klebenow, D. A. 1965. A montane forest winter deer habitat in western Montana. *Journal of Wildlife Management* 29:27–33.

———. 1973. The habitat requirements of sage grouse and the role of fire in management. *Proceedings of the Tall Timbers Fire Ecology Conference* 12:305–15.

Klebenow, D. A., and R. C. Beall. 1978. Fire impacts on birds and mammals on Great Basin rangelands. Pp. 59–62 in *Proceedings of the 1977 rangeland management and fire symposium, Nov. 1–3, 1977, Casper, Wyoming,* ed. C. M. Bourassa and A. P. Brackebusch. Missoula: University of Montana School of Forestry, Montana Forest and Conservation Experiment Station.

Kleinman, L. H. 1973. Community characteristics of six burned aspen-conifer sites and their related animal use. M.S. thesis, Brigham Young University, Provo, Utah.

Knapp, P. A. 1998. Spatio-temporal patterns of large grassland fires in the Intermountain West, U.S.A. *Global Ecology and Biogeography Letters* 7:259–72.

Knapp, P. A., and P. T. Soulé. 2007. Trends in midlatitude cyclone frequency and occurrence during fire season in the northern Rockies, 1900–2004. *Geophysical Research Letters* 34(L20707): 1–5.

Knick, S. T. 1999. Requiem for a sagebrush ecosystem? *Northwest Science* 73:53–57.

Knick, S. T., A. L. Holmes, and R. F. Miller. 2005. The role of fire in structuring sagebrush habitats and bird communities. Pp. 63–75 in *Fire and avian ecology in North America,* ed. V. A. Saab and H. D. W. Powell. Studies in Avian Biology No. 30, Camarillo, California: Cooper Ornithological Society.

Knight, D. H. 1987. Parasites, lightning, and the vegetation mosaic in wilderness landscapes. Pp. 59–83 in *Landscape Heterogeneity and Disturbance,* ed. M. G. Turner. New York: Springer-Verlag.

———. 1994. *Mountains and plains: The ecology of Wyoming landscapes.* New Haven, Connecticut: Yale University Press.

Knight, D. H., and L. L. Wallace. 1989. The Yellowstone fires: Issues in landscape ecology. *BioScience* 39:700–706.

Knowles, P., and M. C. Grant. 1983. Age and size structure analyses of Engelmann spruce, ponderosa pine, lodgepole pine, and limber pine in Colorado. *Ecology* 64:1–9.

Koch, E. 1935. The passing of the Lolo trail. *Journal of Forestry* 33:98–110.

———. 1942. History of the 1910 forest fires in Idaho and western Montana. Unpublished manuscript. Missoula, Montana: USDA Forest Service, Region 1.

Koehler, G. M., and M. G. Hornocker. 1977. Fire effects on marten habitat in the Selway-Bitterroot Wilderness. *Journal of Wildlife Management* 41:500–505.

Kolb, P. F., and R. Robberecht. 1996. Pinus ponderosa seedling establishment and the influence of competition with the bunchgrass *Agropyron spicatum. International Journal of Plant Sciences* 157:509–15.

Komarek, E. V. 1964. The natural history of lightning. *Proceedings of the Tall Timbers Fire Ecology Conference* 3:139–83.

———. 1966. The meteorological basis for fire ecology. *Proceedings of the Tall Timbers Fire Ecology Conference* 5:85–125.

———. 1968. The nature of lightning fires. *Proceedings of the Tall Timbers Fire Ecology Conference* 7:5–41.

————. 1969. Fire and animal behavior. *Proceedings of the Tall Timbers Fire Ecology Conference* 9:161–207.

Koplin, J. R. 1969. The numerical response of woodpeckers to insect prey in a subalpine forest in Colorado. *The Condor* 71:436–38.

Korb, J. E., N. C. Johnson, and W. W. Covington. 2004. Slash pile burning effects on soil biotic and chemical properties and plant establishment: Recommendations for amelioration. *Restoration Ecology* 12:52–62.

Korb, N. T. 2005. Historical fire regimes and structures of Douglas-fir forests in the Centennial Valley of southwest Montana. M.S. thesis, Colorado State University, Fort Collins.

Koski, W. H., and W. C. Fischer. 1979. *Photo series for appraising thinning slash in north Idaho.* USDA Forest Service General Technical Report INT-46. Ogden, Utah: Intermountain Forest and Range Experiment Station.

Kotliar, N. B., S. L. Haire, and C. H. Key. 2003. Lessons from the fires of 2000: Post-fire heterogeneity in ponderosa pine forests. Pp. 277–79 in *Fire, fuel treatments, and ecological restoration: Conference proceedings, April 16–18, 2002, Fort Collins, Colorado,* ed. P. N. Omi and L. A. Joyce. USDA Forest Service Proceedings RMRS-P-29. Fort Collins, Colorado: Rocky Mountain Research Station.

Kotliar, N. B., S. J. Hejl, R. L. Hutto, V. A. Saab, C. P. Melcher, and M. E. McFadzen. 2002. Effects of fire and post-fire salvage logging on avian communities in conifer-dominated forests of the western United States. *Studies in Avian Biology* 25:49–64.

Kotliar, N. B., P. L. Kennedy, and K. Ferree. 2007. Avifaunal responses to fire in southwestern montane forests along a burn severity gradient. *Ecological Applications* 17:491–507.

Kotliar, N. B., E. W. Reynolds, and D. H. Deutschman. 2008. American three-toed woodpecker response to burn severity and prey availability at multiple spatial scales. *Fire Ecology* 4:26–45.

Kou, X., and W. L. Baker. 2006a. Accurate estimation of mean fire interval for managing fire. *International Journal of Wildland Fire* 15:489–95.

————. 2006b. A landscape model quantifies error in reconstructing fire history from scars. *Landscape Ecology* 21:735–45.

Kramer, N. B., and F. D. Johnson. 1987. Mature forest seed banks of three habitat types in central Idaho. *Canadian Journal of Botany* 65:1961–66.

Krasnow, K. 2007. Forest fuel mapping and strategic wildfire mitigation in the montane zone of Boulder County, Colorado. M.S. thesis, University of Colorado, Boulder.

Krebill, R. G. 1972. *Mortality of aspen on the Gros Ventre elk winter range.* USDA Forest Service Research Paper INT-129. Ogden, Utah: Intermountain Forest and Range Experiment Station.

Kruse, R., E. Bend, and P. Bierzychudek. 2004. Native plant regeneration and introduction of non-natives following post-fire rehabilitation with straw mulch and barley seeding. *Forest Ecology and Management* 196:299–310.

Kufeld, R. C. 1983. *Responses of elk, mule deer, cattle, and vegetation to burning, spraying, and chaining of Gambel oak rangeland.* Technical Publication No. 34. Fort Collins: Colorado Division of Wildlife.

Kulakowski, D., and T. T. Veblen. 2002. Influences of fire history and topography on the pattern of a severe wind blowdown in a Colorado subalpine forest. *Journal of Ecology* 90:806–19.

———. 2006. Historical range of variability for forest vegetation of the Grand Mesa National Forest, Colorado. Unpublished report. Lakewood, Colorado: USDA Forest Service, Rocky Mountain Region.

———. 2007. Effect of prior disturbances on the extent and severity of wildfire in Colorado subalpine forests. *Ecology* 88:759–69.

Kulakowski, D., T. T. Veblen, and P. Bebi. 2003. Effects of fire and spruce beetle outbreak legacies on the disturbance regime of a subalpine forest in Colorado. *Journal of Biogeography* 30:1445–56.

Kulakowski, D., T. T. Veblen, and S. Drinkwater. 2004. The persistence of quaking aspen (*Populus tremuloides*) in the Grand Mesa area, Colorado. *Ecological Applications* 14:1603–14.

Kulakowski, D., T. T. Veblen, and B. P. Kurzel. 2006. Influences of infrequent fire, elevation and pre-fire vegetation on the persistence of quaking aspen (*Populus tremuloides* Michx.) in the Flat Tops area, Colorado, USA. *Journal of Biogeography* 33:1397–1413.

Kuntz, D. E. 1982. Plant response following spring burning in an *Artemisia tridentata* subsp. *vaseyana*/*Festuca idahoensis* habitat type. Ph.D. dissertation, University of Idaho, Moscow.

Kunzler, L. M., and K. T. Harper. 1980. Recovery of Gambel oak after fire in central Utah. *Great Basin Naturalist* 40:127–30.

Kunzler, L. M., K. T. Harper, and D. B. Kunzler. 1981. Compositional similarity within the oakbrush type in central and northern Utah. *Great Basin Naturalist* 41:147–53.

Kurzel, B. P., T. T. Veblen, and D. Kulakowski. 2007. A typology of stand structure and dynamics of quaking aspen in northwestern Colorado. *Forest Ecology and Management* 252:176–90.

Lachmund, H. G. 1923. Relative susceptibility of incense cedar and yellow pine to bole injury in forest fires. *Journal of Forestry* 21:815–17.

Landis, A. G., and J. D. Bailey. 2005. Reconstruction of age structure and spatial arrangement of piñon-juniper woodlands and savannas of Anderson Mesa, northern Arizona. *Forest Ecology and Management* 204:221–36.

Landres, P. B., P. Morgan, and F. J. Swanson. 1999. Overview of the use of natural variability concepts in managing ecological systems. *Ecological Applications* 9:1179–88.

Langenheim, J. H. 1962. Vegetation and environmental patterns in the Crested Butte area, Gunnison County, Colorado. *Ecological Monographs* 32:249–85.

Larsen, E. J., and W. J. Ripple. 2003. Aspen age structure in the northern Yellowstone ecosystem: USA. *Forest Ecology and Management* 179:469–82.

Larsen, J. A. 1925a. The forest-fire season at different elevations in Idaho. *Monthly Weather Review* 53:60–63.

———. 1925b. Natural reproduction after forest fires in northern Idaho. *Journal of Agricultural Research* 30:1177–97.

———. 1929. Fires and forest succession in the Bitterroot Mountains of northern Idaho. *Ecology* 10:67–76.

Larsen, J. A., and C. C. Delavan. 1922. Climate and forest fires in Montana and northern Idaho, 1909–1919. *Monthly Weather Review* 50:55–68.

Larson, E. R. 2005. Spatiotemporal variations in the fire regimes of whitebark pine (*Pinus albicaulis* Engelm.) forests, western Montana, USA, and their management implications. M.S. thesis, University of Tennessee, Knoxville.

Latham, D. J., and R. C. Rothermel. 1993. *Probability of fire-stopping precipitation events.* USDA Forest Service Research Note INT-410. Ogden, Utah: Intermountain Research Station.

Latham, D. J., and J. A. Schlieter. 1989. *Ignition probabilities of wildland fuels based on simulated lightning discharges.* USDA Forest Service Research Paper INT-411. Ogden, Utah: Intermountain Research Station.

Latham, D., and E. Williams. 2001. Lightning and forest fires. Pp. 375–418 in *Forest fires: Behavior and ecological effects,* ed. E. A. Johnson and K. Miyanishi. San Diego, California: Academic Press.

Laughlin, D. C., and J. B. Grace. 2006. A multivariate model of plant species richness in forested systems: Old-growth montane forests with a long history of fire. *Oikos* 114:60–70.

Laven, R. D., P. N. Omi, J. G. Wyant, and A. S. Pinkerton. 1980. Interpretation of fire scar data from a ponderosa pine ecosystem in the central Rocky Mountains, Colorado. Pp. 46–49 in *Proceedings of the Fire History Workshop,* ed. M. A. Stokes and J. H. Dieterich. USDA Forest Service General Technical Report RM-81. Fort Collins, Colorado: Rocky Mountain Forest and Range Experiment Station.

Lawrence, G. E. 1966. Ecology of vertebrate animals in relation to chaparral fire in the Sierra Nevada foothills. *Ecology* 47:278–91.

Lawson, B. D. 1972. Fire spread in lodgepole pine stands. Master's thesis, University of Montana, Missoula.

Laycock, W. A. 1967. How heavy grazing and protection affect sagebrush-grass ranges. *Journal of Range Management* 20:206–13.

League, K., and T. Veblen. 2006. Climatic variability and episodic *Pinus ponderosa* establishment along the forest-grassland ecotones of Colorado. *Forest Ecology and Management* 228:98–107.

Lecky, J. H. 2007. *The aerial application of eight long-term fire retardants on all Forest Service lands.* Biological Opinion, Endangered Species Act, Section 7 Consultation. Washington, D.C.: U.S. Department of Commerce, National Marine Fisheries Service.

Leege, T. A. 1968. Prescribed burning for elk in northern Idaho. *Proceedings of the Tall Timbers Fire Ecology Conference* 8:235–53.

———. 1979. Effects of repeated prescribed burns on northern Idaho elk browse. *Northwest Science* 53:107–13.

Leiberg, J. B. 1897. *General report on a botanical survey of the Coeur d'Alene Mountains in Idaho during the summer of 1895.* Contributions from the U.S. National Herbarium, vol. 5, no. 1, pp. 41–85. Washington, D.C.: USDA Division of Botany.

———. 1899a. Bitterroot forest reserve. *Nineteenth annual report of the USDI Geological Survey, Part V—Forest reserves,* pp. 253–82. Washington, D.C.: U.S. Government Printing Office.

————. 1899b. Present conditions of the forested areas in northern Idaho outside the limits of the Priest River forest reserve and north of the Clearwater River. *Nineteenth annual report of the USDI Geological Survey, Part V—Forest reserves*, pp. 373–86. Washington, D.C.: U.S. Government Printing Office.

————. 1900a. Bitterroot forest reserve. House of Representatives, 56th Cong., 1st sess., document no. 5, *Annual Reports of the Department of the Interior for the Fiscal Year Ended June 30, 1899, 20th Annual Report of the United States Geological Survey, Part V—Forest Reserves*, pp. 317–410. Washington, D.C.: U.S. Government Printing Office.

————. 1900b. Sandpoint quadrangle, Idaho. Pp. 583–95 in *Classification of lands, twenty-first annual report of the USDI Geological Survey, Part V—Forest reserves*, ed. H. Gannett. Washington, D.C.: U.S. Government Printing Office.

————. 1904a. *Forest conditions in the Absaroka division of the Yellowstone forest reserve, Montana.* USDI Geological Survey Professional Paper No. 29. Washington, D.C.: U.S. Government Printing Office.

————. 1904b. *Forest conditions in the Little Belt Mountains Forest Reserve, Montana, and the Little Belt Mountains quadrangle.* USDI Geological Survey Professional Paper No. 30. Washington, D.C.: U.S. Government Printing Office.

Lemke, T. O., J. A. Mack, and D. B. Houston. 1998. Winter range expansion by the northern Yellowstone elk herd. *Intermountain Journal of Sciences* 4:1–9.

Lenihan, J. M., D. Bachelet, R. P. Neilson, and R. Drapek. 2008. Simulated response of conterminous United States ecosystems to climate change at different levels of fire suppression, CO_2 emission rate, and growth response to CO_2. *Global and Planetary Change* 64:16–25.

Lentile, L. B. 2004. Causal factors and consequences of mixed-severity fire in Black Hills ponderosa pine forests. M.S. thesis, Colorado State University, Fort Collins.

Lentile, L. B., F. W. Smith, and W. D. Shepperd. 2005. Patch structure, fire-scar formation, and tree regeneration in a large mixed-severity fire in the South Dakota Black Hills, USA. *Canadian Journal of Forest Research* 35:2875–85.

————. 2006. Influence of topography and forest structure on patterns of mixed severity fire in ponderosa pine forests of the South Dakota Black Hills, USA. *International Journal of Wildland Fire* 15:557–66.

Leopold, A. 1924. Grass, brush, timber, and fire in southern Arizona. *Journal of Forestry* 22:1–10.

Leroux, S. J., F. K. A. Schmiegelow, R. B. Lessard, and S. G. Cumming. 2007. Minimum dynamic reserves: A framework for determining reserve size in ecosystems structured by large disturbances. *Biological Conservation* 138:464–73.

Lesica, P., S. V. Cooper, and G. Kudray. 2005. Big sagebrush shrub-steppe postfire succession in southwest Montana. Unpublished report prepared for the USDI Bureau of Land Management, Dillon, Montana. Helena: Montana Natural Heritage Program.

————. 2007. Recovery of big sagebrush following fire in southwest Montana. *Rangeland Ecology and Management* 60:261–69.

Levitt, J. 1980. *Responses of plants to environmental stresses,* 2nd ed, vol. 1. New York: Academic Press.

Liang, L.-M. 2005. True mountain-mahogany sprouting following fires in ponderosa pine forests along the Colorado Front Range. M.S. thesis, Colorado State University, Fort Collins.

Lindenmayer, D. B., and R. F. Noss. 2006. Salvage logging, ecosystem processes, and biodiversity conservation. *Conservation Biology* 20:949–58.

Linn, J. C. 1980. Simulation of fire effects on certain forest rangelands. M.S. thesis, Colorado State University, Fort Collins.

Littell, J. S. 2002. Determinants of fire regime variability in lower elevation forests of the northern greater Yellowstone ecosystem. Master's thesis, Montana State University, Bozeman.

Liu, Y. 2006. North Pacific warming and intense northwestern U.S. wildfires. *Geophysical Research Letters* 33:1–5.

Liu, J., and W. W. Taylor, eds. 2002. *Integrating landscape ecology into natural resource management.* Cambridge, England: Cambridge University Press.

Lo, E. 2005. Gaussian error propagation applied to ecological data: Post-ice-storm-downed woody biomass. *Ecological Monographs* 75:451–66.

Loehle, C. 1988. Tree life history strategies: The role of defenses. *Canadian Journal of Forest Research* 18:209–22.

Logan, J. A., P. White, B. J. Bentz, and J. A. Powell. 1998. Model analysis of spatial patterns in mountain pine beetle outbreaks. *Theoretical Population Biology* 53:236–55.

Long, A. J., D. D. Wade, and F. C. Beall. 2005. Managing for fire in the interface: Challenges and opportunities. Pp. 201–23 in *Forests at the wildland-urban interface: Conservation and management,* ed. S. W. Vince, M. L. Duryea, E. A. Macie, and L. A. Hermansen. Boca Raton, Florida: CRC Press.

Long, D., B. J. Losensky, and D. Bedunah. 2006. Vegetation succession modeling for the LANDFIRE prototype project. Pp. 217–76 in *The LANDFIRE prototype project: National consistent and locally relevant geospatial data for wildland fire management,* ed. M. G. Rollins and C. K. Frame. USDA Forest Service General Technical Report RMRS-GTR-175. Fort Collins, Colorado: Rocky Mountain Research Station.

Long, J. N. 2003. Diversity, complexity and interactions: An overview of Rocky Mountain ecosystems. *Tree Physiology* 23:1091–99.

Longland, W. S., and S. L. Bateman. 2002. Viewpoint: The ecological value of shrub islands on disturbed sagebrush rangelands. *Journal of Range Management* 55:571–75.

Loope, L. L., and G. E. Gruell. 1973. The ecological role of fire in the Jackson Hole area, northwestern Wyoming. *Quaternary Research* 3:425–43.

Lorimer, C. G. 1985. Methodological considerations in the analysis of forest disturbance history. *Canadian Journal of Forest Research* 15:200–213.

Lotan, J. E. 1975. The role of cone serotiny in lodgepole pine forests. Pp. 471–95 in *Management of lodgepole pine ecosystems: Symposium proceedings, Oct. 9–11, 1973, Pullman, Washington,* ed. D. M. Baumgartner. Pullman: Washington State University, Cooperative Extension Service.

———. 1976. Cone serotiny–fire relationships in lodgepole pine. *Proceedings of the Tall Timbers Fire Ecology Conference* 14:267–78.

Lotan, J. E., M. E. Alexander, S. F. Arno, R. E. French, O. G. Langdon, R. M. Loomis, R. A. Norum, R. C. Rothermel, W. C. Schmidt, and J. Van Wagtendonk. 1981. *Effects of fire on flora: A state-of-knowledge review.* USDA Forest Service General Technical Report WO-16. Washington, D.C.: U.S. Government Printing Office.

Lowdermilk, W. C. 1925. Factors affecting reproduction of Engelmann spruce. *Journal of Agricultural Research* 30:995–1009.

Lunan, J. S. 1972. Phytosociology and fuel description of *Pinus ponderosa* communities in Glacier National Park. M.S. thesis, University of Montana, Missoula.

Lundquist, J. E. 1995. Characterizing disturbance in managed ponderosa pine stands in the Black Hills. *Forest Ecology and Management* 74:61–74.

———. 2007. The relative influence of diseases and other small-scale disturbances on fuel loading in the Black Hills. *Plant Disease* 91:147–52.

Lundquist, J. E., and J. F. Negron. 2000. Endemic forest disturbances and stand structure of ponderosa pine (*Pinus ponderosa*) in the upper Pine Creek Research Natural Area, South Dakota, USA. *Natural Areas Journal* 20:126–32.

Lutes, D. C., R. E. Keane, J. F. Caratti, C. H. Key, N. C. Benson, S. Sutherland, and L. J. Gangi. 2006. *FIREMON: Fire effects monitoring and inventory system.* USDA Forest Service General Technical Report RMRS-GTR-164-CD. Fort Collins, Colorado: Rocky Mountain Research Station.

Lyman, C. K. 1947. Slash disposal as related to fire control on the national forests of western Montana and northern Idaho. *Journal of Forestry* 45:259–62.

Lynch, D. 1955. Ecology of the aspen groveland in Glacier County, Montana. *Ecological Monographs* 25:321–44.

Lynch, H. J., R. A. Renkin, R. L. Crabtree, and P. R. Moorcraft. 2006. The influence of previous mountain pine beetle (*Dendroctonus ponderosae*) activity on the 1988 Yellowstone fires. *Ecosystems* 9:1318–27.

Lyon, L. J. 1966. *Initial vegetal development following prescribed burning of Douglas-fir in south-central Idaho.* USDA Forest Service Research Paper INT-29. Ogden, Utah: Intermountain Forest and Range Experiment Station.

———. 1969. Wildlife habitat research and fire in the northern Rockies. *Proceedings of the Tall Timbers Fire Ecology Conference* 9:213–27.

———. 1971. *Vegetal development following prescribed burning of Douglas-fir in south-central Idaho.* USDA Forest Service Research Paper INT-105. Ogden, Utah: Intermountain Forest and Range Experiment Station.

———. 1976. *Vegetal development on the Sleeping Child burn in western Montana, 1961 to 1973.* USDA Forest Service Research Paper INT-184. Ogden, Utah: Intermountain Forest and Range Experiment Station.

———. 1984. *The Sleeping Child burn: 21 years of postfire change.* USDA Forest Service Research Paper INT-330. Ogden, Utah: Intermountain Forest and Range Experiment Station.

Lyon, L. J., K. B. Aubry, W. J. Zielinski, S. W. Buskirk, and L. F. Ruggiero. 1994. The scientific basis for conserving forest carnivores: Considerations for management. Pp. 128–37 in *The scientific basis for conserving forest carnivores: American marten, fisher, lynx, and wolverine in the western United States,* ed. L. F. Ruggiero, K. B. Aubry, S. W. Buskirk, L. J. Lyon, and W. J. Zielinski. USDA Forest Service General Technical Report RM-254. Fort Collins, Colorado: Rocky Mountain Forest and Range Experiment Station.

Lyon, L. J., H. S. Crawford, E. Czuhai, R. L. Frederiksen, R. F. Harlow, L. J. Metz, and H. A. Pearson. 1978. *Effects of fire on fauna: A state-of-knowledge review.* USDA Forest Service General Technical Report WO-6. Washington, D.C.: U.S. Government Printing Office.

Lyon, L. J., and P. F. Stickney. 1974. Early vegetal succession following large northern Rocky Mountain wildfires. *Proceedings of the Tall Timbers Fire Ecology Conference* 14:355–75.

MacDonald, G. M., and R. A. Case. 2005. Variations in the Pacific Decadal Oscillation over the past millennium. *Geophysical Research Letters* 32:L08703.

MacDonald, G. M., C. P. S. Larsen, J. M. Szeicz, and K. A. Moser. 1991. The reconstruction of boreal forest fire history from lake sediments: A comparison of charcoal, pollen, sedimentological, and geochemical indices. *Quaternary Science Reviews* 10:53–71.

Mack, R. N. 1981. Invasion of *Bromus tectorum* L. into western North America: An ecological chronicle. *Agro-Ecosystems* 7:145–65.

MacKenzie, M. D., T. H. DeLuca, and A. Sala. 2004. Forest structure and organic horizon analysis along a fire chronosequence in the low elevation forests of western Montana. *Forest Ecology and Management* 203:331–43.

Madany, M. H., T. W. Swetnam, and N. E. West. 1982. Comparison of two approaches for determining fire dates from tree scars. *Forest Science* 28:856–61.

Madany, M. H., and N. E. West. 1983. Livestock grazing–fire regime interactions within montane forests of Zion National Park, Utah. *Ecology* 64:661–67.

Madole, R. F. 1997. *Fire frequency during the Holocene in a small, forested drainage basin in the southern Rocky Mountains.* Abstract. Rocky Mountain National Park All Scientists Conference, Estes Park, June 9–10.

Maier, A. M., B. L. Perryman, R. A. Olson, and A. L. Hild. 2001. Climatic influences on recruitment of 3 subspecies of *Artemisia tridentata. Journal of Range Management* 54:699–703.

Malanson, G. P., and D. R. Butler. 1984. Avalanche paths as fuel breaks: Implications for fire management. *Journal of Environmental Management* 19:229–38.

Mangan, L., and R. Autenrieth. 1985. Vegetation changes following 2,4-D application and fire in a mountain big sagebrush habitat type. Pp. 61–65 in *Rangeland fire effects: A symposium,* ed. K. Sanders and J. Durham. Boise: Bureau of Land Management, Idaho State Office.

Manier, D. J., N. T. Hobbs, D. M. Theobold, R. M. Reich, M. A. Kalkhan, and M. R. Campbell. 2005. Canopy dynamics and human-caused disturbance on a semi-arid landscape in the Rocky Mountains, USA. *Landscape Ecology* 20:1–17.

Manier, D. J., and R. D. Laven. 2002. Changes in landscape patterns associated with the persistence of aspen (*Populus tremuloides* Michx.) on the western slope of the Rocky Mountains, Colorado. *Forest Ecology and Management* 167:263–84.

Mann, C. C. 2005. *1491.* New York: Alfred A. Knopf.

Mantua, N. J., S. R. Hare, Y. Zhang, J. M. Wallace, and R. C. Francis. 1997. A Pacific interdecadal climate oscillation with impacts on salmon production. *Bulletin of the American Meteorological Society* 78:1069–79.

Margolis, E. Q., T. W. Swetnam, and C. D. Allen. 2007. A stand-replacing fire history in upper montane forests of the southern Rocky Mountains. *Canadian Journal of Forest Research* 37:2227–41.

Marlon, J., P. J. Bartlein, and C. Whitlock. 2006. Fire-fuel-climate linkages in the northwestern USA during the Holocene. *The Holocene* 16:1059–71.

Marshall, R. 1928. The life history of some western white pine stands on the Kaniksu National Forest. *Northwest Science* 2:48–53.

Martell, D. L. 2001. Forest fire management. Pp. 527–83 in *Forest fires: Behavior and ecological effects,* ed. E. A. Johnson and K. Miyanishi. San Diego, California: Academic Press.

Martin, R. C. 1990. Sage grouse responses to wildfire in spring and summer habitats. Master's thesis. University of Idaho, Moscow.

Martin, R. E. 1963. A basic approach to fire injury of tree stems. *Proceedings of the Tall Timbers Fire Ecology Conference* 2:151–62.

Martin, R. E., and C. H. Driver. 1983. Factors affecting antelope bitterbrush reestablishment following fire. Pp. 266–79 in *Research and management of bitterbrush and cliffrose in western North America,* ed. A. R. Tiedemann and K. L. Johnson. USDA Forest Service General Technical Report INT-152. Ogden, Utah: Intermountain Forest and Range Experiment Station.

Martinka, C. J. 1974. Fire and elk in Glacier National Park. *Proceedings of the Tall Timbers Fire Ecology Conference* 14:377–89.

Martinka, R. R. 1972. Structural characteristics of blue grouse territories in southwestern Montana. *Journal of Wildlife Management* 36:498–510.

Martinson, E. J., and P. N. Omi. 2003. Performance of fuel treatments subjected to wildfires. Pp. 7–13 in *Fire, fuel treatments, and ecological restoration: Conference proceedings, April 16–18, 2002, Fort Collins, Colorado,* ed. P. N. Omi and L. A. Joyce. USDA Forest Service Proceedings RMRS-P-29. Fort Collins, Colorado: Rocky Mountain Research Station.

Maser, C., R. G. Anderson, K. Cromack Jr., J. T. Williams, and R. E. Martin. 1979. Dead and down woody material. Pp. 78–95 in *Wildlife habitats in managed forests: The Blue Mountains of Oregon and Washington,* ed. J. W. Thomas. Agricultural Handbook No. 553. Washington, D.C.: USDA Forest Service.

Mason, D. T. 1915. *The life history of lodgepole pine in the Rocky Mountains.* USDA Bulletin No. 154. Washington, D.C.: U.S. Government Printing Office.

Massman, W. J., and J. M. Frank. 2004. Effect of a controlled burn on the thermophysical properties of a dry soil using a new model of soil heat flow and a new high temperature heat flux sensor. *International Journal of Wildland Fire* 13:427–42.

Massman, W. J., J. M. Frank, and N. B. Reisch. 2008. Long term impacts of prescribed burns on soil thermal conductivity and soil heating at a Colorado Rocky Mountain site: A data/model fusion study. *International Journal of Wildland Fire* 17:131–46.

Massman, W. J., J. M. Frank, W. D. Shepperd, and M. J. Platten. 2003. In situ soil temperature and heat flux measurements during controlled surface burns at a southern Colorado forest site. Pp. 69–87 in *Fire, fuel treatments, and ecological restoration: Conference proceedings, April 16–18, 2002, Fort Collins, Colorado,* ed. P. N. Omi and L. A. Joyce. USDA Forest Service Proceedings RMRS-P-29. Fort Collins, Colorado: Rocky Mountain Research Station.

Mast, J. N. 1993. Climatic and disturbance factors influencing *Pinus ponderosa* stand structure near the forest/grassland ecotone in the Colorado Front Range. Ph.D. dissertation, University of Colorado, Boulder.

Mast, J. N., and T. T. Veblen. 1999. Tree spatial patterns and stand development along the pine-grassland ecotone in the Colorado Front Range. *Canadian Journal of Forest Research* 29:575–84.

Mast, J. N., T. T. Veblen, and Y. B. Linhart. 1998. Disturbance and climatic influences on age structure of ponderosa pine at the pine/grassland ecotone, Colorado Front Range. *Journal of Biogeography* 25:743–55.

Mathews, E. E. 1980. The accretion of fuel in lodgepole pine forests of southwest Montana. M.S. thesis, University of Montana, Missoula.

Mattson, D. J., and D. P. Reinhart. 1990. Whitebark pine on the Mount Washburn massif, Yellowstone National Park. Pp. 106–17 in *Proceedings—Symposium on whitebark pine ecosystems: Ecology and management of a high-mountain resource,* ed. W. C. Schmidt and K. J. McDonald. USDA Forest Service General Technical Report INT-270. Ogden, Utah: Intermountain Research Station.

McAdams, A. G. 1995. Changes in ponderosa pine forest structure in the Black Hills, South Dakota, 1874–1995. M.S thesis, Northern Arizona University, Flagstaff.

McBride, J. R. 1983. Analysis of tree rings and fire scars to establish fire history. *Tree-Ring Bulletin* 43:51–67.

McCabe, G. J., and M. A. Palecki. 2006. Multidecadal climate variability of global lands and oceans. *International Journal of Climatology* 26:849–65.

McCabe, G. J., M. A. Palecki, and J. L. Betancourt. 2004. Pacific and Atlantic Ocean influences on multidecadal drought frequency in the United States. *Proceedings of the National Academy of Sciences* 101:4136–41.

McCarthy, M. A., A. M. Gill, and R. A. Bradstock. 2001. Theoretical fire-interval distributions. *International Journal of Wildland Fire* 10:73–77.

McCaughey, W. W., W. C. Schmidt, and R. C. Shearer. 1986. Seed-dispersal characteristics of conifers in the inland mountain West. Pp. 50–62 in *Proceedings: Conifer tree seed in the inland mountain West symposium, August 5–6, 1985, Missoula, Montana,* ed. R. C. Shearer. USDA Forest Service General Technical Report INT-203. Ogden, Utah: Intermountain Research Station.

McCulloch, C. Y. Jr. 1955. Utilization of winter browse on wilderness big game range. *Journal of Wildlife Management* 19:206–15.

———. 1969. Some effects of wildfire on deer habitat in pinyon-juniper woodland. *Journal of Wildlife Management* 33:778–84.

McCullough, D. G., R. A. Werner, and D. Neumann. 1998. Fire and insects in northern and boreal forest ecosystems of North America. *Annual Review of Entomology* 43:107–27.

McCune, B. 1983. Fire frequency reduced two orders of magnitude in the Bitterroot canyons, Montana. *Canadian Journal of Forest Research* 13:212–18.

McCune, B., and T. F. H. Allen. 1985. Forest dynamics in the Bitterroot Canyons, Montana. *Canadian Journal of Botany* 63:377–83.

McCutchan, M. H., and W. A. Main. 1989. The relationship between mean monthly fire potential indices and monthly fire severity. *Proceedings of the Conference on Fire and Forest Meteorology* 10:430–35.

McDowell, M. K. D. 2000. The effects of burning in mountain big sagebrush on key sage grouse habitat characteristics in southeastern Oregon. M.S. thesis, Oregon State University, Corvallis.

McEneaney, T. 1989. The Yellowstone wildfires of 1988: What really happened to the birdlife? *Winging It* 1:1–2, 7.

McGee, J. M. 1982. Small mammal populations in an unburned and early fire successional sagebrush community. *Journal of Range Management* 35:177–80.

McIver, J. D., and R. Ottmar. 2007. Fuel mass and stand structure after post-fire logging of a severely burned ponderosa pine forest in northeastern Oregon. *Forest Ecology and Management* 238:268–79.

McIver, J. D., and L. Starr. 2001. A literature review on the environmental effects of postfire logging. *Western Journal of Applied Forestry* 16:159–68.

McKell, C. M. 1950. A study of plant succession in the oak brush (*Quercus gambelii*) zone after fire. M.S. thesis, University of Utah, Salt Lake City.

McKenzie, C. 2001. Establishment patterns and height growth rates of aspen and conifer species growing on the Gunnison National Forest in western Colorado. M.S. thesis, Colorado: Colorado State University, Fort Collins.

McKenzie, D., Z. Gedalof, D. L. Peterson, and P. Mote. 2004. Climatic change, wildfire, and conservation. *Conservation Biology* 18:890–902.

McLean, A. 1969. Fire resistance of forest species as influenced by root systems. *Journal of Range Management* 22:120–22.

McLean, H. 1993. The Boise quickstep. *American Forests* 99:11–14.

McPherson, G. R. 1997. *Ecology and management of North American savannas.* Tucson: University of Arizona Press.

Meagher, M. 1989. Range expansion by bison of Yellowstone National Park. *Journal of Mammalogy* 70:670–75.

Means, J. A. 1989. Estimating the date of single bole scars by counting tree rings in increment cores. *Canadian Journal of Forest Research* 19:1491–96.

Mehringer, P. J. Jr., S. F. Arno, and K. L. Petersen. 1977. Postglacial history of Lost Trail Pass Bog, Bitterroot Mountains, Montana. *Arctic and Alpine Research* 9:345–68.

Mehus, C. A. 1995. Influences of browsing and fire on sagebrush taxa of the northern Yellowstone winter range. M.S. thesis, Montana State University, Bozeman.

Meinecke, E. P. 1929. *Quaking aspen: A study in applied forest pathology.* USDA Forest Service Technical Bulletin No. 155. Washington, D.C.: U.S. Government Printing Office.

Meisner, B. N., R. A. Chase, M. H. McCutchan, R. Mees, J. W. Benoit, B. Ly, D. Albright, D. Strauss, and T. Ferryman. 1994. A lightning fire ignition assessment model. *Proceedings of the Conference on Fire and Forest Meteorology* 12:172–78.

Melgoza, G., R. S. Nowak, and R. J. Tausch. 1990. Soil water exploitation after fire: Competition between *Bromus tectorum* (cheatgrass) and two native species. *Oecologia* 83:7–13.

Menke, C. A., and P. S. Muir. 2004. Short-term influence of wildfire on canyon grassland plant communities and Spalding's catchfly, a threatened plant. *Northwest Science* 78:192–203.

Mensing, S., S. Livingston, and P. Barker. 2006. Long-term fire history in Great Basin sagebrush reconstructed from macroscopic charcoal in spring sediments, Newark Valley, Nevada. *Western North American Naturalist* 66:64–77.

Merriam, K. E., J. E. Keeley, and J. L. Beyers. 2006. Fuel breaks affect nonnative species abundance in Californian plant communities. *Ecological Applications* 16:515–27.

Merrill, E. H., and H. F. Mayland. 1982. Shrub responses after fire in an Idaho ponderosa pine community. *Journal of Wildlife Management* 46:496–502.

Merrill, E. H., H. F. Mayland, and J. M. Peek. 1980. Effects of a fall wildfire on herbaceous vegetation on xeric sites in the Selway-Bitterroot Wilderness, Idaho. *Journal of Range Management* 33:363–67.

Metlen, K. L., and C. E. Fiedler. 2006. Restoration treatment effects on the understory of ponderosa pine/Douglas-fir forests in western Montana, USA. *Forest Ecology and Management* 222:355–69.

Metlen, K. L., C. E. Fiedler, and A. Youngblood. 2004. Understory response to fuel reduction treatments in the Blue Mountains of northeastern Oregon. *Northwest Science* 78:175–85.

Meyer, C. B., D. H. Knight, and G. K. Dillon. 2005. *Historic range of variability for upland vegetation in the Bighorn National Forest, Wyoming.* USDA Forest Service General Technical Report RMRS-GTR-140. Fort Collins, Colorado: Rocky Mountain Research Station.

Meyer, G. A., and J. L. Pierce. 2003. Climatic controls on fire-induced sediment pulses in Yellowstone National Park and central Idaho: A long-term perspective. *Forest Ecology and Management* 178:89–104.

Meyer, G. A., S. G. Wells, and A. J. T. Jull. 1995. Fire and alluvial chronology in Yellowstone National Park: Climatic and intrinsic controls on Holocene geomorphic processes. *Geological Society of America Bulletin* 107:1211–30.

Milchunas, D. G., and W. K. Lauenroth. 1993. Quantitative effects of grazing on vegetation and soils over a global range of environments. *Ecological Monographs* 63:327–66.

Miller, C. 2006. Wilderness fire management in a changing world. *International Journal of Wilderness* 12:18–21,13.

Miller, J. M., and J. E. Patterson. 1927. Preliminary studies on the relation of fire injury to bark-beetle attack in western yellow pine. *Journal of Agricultural Research* 34:597–613.

Miller, M. 2000. Fire autoecology. Pp. 9–34 in *Wildland fire in ecosystems: Effects of fire on flora,* ed. J. K. Brown and J. K. Smith. USDA Forest Service General Technical Report RMRS-GTR-42, vol. 2. Fort Collins, Colorado: Rocky Mountain Research Station.

Miller, P. C. 1970. Age distributions of spruce and fir in beetle-killed forests on the White River Plateau, Colorado. *American Midland Naturalist* 83:206–12.

Miller, R. F., and J. A. Rose. 1999. Fire history and western juniper encroachment in sagebrush steppe. *Journal of Range Management* 52:550–59.

Miller, R. F., T. J. Svejcar, and N. E. West. 1994. Implications of livestock grazing in the intermountain sagebrush region: Plant composition. Pp. 101–46 in *Ecological implications of livestock herbivory in the West,* ed. M. Vavra, W. A. Laycock, and R. D. Pieper. Denver, Colorado: Society for Range Management.

Miller, R. F., and R. J. Tausch. 2001. The role of fire in juniper and pinyon woodlands: A descriptive analysis. Pp. 15–30 in *Proceedings of the invasive species workshop: The role of fire in the control and spread of invasive species,* ed. K. E. M. Galley and T. P. Wilson. Miscellaneous Publication No. 11. Tallahassee, Florida: Tall Timbers Research Station.

Miller, R. F., R. J. Tausch, E. D. McArthur, D. D. Johnson, and S. C. Sanderson. 2008. *Age structure and expansion of piñon-juniper woodlands: A regional perspective in the intermountain West.* USDA Forest Service Research Paper Report RMRS-RP-69. Fort Collins, Colorado: Rocky Mountain Research Station.

Miller, S. L., T. M. McClean, N. L. Stanton, and S. E. Williams. 1998. Mycorrhization, physiognomy, and first-year survivability of conifer seedlings following natural fire in Grand Teton National Park. *Canadian Journal of Forest Research* 28:115–22.

Millspaugh, S. H., and C. Whitlock. 1995. A 750-year fire history based on lake sediment records in central Yellowstone National Park, USA. *The Holocene* 5:283–92.

Millspaugh, S. H., C. Whitlock, and P. J. Bartlein. 2000. Variations in fire frequency and climate over the past 17000 yr in central Yellowstone National Park. *Geology* 28:211–14.

———. 2004. Postglacial fire, vegetation, and climate history of the Yellowstone-Lamar and central plateau provinces, Yellowstone National Park. Pp. 10–28 in *After the fires: The ecology of change in Yellowstone National Park,* ed. L. L. Wallace. New Haven, Connecticut: Yale University Press.

Minore, D. 1979. *Comparative autecological characteristics of northwestern tree species: A literature review.* USDA Forest Service General Technical Report PNW-87. Portland, Oregon: Pacific Northwest Forest and Range Experiment Station.

Mitchell, J. M. Jr., C. W. Stockton, and D. M. Meko. 1979. Evidence of a 22-year rhythm of drought in the western United States related to the Hale solar cycle since the 17th century. Pp. 125–43 in *Solar-terrestrial influences on weather and climate,* ed. B. M. McCormac and T. A. Seliga. Dordrecht, Netherlands: Reidel.

Mitchell, V. L. 1976. The regionalization of climate in the western United States. *Journal of Applied Meteorology* 15:920–27.

Mohr, F. 1989. Light-hand suppression tactics: A fire management challenge. *Fire Management Notes* 50(1): 21–23.

———. 1992–93. Wildfire suppressed—And the wilderness still looks natural! *Fire Management Notes* 53–54(3): 3–6.

Mohr, F., and K. Curtiss. 1998. U.S. Army firefighters practice "no trace camping" on wilderness wildfires. *Fire Management Notes* 58(1): 4–8.

Mohrle, C. R. 2003. The southwest monsoon and the relation to fire occurrence. M.S. thesis, University of Nevada, Reno.

Moir, W. H. 1969. The lodgepole pine zone in Colorado. *American Midland Naturalist* 81:87–98.

Moir, W. H., and L. S. Huckaby. 1994. Displacement ecology of trees near upper timberline. *International Conference for Bear Research and Management* 9:35–42.

Montana Forest Restoration Committee. 2007. Restoring Montana's national forest lands: Guiding principles and recommended implementation. Unpublished report. http://www.montanarestoration.org.

Mooney, J. 1928. *The aboriginal population of America north of Mexico.* Smithsonian Miscellaneous Collections 80(7). Washington, D.C.: Smithsonian Institution.

Moore, C. T. 1972. Man and fire in the central North American grassland 1535–1890: A documentary historical geography. Ph.D. dissertation, University of California, Los Angeles.

Moore, J., A. Grinsted, and S. Jevrejeva. 2006. Is there evidence for sunspot forcing of climate at multi-year and decadal periods? *Geophysical Research Letters* 33:1–5.

Moore, J. M., and R. W. Wein. 1977. Viable seed populations by soil depth and potential site recolonization after disturbance. *Canadian Journal of Botany* 55:2408–12.

Moore, M. M., C. A. Casey, J. D. Bakker, J. D. Springer, P. Z. Fulé, W. W. Covington, and D. C. Laughlin. 2006. Herbaceous vegetation responses (1992–2004) to restoration treatments in a ponderosa pine forest. *Rangeland Ecology and Management* 59:135–44.

Moore, M. M., D. W. Huffman, P. Z. Fulé, W. W. Covington, and J. E. Crouse. 2004. Comparison of historical and contemporary forest structure and composition on permanent plots in southwestern ponderosa pine forests. *Forest Science* 50:162–76.

Morgan, P., and S. C. Bunting. 1990. Fire effects in whitebark pine forests. Pp. 166–70 in *Proceedings—Symposium on whitebark pine ecosystems: Ecology and management of a high-mountain resource,* ed. W. C. Schmidt and K. J. McDonald. USDA Forest Service General Technical Report INT-270. Ogden, Utah: Intermountain Research Station.

Morgan, P., S. C. Bunting, A. E. Black, T. Merrill, and S. Barrett. 1996. *Fire regimes in the interior Columbia River basin: Past and present.* Final report to the Intermountain Fire Sciences Laboratory. Missoula, Montana: Intermountain Research Station.

Morgan, P., S. C. Bunting, R. E. Keane, and S. F. Arno. 1994. Fire ecology of whitebark pine forests of the northern Rocky Mountains, U.S.A. Pp. 136–41 in *Proceedings—International workshop on subalpine stone pines and their environment: The status of our knowledge, Sept. 5–11, 1992, St. Moritz, Switzerland,* ed. W. C. Schmidt and F.-K. Holtmeier. USDA Forest Service General Technical Report INT-GR-309. Ogden, Utah: Intermountain Research Station.

Morgan, P., E. K. Heyerdahl, and C. E. Gibson. 2008. Multi-season climate synchronized forest fires throughout the 20th century, northern Rockies, USA. *Ecology* 89:717–28.

Morgan, P., and M. P. Murray. 2001. Landscape ecology and isolation implications for conservation of whitebark pine. Pp. 289–309 in *Whitebark pine communities: Ecology and restoration,* ed. D. F. Tomback, S. F. Arno, and R. E. Keane. Washington, D.C.: Island Press.

Morgan, T. A., C. E. Fiedler, and C. Woodall. 2002. Characteristics of dry site old-growth ponderosa pine in the Bull Mountains of Montana, USA. *Natural Areas Journal* 22:11–19.

Moritz, M. A., M. E. Morais, L. A. Summerell, J. M. Carlson, and J. Doyle. 2005. Wildfires, complexity, and highly optimized tolerance. *Proceedings of the National Academy of Sciences* 102:17912–17.

Moritz, M. A., and D. C. Odion. 2004. Prescribed fire and natural disturbance. *Science* 306:1680.

Moritz, W. E. 1988. Wildlife use of fire-disturbed areas in sagebrush steppe on the Idaho National Engineering Laboratory. M.S. thesis, Montana State University, Bozeman.

Morris, W. G., and E. L. Mowat. 1958. Some effects of thinning a ponderosa pine thicket with a prescribed fire. *Journal of Forestry* 56:203–209.

Mueggler, W. F. 1956. *Is sagebrush seed residual in the soil of burns, or is it wind-borne?* USDA Forest Service Research Note No. 35. Ogden, Utah: Intermountain Forest and Range Experiment Station.

———. 1989. Age distribution and reproduction of intermountain aspen stands. *Western Journal of Applied Forestry* 4:41–45.

———. 1994. *Sixty years of change in tree numbers and basal area in central Utah aspen stands.* USDA Forest Service Research Paper INT-RP-478. Ogden, Utah: Intermountain Research Station.

Mueggler, W. F., and J. P. Blaisdell. 1958. Effects on associated species of burning, rotobeating, spraying, and railing sagebrush. *Journal of Range Management* 11:61–66.

Mueller, R. C., C. M. Scudder, M. E. Porter, R. T. Trotter III, C. A. Gehring, and T. G. Whitham. 2005. Differential tree mortality in response to severe drought: Evidence for long-term vegetation shifts. *Journal of Ecology* 93:1085–93.

Muir, P. S. 1993. Disturbance effects on structure and tree species composition of *Pinus contorta* forests in western Montana. *Canadian Journal of Forest Research* 23:1617–25.

Muir, P. S., and J. E. Lotan. 1985. Disturbance history and serotiny of *Pinus contorta* in western Montana. *Ecology* 66:1658–68.

Murphy, J. L., L. R. Green, and J. R. Bentley. 1967. Fuel-breaks—Effective aids, not cure-alls. *Fire Control Notes* 28(1): 4–5.

Murray, M. P. 1996. Landscape dynamics of an island range: Interrelationships of fire and whitebark pine (*Pinus albicaulis*). Ph.D. dissertation, University of Idaho, Moscow.

Murray, M. P., S. C. Bunting, and P. Morgan. 1998. Fire history of an isolated subalpine mountain range of the Intermountain Region, United States. *Journal of Biogeography* 25:1071–80.

———. 2000. Landscape trends (1753–1993) of whitebark pine (*Pinus albicaulis*) forests in the west Big Hole Range of Idaho/Montana, U.S.A. *Arctic, Antarctic, and Alpine Research* 32:412–18.

Murray, R. B. 1983. Response of antelope bitterbrush to burning and spraying in southeastern Idaho. Pp. 142–52 in *Research and management of bitterbrush and cliffrose in western North America,* ed. A. R. Tiedemann and K. L. Johnson. USDA Forest Service General Technical Report INT-152. Ogden, Utah: Intermountain Forest and Range Experiment Station.

Mutlu, M., S. C. Popescu, C. Stripling, and T. Spencer. 2008. Mapping surface fuel models using lidar and multispectral data fusion for fire behavior. *Remote Sensing of the Environment* 112:274–85.

National Interagency Fire Center. 2006. Online data. http://www.nifc.gov.

Neary, D. G., C. C. Klopatek, L. F. DeBano, and P. F. Ffolliott. 1999. Fire effects on belowground sustainability: A review and synthesis. *Forest Ecology and Management* 122:51–71.

Neilson, R. P., J. M. Lenihan, D. Bachelet, and R. J. Drapek. 2005. Climate change implications for sagebrush ecosystems. *Transactions of the North American Wildlife and Natural Resources Conference* 70:145–59.

Nelle, P. J., K. P. Reese, and J. W. Connelly. 2000. Long-term effects of fire on sage grouse habitat. *Journal of Range Management* 53:586–91.

Nelson, J. G., and R. E. England. 1978. Some comments on the causes and effects of fire in the northern grasslands area of Canada and the nearby United States, 1750–1900. Pp. 39–47 in *Proceedings of the 1977 rangeland management and fire symposium, Nov. 1–3, 1977, Casper, Wyoming,* ed. C. M. Bourassa and A. P. Brackebusch. Missoula: University of Montana School of Forestry, Montana Forest and Conservation Experiment Station.

Nelson, R. A. 1934. Further observations of revegetation on the Twin Sisters Burn in Rocky Mountain National Park. *Journal of the Colorado-Wyoming Academy of Sciences* 1:51–52.

———. 1954. The Twinsisters burn after twenty years. *Journal of the Colorado-Wyoming Academy of Sciences* 4:45.

Nelson, R. M. Jr. 1980. *Flame characteristics for fire in southern fuels.* USDA Forest Service Research Paper SE-205. Asheville, North Carolina: Southeastern Forest Experiment Station.

Nelson, R. M. Jr. and C. W. Adkins. 1986. Flame characteristics of wind-driven surface fires. *Canadian Journal of Forest Research* 16:1293–1300.

Nelson, T. C. 1979. Fire management policy in the national forests—A new era. *Journal of Forestry* 77:723–25.

Newland, J. A., and T. H. DeLuca. 2000. Influence of fire on native nitrogen-fixing plants and soil nitrogen status in ponderosa pine–Douglas-fir forests in western Montana. *Canadian Journal of Forest Research* 30:274–82.

Newman, M., G. P. Compo, and M. A. Alexander. 2003. ENSO-forced variability of the Pacific Decadal Oscillation. *Journal of Climate* 16:3853–57.

Nimchuk, N. 1983. *Wildfire behavior associated with upper ridge breakdown.* Forest Service, Report 0701-9920 No. T/50. Edmonton, Alb., Canada: Alberta Energy and Natural Resources.

Nimir, M. B., and G. F. Payne. 1978. Effects of spring burning on a mountain range. *Journal of Range Management* 31:259–63.

Noble, I. R. 1993. A model of the responses of ecotones to climate change. *Ecological Applications* 3:396–403.

Nordyke, K. A., and S. W. Buskirk. 1991. Southern red-backed vole, *Clethrionomys gapperi,* populations in relation to stand succession and old-growth character in central Rocky Mountains. *Canadian Field Naturalist* 105:330–34.

Norland, J. E., F. J. Singer, and L. Mack. 1996. Effects of the Yellowstone firest of 1988 on elk habitats. Pp. 223–32 in *The ecological implications of fire in Greater Yellowstone: Proceedings second biennial conference on the Greater Yellowstone ecosystem, Yellowstone National Park, Sept. 19–21, 1993,* ed. J. M. Greenlee. Fairfield, Washington: International Association of Wildland Fire.

Noss, R. F., J. F. Franklin, W. L. Baker, T. Schoennagel, and P. B. Moyle. 2006. Managing fire-prone forests in the western United States. *Frontiers in Ecology and the Environment* 4:481–87.

Noste, N. V., and J. B. Davis. 1975. A critical look at fire damage appraisal. *Journal of Forestry* 73:715–19.

Nyland, R. D. 1998. Patterns of lodgepole pine regeneration following the 1988 Yellowstone fires. *Forest Ecology and Management* 111:23–33.

Odion, D. C., E. J. Frost, J. R. Strittholt, H. Jiang, D. A. Dellasala, and M. A. Moritz. 2004. Patterns of fire severity and forest conditions in the western Klamath Mountains, California. *Conservation Biology* 18:927–36.

Odion, D. C., and C. T. Hanson. 2008. Fire severity in the Sierra Nevada revisited: Conclusions robust to further analysis. *Ecosystems* 11:12–15.

Ogle, K., and V. DuMond. 1997. *Historical vegetation on national forest lands in the Intermountain region.* Ogden, Utah: USDA Forest Service, Intermountain Region.

Olson, R. A., B. L. Perryman, S. Petersburg, and T. Naumann. 2003. Fire effects on small mammal communities in Dinosaur National Monument. *Western North American Naturalist* 63:50–55.

Omi, P. N., and R. L. Emrick. 1980. *Fire and resource management in Mesa Verde National Park.* Contract CS-1200-9-B015 Final Report. Fort Collins: Colorado State University.

Omi, P. N., E. J. Martinson, and G. W. Chong. 2006. Effectiveness of pre-fire fuel treatments. Unpublished final report to the Joint Fire Science Program. Fort Collins: Colorado State University.

Omphile, U. 1986. Vegetation and soil moisture responses following three big sagebrush control methods. M.S. thesis, University of Wyoming, Laramie.

Ortega, Y. K., and D. E. Pearson. 2005. Weak vs. strong invaders of natural plant communities: Assessing invasibility and impact. *Ecological Applications* 15:651–61.

Orville, R. E., and G. R. Huffines. 2001. Cloud-to-ground lightning in the United States: NLDN results in the first decade, 1989–98. *Monthly Weather Review* 129:1179–93.

Ott, J. E., E. D. McArthur, and B. A. Roundy. 2003. Vegetation of chained and non-chained seedings after wildfire in Utah. *Journal of Range Management* 56:81–91.

Ottmar, R. D., D. V. Sandberg, C. L. Riccardi, and S. J. Prichard. 2007. An overview of the Fuel Characteristic Classification System: Quantifying, classifying, and creating fuelbeds for resource planning. *Canadian Journal of Forest Research* 37:2383–93.

Ottmar, R. D., R. E. Vihnanek, and J. C. Regelbrugge. 2000. *Stereo photo series for quantifying natural fuels.* Vol. 4, *Pinyon-juniper, chaparral, and sagebrush types in the Southwestern United States.* PMS 833, National Wildfire Coordinating Group. Boise, Idaho: National Interagency Fire Center.

Ottmar, R. D., R. E. Vihnanek, and C. S. Wright. 1998. *Stereo photo series for quantifying natural fuels.* Vol. 1, *Mixed-conifer with mortality, western juniper, sagebrush, and grassland types in the Interior Pacific Northwest.* PMS-830, National Wildfire Coordinating Group. Boise, Idaho: National Interagency Fire Center.

————. 2000. *Stereo photo series for quantifying natural fuels.* Vol. 3, *Lodgepole pine, quaking aspen, and Gambel oak types in the Rocky Mountains.* PMS 832, National Wildfire Coordinating Group. Boise, Idaho, National Interagency Fire Center.

————. 2007. *Stereo photo series for quantifying natural fuels.* Vol. 10, *Sagebrush with grass and ponderosa pine–juniper types in central Montana.* USDA Forest Service General Technical Report PNW-GTR-719. Portland, Oregon: Pacific Northwest Research Station.

Page, W. G., and M. J. Jenkins. 2007a. Mountain pine beetle–induced changes to selected lodgepole pine fuel complexes within the Intermountain region. *Forest Science* 53:507–18.

————. 2007b. Predicted fire behavior in selected mountain pine beetle–infested lodgepole pine. *Forest Science* 53:662–74.

Parker, A. J., and K. C. Parker. 1983. Comparative successional roles of trembling aspen and lodgepole pine in the southern Rocky Mountains. *Great Basin Naturalist* 43:447–55.

————. 1994. Structural variability of mature lodgepole pine stands on gently sloping terrain in Taylor Park Basin, Colorado. *Canadian Journal of Forest Research* 24:2020–29.

Parker, A. J., and R. K. Peet. 1984. Size and age structure of conifer forests. *Ecology* 65:1685–89.

Parker, D. L., and L. E. Stipe. 1993. *A sequence of destruction: Mountain pine beetle and wildfire, Yellowstone National Park.* Albuquerque, New Mexico: USDA Forest Service, Southwestern Region.

Parks, C. G., S. R. Radosevich, B. A. Endress, B. J. Naylor, D. Anzinger, L. J. Rew, B. D. Maxwell, and K. A. Dwire. 2005. Natural and land-use history of the Northwest mountain ecoregions (USA) in relation to patterns of plant invasions. *Perspectives in Plant Ecology, Evolution and Systematics* 7:137–58.

Parks, J. M. 2006. True story: A 4-million acre "mega" maximum manageable area. *Fire Management Today* 66(4): 28–32.

Parr, C. L., and A. N. Andersen. 2006. Patch mosaic burning for biodiversity conservation: A critique of the pyrodiversity paradigm. *Conservation Biology* 20:1610–19.

Parsons, D. J., P. B. Landres, and C. Miller. 2003. Wildland fire use: The dilemma of managing and restoring natural fire and fuels in United States wilderness. Pp. 19–26 in *Proceedings of fire conference 2000: The first national congress on fire ecology, prevention, and management,* ed. K. E. M. Galley, R. C. Klinger, and N. G. Sugihara. Miscellaneous Publication No. 13. Tallahassee, Florida: Tall Timbers Research Station.

Parsons, R. A., E. K. Heyerdahl, R. E. Keane, B. Dorner, and J. Fall. 2007. Assessing accuracy of point fire intervals across landscapes with simulation modelling. *Canadian Journal of Forest Research* 37:1605–14.

Patterson, W. A. III, K. J. Edwards, and D. J. Maguire. 1987. Microscopic charcoal as a fossil indicator of fire. *Quaternary Science Reviews* 6:3–23.

Patton, D. R., and H. D. Avant. 1970. *Fire stimulated aspen sprouting in a spruce-fir forest in New Mexico.* USDA Forest Service Research Note RM-159. Fort Collins, Colorado: Rocky Mountain Forest and Range Experiment Station.

Paulson, D. D., and W. L. Baker. 2006. *The nature of southwestern Colorado: Recognizing human legacies and restoring natural places.* Boulder: University Press of Colorado.

Pausas, J. G., R. A. Bradstock, D. A. Keith, J. E. Keeley, and GCTE Fire Network. 2004. Plant functional traits in relation to fire in crown-fire ecosystems. *Ecology* 2004:1085–1100.

Pausas, J. G., and S. Lavorel. 2003. A hierarchical deductive approach for functional types in disturbed ecosystems. *Journal of Vegetation Science* 14:409–16.

Paxon, J. 2000. "Remember Los Alamos": The Cerro Grande fire. *Fire Management Today* 60(4): 9–14.

Pearson, G. A. 1914. The role of aspen in the reforestation of mountain burns in Arizona and New Mexico. *Plant World* 17:249–60.

———. 1934. Grass, pine seedlings and grazing. *Journal of Forestry* 32:545–55.

———. 1942. Herbaceous vegetation a factor in natural regeneration of ponderosa pine in the Southwest. *Ecological Monographs* 12:315–38.

Pearson, J. A., D. H. Knight, and T. J. Fahey. 1987. Biomass and nutrient accumulation during stand development in Wyoming lodgepole pine forests. *Ecology* 68:1966–73.

Pearson, S., M. G. Turner, L. L. Wallace, and W. H. Romme. 1995. Winter habitat use by large ungulates following fire in northern Yellowstone National Park. *Ecological Applications* 5:744–55.

Pechanec, J. F., and G. Stewart. 1944. *Sagebrush burning: Good and bad.* USDA Farmer's Bulletin No. 1948. Washington, D.C.: U.S. Government Printing Office.

Pedersen, E. K., J. W. Connelly, J. R. Hendrickson, and W. E. Grant. 2003. Effect of sheep grazing and fire on sage grouse populations in southeastern Idaho. *Ecological Modelling* 165:23–47.

Peek, J. M., D. A. Demarchi, R. A. Demarchi, and D. E. Stucker. 1985. Bighorn sheep and fire: Seven case studies. Pp. 36–43 in *Fire's effects on wildlife habitat: Symposium proceedings,* ed. J. E. Lotan and J. K. Brown. USDA Forest Service General Technical Report INT-186. Ogden, Utah: Intermountain Research Station.

Peek, J. M., R. A. Riggs, and J. L. Lauer. 1979. Evaluation of fall burning on bighorn sheep winter range. *Journal of Range Management* 32:430–32.

Peet, R. K. 1975. Forest vegetation of the east slope of the northern Colorado Front Range. Ph.D. dissertation, Cornell University, Ithaca, New York.

———. 1978. Forest vegetation of the Colorado Front Range: Patterns of species diversity. *Vegetatio* 37:65–78.

———. 1981. Forest vegetation of the Colorado Front Range: Composition and dynamics. *Vegetatio* 45:3–75.

———. 2000. Forests and meadows of the Rocky Mountains. Pp. 75–121 in *North American terrestrial vegetation,* 2nd ed., ed. M. G. Barbour and W. D. Billings. New York: Cambridge University Press.

Peirce, E. S. 1915. The regeneration of denuded areas in the Bighorn Mountains by Douglas fir. *Forestry Quarterly* 13:300–307.

Pellant, M., and L. Reichert. 1984. Management and rehabilitation of a burned winterfat community in southwestern Idaho. Pp. 281–85 in *Proceedings: Symposium on the biology of Atriplex and related chenopods,* ed. A. R. Tiedemann, E. D. McArthur, H. C. Stutz, R. Stevens, and K. L. Johnson. USDA Forest Service General Technical Report INT-172. Ogden, Utah: Intermountain Forest and Range Experiment Station.

Perry, D. A., and J. E. Lotan. 1979. A model of fire selection for serotiny in lodgepole pine. *Evolution* 33:958–68.

Perryman, B. L., and W. A. Laycock. 2000. Fire history of the Rochelle Hills Thunder Basin National Grasslands. *Journal of Range Management* 53:660–65.

Perryman, B. L., R. A. Olson, S. Petersburg, and T. Naumann. 2002. Vegetation response to prescribed fire in Dinosaur National Monument. *Western North American Naturalist* 62:414–22.

Peters, G., and A. Sala. 2008. Reproductive output of ponderosa pine in response to thinning and prescribed burning in western Montana. *Canadian Journal of Forest Research* 38:844–50.

Petersen, K. L. 1988. *Climate and the Dolores River Anasazi: A paleoenvironmental reconstruction from a 10,000-year pollen record, La Plata Mountains, Southwestern Colorado.* Anthropological Papers No. 113. Salt Lake City: University of Utah.

Petersen, K. L., and L. B. Best. 1987. Effects of prescribed burning on nongame birds in a sagebrush community. *Wildlife Society Bulletin* 15:317–29.

———. 1999. Design and duration of perturbation experiments: Implications for data interpretation. *Studies in Avian Biology* 19:230–36.

Peterson, D. L. 1985. Crown scorch volume and scorch height: Estimates of postfire tree condition. *Canadian Journal of Forest Research* 15:596–98.

Peterson, D. L., and M. J. Arbaugh. 1986. Postfire survival in Douglas-fir and lodgepole pine: Comparing the effects of crown and bole damage. *Canadian Journal of Forest Research* 16:1175–79.

Peterson, D. L., M. J. Arbaugh, G. H. Pollock, and L. J. Robinson. 1991. Postfire growth of *Pseudotsuga menziesii* and *Pinus contorta* in the northern Rocky Mountains, USA. *International Journal of Wildland Fire* 1:63–71.

Peterson, D. L., M. C. Johnson, J. K. Agee, T. B. Jain, D. McKenzie, and E. D. Reinhardt. 2005. *Forest structure and fire hazard in dry forests of the western United States.* USDA Forest Service General Technical Report PNW-GTR-628. Portland, Oregon: Pacific Northwest Research Station.

Peterson, K. T. 1999. Whitebark pine (*Pinus albicaulis*) decline and restoration in Glacier National Park. M.S. thesis, University of North Dakota, Grand Forks.

Petterson, E. S. 1999. Prescribed fire effects on plant communities in Rocky Mountain bighorn sheep habitat. M.S. thesis, Colorado State University, Fort Collins.

Pettit, N. E., and R. J. Naiman. 2007. Fire in the riparian zone: Characteristics and ecological consequences. *Ecosystems* 10:673–87.

Pew, K. L., and C. P. S. Larsen. 2001. GIS analysis of spatial and temporal patterns of human-caused wildfires in the temperate rain forest of Vancouver Island, Canada. *Forest Ecology and Management* 140:1–18.

Pfister, A. R. 1980. Postfire avian ecology in Yellowstone National Park. M.S. thesis, Washington State University, Pullman.

Pickett, S. T. A., S. L. Collins, and J. J. Armesto. 1987. Models, mechanisms and pathways of succession. *Botanical Review* 53:335–71.

Pickett, S. T. A., and M. J. McDonnell. 1989. Changing perspectives in community dynamics: A theory of successional forces. *Trends in Ecology and Evolution* 4:241–45.

Pickford, G. D. 1932. The influence of continued heavy grazing and of promiscuous burning on spring-fall ranges in Utah. *Ecology* 13:159–71.

Pierce, J., and G. Meyer. 2008. Long-term fire history from alluvial sediments: The role of drought and climate variability, and implications for management of Rocky Mountain forests. *International Journal of Wildland Fire* 17:84–95.

Pierce, J. L., G. A. Meyer, and A. J. T. Jull. 2004. Fire-induced erosion and millennial-scale climate change in northern ponderosa pine forests. *Nature* 432:87–90.

Pilliod, D. S., E. L. Bull, J. L. Hayes, and B. C. Wales. 2006. *Wildlife and invertebrate response to fuel reduction treatments in dry coniferous forests of the western United States: A synthesis.* USDA Forest Service General Technical Report RMRS-GTR-173. Fort Collins, Colorado: Rocky Mountain Research Station.

Pilliod, D. S., R. B. Bury, E. J. Hyde, C. A. Pearl, and P. S. Corn. 2003. Fire and amphibians in North America. *Forest Ecology and Management* 178:163–81.

Pinchot, G. 1908. *Forest tables: Western yellow pine.* USDA Forest Service Circular No. 127. Washington, D.C.: U.S. Government Printing Office.

Platt, R. V., T. T. Veblen, and R. L. Sherriff. 2006. Are wildfire mitigation and restoration of historic forest structure compatible? A spatial modeling assessment. *Annals of the Association of American Geographers* 96:455–70.

Plevel, S. R. 1997. Fire policy at the wildland-urban interface: A local responsibility. *Journal of Forestry* 95:12–17.

Plummer, F. G. 1912. *Forest fires: Their causes, extent, and effects, with a summary of recorded destruction and loss.* USDA Forest Service Bulletin 117. Washington, D.C.: U.S. Government Printing Office.

Polakow, D. A., and T. T. Dunne. 1999. Modelling fire-return interval *T:* Stochasticity and censoring in the two-parameter Weibull model. *Ecological Modeling* 121:79–102.

Pollet, J., and P. N. Omi. 2002. Effect of thinning and prescribed burning on crown fire severity in ponderosa pine forests. *International Journal of Wildland Fire* 11:1–10.

Popp, J. B., and J. E. Lundquist. 2006. *Photo series for quantifying forest residues in managed lands of the Medicine Bow National Forest.* USDA Forest Service General Technical Report RMRS-GTR-172. Fort Collins: Rocky Mountain Research Station.

Poreda, S. F. 1992. Vegetation recovery and dynamics following the Wasatch Mountain fire (1990), Midway, Utah. M.S. thesis, University of Utah, Salt Lake City.

Poreda, S. F., and L. H. Wullstein. 1994. Vegetation recovery following fire in an oakbrush vegetation mosaic. *Great Basin Naturalist* 54:380–83.

Potter, L. D., and T. Foxx. 1984. Postfire recovery and mortality of the ponderosa pine forest after La Mesa fire. Pp. 39–55 in *La Mesa fire symnposium, Los Alamos, New Mexico, Oct. 6–7, 1981,* ed. T. S. Foxx. Publication LA-9236-NERP. Los Alamos, New Mexico: Los Alamos National Laboratory.

Potter, M. W., and S. R. Kessell. 1980. Predicting mosaics and wildlife diversity resulting from fire disturbance to a forest ecosystem. *Environmental Management* 4:247–54.

Powell, H. D. W. 2000. The influence of prey density on post-fire habitat use of the black-backed woodpecker. M.S. thesis, University of Montana, Missoula.

Powell, J. W. 1879. *Report on the lands of the arid region of the United States, with a more detailed account of the lands of Utah,* 2nd ed. Washington, D.C.: U.S. Government Printing Office.

Powell, S. L., and A. J. Hansen. 2007. Conifer cover increase in the Greater Yellowstone Eco-system: Frequency, rates, and spatial variation. *Ecosystems* 10:204–16.

Power, M. J., C. Whitlock, P. Bartlein, and L. R. Stevens. 2006. Fire and vegetation history during the last 3800 years in northwestern Montana. *Geomorphology* 75:420–36.

Pratt, D. W., R. A. Black, and B. A. Zamora. 1984. Buried viable seed in a ponderosa pine community. *Canadian Journal of Botany* 62:44–52.

Prestemon, J. P., J. M. Pye, D. T. Butry, T. P. Holmes, and D. E. Mercer. 2002. Understanding broadscale wildfire risks in a human-dominated landscape. *Forest Science* 48:685–93.

Prevedel, D. A. 2007. Linking intense western wildfires with weather patterns and condi-tions. *Fire Management Today* 67(2): 35–38.

Price, C., and D. Rind. 1994. Lightning fires in a 2xCO$_2$ world. *Proceedings of the Conference on Fire and Forest Meteorology* 12:77–84.

Prichard, S. J., R. D. Ottmar, and G. K. Anderson. 2007. Consume 3.0 user's guide. Unpub-lished report, USDA Forest Service, Pacific Wildland Fire Sciences Laboratory. Seattle, Washington: Pacific Northwest Research Station. http://www.fs.fed.us/pnw/fera/research/smoke/consume/consume30_users_guide.pdf.

Prichard, S. J., C. L. Riccardi, D. V. Sandberg, and R. D. Ottmar. 2007. FCCS user's guide: Version 1.1. Unpublished report, Pacific Wildland Fire Sciences Laboratory, USDA Forest Service. Seattle, Washington: Pacific Northwest Research Station. www.fs.fed.us/pnw/fera/fccs/pubs/shtml.

Pyle, W. H., and J. A. Crawford. 1996. Availability of foods of sage grouse chicks following prescribed fire in sagebrush-bitterbrush. *Journal of Range Management* 49:320–24.

Pyne, S. J. 1982. *Fire in America: A Cultural History of Wildland and Rural Fire.* Princeton, New Jersey: Princeton University Press.

———. 1986. These conflagrated prairies: A cultural fire history of the grasslands. Pp. 131–37 in *The prairie: Past, present and future—Proceedings of the ninth North American prairie conference, July 29–August 1, 1984, Moorhead, Minnesota,* ed. G. K. Clambey and R. H. Pemble. Fargo: North Dakota State University, Tri-College University Center for Environmental Studies.

———. 2001. *Year of the fires: The story of the great fires of 1910.* New York: Viking.

———. 2004. *Tending fire: Coping with America's wildland fires.* Washington, D.C.: Island Press.

Radeloff, V. C., R. B. Hammer, S. I. Stewart, J. S. Fried, S. S. Holcomb, and J. F. McKeefry. 2005. The wildland-urban interface in the United States. *Ecological Applications* 15:799–805.

Raffa, K. F., B. H. Aukema, B. J. Bentz, A. L. Carroll, J. A. Hicke, M. G. Turner, and W. H. Romme. 2008. Cross-scale drivers of natural disturbances prone to anthropogenic am-plification: The dynamics of bark beetle eruptions. *BioScience* 58:501–17.

Raper, B., B. Clark, M. Matthews, and A. Aldrich. 1985. Early effects of a fall burn in a west-ern Wyoming mountain big sagebrush–grass community. Pp. 88–92 in *Rangeland fire ef-fects: A symposium,* ed. K. Sanders and J. Durham. Boise: Idaho State Office, Bureau of Land Management.

Rapraeger, E. F. 1936. Effect of repeated ground fires upon stumpage returns in western white pine. *Journal of Forestry* 34:715–18.

Rasmussen, L. A., G. D. Amman, J. C. Vandygriff, R. D. Oakes, A. S. Munson, and K. E. Gibson. 1996. *Bark beetle and wood borer infestation in the Greater Yellowstone Area during four postfire years.* USDA Forest Service Research Paper INT-RP-487. Ogden, Utah: Intermountain Research Station.

Ratzlaff, T. D., and J. E. Anderson. 1995. Vegetal recovery following wildfire in seeded and unseeded sagebrush steppe. *Journal of Range Management* 48:386–91.

Raymond, C. L., and D. L. Peterson. 2005. Fuel treatments alter the effects of wildfire in a mixed-evergreen forest, Oregon, USA. *Canadian Journal of Forest Research* 35:2981–95.

Ream, C. H. 1981. *The effects of fire and other disturbances on small mammals and their predators: An annotated bibliography.* USDA Forest Service General Technical Report INT-106. Ogden, Utah: Intermountain Forest and Range Experiment Station.

Reap, R. M. 1986. Evaluation of cloud-to-ground lightning data from the western United States for the 1983–84 summer seasons. *Journal of Climate and Applied Meteorology* 25:785–99.

Rebertus, A. J., B. R. Burns, and T. T. Veblen. 1991. Stand dynamics of *Pinus flexilis*–dominated subalpine forests in the Colorado Front Range. *Journal of Vegetation Science* 2:445–58.

Reed, R. A., M. E. Finley, W. H. Romme, and M. G. Turner. 1999. Aboveground net primary production and leaf-area index in early postfire vegetation in Yellowstone National Park. *Ecosystems* 2:88–94.

Reed, R. A., J. Johnson-Barnard, and W. L. Baker. 1996. Contribution of roads to forest fragmentation in the Rocky Mountains. *Conservation Biology* 10:1098–1106.

Reed, W. J. 2006. A note on fire frequency concepts and definitions. *Canadian Journal of Forest Research* 36:1884–88.

Reed, W. J., and K. S. McKelvey. 2002. Power-law behaviour and parametric models for the size-distribution of forest fires. *Ecological Modelling* 150:239–54.

Reeves, G. H., P. A. Bisson, B. E. Rieman, and L. E. Benda. 2006. Postfire logging in riparian areas. *Conservation Biology* 20:994–1004.

Reeves, M. C., J. R. Kost, and K. C. Ryan. 2006. Fuels products of the LANDFIRE project. Pp. 239–52 in *Fuels management: How to measure success—Conference proceedings, March 28–30, 2006, Portland, Oregon,* ed. P. L. Andrews and B. W. Butler. USDA Forest Service Proceedings RMRS-P-41. Fort Collins, Colorado: Rocky Mountain Research Station.

Reich, R. M., J. E. Lundquist, and V. A. Bravo. 2004. Spatial models for estimating fuel loads in the Black Hills, South Dakota, USA. *International Journal of Wildland Fire* 13:119–29.

Reid, M. S. 1989. The response of understory vegetation to major canopy disturbance in the subalpine forests of Colorado. M.S. thesis, University of Colorado, Boulder.

Reinhardt, E. 2003. *Using FOFEM 5.0 to estimate tree mortality, fuel consumption, smoke production and soil heating from wildland fire.* Presentation at the 2nd International Wildland Fire Ecology and Fire Management Congress, November 16–20, 2003, Orlando, Florida. http://fire.org/downloads/fofem/5.2/FOFEM5using.pdf.

Reinhardt, E. D., R. E. Keane, and J. K. Brown. 2001. Modeling fire effects. *International Journal of Wildland Fire* 10:373–80.

Reinhardt, E. D., R. E. Keane, J. K. Brown, and D. L. Turner. 1991. Duff consumption from prescribed fire in the U.S. and Canada: A broadly based empirical approach. Pp. 362–70 in *Proceedings of the 11th Conference on Fire and Forest Meteorology, April 16–19, 1991, Missoula, Montana,* ed. P. L. Andrews and D. F. Potts. Bethesda, Maryland: Society of American Foresters.

Reinhardt, E. D., and K. C. Ryan. 1988. *Eight-year tree growth following prescribed underburning in a western Montana Douglas-fir/western larch stand.* USDA Forest Service Research Note INT-387. Ogden, Utah: Intermountain Research Station.

———. 1989. Estimating tree mortality resulting from prescribed fire. Pp. 41–44 in *Prescribed fire in the Intermountain region: Forest site preparation and range improvement—Symposium proceedings,* ed. D. M. Baumgartner, D. W. Breuer, B. A. Zamora, L. F. Neuenschwander, and R. H. Wakimoto. Pullman: Washington State University, Cooperative Extension Service.

———. 1998. Analyzing effects of management actions including salvage, fuel treatment, and prescribed fire on fuel dynamics and fire potential. *Proceedings of the Tall Timbers Fire Ecology Conference* 20:206–9.

Renkin, R. A., and D. G. Despain. 1992. Fuel moisture, forest type, and lightning-caused fire in Yellowstone National Park. *Canadian Journal of Forest Research* 22:37–45.

Rennick, R. B. 1981. Effects of prescribed burning on mixed prairie vegetation in southeastern Montana. M.S. thesis, Montana State University, Bozeman.

Rens, R. J. 2001. Elk effects on sagebrush-grassland after fire on Yellowstone's northern range. M.S. thesis, Montana State University, Bozeman.

Rhodes, J. J. 2007. The watershed impacts of forest treatments to reduce fuels and modify fire behavior. Unpublished report. Eugene, Oregon: Pacific Rivers Council.

Rhodes, J. J., and W. L. Baker. 2008. Fire probability, fuel treatment effectiveness and ecological tradeoffs in western U.S. public forests. *Open Forest Science Journal* 1:1–8.

Riaño, D., E. Meier, B. Allgöwer, E. Chuvieco, and S. L. Ustin. 2003. Modeling airborne laser scanning data for the spatial generation of critical forest parameters in fire behavior modeling. *Remote Sensing of the Environment* 86:177–86.

Riccardi, C. L., R. D. Ottmar, D. V. Sandberg, A. Andreu, E. Elman, K. Kopper, and J. Long. 2007. The fuelbed: A key element of the Fuel Characteristic Classification System. *Canadian Journal of Forest Research* 37:2394–2412.

Riccardi, C. L., S. J. Prichard, D. V. Sandberg, and R. D. Ottmar. 2007. Quantifying physical characteristics of wildland fuels using the Fuel Characteristic Classification System. *Canadian Journal of Forest Research* 37:2413–20.

Rice, C. L. 1983. A literature review of the fire relationships of antelope bitterbrush. Pp. 256–65 in *Research and management of bitterbrush and cliffrose in western North America,* ed. A. R. Tiedemann and K. L. Johnson. USDA Forest Service General Technical Report INT-152. Ogden, Utah: Intermountain Forest and Range Experiment Station.

Rice, P. M. 2005. *INVADERS database system.* Missoula: University of Montana, Division of Biological Sciences. http://invader.dbs.umt.edu.

Richards, R. T., J. C. Chambers, and C. Ross. 1998. Use of native plants on federal lands: Policy and practice. *Journal of Range Management* 51:625–32.

Rickard, W. H. 1970. Ground dwelling beetles in burned and unburned vegetation. *Journal of Range Management* 23:293–94.

Riggs, R. A., and J. M. Peek. 1980. Mountain sheep habitat-use patterns related to post-fire succession. *Journal of Wildlife Management* 44:933–38.

Ripple, W. J. 2001. The role of postfire coarse woody debris in aspen regeneration. *Western Journal of Applied Forestry* 16:61–64.

Ripple, W. J., and E. J. Larsen. 2000. Historic aspen recruitment, elk, and wolves in northern Yellowstone National Park, USA. *Biological Conservation* 95:361–70.

Ripple, W. J., E. J. Larsen, R. A. Renkin, and D. W. Smith. 2001. Trophic cascades among wolves, elk and aspen on Yellowstone National Park's northern range. *Biological Conservation* 102:227–34.

Rivas-Martínez, S., A. Penas, M. A. Luengo, and S. Rivas-Sáenz. 2002. *Worldwide bioclimatic classification system.* CD Series II: Climate and biosphere. Leiden, Netherlands: Backhuys Publishers.

Roberts, M. R. 2004. Response of the herbaceous layer to natural disturbance in North American forests. *Canadian Journal of Botany* 82:1273–83.

Robertson, M. D. 1991. Winter ecology of migratory sage grouse and associated effects of prescribed fire in southeastern Idaho. Master's thesis, University of Idaho, Moscow.

Robertson, P. A., and Y. H. Bowser. 1999. Coarse woody debris in mature *Pinus ponderosa* stands in Colorado. *Journal of the Torrey Botanical Society* 126:255–67.

Robichaud, P. R., J. L. Beyers, and D. G. Neary. 2000. *Evaluating the effectiveness of postfire rehabilitation treatments.* USDA Forest Service General Technical Report RMRS-GTR-63. Fort Collins, Colorado: Rocky Mountain Research Station.

Robichaud, P. R., and R. E. Brown. 2006. Postfire rehabilitation treatments: Are we learning what works? Pp. 1–12 in *Managing watersheds for human and natural impacts: Engineering, ecological, and economic challenges—Proceedings of the 2005 watershed manaement conference, July 19–22, 2005, Williamsburg, VA,* ed. G. E. Moglen. Reston, Virginia: American Society of Civil Engineers.

Robichaud, P. R., T. R. Lillybridge, and J. W. Wagenbrenner. 2006. Effects of postfire seeding and fertilizing on hillslope erosion in north-central Washington, USA. *Catena* 67:56–67.

Rocky Mountain Geographic Science Center. 2007. Historic fire data. U.S. Geological Survey: http://rmgsc.cr.usgs.gov/outgoing/geoMAC/historicfiredata.

Roe, A. L., W. R. Beaufait, L. J. Lyon, and J. L. Oltman. 1971. Fire and forestry in the northern Rocky Mountains: A task force report. *Journal of Forestry* 69:464–70.

Rogers, P. 2002. Using forest health monitoring to assess aspen forest cover change in the southern Rockies ecoregion. *Forest Ecology and Management* 155:223–36.

Rollins, M. G., R. E. Keane, and R. A. Parsons. 2004. Mapping fuels and fire regimes using remote sensing, ecosystem simulation, and gradient modeling. *Ecological Applications* 14:75–95.

Rollins, M. G., P. Morgan, and T. Swetnam. 2002. Landscape-scale controls over 20th century fire occurrence in two large Rocky Mountain (USA) wilderness areas. *Landscape Ecology* 17:539–57.

Rollins, M. G., T. W. Swetnam, and P. Morgan. 2001. Evaluating a century of fire patterns in two Rocky Mountain wilderness areas using digital fire atlases. *Canadian Journal of Forest Research* 31:2107–23.

Romme, W. H. 1980. Fire history terminology: Report of the ad hoc committee. Pp. 135–40 in *Proceedings of the Fire History Workshop,* ed. M. A. Stokes and J. H. Dieterich. USDA Forest Service General Technical Report RM-81. Fort Collins, Colorado: Rocky Mountain Forest and Range Experiment Station.

———. 1982. Fire and landscape diversity in subalpine forests of Yellowstone National Park. *Ecological Monographs* 52:199–221.

Romme, W. H., C. Allen, J. Bailey, W. Baker, B. Bestelmeyer, P. Brown, K. Eisenhart, L. Floyd-Hanna, D. W. Huffman, B. Jacobs, R. Miller, E. Muldavin, T. Swetnam, R. Tausch, and P. Weisberg. 2007. Historical and modern disturbance regimes of piñon–juniper vegetation in the western U.S. Unpublished report. Fort Collins: Colorado Forest Restoration Institute; and Boulder, Colorado: The Nature Conservancy.

———. 2008. Historical and modern disturbance regimes, stand structures, and landscape dynamics in piñon–juniper vegetation of the western U.S. Unpublished report. Fort Collins: Colorado Forest Restoration Institute and Colorado State University.

Romme, W. H., J. Clement, J. Hicke, D. Kulakowski, L. H. MacDonald, T. L. Schoennagel, and T. T. Veblen. 2006. Recent forest insect outbreaks and fire risk in Colorado forests: A brief synthesis of relevant research. Unpublished report. Fort Collins, Colorado: Department of Forest, Rangeland and Watershed Stewardship and Colorado State University.

Romme, W. H., and D. G. Despain. 1989a. Historical perspective on the Yellowstone fires of 1988. *BioScience* 39:695–99.

———. 1989b. The long history of fire in the Greater Yellowstone ecosystem. *Western Wildlands* 15:10–17.

Romme, W. H., L. Floyd-Hanna, D. D. Hanna, and E. Bartlett. 2001. Aspen's ecological role in the West. Pp. 243–59 in *Sustaining aspen in western landscapes: Symposium proceedings,* ed. W. D. Shepperd, D. Binkley, D. L. Bartos, T. J. Stohlgren, and L. G. Eskew. USDA Forest Service Proceedings RMRS-P-18. Fort Collins, Colorado: Rocky Mountain Research Station.

Romme, W. H., D. Hanna, L. Floyd-Hanna, and E. J. Bartlett. 1996. Fire history and successional status in aspen forests of the San Juan National Forest: Final report. Unpublished report to the San Juan National Forest, Durango, Colorado. Durango, Colorado: Fort Lewis College, Department of Biology.

Romme, W. H., M. Kaufmann, T. T. Veblen, R. Sherriff, and C. Regan. 2003. Ecological effects of the Hayman fire. Part 2: Historical (pre-1860) and current (1860–2002) forest and landscape structure. Pp. 196–203 in *Hayman fire case study,* ed. R. T. Graham. USDA Forest Service General Technical Report RMRS-GTR-114. Fort Collins, Colorado: Rocky Mountain Research Station.

Romme, W. H., and D. H. Knight. 1981. Fire frequency and subalpine forest succession along a topographic gradient in Wyoming. *Ecology* 62:319–26.

———. 1982. Landscape diversity: The concept applied to Yellowstone Park. *BioScience* 32:664–70.

Romme, W. H., M. G. Turner, R. H. Gardner, W. W. Hargrove, G. A. Tuskan, D. G. Despain, and R. A. Renkin. 1997. A rare episode of sexual reproduction in aspen (*Populus tremuloides* Michx.) following the 1988 Yellowstone fires. *Natural Areas Journal* 17:17–25.

Romme, W. H., M. G. Turner, G. A. Tuskan, and R. A. Reed. 2005. Establishment, persistence, and growth of aspen (*Populus tremuloides*) seedlings in Yellowstone National Park. *Ecology* 86:404–18.

Romme, W. H., M. G. Turner, L. L. Wallace, and J. S. Walker. 1995. Aspen, elk, and fire in northern Yellowstone National Park. *Ecology* 76:2097–2106.

Romme, W. H., T. T. Veblen, M. R. Kaufmann, R. Sherriff, and C. M. Regan. 2003. Ecological effects of the Hayman fire. Part 1: Historical (pre-1860) and current (1860–2002) fire regimes. Pp. 181–95 in *Hayman fire case study,* ed. R. T. Graham. USDA Forest Service General Technical Report RMRS-GTR-114. Fort Collins, Colorado: Rocky Mountain Research Station.

Rood, S. B., L. A. Goater, J. M. Mahoney, C. M. Pearce, and D. G. Smith. 2007. Floods, fire, and ice: Disturbance ecology of riparian cottonwoods. *Canadian Journal of Botany* 85:1019–32.

Roovers, L. M., and A. J. Rebertus. 1993. Stand dynamics and conservation of an old-growth Engelmann spruce–subalpine fir forest in Colorado. *Natural Areas Journal* 13:256–67.

Ropelewski, C. F., and M. S. Halpert. 1986. North American precipitation and temperature patterns associated with the El Niño–Southern Oscillation (ENSO). *Monthly Weather Review* 114:2352–62.

Roppe, J. A., and D. Hein. 1978. Effects of fire on wildlife in a lodgepole pine forest. *Southwestern Naturalist* 23:279–88.

Rorig, M. L., and S. A. Ferguson. 1999. Characteristics of lightning and wildland fire ignition in the Pacific Northwest. *Journal of Applied Meteorology* 38:1565–75.

———. 2002. The 2000 fire season: Lightning-caused fires. *Journal of Applied Meteorology* 41:786–91.

Ross, A. R. 1947. The prairie fire. *Colorado Magazine* 24:92–94.

Rotenberry, J. T., and S. T. Knick. 1999. Multiscale habitat associations of the sage sparrow: Implications for conservation biology. *Studies in Avian Biology* 19:95–103.

Rothermel, R. C. 1972. *A mathematical model for predicting fire spread in wildland fuels.* USDA Forest Service Research Paper INT-115. Ogden, Utah: Intermountain Forest and Range Experiment Station.

———. 1991a. Crown fire analysis and interpretation. *Proceedings of the Conference on Fire and Forest Meteorology* 11:253–63.

———. 1991b. *Predicting behavior and size of crown fires in the northern Rocky Mountains.* USDA Forest Service Research Paper INT-438. Ogden, Utah: Intermountain Research Station.

———. 1993. *Mann Gulch fire: A race that couldn't be won.* USDA Forest Service General Technical Report INT-299. Ogden, Utah: Intermountain Research Station.

———. 1995. Characterizing severe fire behavior. Pp. 153–158 in *Proceedings: symposium on fire in wilderness and park management,* eds. J. K. Brown, R. W. Mutch, C. W. Spoon and

R. H. Wakimoto. USDA Forest Service General Technical Report INT-GTR-320. Ogden, Utah: Intermountain Research Station.

Rothermel, R. C., and J. E. Deeming. 1980. *Measuring and interpreting fire behavior for correlation with fire effects.* USDA Forest Service General Technical Report INT-93. Ogden, Utah: Intermountain Forest and Range Experiment Station.

Rothermel, R. C., R. A. Hartford, and C. H. Chase. 1994. *Fire growth maps for the 1988 Greater Yellowsone Area fires.* USDA Forest Service General Technical Report INT-304. Ogden, Utah: Intermountain Research Station.

Rothermel, R. C., and R. W. Mutch. 1986. Behavior of the life-threatening Butte fire: August 27–29, 1985. *Fire Management Notes* 47(2): 14–24.

Rowdabaugh, K. M. 1978. The role of fire in the ponderosa pine–mixed conifer ecosystems. M.S. thesis, Colorado State University, Fort Collins.

Rowe, J. S. 1969. Lightning fires in Saskatchewan grassland. *Canadian Field Naturalist* 83:317–27.

Rowland, M. M., A. W. Alldredge, J. E. Ellis, B. J. Weber, and G. C. White. 1983. Comparative winter diets of elk in New Mexico. *Journal of Wildlife Management* 47:924–32.

Roy, D. F. 1956. *Salvage logging may destroy Douglas-fir reproduction.* USDA Forest Service Research Note No. 107. Berkeley: California Forest and Range Experiment Station.

Rubino, D. L., and B. C. McCarthy. 2004. Comparative analysis of dendroecological methods used to assess disturbance events. *Dendrochronologia* 21:97–115.

Rummell, R. S. 1951. Some effects of livestock grazing on ponderosa pine forest and range in central Washington. *Ecology* 32:594–607.

Russell, K. R., D. H. Van Lear, and D. C. Guynn Jr. 1999. Prescribed fire effects on herpetofauna: Review and management implications. *Wildlife Society Bulletin* 27:374–84.

Ryan, K. C. 1976. Forest fire hazard and risk in Colorado. M.S. thesis, Colorado State University, Fort Collins.

————. 1981. *Evaluation of a passive flame-height sensor to estimate forest fire intensity.* USDA Forest Service Research Note PNW-390. Portland, Oregon: Pacific Northwest Forest and Range Experiment Station.

————. 2000. Global change and wildland fire. Pp. 175–83 in *Wildland fire in ecosystems: Effects of fire on flora,* ed. J. K. Brown and J. K. Smith. USDA Forest Service General Technical Report RMRS-GTR-42, vol. 2. Fort Collins, Colorado: Rocky Mountain Research Station.

Ryan, K. C., and W. H. Frandsen. 1991. Basal injury from smoldering fires in mature *Pinus ponderosa* Laws. *International Journal of Wildland Fire* 1:107–18.

Ryan, K. C., and N. V. Noste. 1985. Evaluating prescribed fires. Pp. 230–38 in *Proceedings: Symposium and workshop on wilderness fire, Missoula, Montana, Nov. 15–18, 1983,* ed. J. E. Lotan, B. M. Kilgore, W. C. Fischer, and R. W. Mutch. USDA Forest Service General Technical Report INT-182. Ogden, Utah: Intermountain Forest and Range Experiment Station.

Ryan, K. C., D. L. Peterson, and E. D. Reinhardt. 1988. Modeling long-term fire-caused mortality of Douglas-fir. *Forest Science* 34:190–99.

Ryan, K. C., and E. D. Reinhardt. 1988. Predicting postfire mortality of seven western conifers. *Canadian Journal of Forest Research* 18:1291–97.

Rychert, R. C. 2002. Assessment of cryptobiotic crust recovery. *Western North American Naturalist* 62:223–27.

Ryerson, D. E., T. W. Swetnam, and A. M. Lynch. 2003. A tree-ring reconstruction of western spruce budworm outbreaks in the San Juan Mountains, Colorado, U.S.A. *Canadian Journal of Forest Research* 33:1010–28.

Ryker, R. A. 1975. *A survey of factors affecting regeneration of Rocky Mountain Douglas-fir.* USDA Forest Service Research Paper INT-174. Ogden, Utah: Intermountain Forest and Range Experiment Station.

Rykiel, E. J. Jr. 1996. Testing ecological models: The meaning of validation. *Ecological Modelling* 90:229–44.

Saab, V., L. Bate, J. Lehmkuhl, B. Dickson, S. Story, S. Jentsch, and W. Block. 2006. Changes in downed wood and forest structure after prescribed fire in ponderosa pine forests. Pp. 477–87 in *Fuels management: How to measure success—Conference proceedings,* ed. P. L. Andrews and B. W. Butler. USDA Forest Service Proceedings RMRS-P-41. Fort Collins, Colorado: Rocky Mountain Research Station.

Saab, V. A., W. Block, R. Russell, J. Lehmkuhl, L. Bate, and R. White. 2007. *Birds and burns of the interior West: Descriptions, habitats, and management in western forests.* USDA Forest Service General Technical Report PNW-GTR-712. Portland, Oregon: Pacific Northwest Research Station.

Saab, V. A., J. Dudley, and W. L. Thompson. 2004. Factors influencing occupancy of nest cavities in recently burned forests. *The Condor* 106:20–36.

Saab, V. A., and H. D. W. Powell, eds. 2005a. *Fire and avian ecology in North America.* Studies in Avian Biology No. 30. Camarillo, California: Cooper Ornithological Society.

———. 2005b. Fire and avian ecology in North America: Process influencing pattern. Pp. 1–13 in *Fire and avian ecology in North America,* ed. V. A. Saab and H. D. W. Powell. Studies in Avian Biology No. 30. Camarillo, California: Cooper Ornithological Society.

Saab, V. A., H. D. W. Powell, N. B. Kotliar, and K. R. Newlon. 2005. Variation in fire regimes of the Rocky Mountains: Implications for avian communities and fire management. Pp. 76–96 in *Fire and avian ecology in North America,* ed. V. A. Saab and H. D. W. Powell. Studies in Avian Biology No. 30. Camarillo, California: Cooper Ornithological Society.

Saab, V. A., R. E. Russell, and J. G. Dudley. 2007. Nest densities of cavity-nesting birds in relation to postfire salvage logging and time since wildfire. *The Condor* 109:97–108.

Saab, V. A., and K. T. Vierling. 2001. Reproductive success of Lewis's woodpecker in burned pine and cottonwood riparian forests. *The Condor* 103:491–501.

Safay, R. E. 1981. Intensity of bark beetle attack following prescribed understory burning in seral ponderosa pine stands in northern Idaho. M.S. thesis, University of Idaho, Moscow.

Safford, H. D., J. Miller, D. Schmidt, B. Roath, and A. Parsons. 2008. BAER soil burn severity maps do not measure fire effects to vegetation: A comment on Odion and Hanson (2006). *Ecosystems* 11:1–11.

Sala, A., G. D. Peters, L. R. McIntyre, and M. G. Harrington. 2005. Physiological responses of ponderosa pine in western Montana to thinning, prescribed fire and burning season. *Tree Physiology* 25:339–48.

Sallach, B. K. 1986. Vegetation changes in New Mexico documented by repeat photography. M.S. thesis, New Mexico State University, Las Cruces.

Sandberg, D. V., C. L. Riccardi, and M. D. Schaaf. 2007. Fire potential rating for wildland fuelbeds using the Fuel Characteristic Classification System. *Canadian Journal of Forest Research* 37:2456–63.

Sapsis, D. B., and J. B. Kauffman. 1991. Fuel consumption and fire behavior associated with prescribed fires in sagebrush ecosystems. *Northwest Science* 65:173–79.

Sargent, C. S. 1884. *Tenth census of the United States [1880].* Vol. 9, *Report on the forests of North America (exclusive of Mexico).* Washington, D.C.: U.S. Government Printing Office.

Savage, M. 1991. Structural dynamics of a southwestern pine forest under chronic human influence. *Annals of the Association of American Geographers* 81:271–89.

Savage, M., and J. N. Mast. 2005. How resilient are southwestern ponderosa pine forests after crown fires? *Canadian Journal of Forest Research* 35:967–77.

Savage, M., and T. W. Swetnam. 1990. Early 19th-century fire decline following sheep pasturing in a Navajo ponderosa pine forest. *Ecology* 71:2374–78.

Saveland, J. M., S. R. Bakken, and L. Neuenschwander. 1990. *Predicting mortality and scorch height from prescribed burning for ponderosa pine in northern Idaho.* Forest, Wildlife and Range Experiment Station Bulletin No. 53. Moscow: University of Idaho, College of Forestry.

Saveland, J. M., and L. F. Neuenschwander. 1989. Predicting ponderosa pine mortality from understory prescribed burning. Pp. 45–48 in *Prescribed fire in the Intermountain region: Forest site preparation and range improvement—Symposium proceedings,* ed. D. M. Baumgartner, D. W. Breuer, B. A. Zamora, L. F. Neuenschwander, and R. H. Wakimoto. Pullman: Washington State University, Cooperative Extension Service.

———. 1990. A signal detection framework to evaluate models of tree mortality following fire damage. *Forest Science* 36:66–76.

Schaaf, M. D., D. V. Sandberg, M. D. Schreuder, and C. L. Riccardi. 2007. A conceptual framework for ranking crown fire potential in wildland fuelbeds. *Canadian Journal of Forest Research* 37:2464–78.

Schaefer, V. J. 1957. The relationshp of jet streams to forest wildfires. *Journal of Forestry* 55:419–25.

Scheldt, R. S. 1969. Ecology and utilization of curl-leaf mountain mahogany in Idaho. M.S. thesis, University of Idaho, Moscow.

Schier, G. A., and R. B. Campbell. 1978. Aspen sucker regeneration following burning and clearcutting on two sites in the Rocky Mountains. *Forest Science* 24:303–8.

Schimpf, D. J., J. A. Henderson, and J. A. MacMahon. 1980. Some aspects of succession in the spruce-fir forest zone of northern Utah. *Great Basin Naturalist* 40:1–26.

Schlatterer, E. F. 1960. Productivity and movements of a population of sage-grouse in southeastern Idaho. M.S. thesis, University of Idaho, Moscow.

Schmid, J. M., and S. A. Mata. 1996. *Natural variability of specific forest insect populations and their associated effects in Colorado.* USDA Forest Service General Technical Report RM-GTR-275. Fort Collins, Colorado: Rocky Mountain Forest and Range Experiment Station.

Schmid, J. M., and D. L. Parker. 1990. Fire and forest insect pests. Pp. 232–33 in *Effects of fire management of southwestern natural resources: Proceedings of the symposium, Nov. 15–17, 1988, Tucson, AZ,* ed. J. S. Krammes. USDA Forest Service General Technical Report RM-191. Fort Collins, Colorado: Rocky Mountain Forest and Range Experiment Station.

Schmidt, K. M., J. P. Menakis, C. C. Hardy, W. J. Hann, and D. L. Bunnell. 2002. *Development of coarse-scale spatial data for wildland fire and fuel management.* USDA Forest Service General Technical Report RMRS-87. Fort Collins, Colorado: Rocky Mountain Research Station.

Schmitz, R. F., and A. R. Taylor. 1969. *An instance of lightning damage and infestation of ponderosa pines by the pine engraver beetle in Montana.* USDA Forest Service Research Note INT-88. Ogden, Utah: Intermountain Forest and Range Experiment Station.

Schmoldt, D. L., D. L. Peterson, R. E. Keane, J. M. Lenihan, D. McKenzie, D. R. Weise, and D. V. Sandberg. 1999. *Assessing the effects of fire disturbance on ecosystems: A scientific agenda for research and management.* USDA Forest Service General Technical Report PNW-GTR-455. Portland, Oregon: Pacific Northwest Research Station.

Schoennagel, T., M. G. Turner, D. M. Kashian, and A. Fall. 2006. Influence of fire regimes on lodgepole pine stand age and density across the Yellowstone National Park (USA) landscape. *Landscape Ecology* 21:1281–96.

Schoennagel, T., M. G. Turner, and W. H. Romme. 2003. The influence of fire interval and serotiny on postfire lodgepole pine density in Yellowstone National Park. *Ecology* 84:2967–78.

Schoennagel, T., T. T. Veblen, D. Kulakowski, and A. Holz. 2007. Multidecadal climate variability and climate interactions affect subalpine fire occurrence, western Colorado (USA). *Ecology* 88:2891–2902.

Schoennagel, T., T. T. Veblen, and W. H. Romme. 2004. The interaction of fire, fuels, and climate across Rocky Mountain forests. *BioScience* 54:661–76.

Schoennagel, T., T. T. Veblen, W. H. Romme, J. S. Sibold, and E. R. Cook. 2005. ENSO and PDO variability affect drought-induced fire occurrence in Rocky Mountain subalpine forests. *Ecological Applications* 15:2000–2014.

Schoennagel, T., D. M. Waller, M. G. Turner, and W. H. Romme. 2004. The effect of fire interval on post-fire understory communities in Yellowstone National Park. *Journal of Vegetation Science* 15:797–806.

Schroeder, M. J., and C. C. Buck. 1970. *Fire weather.* USDA Agricultural Handbook No. 360. Washington, D.C.: U.S. Government Printing Office.

Schulz, B. W., R. J. Tausch, and P. T. Tueller. 1991. Size, age, and density relationships in curlleaf mahogany (*Cercocarpus ledifolius*) populations in western and central Nevada: Competitive implications. *Great Basin Naturalist* 51:183–91.

Schulz, B. W., P. T. Tueller, and R. J. Tausch. 1990. Ecology of curlleaf mahogany in western and central Nevada: Community and population structure. *Journal of Range Management* 43:13–20.

Schwecke, D. A. 1989. North Hill fire and rehabilitation. Pp. 95–97 in *Prescribed fire in the Intermountain region: Forest site preparation and range improvement—Symposium proceedings,* ed. D. M. Baumgartner, D. W. Breuer, B. A. Zamora, L. F. Neuenschwander, and R. H. Wakimoto. Pullman: Washington State University, Cooperative Extension Service.

Schwecke, D. A., and W. Hann. 1989. Fire behavior and vegetation response to spring and fall burning on the Helena National forest. Pp. 135–42 in *Prescribed fire in the Intermountain region: Forest site preparation and range improvement—Symposium proceedings,* ed. D. M. Baumgartner, D. W. Breuer, B. A. Zamora, L. F. Neuenschwander, and R. H. Wakimoto. Pullman: Washington State University, Cooperative Extension Service.

Scott, J. H. 1999. NEXUS: A system for assessing crown fire hazard. *Fire Management Notes* 59(2): 20–24.

———. 2006. *Comparison of crown fire modeling systems used in three fire management applications.* USDA Forest Service Research paper RMRS-RP-58. Fort Collins, Colorado: Rocky Mountain Research Station.

Scott, J. H., and R. E. Burgan. 2005. *Standard fire behavior fuel models: A comprehensive set for use with Rothermel's surface fire spread model.* USDA Forest Service General Technical Report RMRS-GTR-153. Fort Collins, Colorado: Rocky Mountain Research Station.

Scott, J. H., and E. D. Reinhardt. 2001. *Assessing crown fire potential by linking models of surface and crown fire behavior.* USDA Forest Service Research Paper RMRS-RP-29. Fort Collins, Colorado: Rocky Mountain Research Station.

———. 2002. Estimating canopy fuels in conifer forests. *Fire Management Today* 62(4): 45–50.

Scott, K., B. Oswald, K. Farrish, and D. Unger. 2002. Fuel loading prediction models developed from aerial photographs of the Sangre de Cristo and Jemez mountains of New Mexico, USA. *International Journal of Wildland Fire* 11:85–90.

Scott, M. D., and H. Geisser. 1996. Pronghorn migration and habitat use following the 1988 Yellowstone fires. Pp. 123–32 in *The ecological implications of fire in Greater Yellowstone: Proceedings second biennial conference on the Greater Yellowstone ecosystem, Yellowstone National Park, Sept. 19–21, 1993,* ed. J. M. Greenlee. Fairfield, Washington: International Association of Wildland Fire.

Seefeldt, S. S., M. Germino, and K. DiCristina. 2007. Prescribed fires in *Artemisia tridentata* ssp. *vaseyana* steppe have minor and transient effects on vegetation cover and composition. *Applied Vegetation Science* 7 19:249–56.

Seefeldt, S. S., and S. D. McCoy. 2003. Measuring plant diversity in the tall threetip sagebrush steppe: Influence of previous grazing management practices. *Environmental Management* 32:234–45.

Seidel, K. W. 1986. Tolerance of seedlings of ponderosa pine, Douglas-fir, grand fir, and Engelmann spruce to high temperatures. *Northwest Science* 60:1–7.

Senft, R. L., M. B. Coughenour, D. W. Bailey, L. R. Rittenhouse, O. E. Sala, and D. M. Swift. 1987. Large herbivore foraging and ecological hierarchies. *BioScience* 37:789–99.

Sexton, T. O. 1998. Ecological effects of post-wildfire management activities (salvage-logging and grass-seeding) on vegetation composition, diversity, biomass, and growth and survival of *Pinus ponderosa* and *Purshia tridentata.* M.S. thesis, Oregon State University, Corvallis.

————. 2006. Forest Service wildland fire use program is expanding. *Fire Management Today* 66(4): 5–6.

Shankman, D. 1984. Tree regeneration following fire as evidence of timberline stability in the Colorado Front Range, U.S.A. *Arctic and Alpine Research* 16:413–17.

Shankman, D., and C. Daly. 1988. Forest regeneration above tree limit depressed by fire in the Colorado Front Range. *Bulletin of the Torrey Botanical Club* 115:272–79.

Shapiro-Miller, L. B., E. K. Heyerdahl, and P. Morgan. 2007. Comparison of fire scars, fire atlases, and satellite data in the northwestern United States. *Canadian Journal of Forest Research* 37:1933–43.

Shariff, A. R. 1988. The vegetation, soil moisture and nitrogen responses to three big sagebrush (*Artemisia tridentata* Nutt.) control methods. M.S. thesis, University of Wyoming, Laramie.

Shaw, J. H., and T. S. Carter. 1990. Bison movements in relation to fire and seasonality. *Wildlife Society Bulletin* 18:426–30.

Shea, K. L. 1985. Demographic aspects of coexistence in Engelmann spruce and subalpine fir. *American Journal of Botany* 72:1823–33.

Shearer, R. C., and W. C. Schmidt. 1970. *Natural regeneration in ponderosa pine forests of western Montana.* USDA Forest Service Research Paper INT-86. Ogden, Utah: Intermountain Forest and Range Experiment Station.

Sheeter, G. R. 1968. Secondary succession and range improvements after wildfire in northeastern Nevada. M.S. thesis, University of Nevada, Reno.

Sheppard, P. R., J. E. Means, and J. P. Lassoie. 1988. Cross-dating cores as a nondestructive method for dating living, scarred trees. *Forest Science* 34:781–89.

Shepperd, W. D. 1981. Stand characteristics of Rocky Mountain aspen. Pp. 22–30 in *Situation management of two intermountain species: Aspen and coyotes,* ed. N. V. DeByle. Logan: Utah State University.

————. 1990. A classification of quaking aspen in the central Rocky Mountains based on growth and stand characteristics. *Western Journal of Applied Forestry* 5:69–75.

Shepperd, W. D., C. B. Edminster, and S. A. Mata. 2006. Long-term seedfall, establishment, survival, and growth of natural and planted ponderosa pine in the Colorado Front Range. *Western Journal of Applied Forestry* 21:19–26.

Sherriff, R. L. 2004. The historic range of variability of ponderosa pine in the northern Colorado Front Range: Past fire types and fire effects. Ph.D. dissertation, University of Colorado, Boulder.

Sherriff, R. L., and T. T. Veblen. 2006. Ecological effects of changes in fire regimes in *Pinus ponderosa* ecosystems in the Colorado Front Range. *Journal of Vegetation Science* 17:705–18.

————. 2007. A spatially-explicit reconstruction of historical fire occurrence in the ponderosa pine zone of the Colorado Front Range. *Ecosystems* 10:311–23.

————. 2008. Variability in fire-climate relationships in ponderosa pine forests in the Colorado Front Range. *International Journal of Wildland Fire* 17:50–59.

Sherriff, R. L., T. T. Veblen, and J. S. Sibold. 2001. Fire history in high elevation subalpine forests in the Colorado Front Range. *Ecoscience* 8:369–80.

Shimkin, D. B. 1986. The introduction of the horse. Pp. 517–24 in *Handbook of North American Indians*. Vol. 11, *Great Basin*, ed. W. L. D'Azevedo. Washington, D.C.: Smithsonian Institution.

Shinneman, D. J. 2006. Determining restoration needs for piñon-juniper woodlands and adjacent ecosystems on the Uncompahgre Plateau, western Colorado. Ph.D. dissertation, University of Wyoming, Laramie.

Shinneman, D. J., and W. L. Baker. 1997. Nonequilibrium dynamics between catastrophic disturbances and old-growth forests in ponderosa pine landscapes of the Black Hills. *Conservation Biology* 11:1276–88.

———. In press. Historical fire and multidecadal drought as context for piñon-juniper woodland restoration in western Colorado. *Ecological Applications*.

Shinneman, D. J., W. L. Baker, and P. Lyon. 2008. Ecological restoration needs derived from reference conditions for a semi-arid landscape in western Colorado, USA. *Journal of Arid Environments* 72:207–27.

Shinneman, D. J., R. McClellan, and R. Smith. 2000. *The state of the southern Rockies ecoregion*. Nederland, Colorado: Southern Rockies Ecosystem Project.

Shiplett, B., and L. F. Neuenschwander. 1994. Fire ecology of the cedar-hemlock zone in Idaho. Pp. 41–52 in *Interior cedar–hemlock–white pine forests: Ecology and management—Symposium proceedings, Mar. 2–4, 1993, Spokane, Washington,* ed. D. M. Baumgartner, J. E. Lotan, and J. R. Tonn. Pullman: Washington State University, Cooperative Extension Service.

Show, S. B., C. A. Abell, R. L. Deering, and P. D. Hanson. 1941. A planning basis for adequate fire control on the Southern California national forests. *Fire Control Notes* 5(1): 1–59.

Sibold, J. S., and T. T. Veblen. 2006. Relationships of subalpine forest fires in the Colorado Front Range with interannual and multidecadal-scale climatic variation. *Journal of Biogeography* 33:833–42.

Sibold, J. S., T. T. Veblen, K. Chipko, L. Lawson, E. Mathis, and J. Scott. 2007. Influences of secondary disturbances on lodgepole pine stand development in Rocky Mountain National Park. *Ecological Applications* 17:1638–55.

Sibold, J. S., T. T. Veblen, and M. E. González. 2006. Spatial and temporal variation in historic fire regimes in subalpine forests across the Colorado Front Range. *Journal of Biogeography* 33:631–47.

Sieg, C. H. 1997. The role of fire in managing for biological diversity on native rangelands of the northern Great Plains. Pp. 31–38 in *Proceedings of the symposium on conserving biodiverstiy on native rangelands, Aug. 17, 1995, Fort Robinson State Park, Nebraska,* ed. D. W. Uresk, G. L. Schenbeck, and J. T. O'Rourke. USDA Forest Service General Technical Report RM-298. Fort Collins, Colorado: Rocky Mountain Research Station.

Sieg, C. H., J. D. McMillin, J. F. Fowler, K. K. Allen, J. F. Negron, L. L. Wadleigh, J. A. Anhold, and K. E. Gibson. 2006. Best predictors for postfire mortality of ponderosa pine trees in the Intermountain West. *Forest Science* 52:718–28.

Sindelar, B. W. 1971. Douglas-fir invasion of western Montana grasslands. Ph.D. dissertation, University of Montana, Missoula.

Singer, F. J. 1975. Wildfire and ungulates in the Glacier National Park area, northwestern Montana. M.S. thesis, University of Idaho, Moscow.

———. 1979. Habitat partitioning and wildfire relationships of cervids in Glacier National Park, Montana. *Journal of Wildlife Management* 43:437–44.

Singer, F. J., M. B. Coughenour, and J. E. Norland. 2004. Elk biology and ecology before and after the Yellowstone fires of 1988. Pp. 117–39 in *After the fires: The ecology of change in Yellowstone National Park,* ed. L. L. Wallace. New Haven, Connecticut: Yale University Press.

Singer, F. J., and M. K. Harter. 1996. Comparative effects of elk herbivory and 1988 fires on northern Yellowstone National Park grasslands. *Ecological Applications* 6:185–99.

Singer, F. J., and J. Mack. 1999. Predicting the effects of wildfire and carnivore predation on ungulates. Pp. 188–237 in *Carnivores in ecosystems: The Yellowstone experience,* ed. T. W. Clark, A. P. Curlee, S. C. Minta, and P. M. Karieva. New Haven, Connecticut: Yale University Press.

Singer, F. J., W. Schreier, J. Oppenheim, and E. O. Garton. 1989. Drought, fires, and large mammals. *BioScience* 39:716–22.

Singer, F. J., and P. Schullery. 1989. Yellowstone wildlife populations in process. *Western Wildlands* 15:18–22.

Skinner, T. V., and R. D. Laven. 1982. Final report: Background data for natural fire management in Rocky Mountain National Park. Unpublished report to Rocky Mountain National Park, Contract CX 1200–1-B020, by Department of Forest and Wood Sciences. Fort Collins: Colorado State University.

———. 1983. A fire history of the Longs Peak region of Rocky Mountain National Park. *Proceedings of the Conference on Fire and Forest Meteorology* 7:71–74.

Skowronski, N., K. Clark, R. Nelson, J. Hom, and M. Patterson. 2007. Remotely sensed measurements of forest structure and fuel loads in the Pinelands of New Jersey. *Remote Sensing of the Environment* 108:123–29.

Sloan, J. P. 1998. Historical density and stand structure of an old-growth forest in the Boise Basin of central Idaho. *Proceedings of the Tall Timbers Fire Ecology Conference* 20:258–66.

Small, R. T. 1957. Relationship of weather factors to rate of spread of the Robie Creek fire. *Fire Control Notes* 18(4): 143–50.

Smith, A. E., and F. W. Smith. 2005. Twenty-year change in aspen dominance in pure aspen and mixed aspen/conifer stands on the Uncompahgre Plateau, Colorado, USA. *Forest Ecology and Management* 213:338–48.

Smith, C. S. 1983. A 4300 year history of vegetation, climate, and fire from Blue Lake, Nez Perce County, Idaho. M.A. thesis, Washington State University, Pullman.

Smith, F. W., and S. C. Resh. 1999. Age-related changes in production and below-ground carbon allocation in *Pinus contorta* forests. *Forest Science* 45:333–41.

Smith, H. Y. 1999. Assessing longevity of ponderosa pine (*Pinus ponderosa*) snags in relation to age, diameter, wood density and pitch content. M.S. thesis, University of Montana, Missoula.

Smith, H. Y., and S. F. Arno. 1999. *Eighty-eight years of change in a managed ponderosa pine forest.* USDA Forest Service General Technical Report RMRS-GTR-23. Fort Collins, Colorado: Rocky Mountain Research Station.

Smith, J. K. 1998. Presettlement fire regimes in northern Idaho. Pp. 83–92 in *Fire management under fire (adapting to change): Proceedings of the 1994 Interior West Fire Council Meeting and Program, November 1–4, 1994, Coeur d'Alene, Idaho,* ed. K. Close and R. A. Bartlette. Fairfield, Washington: International Association of Wildland Fire.

————, ed. 2000. *Wildland fire in ecosystems: Effects of fire on fauna.* USDA Forest Service General Technical Report RMRS-GTR-42, vol. 1. Fort Collins, Colorado: Rocky Mountain Research Station.

Smith, J. K., and W. C. Fischer. 1997. *Fire ecology of the forest habitat types of northern Idaho.* USDA Forest Service General Technical Report INT-GTR-363. Ogden, Utah: Intermountain Research Station.

Smith, J. K., R. D. Laven, and P. N. Omi. 1993. Microplot sampling of fire behavior on *Populus tremuloides* stands in north-central Colorado. *International Journal of Wildland Fire* 3:85–94.

Smith, L. M., and J. A. Kadlec. 1985. Fire and herbivory in a Great Salt Lake marsh. *Ecology* 66:259–65.

Smith, P. T. 1915. A silvicultural system for western yellow pine in the Black Hills. *Proceedings of the Society of American Foresters* 10:294–300.

Smith, S. R., and J. J. O'Brien. 2001. Regional snowfall distributions associated with ENSO: Implications for seasonal forecasting. *Bulletin of the American Meteorological Society* 82:1179–91.

Smucker, K. M., R. L. Hutto, and B. M. Steele. 2005. Changes in bird abundance after wildfire: Importance of fire severity and time since fire. *Ecological Applications* 15:1535–49.

Sneck, K. M. 1977. The fire history: Coram Experimental Forest. Master's thesis, University of Montana, Missoula.

Society for Ecological Restoration. 2004. *The SER International Primer on Ecological Restoration.* Tucson: Society for Ecological Restoration International, International Science & Policy Working Group. http://www.ser.org.

Sparhawk, W. N. 1925. The use of liability ratings in planning forest fire protection. *Journal of Agricultural Research* 30:693–762.

Spyratos, V., P. S. Bourgeron, and M. Ghil. 2007. Development at the wildland-urban interface and the mitigation of forest-fire risk. *Proceedings of the National Academy of Sciences* 104:14272–76.

Stahelin, R. 1943. Factors influencing the natural restocking of high altitude burns by coniferous trees in the central Rocky Mountains. *Ecology* 24:19–30.

Stark, K. E., A. Arsenault, and G. E. Bradfield. 2006. Soil seed banks and plant community assembly following disturbance by fire and logging in interior Douglas-fir forests of south-central British Columbia. *Canadian Journal of Botany* 84:1548–60.

Steele, R., S. F. Arno, and K. Geier-Hayes. 1986. Wildfire patterns change in central Idaho's ponderosa pine–Douglas-fir forest. *Western Journal of Applied Forestry* 1:16–18.

Steele, R., and K. Geier-Hayes. 1995. *Major Douglas-fir habitat types of central Idaho: A summary of succession and management.* USDA Forest Service General Technical Report INT-GTR-331. Ogden, Utah: Intermountain Research Station.

Steelman, T. A., and C. A. Burke. 2007. Is wildfire policy in the United States sustainable? *Journal of Forestry* 105:67–72.

Steinhoff, H. W. 1978. Management of Gambel oak associations for wildlife and livestock. Unpublished report, USDA Forest Service. Lakewood, Colorado: Rocky Mountain Region; and Fort Collins: Colorado State University.

Stephens, S. L. 1998. Evaluation of the effects of silvicultural and fuels treatments on potential fire behavior in Sierra Nevada mixed-conifer forests. *Forest Ecology and Management* 105:21–35.

Stephens, S. L., and J. J. Moghaddas. 2005a. Fuel treatment effects on snags and coarse woody debris in a Sierra Nevada mixed conifer forest. *Forest Ecology and Management* 214:53–64.

———. 2005b. Silvicultural and reserve impacts on potential fire behavior and forest conservation: Twenty-five years of experience from Sierra Nevada mixed conifer forests. *Biological Conservation* 125:369–79.

Stephens, S. L., and L. W. Ruth. 2005. Federal forest-fire policy in the United States. *Ecological Applications* 15:532–42.

Steuter, A. A. 1987. C_3/C_4 production shift on seasonal burns–northern mixed prairie. *Journal of Range Management* 40:27–31.

Steward, F. R., S. Peter, and J. B. Richon. 1990. A method for predicting the depth of lethal heat penetration into mineral soils exposed to fires of various intensities. *Canadian Journal of Forest Research* 20:919–26.

Stewart, L. M. 1958. Planning for smokejumping. *Fire Control Notes* 19(2): 68–71.

Stewart, O. C. 1954. Forest fires with a purpose. *Southwestern Lore* 20:42–45.

———. 1955. Why were the prairies treeless? *Southwestern Lore* 20:59–64.

Stewart, S. I., V. C. Radeloff, R. B. Hammer, and T. J. Hawbaker. 2007. Defining the wildland-urban interface. *Journal of Forestry* 105:201–7.

Stickney, P. F. 1986. *First decade plant succession following the Sundance forest fire, northern Idaho.* USDA Forest Service General Technical Report INT-197. Ogden, Utah: Intermountain Forest and Range Experiment Station.

———. 1990. Early development of vegetation following holocaustic fire in northern Rocky Mountain forests. *Northwest Science* 64:243–46.

Stickney, P. F., and R. B. Campbell Jr. 2000. *Data base for early postfire succession in northern Rocky Mountain forests.* USDA Forest Service General Technical Report RMRS-GTR-61-CD. Fort Collins, Colorado: Rocky Mountain Research Station.

Stockstad, D. S. 1979. *Spontaneous and piloted ignition of rotten wood.* USDA Forest Service Research Note INT-267. Ogden, Utah: Intermountain Forest and Range Experiment Station.

Stokes, M. A., and T. L. Smiley. 1968. *Tree-ring dating.* Chicago: University of Chicago Press.

Stott, P. A., S. F. B. Tett, G. S. Jones, M. R. Allen, J. F. B. Mitchell, and G. J. Henkins. 2000. External control of 20th century temperature by natural and anthropogenic forcings. *Science* 290:2133–37.

Stout, J., A. L. Farris, and V. L. Wright. 1971. Small mammal populations of an area in northern Idaho severely burned in 1967. *Northwest Science* 45:219–26.

Strauss, D., L. Bednar, and R. Mees. 1989. Do one percent of forest fires cause ninety-nine percent of the damage? *Forest Science* 35:319–28.

Striffler, W. D., and E. W. Mogren. 1971. Erosion, soil properties, and revegetation following a severe burn in the Colorado Rockies. Pp. 25–36 in *Fire in the northern environment: A symposium,* ed. C. W. Slaughter, R. J. Barney, and G. M. Hansen. Portland, Oregon: USDA Forest Service, Pacific Northwest Forest and Range Experiment Station.

Strom, B. A., and P. Z. Fulé. 2007. Pre-wildfire fuel treatments affect long-term ponderosa pine forest dynamics. *International Journal of Wildland Fire* 16:128–38.

Sturman, A. P. 1987. Thermal influences on airflow in mountainous terrain. *Progress in Physical Geography* 11:183–206.

Sudworth, G. B. 1900a. Battlement Mesa forest reserve. House of Representatives, 56th Cong., 1st sess., document no. 5, *Annual Reports of the Department of the Interior for the Fiscal Year Ended June 30, 1899, 20th Annual Report of the United States Geological Survey, Part V—Forest Reserves,* pp. 181–243. Washington, D.C.: U.S. Government Printing Office.

———. 1900b. White River Plateau timber land reserve. House of Representatives, 56th Cong., 1st sess., document no. 5, *Annual Reports of the Department of the Interior for the Fiscal Year Ended June 30, 1899, 20th Annual Report of the United States Geological Survey, Part V—Forest Reserves,* pp. 117–79. Washington, D.C.: U.S. Government Printing Office.

Suh, Y.-C., and G.-H. Lim. 2006. Effects of the 11-year solar cycle on the earth atmosphere revealed in ECMWF reanalyses. *Geophysical Research Letters* 33:1–4.

Sund, S. K. 1988. Post-fire regeneration of *Pinus albicaulis* in western Montana: Patterns of occurrence and site characteristics. M.S. thesis, University of Colorado, Denver.

Sutton, R. T., and D. L. R. Hodson. 2005. Atlantic ocean forcing of North American and European summer climate. *Science* 309:115–18.

Swetnam, T. W., and A. M. Lynch. 1989. A tree-ring reconstruction of western spruce budworm history in the southern Rocky Mountains. *Forest Science* 35:962–86.

Swezy, D. M., and J. K. Agee. 1991. Prescribed-fire effects on fine-root and tree mortality in old-growth ponderosa pine. *Canadian Journal of Forest Research* 21:626–34.

Sydoriak, C. A., C. D. Allen, and B. F. Jacobs. 2000. Would ecological landscape restoration make the Bandelier Wilderness more or less of a wilderness? Pp. 209–15 in *Wilderness science in a time of change conference.* Vol. 5, *Wilderness ecosystems, threats, and management,* ed. D. N. Cole, S. F. McCool, W. T. Borrie, and J. O'Loughlin. USDA Forest Service Proceedings RMRS-P-15-VOL-5. Fort Collins, Colorado: Rocky Mountain Research Station.

Syphard, A. D., V. C. Radeloff, J. E. Keeley, T. J. Hawbaker, M. K. Clayton, S. I. Stewart, and R. B. Hammer. 2007. Human influence on California fire regimes. *Ecological Applications* 17:1388–1402.

Tansley, A. G. 1935. The use and abuse of vegetational concepts and terms. *Ecology* 16:284–307.

Taylor, A. H. 2000. Fire regimes and forest changes in mid and upper montane forests of the southern Cascades, Lassen Volcanic National Park, California, U.S.A. *Journal of Biogeography* 27:87–104.

Taylor, A. H., and C. N. Skinner. 1998. Fire history and landscape dynamics in a late-successional reserve, Klamath Mountains, California, USA. *Forest Ecology and Management* 111:285–301.

Taylor, A. R. 1964. Lightning damage to forest trees in Montana. *Weatherwise* 17:61–65.

———. 1974. Ecological aspects of lightning in forests. *Proceedings of the Tall Timbers Fire Ecology Conference* 13:455–82.

Taylor, D. L. 1969. Biotic succession of lodgepole pine forests of fire origin in Yellowstone National Park. Ph.D. dissertation, University of Wyoming, Laramie.

———. 1971. Biotic succession of lodgepole-pine forests of fire origin in Yellowstone National Park. *National Geographic Research Reports* 12:693–702.

———. 1973. Some ecological implications of forest fire control in Yellowstone National Park, Wyoming. *Ecology* 54:1394–96.

Taylor, D. L., and W. J. Barmore Jr. 1980. Post-fire succession of avifauna in coniferous forests of Yellowstone and Grant Teton National Parks, Wyoming. Pp. 130–45 in *Workshop proceedings: Management of western forests and grasslands for nongame birds,* ed. R. M. DeGraff. USDA Forest Service General Technical Report INT-GTR-86. Ogden, Utah: Intermountain Forest and Range Experiment Station.

Tesch, S. D. 1981. Comparative stand development in an old-growth Douglas-fir (*Pseudotsuga menziesii* var. *glauca*) forest in western Montana. *Canadian Journal of Forest Research* 11:82–89.

Theobald, D. M., and W. H. Romme. 2007. Expansion of the US wildland-urban interface. *Landscape and Urban Planning* 83:340–54.

Thomas, D. A. 1991. The Old Faithful fire run of September 7, 1988. *Proceedings of the Conference on Fire and Forest Meteorology* 11:272–80.

Thomas, J. W. 1979. Snags. Pp. 60–77 in *Wildlife habitats in managed forests: The Blue Mountains of Oregon and Washington,* ed. J. W. Thomas. Agricultural Handbook No. 553. Washington, D.C.: USDA Forest Service.

Thomas, P. H. 1963. The size of flames from natural fires. Pp. 844–59 in *Proceedings of the symposium on combustion,* vol. 9. New York: Academic Press.

Thompson, A. W. 1947. My first prairie fire fight. *Colorado Magazine* 24:220–22.

Thompson, G. A. 1964. Fires in wilderness areas. *Proceedings of the Tall Timbers Fire Ecology Conference* 3:104–10.

Thompson, T. W., B. A. Roundy, E. D. McArthur, B. D. Jessop, B. Waldron, and J. N. Davis. 2006. Fire rehabilitation using native and introduced species: A landscape trial. *Rangeland Ecology and Management* 59:237–48.

Thornbury, W. D. 1965. *Regional geomorphology of the United States.* New York: John Wiley.

Thornton, R. 1987. *American Indian holocaust and survival: A population history since 1492.* Norman: University of Oklahoma Press.

Tiedemann, A. R., W. P. Clary, and R. J. Barbour. 1987. Underground systems of Gambel oak (*Quercus gambelii*) in central Utah. *American Journal of Botany* 74:1065–71.

Tiedemann, A. R., J. O. Klemmedson, and E. L. Bull. 2000. Solution of forest health problems with prescribed fire: Are forest productivity and wildlife at risk? *Forest Ecology and Management* 127:1–18.

Tinker, D. B., and D. H. Knight. 2000. Coarse woody debris following fire and logging in Wyoming lodgepole pine forests. *Ecosystems* 3:472–83.

———. 2001. Temporal and spatial dynamics of coarse woody debris in harvested and unharvested lodgepole pine forests. *Ecological Modeling* 141:125–49.

————. 2004. Snags and coarse woody debris: An important legacy of forests in the Greater Yellowstone ecosystem. Pp. 279–98 in *After the fires: The ecology of change in Yellowstone National Park,* ed. L. L. Wallace. New Haven, Connecticut: Yale University Press.

Tinker, D. B., W. H. Romme, W. W. Hargrove, R. H. Gardner, and M. G. Turner. 1994. Landscape-scale heterogeneity in lodgepole pine serotiny. *Canadian Journal of Forest Research* 24:897–903.

Tomback, D. F. 1994. Effects of seed dispersal by Clark's nutcracker on early postfire regeneration of whitebark pine. Pp. 193–98 in *Proceedings—International workshop on subalpine stone pines and their environment: The status of our knowledge, Sept. 5–11, 1992, St. Moritz, Switzerland,* ed. W. C. Schmidt and F.-K. Holtmeier. USDA Forest Service General Technical Report INT-GR-309. Ogden, Utah: Intermountain Research Station.

————. 2001. Clark's nutcracker: Agent of regeneration. Pp. 89–104 in *Whitebark pine communities: Ecology and restoration,* ed. D. F. Tomback, S. F. Arno, and R. E. Keane. Washington, D.C.: Island Press.

Tomback, D. F., A. J. Anderies, K. S. Carsey, M. L. Powell, and S. Mellmann-Brown. 2001. Delayed seed germination in whitebark pine and regeneration patterns following the Yellowstone fires. *Ecology* 82:2587–2600.

Tomback, D. F., S. F. Arno, and R. E. Keane. 2001. The compelling case for management intervention. Pp. 3–25 in *Whitebark pine communities: Ecology and restoration,* ed. D. F. Tomback, S. F. Arno, and R. E. Keane. Washington, D.C.: Island Press.

Tomback, D. F., J. K. Clary, J. Koehler, R. J. Hoff, and S. F. Arno. 1995. The effects of blister rust on post-fire regeneration of whitebark pine: The Sundance burn of northern Idaho (U.S.A.). *Conservation Biology* 9:654–64.

Tomback, D. F., L. A. Hoffman, and S. K. Sund. 1990. Coevolution of whitebark pine and nutcrackers: Implications for forest regeneration. Pp. 118–29 in *Proceedings—Symposium on whitebark pine ecosystems: Ecology and management of a high-mountain resource,* ed. W. C. Schmidt and K. J. McDonald. USDA Forest Service General Technical Report INT-270. Ogden, Utah: Intermountain Research Station.

Tomback, D. F., S. K. Sund, and L. A. Hoffman. 1993. Post-fire regeneration of *Pinus albicaulis*: Height-age relationships, age structure, and microsite characteristics. *Canadian Journal of Forest Research* 23:113–19.

Toney, J. L., and R. S. Anderson. 2006. A postglacial palaeoecological record from the San Juan Mountains of Colorado USA: Fire, climate and vegetation history. *The Holocene* 16:505–17.

Tonnesen, A. S., and J. J. Ebersole. 1997. Human trampling effects on regeneration and age structures of *Pinus edulis* and *Juniperus monosperma*. *Great Basin Naturalist* 57:50–56.

Toth, B. L. 1992. Factors affecting conifer regeneration and community structure after a wildfire in western Montana. M.S. thesis, Oregon State University, Corvallis.

Touchan, R., C. D. Allen, and T. W. Swetnam. 1996. Fire history and climatic patterns in ponderosa pine and mixed-conifer forests of the Jemez Mountains, northern New Mexico. Pp. 33–46 in *Fire effects in southwestern forests: Proceedings of the second La Mesa fire symposium,* ed. C. D. Allen. USDA Forest Service General Technical Report RM-GTR-286. Fort Collins, Colorado: Rocky Mountain Forest and Range Experiment Station.

Touchan, R., T. W. Swetnam, and H. D. Grissino-Mayer. 1995. Effects of livestock grazing on pre-settlement fire regimes in New Mexico. Pp. 268–72 in *Proceedings: Symposium on fire in wilderness and park management,* ed. J. K. Brown, R. W. Mutch, C. W. Spoon, and R. H. Wakimoto. USDA Forest Service General Technical Report INT-GTR-320. Ogden, Utah: Intermountain Research Station.

Town, F. E. 1899. Bighorn Forest Reserve. Pp. 165–90 in *USDI Geological Survey, Nineteenth Annual Report, Part V—Forest Reserves.* Washington, D.C.: U.S. Government Printing Office.

Tracy, B. F. 1997. Fire effects in Yellowstone's grasslands. *Yellowstone Science* 5:2–5.

———. 2004. Fire effects, elk, and ecosystem resilience in Yellowstone's sagebrush grasslands. Pp. 102–16 in *After the fires: The ecology of change in Yellowstone National Park,* ed. L. L. Wallace. New Haven, Connecticut: Yale University Press.

Tracy, B. F., and S. J. McNaughton. 1997. Elk grazing and vegetation responses following a late season fire in Yellowstone National Park. *Plant Ecology* 130:111–19.

Trimble, G. R. Jr., and N. R. Tripp. 1949. Some effects of fire and cutting on forest soils in the lodgepole pine forest of the northern Rocky Mountains. *Journal of Forestry* 47:640–42.

Trouet, V., A. H. Taylor, A. M. Carleton, and C. N. Skinner. 2006. Fire-climate interactions in forests of the American Pacific coast. *Geophysical Research Letters* 31:1–5.

Turner, D. P. 1985. Successional relationships and a comparison of biological characteristics among six northwestern conifers. *Bulletin of the Torrey Botanical Club* 112:421–28.

Turner, M. G., ed. 1987. *Landscape heterogeneity and disturbance.* New York: Springer.

Turner, M. G., W. W. Hargrove, R. H. Gardner, and W. H. Romme. 1994. Effects of fire on landscape heterogeneity in Yellowstone National Park, Wyoming. *Journal of Vegetation Science* 5:731–42.

Turner, M. G., and W. H. Romme. 1994. Landscape dynamics in crown fire ecosystems. *Landscape Ecology* 9:59–77.

Turner, M. G., W. H. Romme, and R. H. Gardner. 1999. Prefire heterogeneity, fire severity, and early postfire plant reestablishment in subalpine forests of Yellowstone National Park, Wyoming. *International Journal of Wildland Fire* 9:21–36.

Turner, M. G., W. H. Romme, R. H. Gardner, and W. W. Hargrove. 1997. Effects of fire size and pattern on early succession in Yellowstone National Park. *Ecological Monographs* 67:411–33.

Turner, M. G., W. H. Romme, and R. A. Reed. 2003. Postfire aspen seedling recruitment across the Yellowstone (USA) landscape. *Landscape Ecology* 18:127–40.

Turner, M. G., D. B. Tinker, W. H. Romme, D. M. Kashian, and C. M. Litton. 2004. Landscape patterns of sapling density, leaf area, and aboveground net primary production in postfire lodgepole pine forests, Yellowstone National Park (USA). *Ecosystems* 7:751–75.

Turner, M. G., Y. Wu, L. L. Wallace, W. H. Romme, and A. Brenkert. 1994. Simulating winter interactions among ungulates, vegetation, and fire in northern Yellowstone Park. *Ecological Applications* 4:472–96.

Tyers, D. B., and L. R. Irby. 1995. Shiras moose winter habitat use in the upper Yellowstone River valley prior to and after the 1988 fires. *Alces* 31:35–43.

Umbanhowar, C. E. Jr. 1996. Recent fire history of the northern Great Plains. *American Midland Naturalist* 135:115–21.

———. 2004. Interactions of climate and fire at two sites in the northern Great Plains, USA. *Palaeogeography, Palaeoclimatology, Palaeoecology* 208:141–52.

USDA Forest Service. 1958. *Timber resources for America's future.* Forest Resource Report No. 14. Washington, D.C.: U.S. Government Printing Office.

———. 1966. *Forest fire statistics.* Washington, D.C.: U.S. Government Printing Office, Division of Cooperative Forest Fire Control.

———. 1973. *The outlook for timber in the United States.* USDA Forest Service Forest Resource Report No. 20. Washington, D.C.: U.S. Government Printing Office.

———. 2000. Forest Service roadless area conservation. *Draft Environmental Impact Statement,* vol. 1. Washington, D.C.: USDA Forest Service.

———. 2005. *A strategic assessment of forest biomass and fuel reduction treatments in western states.* USDA Forest Service General Technical Report RMRS-GTR-149. Fort Collins, Colorado: Rocky Mountain Research Station.

———. 2006. Fire effects information system. http://www.fs.fed.us/database/feis/index.html.

USDA and USDI. 2006. *Wildland fire use implementation procedures reference guide.* Washington, D.C.: U.S. Department of Agriculture and Department of Interior.

U.S. Department of Agriculture. 2006. *Audit report: Forest Service large fire suppression costs.* Report No. 08601–44-SF, Office of Inspector General, Western Region. Washington, D.C.: U.S. Government Printing Office.

USDI Bureau of Land Management. 2004. *Fire history database.* ftp://ftp.blm.gov/pub/gis/wildfire/firehistory2003.

USDI Bureau of Land Management, USDA Forest Service, USDI National Park Service, USDI Fish & Wildlife Service, and USDI Bureau of Indian Affairs. 2006. *Interagency burned area emergency response guidebook.* Interpretation of Department of the Interior 620 DM 3 and USDA Forest Service Manual 2523 for the emergency stabilization of federal and tribal trust lands, Version 4.0. Washingon, D.C.: U.S. Government Printing Office.

Vaillancourt, D. A. 1995. Structural and microclimatic edge effects associated with clearcutting in a Rocky Mountain forest. M.S. thesis, University of Wyoming, Laramie.

Vale, T. R. 1998. The myth of the humanized landscape: An example from Yosemite National Park. *Natural Areas Journal* 18:231–36.

———, ed. 2002. *Fire, native peoples, and the natural landscape.* Washington, D.C.: Island Press.

Vales, D. J., and J. M. Peek. 1996. Responses of elk to the 1988 Yellowstone fires and drought. Pp. 159–67 in *The ecological implications of fire in Greater Yellowstone: Proceedings second biennial conference on the Greater Yellowstone ecosystem, Yellowstone National Park, Sept. 19–21, 1993,* ed. J. M. Greenlee. Fairfield, Washington: International Association of Wildland Fire.

Van Dyke, F., and J. A. Darragh. 2006. Short- and longer-term effects of fire and herbivory on sagebrush communities in south-central Montana. *Environmental Management* 38:365–76.

Van Dyke, F., M. J. Deboer, and G. M. Van Beek. 1996. Winter range plant production and elk use following prescribed burning. Pp. 193–200 in *The ecological implications of fire in Greater Yellowstone: Proceedings second biennial conference on the Greater Yellowstone ecosystem, Yellowstone National Park, Sept. 19–21, 1993,* ed. J. M. Greenlee. Fairfield, Washington: International Association of Wildland Fire.

Van Horne, M. L., and P. Z. Fulé. 2006. Comparing methods of reconstructing fire history using fire scars in a southwestern United States ponderosa pine forest. *Canadian Journal of Forest Research* 36:855–67.

Van Wagner, C. E. 1973. Height of crown scorch in forest fires. *Canadian Journal of Forest Research* 3:373–78.

———. 1977. Conditions for the start and spread of crown fire. *Canadian Journal of Forest Research* 7:23–34.

———. 1978. Age-class distribution and the forest fire cycle. *Canadian Journal of Forest Research* 8:220–27.

Veblen, T. T. 1986. Age and size structure of subalpine forests in the Colorado Front Range. *Bulletin of the Torrey Botanical Club* 113:225–40.

Veblen, T. T., and J. A. Donnegan. 2005. Historic range of variability for forest vegetation of the national forests of the Colorado Front Range. Unpublished report, USDA Forest Service. Lakewood, Colorado: Rocky Mountain Region.

Veblen, T. T., K. S. Hadley, E. M. Nel, T. Kitzberger, M. Reid, and R. Villalba. 1994. Disturbance regime and disturbance interactions in a Rocky Mountain subalpine forest. *Journal of Ecology* 82:125–35.

Veblen, T. T., K. S. Hadley, and M. S. Reid. 1991. Disturbance and stand development of a Colorado subalpine forest. *Journal of Biogeography* 18:707–16.

Veblen, T. T., K. S. Hadley, M. S. Reid, and A. J. Rebertus. 1991. Methods of detecting past spruce beetle outbreaks in Rocky Mountain subalpine forest. *Canadian Journal of Forest Research* 21:242–54.

Veblen, T. T., T. Kitzberger, and J. Donnegan. 2000. Climatic and human influences on fire regimes in ponderosa pine forests in the Colorado Front Range. *Ecological Applications* 10:1178–95.

Veblen, T. T., and D. C. Lorenz. 1986. Anthropogenic disturbance and recovery patterns in montane forests, Colorado Front Range. *Physical Geography* 7:1–24.

———. 1991. *The Colorado Front Range: A century of ecological change.* Salt Lake City: University of Utah Press.

Vesk, P. A., D. I. Warton, and M. Westoby. 2004. Sprouting by semi-arid plants: Testing a dichotomy and predictive traits. *Oikos* 107:72–89.

Vierling, K. T., L. B. Lentile, and N. Nielsen-Pincus. 2008. Preburn characteristics and woodpecker use of burned coniferous forests. *Journal of Wildlife Management* 72:422–27.

Vines, R. G. 1968. Heat transfer through bark, and the resistance of trees to fire. *Australian Journal of Botany* 16:499–514.

Vinton, M. A., D. C. Hartnett, E. J. Finck, and J. M. Briggs. 1993. Interactive effects of fire, bison (*Bison bison*) grazing and plant community composition in tallgrass prairie. *American Midland Naturalist* 129:10–18.

Vissage, J. S., and P. D. Miles. 2003. Fuel–reduction treatment: A west-wide assessment of opportunities. *Journal of Forestry* 101:5–6.

Wadleigh, L. L., and M. J. Jenkins. 1996. Fire frequency and vegetative mosaic of a spruce-fir forest in northern Utah. *Great Basin Naturalist* 56:28–37.

Wadleigh, L. L., C. Parker, and B. Smith. 1998. A fire frequency and comparative fuel load analysis in Gambel oak of northern Utah. *Proceedings of the Tall Timbers Fire Ecology Conference* 20:267–72.

Wahlenberg, W. G. 1930. Effect of Ceanothus brush on western yellow pine plantations in the northern Rocky Mountains. *Journal of Agricultural Research* 41:601–12.

Wakelyn, L. A. 1987. Changing habitat conditions on bighorn sheep ranges in Colorado. *Journal of Wildlife Management* 51:904–12.

Walhof, K. S. 1997. A comparison of burned and unburned big sagebrush communities in southwest Montana. M.S. thesis, Montana State University, Bozeman.

Walker, S. C. 1993. Effects of cattle and big game on the secondary succession of aspen-conifer understudy following fire. M.S thesis, Brigham Young University, Provo, Utah.

Walker, S. C., D. K. Mann, and E. D. McArthur. 1996. Plant community changes over 54 years within the Great Basin Experimental Range, Manti-La Sal National Forest. Pp. 66–68 in *Proceedings: Shrubland ecosystem dynamics in a changing environment,* ed. J. R. Barrow, E. D. McArthur, R. E. Sosebee, and R. J. Tausch. USDA Forest Service General Technical Report INT-GTR-338. Ogden, Utah: Intermountain Research Station.

Wallace, H. A. 1936. *The western range.* 74th Cong., 2nd sess., Senate document no. 199. Washington, D.C.: U.S. Government Printing Office.

Wallace, L. L., M. B. Coughenour, M. G. Turner, and W. H. Romme. 2004. Fire patterns and ungulate survival in northern Yellowstone Park: The results of two independent models. Pp. 299–317 in *After the fires: The ecology of change in Yellowstone National Park,* ed. L. L. Wallace. New Haven, Connecticut: Yale University Press.

Wallace, L. L., M. G. Turner, W. H. Romme, R. V. O'Neill, and Y. Wu. 1995. Scale of heterogeneity of forage production and winter foraging by elk and bison. *Landscape Ecology* 10:75–83.

Walsh, J. R. 2005. Fire regimes and stand dynamics of whitebark pine (*Pinus albicaulus*) communities in the Greater Yellowstone Ecosystem. M.S. thesis, Colorado State University, Fort Collins.

Wambolt, C. L. 1998. Sagebrush and ungulate relationships on Yellowstone's northern range. *Wildlife Society Bulletin* 26:429–37.

Wambolt, C. L., T. Hoffman, and C. A. Mehus. 1999. Response of shrubs in big sagebrush habitats to fire on the northern Yellowstone winter range. Pp. 238–42 in *Proceedings: Shrubland ecotones,* ed. E. D. McArthur, W. K. Ostler, and C. L. Wambolt. USDA Forest Service Proceedings RMRS-P-11. Fort Collins, Colorado: Rocky Mountain Research Station.

Wambolt, C. L., and G. F. Payne. 1986. An 18-year comparison of control methods for Wyoming big sagebrush in southwestern Montana. *Journal of Range Management* 39:314–19.

Wambolt, C. L., K. S. Walhof, and M. R. Frisina. 2001. Recovery of big sagebrush communities after burning in south-western Montana. *Journal of Environmental Management* 61:243–52.

Wang, G. G., and K. J. Kemball. 2005. Effects of fire severity on early development of understory vegetation. *Canadian Journal of Forest Research* 35:254–62.

Watt, A. S. 1947. Pattern and process in the plant community. *Journal of Ecology* 35:1–22.

Watts, M. J., and C. L. Wambolt. 1996. Long-term recovery of Wyoming big sagebrush after four treatments. *Journal of Environmental Management* 46:95–102.

Wauer, R. H., and T. Johnson. 1984. La Mesa fire effects on avifauna. Pp. 145–72 in *La Mesa fire symposium, Los Alamos, New Mexico, Oct. 6–7, 1981,* ed. T. S. Foxx. Publication LA-9236-NERP. Los Alamos, New Mexico: Los Alamos National Laboratory.

Weatherspoon, C. P. 1996. Fire-silviculture relationships in Sierra forests. Pp. 1167–76 in *Sierra Nevada Ecosystem Project, Final Report to Congress.* Vol. 2, *Assessments and scientific basis for management options.* Wildland Resources Center Report No. 37. Davis: University of California, Centers for Water and Wildland Resources.

Weaver, H. A. 1943. Fire as an ecological and silvicultural factor in the ponderosa pine region of the Pacific slope. *Journal of Forestry* 41:7–15.

Weaver, T., and D. Dale. 1974. *Pinus albicaulis* in central Montana: Environment, vegetation and production. *American Midland Naturalist* 92:222–30.

Weaver, T., F. Forcella, and D. Dale. 1990. Stand development in whitebark pine woodlands. Pp. 151–55 in *Proceedings—Symposium on whitebark pine ecosystems: Ecology and management of a high-mountain resource,* ed. W. C. Schmidt and K. J. McDonald. USDA Forest Service General Technical Report INT-270. Ogden, Utah: Intermountain Research Station.

Weidman, R. H. 1921. Forest succession as a basis of the silviculture of western yellow pine. *Journal of Forestry* 19:877–85.

Weisberg, P. J., and W. L. Baker. 1995. Spatial variation in tree seedling and krummholz growth in the forest-tundra ecotone of Rocky Mountain National Park, Colorado, U.S.A. *Arctic and Alpine Research* 27:116–29.

Welch, B. L. 2002. *Bird counts of burned versus unburned big sagebrush sites.* USDA Forest Service Research Note RMRS-RN-16. Ogden, Utah: Rocky Mountain Research Station.

Welch, B. L., and C. Criddle. 2003. *Countering misinformation concerning big sagebrush.* USDA Forest Service Research Paper RMRS-RP-40. Fort Collins, Colorado: Rocky Mountain Research Station.

Wellner, C. A. 1970. Fire history in the northern Rocky Mountains. Pp. 42–64 in *The role of fire in the Intermountain West: Proceedings of a symposium, Missoula, Montana, Oct. 27–29, 1970,* ed. R. Taber. Missoula, Montana: Intermountain Fire Research Council and University of Montana.

Wendtland, K. J., and J. L. Dodd. 1992. The fire history of Scotts Bluff National Monument. Pp. 141–43 in *Proceedings of the 12th North American Prairie Conference,* ed. D. D. Smith and C. A. Jacobs. Cedar City: University of Northern Iowa.

Werth, P. 2007. Fire weather case study: Mann Gulch fire, Montana, August 5, 1949. Unpublished report. Battle Ground, Washington: Weather Research and Consulting

Services.   http://www.fireweather.com/res/419200783514PMMann%20Gulch%20
Fire.pdf.

West, N. E. 1983. Intermountain salt-desert shrubland. Pp. 375–97 in *Ecosystems of the world 5: Temperate deserts and semi-deserts,* ed. N. E. West. Amsterdam: Elsevier Scientific Publishing Company.

————. 1994. Effects of fire on salt-desert shrub rangelands. Pp. 71–74 in *Proceedings: Ecology and management of annual rangelands, May 18–22, 1992, Boise, Idaho,* ed. S. B. Monsen and S. G. Kitchen. USDA Forest Service General Technical Report INT-GTR-313. Ogden, Utah: Intermountain Research Station.

————. 1999. Juniper-piñon savannas and woodlands of western North America. Pp. 288–308 in *Savannas, barrens, and rock outcrop plant communities of North America,* ed. R. C. Anderson, J. S. Fralish, and J. M. Baskin. Cambridge, England: Cambridge University Press.

West, N. E., and M. A. Hassan. 1985. Recovery of sagebrush-grass vegetation following wildfire. *Journal of Range Management* 38:131–34.

West, N. E., and T. P. Yorks. 2002. Vegetation responses following wildfire on grazed and ungrazed sagebrush semi-desert. *Journal of Range Management* 55:171–81.

Westerling, A. L., A. Gershunov, T. J. Brown, D. R. Cayan, and M. D. Dettinger. 2003. Climate and wildfire in the western United States. *Bulletin of the American Meteorological Society* 84:595–604.

Westerling, A. L., H. G. Hidalgo, D. R. Cayan, and T. W. Swetnam. 2006. Warming and earlier spring increase western U.S. forest wildfire activity. *Science* 313:940–43.

Westerling, A. L., and T. W. Swetnam. 2003. Interannual to decadal drought and wildfire in the western United States. *EOS* 84:545, 554–55.

Whelan, R. J. 1995. *The ecology of fire.* Cambridge, England: Cambridge University Press.

Whicker, J. J., J. E. Pinder III, and D. D. Breshears. 2008. Thinning semiarid forests amplifies wind erosion comparably to wildfire: Implications for restoration and soil stability. *Journal of Arid Environments* 72:494–508.

Whipple, S. A. 1978. The relationship of buried, germinating seeds to vegetation in an old-growth Colorado subalpine forest. *Canadian Journal of Botany* 56:1505–09.

Whipple, S. A., and R. L. Dix. 1979. Age structure and successional dynamics of a Colorado subalpine forest. *American Midland Naturalist* 101:142–58.

Whisenant, S. G. 1990. Changing fire frequencies on Idaho's Snake River plains: Ecological and management implications. Pp. 4–10 in *Proceedings: Symposium on cheatgrass invasion, shrub die-off, and other aspects of shrub biology and management,* ed. E. D. McArthur, E. M. Romney, S. D. Smith, and P. T. Tueller. USDA Forest Service General Technical Report INT-276. Ogden, Utah: Intermountain Forest and Range Experiment Station.

White, R. S., and P. O. Currie. 1983. The effects of prescribed burning on silver sagebrush. *Journal of Range Management* 36:611–13.

Whitford, H. N. 1905. The forests of the Flathead Valley, Montana. II. Edaphic formations in Flathead Valley. *Botanical Gazette* 39:194–218, 276–96.

Whitlock, C., J. Marlon, C. Briles, A. Brunelle, C. Long, and P. Bartlein. 2008. Long-term relations among fire, fuel, and climate in the north-western US based on lake-sediment studies. *International Journal of Wildland Fire* 17:72–83.

Whittaker, R. H., and S. A. Levin. 1977. The role of mosaic phenomena in natural communities. *Theoretical Population Biology* 12:117–39.

Wicker, E. F., and C. D. Leaphart. 1976. Fire and dwarf mistletoe (*Arceuthobium* spp.) relationships in the northern Rocky Mountains. *Proceedings of the Tall Timbers Fire Ecology Conference* 14:279–98.

Wickham, J. D., T. G. Wade, K. B. Jones, K. H. Riiters, and R. V. O'Neill. 1995. Diversity of ecological communities of the United States. *Vegetatio* 119:91–100.

Wieder, W. R., and N. W. Bower. 2004. Fire history of the Aiken Canyon grassland–woodland ecotone in the southern foothills of the Colorado Front Range. *Southwestern Naturalist* 49:239–98.

Wienk, C. L., C. H. Sieg, and G. R. McPherson. 2004. Evaluating the role of cutting treatments, fire and soil seed banks in an experimental framework in ponderosa pine forests of the Black Hills, South Dakota. *Forest Ecology and Management* 192:375–93.

Willard, E. E., R. H. Wakimoto, and K. C. Ryan. 1994. Effects of wildfire on survival and regeneration of ponderosa pine in Glacier National Park. *Proceedings of the Conference on Fire and Forest Meteorology* 12:723–28.

———. 1995. Vegetation recovery in sedge meadow communities within the Red Bench fire, Glacier National Park. *Proceedings of the Tall Timbers Fire Ecology Conference* 19:102–10.

Williams, C. S. 1963. Ecology of bluebunch wheatgrass in northwestern Wyoming. Ph.D. dissertation, University of Wyoming, Laramie.

Williams, G. W. 2004. American Indian fire use in the arid west. *Fire Management Today* 64(3): 10–14.

Williams, M. 1989. *Americans and their forests: A historical geography.* Cambridge, England: Cambridge University Press.

Wilmer, B., and G. H. Aplet. 2005. *Targeting the community fire planning zone: Mapping matters.* Washington, D.C.: Wilderness Society.

Wilmes, L., D. Martinez, L. Wadleigh, C. Denton, and D. Geisler. 2002. *Rodeo-Chediski fire effects summary report.* Springerville, Arizona: Apache-Sitgreaves National Forest.

Wilson, C. C. 1979. Roadsides: Corridors with high fuel hazard and risk. *Journal of Forestry* 77:576–80.

Wilson, C. C., and J. D. Dell. 1971. The fuels buildup in American forests: A plan of action and research. *Journal of Forestry* 69:471–75.

Wilson, J. F. 1989. Major transitions in firefighting: 1950 to 1990. *Fire Management Notes* 50: 6–8.

Wilson, R. A. Jr. 1990. *Reexamination of Rothermel's fire spread equations in no-wind and no-slope conditions.* USDA Forest Service Research Paper INT-434. Ogden, Utah: Intermountain Research Station.

Winward, A. H. 1991. A renewed commitment to management of sagebrush grasslands. Pp. 2–7 in *Research in rangeland management.* Agricultural Experiment Station Special Report 880. Corvallis: Oregon State University.

Wolfson, B. A. S., T. E. Kolb, C. H. Sieg, and K. M. Clancy. 2005. Effects of post-fire conditions on germination and seedling success of diffuse knapweed in northern Arizona. *Forest Ecology and Management* 216:342–58.

Wood, M. A. 1981. Small mammal communities after two recent fires in Yellowstone National Park. M.S. thesis, Montana State University, Bozeman.

Wright, C. S., and S. J. Prichard. 2006. Biomass consumption during prescribed fires in big sagebrush ecosystems. Pp. 489–500 in *Fuels management: How to measure success— Conference proceedings, 28–30 March 2006, Portland, Oregon,* ed. P. L. Andrews and B. W. Butler. USDA Forest Service Proceedings RMRS-P-41. Fort Collins, Colorado: Rocky Mountain Research Station.

Wright, H. A. 1971. Why squirreltail is more tolerant to burning than needle-and-thread. *Journal of Range Management* 24:277–84.

Wright, H. A., and A. W. Bailey. 1980. *Fire ecology and prescribed burning in the Great Plains: A research review.* USDA Forest Service General Technical Report INT-77. Ogden, Utah: Intermountain Forest and Range Experiment Station.

Wright, H. A., and J. O. Klemmedson. 1965. Effect of fire on bunchgrasses of the sagebrush-grass region in southern Idaho. *Ecology* 46:680–88.

Wrobleski, D. W., and J. B. Kauffman. 2003. Initial effects of prescribed fire on morphology, abundance, and phenology of forbs in big sagebrush communities in southeastern Oregon. *Restoration Ecology* 11:82–90.

Wu, R. 1999. Fire history and forest structure in the mixed conifer forests of southwest Colorado. Master's thesis, Colorado State University, Fort Collins.

Wyant, J. G., R. D. Laven, and P. N. Omi. 1983. Fire effects on shoot growth characteristics of ponderosa pine in Colorado. *Canadian Journal of Forest Research* 13:620–25.

Wyant, J. G., P. N. Omi, and R. D. Laven. 1986. Fire induced tree mortality in a Colorado ponderosa pine/Douglas-fir stand. *Forest Science* 32:49–59.

Yang, J., H. S. He, S. R. Shifley, and E. J. Gustafson. 2007. Spatial patterns of modern period human-caused fire occurrence in the Missouri Ozark Highlands. *Forest Science* 53:1–15.

Yensen, E., D. L. Quinney, K. Johnson, K. Timmerman, and K. Steenhof. 1992. Fire, vegetation changes, and population fluctuations of Townsend's ground squirrels. *American Midland Naturalist* 128:299–312.

Yoder, J., D. Engle, and S. Fuhlendorf. 2004. Liability, incentives, and prescribed fire for ecosystem management. *Frontiers in Ecology and the Environment* 2:361–66.

Young, D. L., and J. A. Bailey. 1975. Effects of fire and mechanical treatment on *Cercocarpus montanus* and *Ribes cereum. Journal of Range Management* 28:495–97.

Young, J. A., and R. A. Evans. 1978. Population dynamics after wildfires in sagebrush grasslands. *Journal of Range Management* 31:283–89.

Young, J. A., and C. G. Young. 1992. *Seeds of woody plants in North America.* Portland, Oregon: Dioscorides Press.

Young, R. P. 1987. Fire ecology and management in plant communities of Malheur National Wildlife Refuge, southeastern Oregon. Ph.D. dissertation, Oregon State University, Corvallis.

Youngblood, A., K. L. Metlen, and K. Coe. 2006. Changes in stand structure and composition after restoration treatments in low elevation dry forests of northeastern Oregon. *Forest Ecology and Management* 234:143–63.

Zager, P. E. 1980. The influence of logging and wildfire on grizzly bear habitat in north-western Montana. M.S. thesis, University of Montana, Missoula.

Zedler, P. H., C. R. Gautier, and G. S. McMaster. 1983. Vegetation change in response to extreme events: The effect of a short interval between fires in California chaparral and coastal scrub. *Ecology* 64:809–18.

Zier, J. L., and W. L. Baker. 2006. A century of vegetation change in the San Juan Mountains, Colorado: An analysis using repeat photography. *Forest Ecology and Management* 228:251–62.

Zimmerman, G. T., and D. L. Bunnell. 2000. The federal wildland fire policy: Opportunities for wilderness fire management. Pp. 288–97 in *Wilderness science in a time of change conference.* Vol. 5, *Wilderness ecosystems, threats, and management,* ed. D. N. Cole, S. F. McCool, W. T. Borrie, and J. O'Loughlin. USDA Forest Service Proceedings RMRS-P-15. Fort Collins, Colorado: Rocky Mountain Research Station.

Zimmerman, G. T., and R. Lasko. 2006. The changing face of wildland fire use. *Fire Management Today* 66(4): 7–12.

Zimmerman, G. T., and R. D. Laven. 1984. Ecological implications of dwarf mistletoe and fire in lodgepole pine forests. Pp. 123–31 in *Proceedings of the symposium on the biology of dwarf mistletoes,* ed. F. G. Hawksworth and R. F. Scharpf. USDA Forest Service General Technical Report RM-111. Fort Collins, Colorado: Rocky Mountain Forest and Range Experiment Station.

Zimmerman, G. T., and L. F. Neuenschwander. 1984. Livestock grazing influences on community structure, fire intensity, and fire frequency within the Douglas-fir/ninebark habitat type. *Journal of Range Management* 37:104–10.

Ziska, L. H., J. B. Reeves III, and B. Blank. 2005. The impact of recent increases in atmospheric CO_2 on biomass production and vegetative retention of cheatgrass (*Bromus tectorum*): Implications for fire disturbance. *Global Change Biology* 11:1325–32.

Zouhar, K., J. K. Smith, S. Sutherland, and M. L. Brooks. 2008. *Wildland fire in ecosystems: Fire and nonnative invasive plants.* USDA Forest Service General Technical Report RMRS GTR-42, vol. 6. Ogden, Utah: Rocky Mountain Research Station.

About the Author

William L. Baker is a professor in the Ecology Program and in the Department of Geography at the University of Wyoming in Laramie, where he teaches courses on fire ecology, landscape ecology, and biogeography. He and an energetic group of graduate and undergraduate students have studied fire ecology, landscape ecology, and human-environment interactions in the Rocky Mountains and the adjoining Colorado Plateau for three decades. He is the winner of the 2009 James J. Parsons Distinguished Career Award from the Association of American Geographers.

Index